Strategic Management of Sustainable Manufacturing Operations

Rameshwar Dubey
Symbiosis International University, India

Angappa Gunasekaran
University of Massachuesetts, Dartmouth, USA

A volume in the Advances in Logistics, Operations, and Management Science (ALOMS) Book Series

BUSINESS SCIENCE Reference
An Imprint of IGI Global

Published in the United States of America by
Business Science Reference (an imprint of IGI Global)
701 E. Chocolate Avenue
Hershey PA, USA 17033
Tel: 717-533-8845
Fax: 717-533-8661
E-mail: cust@igi-global.com
Web site: http://www.igi-global.com

Library of Congress Cataloging-in-Publication Data

Library of Congress Cataloging-in-Publication Data

Names: Dubey, Rameshwar, 1977- editor. | Gunasekaran, A., editor.
Title: Strategic management of sustainable manufacturing operations / Rameshwar Dubey and Angappa Gunasekaran, editors.
Description: Hershey : Business Science Reference, 2016. | Includes bibliographical references and index.
Identifiers: LCCN 2016006120| ISBN 9781522503507 (hardcover : alk. paper) | ISBN 9781522503514 (ebook)
Subjects: LCSH: Business logistics--Environmental aspects. | Organizational change. | Sustainable development.
Classification: LCC HD38.5 .S767 2016 | DDC 658.4/012--dc23 LC record available at https://lccn.loc.gov/2016006120

This book is published in the IGI Global book series Advances in Logistics, Operations, and Management Science (ALOMS) (ISSN: 2327-350X; eISSN: 2327-3518)

British Cataloguing in Publication Data
A Cataloguing in Publication record for this book is available from the British Library.

Advances in Logistics, Operations, and Management Science (ALOMS) Book Series

John Wang
Montclair State University, USA

ISSN: 2327-350X
EISSN: 2327-3518

Mission

Operations research and management science continue to influence business processes, administration, and management information systems, particularly in covering the application methods for decision-making processes. New case studies and applications on management science, operations management, social sciences, and other behavioral sciences have been incorporated into business and organizations real-world objectives.

The **Advances in Logistics, Operations, and Management Science** (ALOMS) Book Series provides a collection of reference publications on the current trends, applications, theories, and practices in the management science field. Providing relevant and current research, this series and its individual publications would be useful for academics, researchers, scholars, and practitioners interested in improving decision making models and business functions.

Coverage

- Decision analysis and decision support
- Production management
- Computing and information technologies
- Services management
- Networks
- Operations Management
- Political Science
- Finance
- Organizational Behavior
- Marketing engineering

IGI Global is currently accepting manuscripts for publication within this series. To submit a proposal for a volume in this series, please contact our Acquisition Editors at Acquisitions@igi-global.com or visit: http://www.igi-global.com/publish/.

The Advances in Logistics, Operations, and Management Science (ALOMS) Book Series (ISSN 2327-350X) is published by IGI Global, 701 E. Chocolate Avenue, Hershey, PA 17033-1240, USA, www.igi-global.com. This series is composed of titles available for purchase individually; each title is edited to be contextually exclusive from any other title within the series. For pricing and ordering information please visit http://www.igi-global.com/book-series/advances-logistics-operations-management-science/37170. Postmaster: Send all address changes to above address.

Titles in this Series

For a list of additional titles in this series, please visit: www.igi-global.com

Supply Chain Strategies and the Engineer-to-Order Approach
Richard Addo-Tenkorang (University of Vaasa, Finland) Jussi Kantola (University of Vaasa, Finland) Petri Helo (University of Vaasa, Finland) and Ahm Shamsuzzoha (Sultan Qaboos University, Oman)
Business Science Reference • copyright 2016 • 283pp • H/C (ISBN: 9781522500216) • US $200.00 (our price)

Optimal Inventory Control and Management Techniques
Mandeep Mittal (Amity School of Engineering and Technology, India) and Nita H. Shah (Gujarat University, India)
Business Science Reference • copyright 2016 • 406pp • H/C (ISBN: 9781466698888) • US $170.00 (our price)

Stochastic Processes and Models in Operations Research
Neelamegam Anbazhagan (Alagappa University, India)
Business Science Reference • copyright 2016 • 338pp • H/C (ISBN: 9781522500445) • US $185.00 (our price)

Sustainable Logistics and Strategic Transportation Planning
Tomaž Kramberger (University of Maribor, Slovenia) Vojko Potočan (University of Maribor, Slovenia) and Vesna Mia Ipavec (University of Maribor, Slovenia)
Business Science Reference • copyright 2016 • 511pp • H/C (ISBN: 9781522500018) • US $225.00 (our price)

Handbook of Research on Strategic Supply Chain Management in the Retail Industry
Narasimha Kamath (JDA Software, India) and Swapnil Saurav (JDA Software, India)
Business Science Reference • copyright 2016 • 429pp • H/C (ISBN: 9781466698949) • US $235.00 (our price)

Handbook of Research on Military, Aeronautical, and Maritime Logistics and Operations
Alberto Ochoa-Zezzatti (Juarez City University, Mexico) Jöns Sánchez (Consejo Nacional De Ciencie Y Tecnologia (CONACYT), Mexico) Miguel Gastón Cedillo-Campos (Transportation Systems and Logistics National Laboratory, Mexican Institute of Transportation, Mexico) and Margain de Lourdes (Polytechnic University of Aguascalientes, Mexico)
Business Science Reference • copyright 2016 • 587pp • H/C (ISBN: 9781466697799) • US $310.00 (our price)

Innovative Solutions for Implementing Global Supply Chains in Emerging Markets
Ashish Dwivedi (University of Hull Business School, UK)
Business Science Reference • copyright 2016 • 313pp • H/C (ISBN: 9781466697959) • US $195.00 (our price)

IGI GLOBAL
DISSEMINATOR OF KNOWLEDGE
www.igi-global.com

701 E. Chocolate Ave., Hershey, PA 17033
Order online at www.igi-global.com or call 717-533-8845 x100
To place a standing order for titles released in this series, contact: cust@igi-global.com
Mon-Fri 8:00 am - 5:00 pm (est) or fax 24 hours a day 717-533-8661

Table of Contents

Detailed Table of Contents

Operational practices of 'green' supply chain processes within as well as across the firm boundaries are strongly linked with the firm performance. However, the desire for such practices is governed by external and/or internal factors. The challenge, therefore, for supply chain and operations managers is to design and develop an operations strategy that ensures the firms investments in such 'green' initiatives. We develop a theoretical framework for resource allocation strategies for sustainable manufacturing operations that can be characterized as speculative, responsible, efficient, and sustainable resource allocation. This can aid managers in characterizing and controlling the firm's underlying operational processes and guide resource allocation decisions at strategic level.

The purpose of this paper is to develop an understanding of employee's retention in manufacturing organizations. The factors identified for the study are work culture, training, compensation, motivation of employee's for their profession, and employee's personality factors which are mediating through job satisfaction of employees. The study established statistically significant relationship between the factors and employee's retention with mediation effect of job satisfaction. Results revealed the importance of job satisfaction as a mediator in predicting employee's retention in manufacturing industry.

The aim of this paper is to develop a conceptual sustainable manufacturing (SM) framework, which is the pressing need of highly competitive and natural resource scarce world. Systematic literature review has been adopted to identify enablers of SM. The enablers identified through systematic literature review includes product design, material selection & procurement, customer requirements, regulatory norms, social values & ethics, environmental priorities, energy costs, natural resource scarcity, supply chain &

mining and minerals industry in the South African region. The approach extends the domain of green supply chain research and produce theories that have greater explanatory power than the current practices. For this purpose a conceptual model was developed from literature review and data collected using a structured questionnaire mailed to a sample of 174 firms which are leading edge ISO14001 certified mining companies in South Africa. Data analysis is further performed through exploratory factor analysis and regression analysis. This paper presents the first empirical evaluation of the link between green supply chain management practices amongst a sample of mining and minerals companies in South Africa.

Green procurement demands a creative attitude and innovation oriented culture which is still missing in many organizations. It is noted that most of the procurement professionals lack essential soft skills which are more important than hard skills. Human resource managers need to develop appropriate teaching methodologies for procurement professionals to enhance the essential soft skills. The purpose of this study is to identify the essential set of soft skills required for green procurement practices. Interpretive structural modeling approach was applied to find the interrelationships. This study also provides deep insight to human resource managers and will be helpful in developing training programs for procurement professionals.

Sustainable Procurement is an emerging concept in India in the recent times. It is imperative for the production firms to look the procurement activity as one of the strategic enabler for sustaining the business in the competitive global environment. Moreover, sustainability in India has been understood in multiple meanings and it is essential to bring the awareness on sustainable procurement and its impact on manufacturing. This chapter elucidates the key drivers of sustainable procurement along with a complete discussion on the importance of those practices in the Indian context. It also lists some of the key trends in the current business environment. The study endorses that manufacturing industry dominated by SMEs are highly reluctant to adopt sustainability as they lack awareness and perceive them as a cost escalation activity. It also reveals challenges faced by the firms and gives a conclusion on the future directions for sustainable procurement research in Indian context.

This chapter gives the overview of social sustainability in the supply chain of manufacturing industry. The author through literature survey identifies various definitions to social sustainability and suggest the

Sustainable operations strategy, an important field of research for engineering optimization, is posing an increasing challenge to operations managers. Sustainable operations can become a key to competitive advantage. Consequently, there is a need for research for the complex interfaces between the dimensions of sustainable operations strategy. The companies need to accommodate environmental demands, emphasized by media and government, along with social aspect in day to day operations for formulating strategies. Non-traditional manufacturing processes involves mechanical, thermal, electrical or chemical energy or combinations of these energies but do not use a sharp cutting tools as required for traditional manufacturing processes. In this chapter a non-traditional machining process namely Electro chemical Machining (ECM) has been considered. Material Removal Rate (MRR) and Surface Roughness (SR) has been optimized using Taguchi, Grey Relation, Artificial Neural Network, Fuzzy Logic and Genetic Algorithm to achieve the best.

Chapter 12
Sustainable Supply Chain: An Institutional Perspective..272
Susheela Girisaballa, Symbiosis International University, India
Sonali Bhattacharya, Symbiosis International University, India

Sustainable supply chain management (SSCM) is an evolving discipline, for which many theories are being proposed. It is dependent on other disciplines, of which organizational theories are of tremendous importance. One such theory that has great relevance to SSCM is Institutional Theory. The purpose of this study is to develop a model that relates the enablers of SSCM to Institutional Theory. This is achieved in three steps: First, SSCM enablers were identified through a literature review. These internal enablers go through certain external challenges when supply chain players practice Sustainability. Second, an attempt is made to explain these challenges from the perspective of Institutional Theory. The role of Institutional Isomorphism in regulating and creating a direction for achieving the sustainability goal is identified. Finally, a theoretical model is developed to describe the link between supply chain players, institutional isomorphism, and the support and participation of the top management in creating an enabling environment for legitimization of sustainable supply chain.

Chapter 13
Sustainable Supply Chain Management: Literature Review...303
Shirish C. Jeble, SIU, India

In the recent years, interest in Sustainable Supply Chain Management has grown as the effects of global warming can be seen through various scientific researches. In a global economy as supply chains are spreading across continents – economic, social and environmental sustainability – commonly referred as triple bottom line (TBL) assumes significance. Focal firms in supply chain are held accountable for their economic performance by shareholders, social and environmental performance by society. Even the actions of members of their extended supply chain come under scrutiny. Growing global supply chains facilitate economic growth, improve standard of living of people and provide business opportunities to distant locations across the world. However, they have unfavorable environmental impact. In order to preserve natural resources for future generations, sustainability needs to be embedded across the supply chains. This literature review - covers different aspects of greening the supply chain. Supply chain is viewed as a system consisting of focal firm, chains of suppliers, distributors, logistics partners, retailers etc. Role of each member of supply chain in sustainability goals is discussed in detail. Extant literature has covered individual parts of supply chain, whereas this paper integrates this knowhow to facilitate learning all interrelated concepts in SSCM. Some of the supply chain companies have already taken

Preface

In recent years, the sustainable manufacturing operations have invited significant attention from academia and practitioners. Sustainable manufacturing is a strategy for the development of new products. It is defined by the U.S. Department of Commerce (2007) as "the creation of manufactured products that use processes that minimize negative environmental impacts, conserve energy and natural resources, are safe for employees, communities, and consumers and are economically sound." The integration of environmental requirements throughout the entire lifetime of products needs a new way of thinking and new decision tools need to be applied (Kaebernick et al. 2003; Jovane et al. 2008; Garetti and Taisch, 2012). Thus sustainable manufacturing involves green product design, green procurement, green technology and green production (e.g. Noci, 1997; Azzone and Noci, 1998; Gunasekaran and Spalanzani, 2012).

The manufacturing practices has evolved in last 2 decades from traditional manufacturing, which was revolving around cost, quality, delivery and flexibility (Sanchez and Perez, 2001) to sustainable manufacturing which aims at achieving a balance between environmental, social and economic dimensions to satisfy stakeholders (Flammer, 2013) and to achieve a competitive advantage (Rusinko, 2007; Carter and Rogers, 2008; Kannegiesser and Gunther, 2014). Molamohamadi and Ismail (2013) have argued that technology, education, ethnic background and accountability are the key enablers of sustainable manufacturing. Prabhu and his coauthors (Prabhu et al., 2012) have argued that the minimization of energy consumption and waste produced are two key aspects of sustainable manufacturing. Gunasekaran and his coauthors (Gunasekaran et al., 2013) have argued that operational strategies, tactics & techniques and operational policies are the foundation of sustainable manufacturing. Garbie (2013, 2014) has further argued that to implement sustainable manufacturing, an organization needs to focus on key enablers like international issues, contemporary issues, innovative products, reconfigurable manufacturing systems, complexity analysis, lean production, agile manufacturing, performance measurement and flexible organization. Dubey and his coauthors (Dubey et al., 2015) have further attempted to take the sustainable manufacturing practices to a world-class sustainable manufacturing level. The pillars identified are: leadership, regulatory pressures, supplier relationship management, employee involvement, reconfigurable manufacturing systems, lean production, and agile manufacturing. Sustainable manufacturing has attracted a lot of attention academia and practitioners in recent years. Dubey et al. (2015) have further attempted to explain sustainable manufacturing using a big data concept.

Thus, the existing literature has discussed sustainable manufacturing (e.g. Garbie, 2013, 2014) and sustainable practices such as waste minimization and energy efficiency through monitoring or technology (Despeisse et al., 2013). Garetti and Taisch (2012), in their review of sustainable manufacturing, highlighting the role of data and BDA suggesting that there is a need for methods that will be able to process large amounts of data related to its environmental, social, and economic implications.

In response to pressing need of the time, we have come out with this edited book of research on the strategic management of sustainable manufacturing operations. This edited book has 15 chapters, contributed by worldwide researchers and practitioners, of the sustainable operations field intended to advance the theories and practices of sustainable manufacturing operations.

Rameshwar Dubey
Symbiosis Institute of Operations Management, Symbiosis International University, India

Angappa Gunasekaran
Charlton College of Business, University of Massachusetts Dartmouth, USA

REFERENCES

Azzone, G., & Noci, G. (1998). Identifying effective PMSs for the deployment of "green" manufacturing strategies. *International Journal of Operations & Production Management*, *18*(4), 308–335. doi:10.1108/01443579810199711

Carter, C. R., & Rogers, D. S. (2008). A framework of sustainable supply chain management: Moving toward new theory. *International Journal of Physical Distribution & Logistics Management*, *38*(5), 360–387. doi:10.1108/09600030810882816

Despeisse, M., Ball, P. D., Evans, S., & Levers, A. (2012). Industrial Ecology at Factory Level – a Conceptual Model. *Journal of Cleaner Production*, *31*(3–4), 30–39. doi:10.1016/j.jclepro.2012.02.027

Dubey, R., Gunasekaran, A., & Chakrabarty, A. (2015). World-class sustainable manufacturing: Framework and a performance measurement system. *International Journal of Production Research*, *53*(17), 5207–5223. doi:10.1080/00207543.2015.1012603

Dubey, R., Gunasekaran, A., Childe, S. J., Wamba, S. F., & Papadopoulos, T. (2015). The impact of big data on world-class sustainable manufacturing. *International Journal of Advanced Manufacturing Technology*, *2015*, 1–15.

Flammer, C. (2013). Corporate Social Responsibility and Shareholder Reaction: The Environmental Awareness of Investors. *Academy of Management Journal*, *56*(3), 758–781. doi:10.5465/amj.2011.0744

Garbie, I. H. (2013). DFSME: Design for sustainable manufacturing enterprises (an economic viewpoint). *International Journal of Production Research*, *51*(2), 479–503. doi:10.1080/00207543.2011.652746

Garbie, I. H. (2014). An analytical technique to model and assess sustainable development index in manufacturing enterprises. *International Journal of Production Research*, *52*(16), 4876–4915. doi:10.1080/00207543.2014.893066

Garetti, M., & Taisch, M. (2012). Sustainable manufacturing: Trends and research challenges. *Production Planning and Control*, *23*(2-3), 83–104. doi:10.1080/09537287.2011.591619

Gunasekaran, A., Irani, Z., & Papadopoulos, T. (2013). Modelling and analysis of sustainable operations management: Certain investigations for research and applications. *The Journal of the Operational Research Society*, *65*(6), 806–823. doi:10.1057/jors.2013.171

Gunasekaran, A., & Spalanzani, A. (2012). Sustainable of manufacturing services: Investigation for research and applications. *International Journal of Production Economics*, *140*(1), 35–47. doi:10.1016/j.ijpe.2011.05.011

Jovane, F., Yoshikawa, H., Alting, L., Boër, C. R., Westkamper, E., Williams, D., & Paci, A. M. et al. (2008). The incoming global technological and industrial revolution towards competitive sustainable manufacturing. *CIRP Annals-Manufacturing Technology*, *57*(2), 641–659. doi:10.1016/j.cirp.2008.09.010

Kaebernick, H., Kara, S., & Sun, M. (2003). Sustainable product development and manufacturing by considering environmental requirements. *Robotics and Computer-integrated Manufacturing*, *19*(6), 461–468. doi:10.1016/S0736-5845(03)00056-5

Kannegiesser, M., & Günther, H. O. (2014). Sustainable development of global supply chains—part 1: Sustainability optimization framework. *Flexible Services and Manufacturing Journal*, *26*(1-2), 24–47. doi:10.1007/s10696-013-9176-5

Molamohamadi, Z., & Ismail, N. (2013). Developing a New Scheme for Sustainable Manufacturing. *International Journal of Materials. Mechanics and Manufacturing*, *1*(1), 1–5. doi:10.7763/IJMMM.2013.V1.1

Noci, G. (1997). Designing 'green'vendor rating systems for the assessment of a supplier's environmental performance. *European Journal of Purchasing & Supply Management*, *3*(2), 103–114. doi:10.1016/S0969-7012(96)00021-4

Prabhu, V. V., Jeon, H. W., & Taisch, M. (2012, August). Modeling green factory physics—An analytical approach. In *Automation Science and Engineering (CASE), 2012 IEEE International Conference on* (pp. 46-51). IEEE.

Rusinko, C. A. (2007). Green Manufacturing: An evaluation of environmentally sustainable manufacturing practices and their impact on competitive outcomes. *IEEE Transactions on Engineering Management*, *54*(3), 445–454. doi:10.1109/TEM.2007.900806

Chapter 1
'Green' Supply Chain Management:
Resource Allocation Strategies for Sustainable Operations

Padmanabha Aital
Symbiosis Institute of Operations Management, India

Prince Vijai
ICFAI Business School, India

ABSTRACT

Operational practices of 'green' supply chain processes within as well as across the firm boundaries are strongly linked with the firm performance. However, the desire for such practices is governed by external and/or internal factors. The challenge, therefore, for supply chain and operations managers is to design and develop an operations strategy that ensures the firms investments in such 'green' initiatives. We develop a theoretical framework for resource allocation strategies for sustainable manufacturing operations that can be characterized as speculative, responsible, efficient, and sustainable resource allocation. This can aid managers in characterizing and controlling the firm's underlying operational processes and guide resource allocation decisions at strategic level.

1. INTRODUCTION

Increased awareness about 'green' products and processes among customers as well as firms challenges supply chain and operations managers to examine such activities and its impact on firm performance. There are several studies that examine the impact of 'green' supply chain practices on firm performance. For more details refer Porter and Linde, 1995; Hart, 1997; Srivastava, 2007; Unruh, 2010; Sarkis et al., 2011; Eccles et al., 2012; Zhu et al., 2013; Souza, 2013; Dubey et al., 2015a; Dubey et al., 2015b.

Kleindorfer et al. (2005) provides the review of sustainable operations management and challenges that poses to the managers to integrate environment, health, and safety issues with the objective of achieving

DOI: 10.4018/978-1-5225-0350-7.ch001

triple bottom line (TBL) results. According to Srivastava (2007), 'green' supply chain management is defined as "integrating environmental thinking into supply chain management, including product design, material sourcing and selection, manufacturing processes, delivery of final product to the consumers a well as end-of-life management of product after its useful life."

Linton et al. (2007) found that there is a strong linkage between supply chains and sustainability in order to design and manage sustainable supply chains that to make use of current available resources without affecting the future generations needs. Carter and Rogers (2008) developed a framework for sustainable supply chain, and further identified its relationship with environmental, social, and economic related firm performance measures. Gupta and Pasule-Desai (2011) developed a framework for sustainable supply chain management based on four categories, namely, strategic considerations, decision at functional interfaces, regulation and government policies, and integrative models and decision support tools.

Liu et al. (2012) integrated the concepts 'Green' marketing and sustainable supply chain management through six dimensions, namely, product, promotion, planning, process, people, and project, to improve the firm's overall business performance. Kang et al. (2012) developed a framework for implementing sustainable supply chain management, and further identified factors and strategies for successful implementation. Hassini et al. (2012) developed a framework for sustainable supply chain management and identified the related performance measures. Gunasekaran and Spalanzani (2012) developed a research framework for sustainable business development in manufacturing and services, and further provide future research directions. Ageron et al. (2012) developed a theoretical framework on sustainable supply management in the context of French firms and found that greening the supply chains is one of the most critical success factors for achieving sustainable supply management. Baskaran et al. (2012) studied the Indian textile industry suppliers to evaluate and categorize by adopting grey approach on sustainability criteria such as discrimination, abuse of human right, child labor, long working hours, unfair competition, and pollution.

Sustainability plays an important role in achieving competitive advantage for any business firms (Berns et al., 2009). Further, there is a growing interest of government agencies in institutionalizing the corporate social responsibility towards sustainable resource development (Musa et al., 2013). Vimal and Vinodh (2013) developed a framework that consists of checklists to assess the sustainability from manufacturing processes perspective in terms of ecology, equity and economy. Reefke and Trocchi (2013) developed a comprehensive performance measurement system for the sustainable supply chains by integrating sustainable supply chains performance measures and balanced scorecard (BSC) performance management system (Table 1).

This chapter aims to address the following questions:

- What is 'green' supply chain management?
- What are the drivers of 'green' initiatives across the supply chain?
- What are the prevailing 'green' practices in the firms?
- What are the performance measures for firm's sustainability development?
- How does 'green' practices are linked to firm performance?
- How does firm allocate resources for sustainable operations?

Table 1. Summary of Recent GSCM Studies at Various Industries

Industry (Country)	Studies
Manufacturing industry (China)	Zhu and Sarkis (2004) Zhu et al. (2005 , 2008a)
Manufacturing industry Japan)	Zhu et al. (2010)
Manufacturing industry (US)	Green et al. (2012)
Manufacturing industry (China)	Zhu et al. (2013)
Manufacturing industry (US)	Clark et al. (2014)
Manufacturing industry (India)	Digalwar et al. (2013)
Manufacturing industry (India)	Dubey et al. (2013)
Manufacturing industry (Thailand)	Laosirihongthong et al. (2013)
Manufacturing industry (UK)	Holt and Ghobadian (2009)
Automotive industry (UK)	Hoejmose et al. (2012)
Automobile industry (Germany)	Caniels et al. (2013)
Electronic appliance industry (China/Japan)	Lau (2011)
Electrical and electronics industries (Taiwan)	Hu and Hsu (2010)
Textile industry (Denmark)	Jorgensen et al. (2010)
Textile industry (India)	Baskaran et al. (2012)
Manufacturing, power generation and distribution, medical and pharmaceutical, sales and distribution, logistic services provider industries (French)	Ageron et al. (2012)
Rubber industry (India)	Dubey et al. (2014)
Oil and gas industry (Nigeria)	Musa et al. (2013)
Power generating, chemical/petroleum, electrical/electronic and automobile industries (China)	Zhu et al. (2008b)
Chemical industry (Jordan)	Eiadat et al. (2008)

Note: US – United States; UK – United Kingdom

2. 'GREEN' DRIVERS/PRESSURES

Anderson et al. (1999) investigated the reasons for firms to seek ISO certification and found that since it also provides competitive advantage to firms. The environmental framework of supply chain shows that the orientation of the supply chain has been molded by four forces such as market, competitive, cultural, and regulatory environment. The major drivers of 'green' supply chain are (Mollenkopf et al., 2010; Jorgensen et al., 2010; Holt & Ghobadian, 2010):

1. Environmental Standards and Reputations,
2. Corporate Environmental Policies and Culture,
3. Government Regulations and Standards,
4. Stakeholder Pressures,
5. Cost Reduction and Profitability,
6. Supplier Readiness and Market Expectations,
7. Non-Governmental Organizations,

8. Competition and Competitors,
9. Risk Management, and
10. Innovation and Corporate Brands.

The institutional theory proposes three kinds of pressures that firm faces towards 'green' initiatives. They are (DiMaggio & Powell, 1983):

1. Normative Pressure (e.g., corporative social responsibility and environment image, societal factors),
2. Coercive Pressure (e.g., strict government environmental laws, legislations), and
3. Mimetic Pressure (e.g., simply copying practices from other firms).

Given the nature and orientation of supply chain, it operates under four environments namely market, competitive, cultural, and regulatory environment. Further, these forces shape the nature and approach of the supply chain management. In addition, the drivers of green supply chain management in connection with lean as well as global supply chain are common global environmental standards, corporate headquarters' environmental policies and corporate culture, global environmental reputations, cost reduction, customers and stakeholder pressures, non-governmental organizations, competitors, commodity risk management, government regulations and standards, profitability, and focus on innovation (Mollenkopf et al., 2010).

It was identified that the major driving forces behind the 'green' initiatives in Danish transnational textile product supply chains are governmental regulation, customer demands, market expectations and protection of corporate brands (Jorgensen et al., 2010).

Sarkis (2003) developed a generic framework for evaluating 'green' supply chain management practices at strategic level. The framework considers product life cycle, operational life cycle, performance measures, and environmentally influential organizational policy elements apart from other decision variables. Further, the extant literature of 'green' supply chain management that was categorized to understand and identify research directions based on organizational theories. The theories that helps understand and apply in the 'green' practices are as follows (Sarkis et al., 2011):

- Complexity theory, Crozier and Thoenig (1976)
- Ecological modernization theory (EMT), Janicke (2008); Murphy and Gouldson (2000)
- Information asymmetry and signaling theory, Simpson et al. (2007)
- Institutional theory, Hirsch (1975); DiMaggio and Powell (1983)
- Resource-based view (RBV), Barney (1991); Barney (2012)
- Resource dependence theory (RDT), Pfeffer and Salancik (1978)
- Social network theory (SNT), Jones et al. (1997); Rowley (1997)
- Stakeholder theory, Freeman (1984)
- Transaction cost economics (TCE), Williamson (1981)
- Diffusion of innovation theory, Rogers (2003)
- Path dependency theory, Pierson (2000)
- Social embeddedness theory (SET), Granovetter (1985)
- Structuration theory, and Giddens (1984)
- Agency theory, Eisenhardt (1989)

Hart (1995) proposed a new theory called natural resource-based view that build upon resource-based view to explain the importance of firm's relationship with natural resources in achieving sustainable competitive advantage by developing strategies for pollution prevention, product stewardship, and sustainable development. Zsidisin and Siferd (2001) developed a theoretical framework to study the purchasing ability to affect environmental performance by linking environmental purchasing and transaction cost analysis. Gavronski et al. (2011) examined the relationship among firm resources, 'green' manufacturing capabilities and 'green' supply management from resource-based view.

The DiMaggio and Powell's (1983) institutional theory postulates that normative, coercive, and mimetic pressures align firms with environmental management practices because the firms adopt new initiatives to simply gain legitimacy. It was found that Chinese firms adopted 'green' supply chain practices due to government regulatory, competitive, and marketing pressures. However, it was also found that these pressures were not strongly translated into practices that in turn lead to firm performance (Zhu et al., 2005). Moreover, it was found that the Japanese firms are largely influenced by normative (i.e., corporative social responsibility and environment image) and coercive (i.e., strict government environmental laws) pressures but not by mimetic pressures (Zhu et al., 2010). It was also found that the coercive pressure (i.e., legislation) plays an important role in driving the diffusion of 'green' supply chain management in UK manufacturing sector whereas least the normative pressure (i.e., societal drivers) (Holt & Ghobadian, 2009).

An institutional theory perspective was adopted to study the role of supermarkets in developing the legitimate of sustainable practices of dairy supply chain in UK; and found interestingly, that the supermarkets exerts pressure on its suppliers in pursing the sustainable agenda (Glover et al., 2014). Rothenberg (2007) studied the influence of both institutional and technical pressure on environmental management by comparing two similar manufacturing plants' practices that operates in different environmental regulation. Schneider and Wallenburg (2012) studied implementation of sustainable sourcing along three dimensions, namely, economic, environmental, and social aspects, from stakeholder theory perspective. Eiadat et al. (2008) examined the mediating role of environmental innovation strategy between environmental pressures, such as government regulations, stakeholders' pressure, and management environmental policies, and firm's business performance in Jordan's chemical industry.

Hall (2000) investigated a phenomenon called environmental supply chain dynamics that is nothing diffusion of environmental innovation from base firm to a supplier firm, in British food retail and aerospace industries as well as in Japanese food retail industry. It was shown that the phenomenon will emerge when a channel leader is under environmental pressure then the channel leader will exercise the power over its suppliers.

Walker et al. (2008) explored the factors that drives and hinders firms to implement 'green' supply chain management practices and found that firms are more influenced by external drivers, such as regulation, competition, customers, suppliers, and society, rather than internal organizational-related drivers. Reuter et al. (2012) examined the relationship between adoption of ethical business culture and sustainability and cost prevalence from stakeholder approach in Germany firms.

Clemens and Douglas (2005) attempted to understand the relationship between institutional pressures and firms strategic responses in Steel industry. Further, Clemens and Douglas (2006) examined the relationship among coercive pressures, superior firm resources, and voluntary 'green' initiatives and found that coercion is positively related to voluntary 'green' initiatives depends upon firm's superior resources and strategy.

Table 2. Drivers/Pressures of 'Green' Initiatives

Drivers/Pressures	Studies
Market pressure; leadership; regulation; and product quality	Dubey et al. (2013)
Institutional and stakeholder pressures	Hoejmose et al. (2012)
Market; competition; cultural; regulatory environment	Mollenkopf et al. (2010)
Governmental regulation; customer demands; market expectations; and protection of corporate brands	Jorgensen et al. (2010)
Normative (i.e., corporative social responsibility and environment image); coercive (i.e., strict government environmental laws); and mimetic pressures	Holt and Ghobadian (2009) Musa et al. (2013)
Government regulatory; competition; and marketing pressures	Zhu et al. (2010) Zhu et al. (2008a , 2008b) Zhu et al. (2005) Zhu and Sarkis (2004)

Hitchcock (2012) discussed the legal and commercial pressures on UK and Chinese supply chains from industry point of view. Zhu et al. (2013) developed a theoretical model to understand the antecedents of 'green' supply chain management practices both internally and externally in the context of Chinese manufacturers. They found that institutional pressures serve as a major driver in the adoption of internal practices which in turn further drive the external practices. Shi et al. (2012) developed a conceptual model for 'green' supply chain management from natural resource based view in order to understand the institutional drivers and its relationship with performance measures.

A theory was proposed by employing a case study approach for making environmental investment decisions, while balancing the firm's economic and environmental priorities (Pagell & Wu, 2009; Wu & Pagell, 2011). The role of trust also plays a significant role among supply chain partners and top management in addition to the institutional and stakeholder pressures in green supply chain management. In the United Kingdom's (UK) firms, it was found that the role of trust and top management support are the most important drivers of green supply chain management practices in business-to-business sector against least important in business-to-customer sector (Hoejmose et al., 2012).

Barari et al. (2012) developed a framework for analyzing the supply chain contracts between producer and retailer by employing the evolutionary game theory approach. They showed that there is an agreement between producer and retailer effort in marketing the green product; and also there is a trade-off between 'green' cost and 'green' tax. The suppliers' participation in the 'green' initiatives plays an important role. In the context of German automotive suppliers, it was found that supplier readiness and customer requirements were the important drivers of the supplier participations in green initiatives (Caniels et al., 2013). It was found that that market pressure, leadership, regulation, and quality play important roles in determining the 'green' purchasing strategies (Dubey et al., 2013). Ghosh and Shah (2012) analyzed the greening policies across the apparel supply chain structures using game-theoretic approach (see Table 2).

3. 'GREEN' SUPPLY CHAIN PRACTICES

Bowen et al. (2001) studied the role of capabilities in supply management in implementing 'green' supply or purchasing. They found that the capabilities are jointly developed by corporate environmental policy and strategic purchasing behavior towards 'green' supply. Bjorklund (2011) explored the factors that influencing 'green' purchasing of transportation services in Swedish firms and found that factors like management, company image, customers, carriers, and means of control play major role. Walker and Brammer (2012) examined the relationship between sustainable procurement and e-procurement and found that e-procurement improves sustainable development along environmental, health, and safety aspects. Further, Walker et al. (2012) reviewed the extant literature and developed a framework for sustainable procurement. Giunipero et al. (2012) explored the factors that influence and hinder sustainable procurement and supply management and found that top management support and government regulations drive the sustainable practices, whereas, investments in sustainability and economic uncertainty impede. Oruezebala and Rico (2012) studied the impact of sustainable public procurement practices on supplier management performance in French hospitals. Tate et al. (2012) explored the factors in environmental purchasing and supplier management practices, such as general practices and philosophy, supplier involvement, supplier deployment, supplier selection criteria, and supplier environmental outcomes. Hoejmose and Adrien-Kirby (2012) developed a theoretical framework for socially and environmentally responsible procurement and its relationship with firm performance. Crespin-Mazet and Dontenwill (2012) developed a conceptual framework for sustainable sourcing and legitimacy in the supply network. They highlighted product, corporate, and cause legitimacies in the supply network and the role of business and non-business actors. Goebel et al. (2012) examined the impact of firms' ethical culture on socially and environmentally sustainable supplier selection. Blengini and Shields (2010) studied the building products supply chain in Italy and reported to European Commission's and Member States' policy makers about the 'green' labels and sustainability towards 'Go Green.'

Lau (2011) studied and compared the 'green' logistics practices in electronic appliance industry between China and Japan. Choi et al. (2012) discussed the 'green' manufacturing and distribution issues and challenges in the fashion and apparel industries. Kudla and Klaas-Wissing (2012) studied the sustainability aspects of logistics service providers from dyadic buyer-supplier relationships. Large et al. (2013) examined the relationship between availing logistics services and sustainable development in European firms and found that procurement of logistics services have weak influence on environmental and social performance of sustainability. Dekker et al. (2012) provided a review of operations research models for designing 'green' logistics.

Zhu et al. (2005) explored the factors that influencing 'green' supply chain management practices in Chinese manufacturing firms, namely, internal environmental management, external 'green' supply chain management, eco-design, and investment recovery. Further, the 'green' supply chain management practices construct was validated in the context of Chinese manufacturing firms (Zhu et al., 2008a). Johnstone and Labonne (2009) explored the motivation behind firms adopting environmental management systems and certification in seven OCED countries and found that both factors improvement in performance and indication to market play important role but their relative importance depends upon firm size.

In addition, Holt and Ghobadian (2009) explored the factors that influencing 'green' supply chain practices in UK's manufacturing sector, namely, internal environmental management practices, logistics, supplier assessment and evaluation, green procurement and logistics policy, supplier education, coaching and mentoring, and industrial networks. Further, it was found that the environmental attitude was the key predictor for 'green' practices. Furthermore, Hu and Hsu (2010) explored the factors that influencing 'green' supply chain practices in Taiwan's electrical and electronics industries. They are supplier management, product recycling, organization involvement, and life cycle management. Caniato et al. (2012) conducted an exploratory case-based research to explore the factors that drives firms in implementing 'green' practices towards environmental sustainability in fashion industry. Caniels et al.

Table 3. Measures of 'Green' Supply Chain Practices

'Green' Supply Chain Practices	Studies
Technology and best practices; supplier relationship management; waste reduction through collaboration and regulatory norms	Dubey et al. (2014)
Internal environmental management; sustainable practices related to suppliers and customers and environmentally conscious product and process design	Hasan (2013)
Green purchasing; product related eco-design; packaging related eco-design; reverse logistics; and legislations and regulations	Laosirihongthong et al. (2013)
Internal environmental management program; internal social-oriented program; supply chain assessment; and supply chain collaboration	Gimenez et al. (2012)
Green purchasing; waste elimination and toxic reduction; supplier as well as customer collaboration; certification; green packaging; and reverse logistics	Azevedo et al. (2011)
Supplier management; product recycling; organization involvement; and life cycle management	Hu and Hsu (2010)
Internal environmental management practices; logistics; supplier assessment and evaluation; green procurement and logistics policy; supplier education; coaching and mentoring; and industrial networks	Holt and Ghobadian (2009)
Internal environmental management; external GSCM – green purchasing and customer cooperation with environmental considerations; eco-design; and investment recovery	Zhu et al. (2010) Zhu et al. (2008a , 2008b) Zhu et al. (2005) Zhu and Sarkis (2004)

(2013) studied the drivers of suppliers participation in 'greening' supply chains in German automotive suppliers and found that supplier readiness and customer requirements are significant drivers, whereas, relational norms and customer investments are not. Chen et al. (2012) presented an approach for component selection system for 'green' supply chain management.

Bierma and Waterstraat (1999) studied the chemical suppliers and developed shared savings contracts to reduce waste and total costs in the supply chain. Boltic et al. (2013) conducted a case study for implementation of cleaner production process to minimize emissions in pharmaceutical industry. Brown and Stone (2007) discussed the environmental policies and 'green' initiatives that had undertaken in New Zealand for the past 15 years.

The 'green' supply chain practices of the firms can be measured as follows (see Table 3):

1. **Internal Environmental Management:** It consists of commitment from senior managers, support from mid-level managers, cross-function cooperation, total quality environmental management, environmental compliance and auditing programs, ISO 14001 certification and environmental management systems.
2. **'Green' Purchasing:** It consists of cooperation from suppliers, environmental audit at supplier's firm, supplier's ISO 14001 certification, tier-two supplier's practice evaluation and environmental obligations for purchasing items.
3. **Customer Cooperation:** It consists of customer cooperation for eco-friendly, cleaner production, 'green' packaging, 'green' labels, disposal and end-of-life usage.
4. **Eco-Design or 'Green' Design:** It is also widely known as 'design for environment.' It consists of designing products for reduced consumption of materials/energy, reuse, recycle, recovery of materials/components and reduce use of hazardous products.

5. **'Green' Logistics:** It consists of cooperation with logistics service providers for environmental conscious transportation operation while procuring and delivering raw materials/products in terms of carbon emissions and fuel consumption.
6. **Investment Recovery:** It consists of recovery or sales option of excess inventories/materials, scraps, used materials and excess capital equipment.

4. FIRM PERFORMANCE

Zhu et al. (2005) explored the factors that determine the firm performance, namely, environmental, operational and economic performance. Bai et al. (2012) developed sustainable performance measures for supply chains with reference to supply chain operations reference model (SCOR) that integrates business and environmental performance. Bjorklund et al. (2012) developed a framework for performance measurement by considering environmental logistics and 'green' supply chain management. Recently, Digalwar et al. (2013) identified the performance measures of 'green' manufacturing practicing firms in Indian manufacturing firms, namely, top management commitment, knowledge management, employee training, green product and process design, employee empowerment, environmental, health and safety, supplier and materials management, production planning and control, quality, cost, customer environmental performance requirement, and customer responsiveness and company growth. Brandenburg et al. (2014) provided a review of quantitative models in forward supply chain management that considered environmental or social factors. Mirhedayatian et al. (2014) presented a mathematical model, data envelopment analysis, for evaluating 'green' supply chain management. Sahu et al. (2013) presented a mathematical model to benchmark and evaluate 'green' supply chain performance and firm's environmental sustainability. Seuring (2013) provided a review of mathematical modeling approaches for sustainable supply chain management by considering environmental and social aspects of sustainability. Chaabane et al. (2012) presented a mathematical model for sustainable supply chain design under emission trading scheme.

Table 4. Measures of Firm Performance

Firm Performance Measures	Studies
Business; and environmental performance	Dubey et al. (2014)
Environmental; and operational performance	Hasan (2013)
Environmental; economic; and intangible performance	Laosirihongthong et al. (2013)
Top management commitment; knowledge management; employee training; green product and process design; employee empowerment; environmental, health and safety; supplier and materials management; production planning and control; quality; cost; customer environmental performance requirement; and customer responsiveness and company growth	Digalwar et al. (2013)
Environmental; social; and economic performance	Gimenez et al. (2012)
Economic; operational; and environmental performance	Azevedo et al. (2011)
Environmental; operational; and economic performances	Zhu et al. (2010) Zhu et al. (2008a , 2008b) Zhu et al. (2005) Zhu and Sarkis (2004)

The outcome of 'green' supply chain management practices of the firm can be measured in terms of sustainability development as follows (see Table 4) (Zhu et al., 2005 & 2010; Gimenez et al., 2012):

1. **Environmental Performance:** It consists of reduction in air emission, waste water and solid wastes, decrease in consumption of hazardous materials and environmental accidents, and also improvement in firm's environmental situation.
2. **Financial Performance:** It consists of decrease in cost for materials purchasing, energy consumption, waste treatment, waste discharge and environmental accidents.
3. **Operational Performance:** It consists of decrease in inventory levels and scrap rate, increase in amount of goods delivered on time and number of product line, and also improvement in product's quality and capacity utilization.
4. **Economic Performance:** It consists of decrease in cost per unit of manufacturing, material cost, and direct labor cost.
5. **Social Performance:** It consists of social obligation, social expectation and social reputation of the firm.

5. LINKING 'GREEN' SUPPLY CHAIN PRACTICES AND FIRM PERFORMANCE

Klassen and McLaughlin (1996) conducted an event study to investigate the impact of environmental performance on firm performance and found differences exist between first-time environmental award announcements and others. Carter et al. (2000) studied the impact of environmental purchasing on firm performance and found that the relationship positively exists. King and Lenox (2001) examined the relationship between lean production practices and environmental performance and found that the link had not hold for improvement in performance. Rao (2002) studied the 'green' supply chain initiatives and its impact on environmental and economic performance of the firms in the South-East Asia. Further, Rao (2004) explored the factors that determine 'greening' production in the region of South-East Asia, namely, corporate environmental responsibility, supplier integration, total quality environmental management, cleaner production, and worker involvement. Furthermore, Rao and Holt (2005) studied the impact of 'green' supply chain management initiatives in the South-East Asia's firms on environmental performance, economic performance and competitiveness.

Pujari et al. (2003) examined the relationship between environmental-friendly new product development activities and market and economic performance of such new products. They found that environmental paradigm of new product development provides synergy than conflict over conventional paradigm. Further, Pujari et al. (2004) explored the organizational antecedents of environmental-friendly new product development. Furthermore, Pujari (2006) studied the impact of environmental-friendly new product development on market performance in North American firms. They found that factors like cross-functional coordination, supplier involvement, market focus, and life cycle analysis influence market performance of 'green' products. Melnyk et al. (2003) studied the impact of environmental management systems on corporate and environmental performance in North American firms. Chen et al. (2004) examined the relationship among strategic purchasing, supply management, and firm performance in terms of customer responsiveness and financial performance in US manufacturing firms. Frondel et al. (2008) examined the relationship between environmental management systems and firm's environmental innovation performance in German firms. Darnall et al. (2008) studied the impact of environmental management systems on business performance in OCED countries. They found that complementary resources and capabilities, such as export orientation, employee commitment, and environmental research and development, motivate firms to adopt and further improve firm performance than merely institutional pressures.

Zhu and Sarkis (2004) studied the relationships between 'green' supply chain management practices and firm performance in Chinese manufacturing enterprises and found that the operational practices of 'green' supply chain management have influence on the firm's environmental and economic performances. In addition, the relationship was further examined in the presence of quality management and just-in-time or lean management practices. Further, the relationship between 'green' supply chain practices and closed-loop supply chain management was studied in Chinese industries and found that 'green' practices differ across industries – namely, power generating, chemical/petroleum, electrical/electronic, and automobile (Zhu et al., 2008b). Furthermore, the relationships between 'green' supply chain management practices and firm performance in Japanese large firms was studied and found that the Japanese firms more actively practicing internal environmental management than any other 'green' practices in comparison to Chinese firms. It was also found that Japanese firms made significant environmental and economic performances (Zhu et al., 2005; Zhu et al., 2010). In addition, Bai and Sarkis (2010) investigated the relationship between organizational characteristics, 'green' supplier development, and environmental and firm performance using rough set theory approach.

Azevedo et al. (2011) developed a conceptual model for evaluating the influence of 'green' practices on supply chain performance in Portuguese automotive industry by conducting a case study. They found that the green practices had strong influence on quality, customer satisfaction and efficiency. Gimenez et al. (2012) studied the impact of environmental and social initiatives on environmental, social and economic performance of the firm and found that the environmental initiatives such as design to recycle and certification had strong impact on the firm performance compared to social initiatives. Bose and Pal (2012) conducted an event study to understand the impact of green supply chain management initiatives on firms' stock prices, and found that adoption of 'green' supply chain management initiatives have strong influence for manufacturing firms compared to non-manufacturing firms. Green et al. (2012) studied the impact of 'green' supply chain management practices on firm performance in US manufacturing firms and found adoption of 'green' practices improves environmental and economic performance, which, in turn improves operational performance of the firm. Lo et al. (2012) studied the impact of the adoption of environmental management systems on firm financial performance, namely, return on assets (ROA), return on sales (ROS), and sales by total assets (SOA), in fashion and textile industries. Wiengarten et al. (2012) examined the relationship between environmental investments in supply chains and firm performance by comparing dynamic versus static industries. They found that in static industries the investments lead to improvement in operational performance in terms of cost, quality, and flexibility, whereas, in dynamic industries do not.

Laosirihongthong et al. (2013) studied the impact of 'green' supply chain management practices on Thailand manufacturing firm performance and found that reactive practices, such as legislation and regulation, had played major role in firm performance, whereas, proactive practices, such as reverse logistics had no role to play. Hasan (2013) studied the relationship between 'green' supply chain management practices and firm performance among the Austrian firms and established by adopting a case study approach. Here, the firm performance was further categorized into two dimensions, namely environmental-related and operational-related firm performance. Dubey et al. (2013) studied the drivers of 'green' purchasing strategies among the Indian firms that improve the firm performance. They found that that market pressure, leadership, regulation, and quality play important roles in determining the 'green' purchasing strategies. Further, the impact of 'green' supply chain management practices on Indian firm performance was studied and found that 'green' purchasing, supplier relationship, green logistics, and regulatory frameworks have strong influence on the firm performance (Dubey and Bag, 2014). Furthermore, specifically, the impact of 'green' supply chain practices was studied on Indian rubber industry's firm performance. It was found that best-in-class technology and waste reduction practices improve the firm's environmental performance (Dubey et al., 2014). Clark et al. (2014) studied the moderating role of stakeholder involvement between organization culture and firm performance from the perspective of market-oriented sustainability framework among US manufacturing firms. They found that market-orientation interacts with 'green' purchasing to improve the logistics performance.

Klassen and Vachon (2003) explored the supply chain activities of collaboration and evaluation and its impact on plant environmental performance in Canadian firms. They found that customer-initiated collaboration have significant impact on level of environmental investment in the plants than plant-initiated collaboration. Further, Vachon and Klassen (2006a) examined the antecedents of 'green' supply chain practices in North American firms and found that both logistical and technological integration related with environmental monitoring and collaboration. Furthermore, the role of collaboration in the relationship between environmental management and firm performance was studied in North American firms (Vachon and Klassen, 2008). Furthermore, Vachon and Klassen (2006b) investigated the relationship between 'green' project partnership and operational performance in Canada and US packaging printing industry. They found that partnership with customers were strongly associated with quality, flexibility, and environmental performance, while partnership with suppliers was associated with delivery performance. Moreover, Vachon and Mao (2008) examined the linkage between supply chain strength - measured in terms of quantity and quality of suppliers and customers - and sustainable development – measured in terms of environmental performance, corporate environmental practices, and social performance- at country-level.

6. PUTTING-IT-ALL-TOGETHER: RESOURCE ALLOCATION STRATEGIES FOR SUSTAINABLE OPERATIONS

Azzone and Noci (1998) developed a framework to design environmental performance measurement systems deploying 'green' manufacturing strategies. Reinhardt (1999) described the importance of environmental investments to reduce risks and earn positive returns as like other investment decisions in the firms. Klassen and Whybark (1999) studied the impact of environmental technologies on manufacturing performance. They found that investments in environmental technologies over time have significant affect on environmental and manufacturing performance. Vachon (2007) examined the relationship between 'green' supply chain practices and section of environmental technologies in Canada and US packaging printing industry. They found that collaboration with suppliers was strongly associated with investments in environmental technologies compared to collaboration with customers. Vachon and Klassen (2007) examined the relationship between supply chain characteristics and investments in environmental technologies. They found that integration with suppliers increase investments in environmental technologies but not with customers. Wang et al. (2011) presented an optimization model, investment decision problem, for designing 'green' supply chain network.

The operations strategies are nothing but designing and planning of the operational processes such that the resulting performances ensure the firms' investments towards 'green' initiatives. Resource allocation strategies can be characterized into four distinct strategies based upon 'green' practices and firm performance that capture the firm's underlying operational processes as follows:

1. **Speculative Allocation:** If the firm is not sure about the resulting firm performance then it always speculate about its 'green' initiatives and therefore make less investments. Here, the focus is more on financial performance rather than environmental or operational performance.
2. **Responsible Allocation:** If the firm is more responsible towards 'green' stimulus then it may invest more in such practices without much bothering about its financial performance. Here, the focus is more on environmental and operational performance.
3. **Efficient Allocation:** If the firm is more concerned about its cost efficiency in operations then it may invest less in 'green' initiatives. Here, the focus is more on financial performance rather than environmental or operational performance.
4. **Sustainable Allocation:** If the firm is more responsible towards 'green' campaign and also concerned about its cost efficiency in operations then it may invest less but practices more. Here, the focus is more holistic in nature and on all the firm performances.

In conclusion, the framework helps operations and supply chain managers to identify opportunities that ensure its supply chain is really 'green.' Lee (2010) also suggested "identifying opportunities that span the supply chain, reinventing manufacturing processes, and even linking up with competitors to tackle challenges of scale. The result can be a *greener supply chain* that requires less capital, has much lower operating costs, and provides a competitive advantage."

REFERENCES

Ageron, B., Gunasekaran, A., & Spalanzani, A. (2012). Sustainable supply management: An empirical study. *International Journal of Production Economics*, *140*(1), 168–182. doi:10.1016/j.ijpe.2011.04.007

Anderson, S. W., Daly, J. D., & Johnson, M. F. (1999). 'Why firms seek ISO 9000 certification: Regulatory compliance advantage? *Production and Operations Management*, *8*(1), 28–43. doi:10.1111/j.1937-5956.1999.tb00059.x

Azevedo, S. G., & Carvalho, H., & CruzMachado, V. (2011). The influence of green practices on supply chain performance: A case study approach. *Transportation Research Part E, Logistics and Transportation Review*, *47*(6), 850–871. doi:10.1016/j.tre.2011.05.017

Azzone, G., & Noci, G. (1998). Identifying effective PMSs for the deployment of 'green' manufacturing strategies. *International Journal of Operations & Production Management*, *18*(4), 308–335. doi:10.1108/01443579810199711

Bai, C., & Sarkis, J. (2010). Greener supplier development: Analytical evaluation using rough Set theory. *Journal of Cleaner Production*, *18*(12), 1200–1210. doi:10.1016/j.jclepro.2010.01.016

Bai, C., Sarkis, J., Wei, X., & Koh, L. (2012). 'Evaluating ecological sustainable performance measures for supply chain management', *Supply Chain Management. International Journal (Toronto, Ont.)*, *17*(1), 78–92.

Barari, S., Agarwal, G., Zhang, W. J., Mahanty, B., & Tiwari, M. K. (2012). A decision framework for the analysis of green supply chain contracts: An evolutionary game approach. *Expert Systems with Applications*, *39*(3), 2965–2976. doi:10.1016/j.eswa.2011.08.158

Barney, J. B. (1991). Firm resources and sustained competitive advantage. *Journal of Management*, *17*(1), 99–120. doi:10.1177/014920639101700108

Barney, J. B. (2012). Purchasing, supply chain management and sustained competitive advantage: The relevance of resource-based theory. *Journal of Supply Chain Management*, *48*(2), 3–6. doi:10.1111/j.1745-493X.2012.03265.x

Baskaran, V., Nachiappan, S., & Rahman, S. (2012). 'Indian textile suppliers' sustainability evaluation using the grey approach'. *International Journal of Production Economics*, *135*(2), 647–658. doi:10.1016/j.ijpe.2011.06.012

Berns, M., Townend, A., Khayat, Z., Balagopal, B., Reeves, M., Hopkins, M. S., & Kruschwitz, N. (2009). Sustainability and competitive advantage. *MIT Sloan Management Review*, *51*(2), 19–26.

Bierma, T. J., & Waterstraat, F. L. (1999). Cleaner production from chemical suppliers: Understanding shared savings contracts. *Journal of Cleaner Production*, *7*(2), 145–158. doi:10.1016/S0959-6526(98)00073-0

Bjorklund, M. (2011). Influence from the business environment to environmental purchasing – drivers and hinders of purchasing green transportation services. *Journal of Purchasing and Supply Management*, *17*(1), 11–22. doi:10.1016/j.pursup.2010.04.002

Bjorklund, M., Martinsen, U., & Abrahamsson, M. (2012). 'Performance measurements in the greening of the supply chains', *Supply Chain Management. International Journal (Toronto, Ont.)*, *17*(1), 29–39.

Blengini, G. A., & Shields, D. J. (2010). Green labels and sustainability reporting: Overview of the building products supply chain in Italy. *Management of Environmental Quality: An International Journal*, *21*(4), 477–493. doi:10.1108/14777831011049115

Boltic, Z., Ruzic, N., Jovanovic, M., Savic, M., Jovanovic, J., & Petrovic, S. (2013). Cleaner production aspects of tablet coating process in pharmaceutical industry: Problem of VOCs emission. *Journal of Cleaner Production*, *44*(1), 123–132. doi:10.1016/j.jclepro.2013.01.004

Bose, I., & Pal, R. (2012). Do green supply chain management initiatives impact stock prices of firms? *Decision Support Systems*, *52*(3), 624–634. doi:10.1016/j.dss.2011.10.020

Bowen, F. E., Cousins, P. D., Lamming, R. C., & Farukt, A. C. (2001). The role of supply management capabilities in green supply. *Production and Operations Management*, *10*(2), 174–189. doi:10.1111/j.1937-5956.2001.tb00077.x

Brandenburg, M., Govindan, K., Sarkis, J., & Seuring, S. (2014). Quantitative models for supply chain management: Developments and directions. *European Journal of Operational Research*, *233*(2), 299–312. doi:10.1016/j.ejor.2013.09.032

Brown, G., & Stone, L. (2007). Cleaner production in New Zealand: Taking stock. *Journal of Cleaner Production*, *15*(8-9), 716–728. doi:10.1016/j.jclepro.2006.06.025

Caniato, F., Caridi, M., Crippa, L., & Moretto, A. (2012). Environmental sustainability in fashion supply chains. *International Journal of Production Economics*, *135*(2), 559–570. doi:10.1016/j.ijpe.2011.06.001

Caniels, M. C. J., Gehrsitz, M. H., & Semeijn, J. (2013). Participation of suppliers in greening the supply chains: An empirical analysis of German automotive suppliers. *Journal of Purchasing and Supply Management*, *19*(3), 134–143. doi:10.1016/j.pursup.2013.02.005

Carter, C. R., Kale, R., & Grimm, C. M. (2000). Environmental purchasing and firm performance: An empirical investigation. *Transportation Research Part E, Logistics and Transportation Review*, *36*(3), 219–228. doi:10.1016/S1366-5545(99)00034-4

Carter, C. R., & Rogers, D. S. (2008). A framework for sustainable supply chain management: Moving toward new theory. *International Journal of Physical Distribution & Logistics Management*, *38*(5), 360–387. doi:10.1108/09600030810882816

Chaabane, A., Ramudhin, A., & Paquet, M. (2012). Design of sustainable supply chains under the emission trading scheme. *International Journal of Production Economics*, *135*(1), 37–49. doi:10.1016/j.ijpe.2010.10.025

Chen, I. J., Paulraj, A., & Lado, A. A. (2004). Strategic purchasing, supply management, and firm performance. *Journal of Operations Management*, *22*(5), 505–523. doi:10.1016/j.jom.2004.06.002

Chen, M.-K., Tai, T.-W., & Hung, T.-Y. (2012). Component selection system for green supply chain. *Expert Systems with Applications*, *39*(5), 5687–5701. doi:10.1016/j.eswa.2011.11.102

Choi, T. M., Lo, C. K., Wong, C. W., & Yee, R. W. (2012). Green manufacturing and distribution in the fashion and apparel industries. *International Journal of Production Economics*, *135*(2), 531. doi:10.1016/j.ijpe.2011.07.012

Clark, J. W., Toms, L. C., & Green, K. W. (2014). Market-oriented sustainability: Moderating impact of stakeholder involvement. *Industrial Management & Data Systems*, *114*(1), 21–36. doi:10.1108/IMDS-04-2013-0194

Clemens, B., & Douglas, T. J. (2005). Understanding strategic responses to institutional pressures. *Journal of Business Research*, *58*(9), 1205–1213. doi:10.1016/j.jbusres.2004.04.002

Clemens, B., & Douglas, T. J. (2006). 'Does coercion drive firms to adopt 'voluntary' green initiatives? Relationships among coercion, superior firm resources, and voluntary green initiatives'. *Journal of Business Research*, *59*(4), 483–491. doi:10.1016/j.jbusres.2005.09.016

Crespin-Mazet, F., & Dontenwill, E. (2012). Sustainable procurement: Building legitimacy in the supply network. *Journal of Purchasing and Supply Management*, *18*(4), 207–217. doi:10.1016/j.pursup.2012.01.002

Crozier, M., & Thoenig, J.-C. (1976). The regulation of complex organized systems. *Administrative Science Quarterly*, *21*(4), 547–570. doi:10.2307/2391716

Darnall, N., Henriques, I., & Sadorsky, P. (2008). Do environmental management systems improve business performance in an international setting? *Journal of International Management*, *14*(4), 364–376. doi:10.1016/j.intman.2007.09.006

Dekker, R., Bloemhof, J., & Mallidis, I. (2012). Operations research for green logistics – an overview of aspects, issues, contributions and challenges. *European Journal of Operational Research, 219*(3), 671–679. doi:10.1016/j.ejor.2011.11.010

Digalwar, A.K., Tagalpallewar, A.R. & Sunnapwar, V.K. (2013) Green manufacturing performance measures: an empirical investigation from Indian manufacturing industries. *Measuring Business Excellence*, 17(4), 59-75.

DiMaggio, P. J., & Powell, W. W. (1983). The iron cage revisited: Industrial isomorphism and collective rationality in organizational fields. *American Sociological Review*, *48*(2), 147–160. doi:10.2307/2095101

Dubey, R., Bag, S., & Ali, S. S. (2014). Green supply chain practices and its impact on organizational performance: An insight from Indian rubber industry. *International Journal of Logistics Systems and Management*, *19*(1), 20–42. doi:10.1504/IJLSM.2014.064029

Dubey, R., Bag, S., Ali, S. S., & Venkatesh, V. G. (2013). Green purchasing is key to superior performance: An empirical study. *International Journal of Procurement Management*, *6*(2), 187–210. doi:10.1504/IJPM.2013.052469

Dubey, R., & Bag, T. (2014) Antecedents of green manufacturing practices: a journey towards manufacturing sustainability. In Smart Manufacturing Innovation and Transformation: Interconnection and Intelligence (Ch. 13, pp. 333-354). doi:10.4018/978-1-4666-5836-3.ch013

Dubey, R., Gunasekaran, A., & Ali, S. S. (2015). Exploring the relationship between leadership, operational practices, institutional pressures and environmental pressure: A framework for green supply chain. *International Journal of Production Economics*, *160*(1), 120–132. doi:10.1016/j.ijpe.2014.10.001

Dubey, R., Gunasekaran, A., Singh, S., & Singh, T. (2015). Building theory of sustainable manufacturing using total interpretive structural modelling. *International Journal of Production Research*.

Eccles, R. G., Perkins, K. M., & Serafeim, G. (2012). How to become a sustainable company. *MIT Sloan Management Review*, *53*(4), 43–50.

Eiadat, Y., Kelly, A., Roche, F., & Eyadat, H. (2008). Green and competitive? An empirical test of the mediating role of environmental innovation strategy. *Journal of World Business*, *43*(2), 131–145. doi:10.1016/j.jwb.2007.11.012

Eisenhardt, K. M. (1989). Agency theory: An assessment and review. *Academy of Management Review*, *14*(1), 57–74.

Freeman, R. E. (1984). *Strategic Management: A Stakeholder Approach*. Boston: Pitman.

Frondel, M., Horbach, J., & Rennings, K. (2008). What triggers environmental management and innovation? Empirical evidence from Germany. *Ecological Economics*, *66*(1), 153–160. doi:10.1016/j.ecolecon.2007.08.016

Gavronski, I., Klassen, R. D., Vachon, S., & do Nascimento, L. S. M. (2011). A resource-based view of green supply management. *Transportation Research Part E, Logistics and Transportation Review*, *47*(6), 872–885. doi:10.1016/j.tre.2011.05.018

Ghosh, D., & Shah, J. (2012). A comparative analysis of greening polices across supply chain structures. *International Journal of Production Economics*, *135*(2), 568–583. doi:10.1016/j.ijpe.2011.05.027

Giddens, A. (1984). *The Constitution of Society: Outline of the Theory of Structuration*. Berkeley, CA: University of California Press.

Gimenez, C., Sierra, V., & Rodon, J. (2012). Sustainable operations: Their impact on the triple bottom line. *International Journal of Production Economics*, *140*(1), 149–159. doi:10.1016/j.ijpe.2012.01.035

Giunipero, L. C., Hooker, R. E., & Denslow, D. (2012). Purchasing and supply management sustainability. *Journal of Purchasing and Supply Management*, *18*(4), 258–269. doi:10.1016/j.pursup.2012.06.003

Glover, J. L., Champion, D., Daniels, K. J., & Dainty, A. D. J. (2014). An institutional theory perspective on sustainable practices across the diary supply chain. *International Journal of Production Economics*, *152*(1), 102–111. doi:10.1016/j.ijpe.2013.12.027

Goebel, P., Reuter, C., Pibernik, R., & Sichtmann, C. (2012). The influence of ethical culture on supplier selection in the context of sustainable sourcing. *International Journal of Production Economics*, *140*(1), 7–17. doi:10.1016/j.ijpe.2012.02.020

Granovetter, M. (1985). Economic action and social structure: The problem of embeddedness. *American Journal of Sociology*, *91*(3), 481–510. doi:10.1086/228311

Green, K. W. Jr, Zelbst, P. J., Meacham, J., & Bhadauria, V. S. (2012). Green supply chain management practices: Impact on performance. *Supply Chain Management: An International Journal*, *17*(3), 290–305. doi:10.1108/13598541211227126

Gunasekaran, A., & Spalanzani, A. (2012). Sustainability of manufacturing and services: Investigations for research and applications. *International Journal of Production Economics*, *140*(1), 35–47. doi:10.1016/j.ijpe.2011.05.011

Gupta, S., & Pasule-Desai, O. D. (2011). Sustainable supply chain management: Review and research opportunities. *IIMB Management Review*, *23*(4), 234–245. doi:10.1016/j.iimb.2011.09.002

Hall, J. (2000). Environmental supply chain dynamics. *Journal of Cleaner Production*, *8*(6), 455–471. doi:10.1016/S0959-6526(00)00013-5

Hart, S. L. (1995). A natural resource-based view of the firm. *Academy of Management Review*, *20*(4), 986–1014.

Hart, S. L. (1997). 'Beyond greening: Strategies for a sustainable world. *Harvard Business Review*, *75*, 66–77. PMID:10174798

Hasan, M. (2013). 'Sustainable supply chain management practices and operational performance. *American Journal of Industrial and Business Management*, *3*(1), 42–48. doi:10.4236/ajibm.2013.31006

Hassini, E., Surti, C., & Searcy, C. (2012). A literature review and a case study of sustainable supply chains with a focus on metrics. *International Journal of Production Economics*, *140*(1), 69–82. doi:10.1016/j.ijpe.2012.01.042

Hirsch, P. M. (1975). Organizational effectiveness and the institutional environment. *Administrative Science Quarterly*, *20*(3), 327–344. doi:10.2307/2391994

Hitchcock, T. (2012). Low carbon and green supply chains: The legal drivers and commercial pressures. *Supply Chain Management, International Journal (Toronto, Ont.)*, *17*(1), 98–101.

Hoejmose, S., Brammer, S., & Millington, A. (2012). Green supply chain management: The role of trust and top management in B2B and B2C markets. *Industrial Marketing Management*, *12*(4), 609–620. doi:10.1016/j.indmarman.2012.04.008

Hoejmose, S. U., & Adrien-Kirby, A. J. (2012). Socially and environmentally responsible procurement: A literature review and future research agenda of a managerial issue in the 21st century. *Journal of Purchasing and Supply Management*, *18*(4), 232–242. doi:10.1016/j.pursup.2012.06.002

Holt, D., & Ghobadian, A. (2009). An empirical study of green amongst UK manufacturers. *Journal of Manufacturing Technology Management*, *20*(7), 933–956. doi:10.1108/17410380910984212

Hu, A. H., & Hsu, C. (2010). Critical factors for implementing green supply chain management practice: An empirical study of electrical and electronics industries in Taiwan. *Management Research Review*, *33*(6), 586–608. doi:10.1108/01409171011050208

Janicke, M. (2008). Ecological modernization: New perspectives. *Journal of Cleaner Production*, *16*(5), 557–565. doi:10.1016/j.jclepro.2007.02.011

Johnstone, N., & Labonne, J. (2009). Why do manufacturing facilities introduce environmental management systems? Improving and/or signaling performance. *Ecological Economics*, *68*(3), 719–730. doi:10.1016/j.ecolecon.2008.06.003

Jones, C., Hesterly, W. S., & Borgatti, S. P. (1997). A general theory of network governance: Exchange conditions and social mechanisms. *Academy of Management Review*, *22*(4), 911–945.

Jorgensen, M. S., Jorgensen, U., Hendriksen, K., Hirsbak, S., Thomsen, H. H., & Thorsen, N. (2010). Environmental management in Danish transnational textile product chains. *Management Research Review*, *33*(4), 357–379. doi:10.1108/01409171011030462

Kang, S.-H., Kang, B., Shim, K., Kim, D., & Han, J. (2012). A theoretical framework for strategy development to introduce sustainable supply chain management. *Procedia: Social and Behavioral Sciences*, *40*(1), 631–635. doi:10.1016/j.sbspro.2012.03.241

King, A. A., & Lenox, M. J. (2001). Lean and green? And empirical examination of the relationship between lean production and environmental performance. *Production and Operations Management*, *10*(3), 244–256. doi:10.1111/j.1937-5956.2001.tb00373.x

Klassen, R. D., & McLaughlin, C. (1996). The impact of environmental management on firm performance. *Management Science*, *42*(8), 1199–1214. doi:10.1287/mnsc.42.8.1199

Klassen, R. D., & Vachon, S. (2003). Collaboration and evaluation in the supply chain: Their impact on plant-level environmental investment. *Production and Operations Management*, *12*(3), 336–352. doi:10.1111/j.1937-5956.2003.tb00207.x

Klassen, R. D., & Whybark, D. C. (1999). The impact of environmental technologies on manufacturing performance. *Academy of Management Journal*, *42*(6), 599–615. doi:10.2307/256982

Kleindorfer, P. R., Singhal, K., & Wassenhove, L. N. (2005). Sustainable operations management. *Production and Operations Management*, *14*(4), 482–492. doi:10.1111/j.1937-5956.2005.tb00235.x

Kudla, N. L., & Klaas-Wissing, T. (2012). Sustainability in shipper-logistics service provider relationships: A tentative taxonomy based on agency theory and stimulus-response analysis. *Journal of Purchasing and Supply Management*, *18*(4), 218–231. doi:10.1016/j.pursup.2012.04.001

Laosirihongthong, T., & Adebanjo, D., & ChoonTan, K. (2013). Green supply chain management practices and performances. *Industrial Management & Data Systems*, *113*(8), 1088–1109. doi:10.1108/IMDS-04-2013-0164

Large, R. O., Kramer, N., & Hartmann, R. K. (2013). Procurement of logistics services and sustainable development in Europe: Fields of activity and empirical results. *Journal of Purchasing and Supply Management*, *19*(3), 122–133. doi:10.1016/j.pursup.2013.05.002

Lau, K. H. (2011). Benchmarking green logistics performance with a composite score. *Benchmarking International Journal (Toronto, Ont.)*, *18*(6), 873–896.

Lee, L. H. (2010). 'Don't tweak your supply chain - rethink it end to end'. *Harvard Business Review*, *88*, 46–53.

Linton, J., Klassen, R., & Jayaraman, V. (2007). Sustainable supply chains: An introduction. *Journal of Operations Management*, *25*(6), 1075–1082. doi:10.1016/j.jom.2007.01.012

Liu, S., Kasturiratne, D., & Moizer, J. (2012). A hub-and-spoke model for multi-dimensional integration of green marketing and sustainable supply chain management. *Industrial Marketing Management*, *41*(4), 581–588. doi:10.1016/j.indmarman.2012.04.005

Lo, C. K. Y., Yeung, A. C. L., & Cheng, T. C. E. (2012). The impact of environmental management systems on financial performance in fashion and textile industries. *International Journal of Production Economics*, *135*(2), 561–567. doi:10.1016/j.ijpe.2011.05.010

Melnyk, S. A., Sroufe, R. P., & Calantone, R. (2003). Assessing the impact of environmental management systems on corporate and environmental performance. *Journal of Operations Management*, *21*(3), 329–351. doi:10.1016/S0272-6963(02)00109-2

Mirhedayatian, S.M., Azadi, M. & Saen, R.F. (2014). A novel network data envelopment analysis model for evaluating green supply chain management. *International Journal of Production Economics*, 147(1 Part B), 544-554.

Mollenkopf, D., Stolze, H., Tate, W. L., & Ueltschy, M. (2010). Green, lean, and global supply chains. *International Journal of Physical Distribution & Logistics Management*, *40*(1/2), 14–41. doi:10.1108/09600031011018028

Murphy, J., & Gouldson, A. (2000). Environmental policy and industrial innovation: Integrating environment and economy through ecological modernization. *Geoforum*, *31*(1), 33–44. doi:10.1016/S0016-7185(99)00042-1

Musa, A., Yusuf, Y., McArdle, L., & Banjoko, G. (2013). Corporate social responsibility in Nigeria' oil and gas industry: The perspectives of the industry. *International Journal of Process Management and Benchmarking*, *3*(2), 101–135. doi:10.1504/IJPMB.2013.057722

Oruezabala, G., & Rico, J.-C. (2012). The impact of sustainable public procurement on supplier management – the case of French public hospitals. *Industrial Marketing Management*, *41*(4), 573–580. doi:10.1016/j.indmarman.2012.04.004

Pagell, M., & Wu, Z. (2009). Building a more complete theory of sustainable supply chain management using case studies of 10 exemplars. *Journal of Supply Chain Management*, *45*(2), 37–56. doi:10.1111/j.1745-493X.2009.03162.x

Pfeffer, J., & Salancik, G. (1978). *The External Control of Organizations: A Resource Dependence Perspective*. New York, NY: Harper & Row.

Pierson, P. (2000). Increasing returns, path dependence, and study of politics. *The American Political Science Review*, *94*(2), 251–267. doi:10.2307/2586011

Porter, M. E., & Linde, C. (1995). Green and competitive: Ending the stalemate. *Harvard Business Review*, *73*(5), 120–134.

Pujari, D. (2006). Eco-innovation and new product development: Understanding the influences on market performance. *Technovation*, *26*(1), 76–85. doi:10.1016/j.technovation.2004.07.006

Pujari, D., Peattie, K., & Wright, G. (2004). Organizational antecedents of environmental responsiveness in industrial new product development. *Industrial Marketing Management*, *33*(5), 381–391. doi:10.1016/j.indmarman.2003.09.001

Pujari, D., Wright, G., & Peattie, K. (2003). Green and competitive: Influences on environmental new product development performance. *Journal of Business Research*, *56*(8), 657–671. doi:10.1016/S0148-2963(01)00310-1

Rao, P. (2002). Greening the supply chain: A new initiative in South East Asia. *International Journal of Operations & Production Management*, *22*(6), 632–655. doi:10.1108/01443570210427668

Rao, P. (2004). Greening production: A South-East Asian experience'. *International Journal of Operations & Production Management*, *24*(3), 289–320. doi:10.1108/01443570410519042

Rao, P., & Holt, D. (2005). Do green supply chains lead to competitiveness and economic performance? *International Journal of Operations & Production Management*, *25*(5), 898–916. doi:10.1108/01443570510613956

Reefke, H., & Trocchi, M. (2013). Balanced scorecard for sustainable supply chains: Design and development guidelines. *International Journal of Productivity and Performance Management*, *62*(8), 805–826. doi:10.1108/IJPPM-02-2013-0029

Reinhardt, F. L. (1999). Bringing the environment down to earth. *Harvard Business Review*, *77*(4), 149–157. PMID:10539206

Reuter, C., Goebel, P., & Foerstl, K. (2012). The impact of shareholder orientation on sustainability and cost prevalence in supplier selection decisions. *Journal of Purchasing and Supply Management*, *18*(4), 270–281. doi:10.1016/j.pursup.2012.06.004

Rogers, E. M. (2003). *Diffusion of Innovations* (5th ed.). New York, NY: Free Press.

Rothenberg, S. (2007). Environmental managers as institutional entrepreneurs: The influence of institutional and technical pressures on waste management. *Journal of Business Research*, *60*(7), 749–757. doi:10.1016/j.jbusres.2007.02.017

Rowley, T. (1997). Moving beyond dyadic ties: A network theory of stakeholder influences. *Academy of Management Review*, *22*(4), 887–910.

Sahu, A. K., Datta, S., & Mahapatra, S. S. (2013). Green supply chain performance benchmarking using integrated IVFN-TOPSIS methodology. *International Journal of Process Management and Benchmarking*, *3*(4), 511–551. doi:10.1504/IJPMB.2013.058272

Sarkis, J. (2003). A strategic framework for green supply chain management. *Journal of Cleaner Production*, *11*(4), 397–409. doi:10.1016/S0959-6526(02)00062-8

Sarkis, J., Zhu, Q., & Lai, K.-H. (2011). An organizational theoretic review of green supply chain management literature. *International Journal of Production Economics*, *130*(1), 1–15. doi:10.1016/j.ijpe.2010.11.010

Schneider, L., & Wallenburg, C. M. (2012). Implementing sustainable sourcing – does purchasing need to change? *Journal of Purchasing and Supply Management*, *18*(4), 243–257. doi:10.1016/j.pursup.2012.03.002

Seuring, S. (2013). A review of modeling approaches for sustainable supply chain management. *Decision Support Systems*, *54*(4), 1513–1520. doi:10.1016/j.dss.2012.05.053

Shi, V.G., Koh, S.C.L., Baldwin, J. & Cucchiella, F. (2012). *Natural resource based green supply chain management*, 12(1), 54-67.

Simpson, D., Power, D., & Samson, D. (2007). Greening the automotive supply chain: A relationship perspective. *International Journal of Operations & Production Management*, *27*(1), 28–48. doi:10.1108/01443570710714529

Souza, G. C. (2013). Closed-loop supply chains: A critical review, and future research. *Decision Sciences*, *44*(1), 7–38. doi:10.1111/j.1540-5915.2012.00394.x

Srivastava, S. K. (2007). Green supply chain management: A state-of-the-art- literature review. *International Journal of Management Reviews*, *9*(1), 53–80. doi:10.1111/j.1468-2370.2007.00202.x

Tate, W. L., Ellram, L. M., & Dooley, K. J. (2012). Environmental purchasing and supplier management: Theory and practice. *Journal of Purchasing and Supply Management*, *18*(3), 173–188. doi:10.1016/j.pursup.2012.07.001

Unruh, G. (2010). *Earth, Inc.: Using Nature's Rules to Build Sustainable Profits*. Boston, Massachusetts: Harvard Business Press.

Vachon, S. (2007). Green supply chain practices and the selection of environmental technologies. *International Journal of Production Research*, *45*(18-19), 4357–4379. doi:10.1080/00207540701440303

Vachon, S., & Klassen, R. D. (2006a). Extending green practices across the supply chain: The impact of upstream and downstream integration. *International Journal of Operations & Production Management*, *26*(7), 795–821. doi:10.1108/01443570610672248

Vachon, S., & Klassen, R. D. (2006b). Green project partnership in the supply chain: The case of the package printing industry. *Journal of Cleaner Production*, *14*(6-7), 661–671. doi:10.1016/j.jclepro.2005.07.014

Vachon, S., & Klassen, R. D. (2007). Supply chain management and environmental technologies: The role of integration. *International Journal of Production Research*, *45*(2), 401–423. doi:10.1080/00207540600597781

Vachon, S., & Klassen, R. D. (2008). Environmental management and manufacturing performance: The role of collaboration in the supply chain. *International Journal of Production Economics*, *111*(2), 299–315. doi:10.1016/j.ijpe.2006.11.030

Vachon, S., & Mao, Z. (2008). Linking supply chain strength to sustainable development: A country-level analysis. *Journal of Cleaner Production*, *16*(15), 1552–1560. doi:10.1016/j.jclepro.2008.04.012

Vimal, K. E. K., & Vinodh, S. (2013). Development of checklist for evaluating sustainability characteristics of manufacturing processes. *International Journal of Process Management and Benchmarking*, *3*(2), 213–232. doi:10.1504/IJPMB.2013.057726

Walker, H., DiSisto, L., & McBain, D. (2008). Drivers and barriers to environmental supply chain management practices: Lessons from the public and private sectors. *Journal of Purchasing and Supply Management*, *14*(1), 69–85. doi:10.1016/j.pursup.2008.01.007

Walker, H., & Brammer, S. (2012). The relationship between sustainable procurement and e-procurement in the public sector. *International Journal of Production Economics*, *140*(1), 256–268. doi:10.1016/j.ijpe.2012.01.008

Walker, H., Miemczyk, J., Johnsen, T., & Spencer, R. (2012). Sustainable procurement: Past, present and future. *Journal of Purchasing and Supply Management*, *18*(4), 201–206. doi:10.1016/j.pursup.2012.11.003

Wang, F., Lai, X., & Shi, N. (2011). A multi-objective optimization for green supply chain network design. *Decision Support Systems*, *51*(2), 262–269. doi:10.1016/j.dss.2010.11.020

Wiengarten, F., Pagell, M., & Fynes, B. (2012). Supply chain environmental investments in dynamic industries: Comparing investment and performance differences with static industries. *International Journal of Production Economics*, *135*(2), 541–551. doi:10.1016/j.ijpe.2011.03.011

Williamson, O. E. (1981). The economics of organization: The transaction cost approach. *American Journal of Sociology*, *87*(3), 548–577. doi:10.1086/227496

Wu, Z., & Pagell, M. (2011). Balancing priorities: Decision-making in sustainable supply chain management. *Journal of Operations Management*, *29*(6), 577–590. doi:10.1016/j.jom.2010.10.001

Zhu, Q., Geng, Y., Fujita, T., & Hashimoto, S. (2010). Green supply chain management in leading manufacturers: Case studies in Japanese large companies. *Management Research Review*, *33*(4), 380–392. doi:10.1108/01409171011030471

Zhu, Q., Geng, Y., Fujita, T., & Hashimoto, S. (2010). Green supply chain management in leading manufacturers: Case studies in Japanese large companies. *Management Research Review*, *33*(4), 380–392. doi:10.1108/01409171011030471

Zhu, Q., & Sarkis, J. (2004). Relationships between operational practices and performance among early adopters of green supply chain management practices in Chinese manufacturing enterprises. *Journal of Operations Management*, *22*(3), 265–289. doi:10.1016/j.jom.2004.01.005

Zhu, Q., & Sarkis, J. (2004, June). Supply chain management practices in Chinese manufacturing enterprises. *Journal of Operations Management*, *22*(3), 265–289. doi:10.1016/j.jom.2004.01.005

Zhu, Q., Sarkis, J., & Geng, Y. (2005). Green supply chain management in China: Pressures, practices, and performance. *International Journal of Operations & Production Management*, *25*(5), 449–468. doi:10.1108/01443570510593148

Zhu, Q., Sarkis, J., & Geng, Y. (2005). Green supply chain management in China: Pressures, practices, and performance. *International Journal of Operations & Production Management*, *25*(5), 449–468. doi:10.1108/01443570510593148

Zhu, Q., Sarkis, J., & Lai, K.-H. (2008a). Confirmation of a measurement model for green supply chain management practices implementation. *International Journal of Production Economics*, *111*(2), 261–273. doi:10.1016/j.ijpe.2006.11.029

Zhu, Q., Sarkis, J., & Lai, K.-H. (2008a). Confirmation of a measurement model for green supply chain management practices implementation. *International Journal of Production Economics*, *111*(2), 261–273. doi:10.1016/j.ijpe.2006.11.029

Zhu, Q., Sarkis, J., & Lai, K.-H. (2008b). Green supply chain management implications for closing the loop. *Transportation Research Part E, Logistics and Transportation Review*, *44*(1), 1–18. doi:10.1016/j.tre.2006.06.003

Zhu, Q., Sarkis, J., & Lai, K.-H. (2008b). Green supply chain management implications for closing the loop. *Transportation Research Part E, Logistics and Transportation Review*, *44*(1), 1–18. doi:10.1016/j.tre.2006.06.003

Zhu, Q., Sarkis, J., & Lai, K.-H. (2013). Institutional-based antecedents and performance outcomes of internal and external green supply chain management practices. *Journal of Purchasing and Supply Management*, *19*(2), 106–117. doi:10.1016/j.pursup.2012.12.001

Zhu, Q., Sarkis, J., & Lai, K.-H. (2013). Institutional-based antecedents and performance outcomes of internal and external green supply chain management practices. *Journal of Purchasing and Supply Management*, *19*(2), 106–117. doi:10.1016/j.pursup.2012.12.001

Zsidisin, G. A., & Siferd, S. P. (2001). Environmental purchasing: A framework for theory development. *European Journal of Purchasing and Supply Management*, *7*(1), 61–73. doi:10.1016/S0969-7012(00)00007-1

Chapter 2
An Empirical Study on Factors Leading to Employee Retention for Manufacturing Industry in India

Rupali Singh
Motilal Nehru National Institute of Technology, India

Anil Sankhi
Constantia Parikh Packaging, India

ABSTRACT

The purpose of this paper is to develop an understanding of employee's retention in manufacturing organizations. The factors identified for the study are work culture, training, compensation, motivation of employee's for their profession, and employee's personality factors which are mediating through job satisfaction of employees. The study established statistically significant relationship between the factors and employee's retention with mediation effect of job satisfaction. Results revealed the importance of job satisfaction as a mediator in predicting employee's retention in manufacturing industry.

1. INTRODUCTION

Talent Management (TM) is widely accepted in corporate and its subsystems are practiced extensively. TM as it is used is a term without value unless it is grounded in a strategic decision framework (Lewis & Heckman, 2006). Retention is one of the sub systems of TM which has substituted the concept of turnover in industry. Employee retention is considered to be proactive approach to increase the efficiency of an organization.

TM becomes all the more complex when used in context of manufacturing industry. A decade into the 21st century, the role of manufacturing in the global economy continues to evolve (Manyika, 2012). The new era of manufacturing will be marked by highly agile, networked enterprises that employ talent as skillfully as machinery to deliver products to diverse global markets. In advanced economies, manu-

DOI: 10.4018/978-1-5225-0350-7.ch002

facturing will continue to drive innovation, exports, and productivity growth. In developing economies, manufacturing will continue to providing a pathway to higher living standards. As long as companies and countries understanding the evolving nature of manufacturing and act on the powerful trends shaping the global competitive environment, they can thrive in the promising future.

We will try to explore the possible linkage between the factors leading to employee retention in mediating effect of employee job satisfaction in context of manufacturing industries.

2. LITERATURE REVIEW

In this section we will examine the existing research of various scholars in the related fields. A Systematic Literature Review (SLR) has been used as suggested by Tranfield et al., 2003. The objectives of the literature review are:

1. To define factors that leads to executive level employee's retention in manufacturing organizations in India.
2. To develop and statistically validate a research model showing relationship between the factors that leads to employee's retention in manufacturing organizations in India.

The literature review is further divided into two stages namely planning the review and conducting review which is duly presented in subsequent section.

Stage 1: Planning the Review

Planning is further divided into three phases as follows:

Phase 1: Identification for the Need of Review

In our present study we want to investigate the impact of work culture, training and development, compensation, motivation for profession, and personality factors on employee's retention under the mediating effect of employee's job satisfaction.

Phase 2: Proposal for Review

The literature review is an integral component of any scientific works. To begin with this section, we first identified reputable journals in the field of Talent Management such as *Human Resource Management Review, The Social Science Journal, The Industrial and Commercial Training, The TQM Magazine, Journal of Managerial Psychology, Journal of Applied Psychology, Strategic HR Review, Strategic Management Journal, Human Resource Planning, Journal of Management, Harvard Business Review, Human Resource Planning, Compensation and Benefits Management, Journal of Management, International Journal of Humanities and Social Science, International Journal of Social Science and Interdisciplinary Research* and other related journals. We used key words like Talent Management, Employee Retention, Executive Employee Retention, Manufacturing Organizations, Job Satisfaction, Work Culture, Training and Development, Employee Compensation, Motivation in manufacturing organizations, Personality factors of executives. In this way, we identified over 100 articles published in mentioned journals.

Phase 3: Development of Review Protocol

We decided initially to review published literatures. In order to eliminate our biasness towards any work, we reviewed all possible articles published in past twenty years in all mentioned journals related to talent management, employee retention, employee job satisfaction and its related concepts so that we do not end up with sketchy literature.

Stage 2: Conducting Review

Execution of the review is further divided into three phases:

Phase 4: Identification of Research

We initially reviewed some of the articles that deal with Talent Management and its related issues to identify our research and define the scope of the present study (e.g., Buckingham & Vosburgh, 2001; Axelrod et al., 2002; Ashton & Morton, 2005; Boudreau & Ramstad, 2005; Cheloha & Swain, 2005; Lewis & Heckman, 2006; Smilansky, 2006; Garg & Rastogi, 2006; Pasewark & Viator, 2006; Freyermuth, 2007; Cheese et al., 2008; Hytter, 2008; Sinha & Sinha, 2008; Gaan, 2008; Kumar, Udayasuriyan & Vimala, 2008; Schiemann, 2009; Gaiduk & Gaiduk, 2009; Makhijani et at., 2009; Ihsan & Naeem, 2009; Yanadoria & Katob, 2010; Iles, Chuai, & Preece, 2010; Irshad, 2012).

The issues identified from literature are duly presented in Table 1: Issues identified from the literature. The horizontal rows of table indicate the authors while vertical columns denotes the variables they have researched over a period of time signified by the alphabet 'A' across each reference.

We derive following issues relevant to the present study:

1. **Work Culture, Employee Job Satisfaction, and Employee Retention** (Lambert et al., 2001; Chuai, & Preece, 2010): Study opined that the work culture is more important in shaping employee job satisfaction than demographic characteristics and that job satisfaction is a highly salient antecedent of turnover intent. Mostly studies have been done on turnover and not on retention of employees, thus, it is pertinent to observe the impact of work culture on employee retention with mediating effect of employee job satisfaction.
2. **Training and development, Job Satisfaction, and Employee Retention** (Hall, 1996; Prince, 2005): Training and development is vital for both employee and employer. To gain and maintain competitive advantage, organizations require talented and productive employees. These employees need training and development opportunities to enhance and cultivate their competencies. Thus, it is important to study that what is the impact of training and development of an executive employee has on its job satisfaction and retention in organization.
3. **Compensation, Job Satisfaction, and Employee Retention** (Ihsan & Naeem, 2009): Study opined that the pay and fringe benefits is the most important retention factor which is highly valued by the sales force of all demographic backgrounds. With a distinct view, Highhouse et al., 1999 suggested that pay is not sufficient to retain employees. A low paid employee can be oriented to switch job

Table 1. Issues identified from the literature

S. No.	Reference	Issues/ Focus of the Study						
		Work culture	Training and Development	Compen-sation	Motivation for Profession	Personality factors	Job Satisfaction	Employee Retention
1	Steel and Ovalle, 1984						A	A
2	Garg, 1985	A						
3	Price and Mueller, 1986						A	A
4	Williams and Hazer, 1986						A	A
5	Verma, 1986						A	
6	Friesen, 1986					A		
7	Wildman and Niles, 1987		A					
8	Lee and Mowday, 1987						A	A
9	Lachman and Diamant, 1987							A
10	Verma, 1988				A		A	
11	Cascio, 1987			A				
12	Graddick, 1988					A		
13	Putti, 1989	A					A	
14	Eisenberger and Associates, 1990	A						
15	Lawler 1990			A				
16	Reilly et al., 1991				A			
17	Goldstein, 1991		A					
18	Roznowski and Hulin 1992						A	
19	Cranny, Smith and Stone, 1992						A	
20	Williams and Dreher, 1992			A				
21	Storey and Sisson 1993		A					
22	Arthur, 1994			A				
23	Greenhaus, 1994	A			A			
24	Huselid, 1995			A				
25	Judge, Hanisch, and Drankoski 1995						A	
26	Gomez et al, 1995		A					
27	Lauri, Benson and Cheney, 1996		A					
28	Black and Lynch, 1996		A					
29	Hall, 1996					A		
30	Koustelios and Bagiatis, 1997						A	
31	Steers, 1997				A			
32	Netemeyer et al., 1997				A			
33	Cable and Judge 1997				A			
34	Hall & Moss, 1998					A		
35	Pfeffer, 1998			A				
36	Bishop, 1998		A					
37	Frazis et at, 1998		A					
38	Lum et al., 1998						A	

continued on following page

Table 1. Continued

S. No.	Reference	Issues/ Focus of the Study						
		Work culture	Training and Development	Compen-sation	Motivation for Profession	Personality factors	Job Satisfaction	Employee Retention
39	Edward, 1999				A			
40	Highhouse et al, 1999			A				
41	Noe, 1999		A					
42	Brannick, 1999			A				
43	Watson Wyatt, 1999			A				
44	Woodruffe, 1999					A		
45	Bassi and Van Buren, 1999		A					
46	Sjoberg and Sverke, 2000	A						
47	Olajide, 2000					A	A	
48	Eskildesen and Hahlgaard, 2000						A	
49	Messmer, 2000		A					
50	Kirby, 2000						A	
51	Konovsky, 2000	A						
52	Willis, 2000			A				
53	Kim, 2000						A	
54	Lambert, Hogan and Barton, 2001	A					A	A
55	Walker, 2001			A				
56	Davies, 2001			A				
57	Gold, 2001			A				
58	Clarke, 2001		A				A	
59	Butler and Waldrop, 2001					A		
60	Smith, 2001			A				
61	Michaels et al., 2001							A
62	Parker and Wright, 2001			A				
63	Accenture, 2001		A					
64	Allen, 2001	A						
65	Buckingham and Vosburgh, 2001				A			
66	Steel et al., 2002					A		
67	Shields and Price, 2002						A	
68	Hsu, Jiang, Klein and Tang, 2003					A		
69	Tower, 2003			A				
70	Wetland, 2003		A					
71	Ramlall, 2003	A						
72	Robertson and Abbey, 2003				A			
73	Mercer, 2003			A				
74	Amar 2004				A			
75	Kyriakidou and Ozbilgin, 2004					A		
76	Ontario, 2004	A						

continued on following page

Table 1. Continued

S. No.	Reference	Issues/ Focus of the Study						
		Work culture	Training and Development	Compen-sation	Motivation for Profession	Personality factors	Job Satisfaction	Employee Retention
77	Bockerman and Iimakunnas, 2004						A	
78	Kristensen et al., 2004						A	
79	Light, 2004						A	
80	Silbert, 2005	A		A				
81	Prince, 2005					A		
82	Zeytinoglu and Denton, 2005	A						
83	Thompson and Prottas, 2005	A						
84	Smilansky, 2006		A					
85	Garg and Rastogi, 2006		A					
86	Pasewark and Viator, 2006	A						
87	Freyermuth, 2007	A						
88	Cheese et al., 2008				A			
89	Hytter, 2008	A						
90	Sinha and Sinha, 2008	A	A	A				A
91	Gaan, 2008	A						
92	Kumar, Udayasuriyan and Vimala, 2008				A		A	
93	Schiemann, 2009				A			
94	Gaiduk and Gaiduk, 2009	A	A					
95	Makhijani et at., 2009		A					
96	Ihsan and Naeem, 2009			A				
97	Yanadoria and Katob, 2010	A						
98	Iles, Chuai, and Preece, 2010		A					
	Total	20	20	20	13	11	22	08

Source: Compiled by author after reviewing various sources

but a high pay may not bring and keep the employees in organization. Thus, it becomes relevant to study impact of compensation on employee retention with mediating effect of employee job satisfaction.

4. **Motivation for Profession, Job Satisfaction, and Employee Retention** (Gaan, 2008; Yanadoria & Katob, 2010): Studies suggests that employee motivation for profession refers to one's basic instinct to choose and carry forward their occupation and feel committed towards organization. Motivational factors bring in satisfaction and helps in reducing turnover (Reilly et al, 1991). It includes views of employees about their profession and hence it becomes pertinent to observe that what motivates in getting and staying of an employee to its profession (Huberman, 1989).
5. **Personality Factors, Job Satisfaction, and Employee Retention** (Makhijani et at., 2009): It is not only the organizational factors which have impact on employee's retention but also individual inherent factors (Kyriakidou & Ozbilgin, 2004). Personality factors are inherent factors which belong to emotional and mental health (Coates & Thoresen, 1976; Farber, 1991) and have a significant

impact on employee's job satisfaction and intention to stay with their organizations. Personality factors can be both, positive or negative. Thus, it is pertinent to observe the impact of employee's personality factors on employee's job satisfaction and retention.

Phase 5: Selection of Studies

We observed that mostly literature is related to employee turnover and not retention with mediation effect of employee's job satisfaction. Also, organizations incur high cost with high employee turnover. Thus, we decided to study retention in manufacturing organizations from employee's perspective.

Phase 6: Study Quality Assessment

We reviewed the articles related to research design adopted, data collection procedures and data analysis. We observed lot of variation in approaches adopted by researchers in terms of research objectives, research design and scope of the study.

The variables we identified through literature review are work culture, training and development, compensation, motivation for profession, personality factors, employee's job satisfaction, and employee's retention.

In the next section, we will discuss our theoretical framework, hypotheses development and research design to test our theoretical framework.

3. THEORETICAL FRAMEWORK, RESEARCH HYPOTHESIS AND RESEARCH DESIGN

This section includes the framework for the study, corresponding research hypothesis and the structure to test these hypotheses under research design:

3.1. Theoretical Framework

As discussed in literature review, the theoretical framework includes work culture, training and development, compensation, motivation for profession, and personality factors as independent factors leading to employee's retention in organization with a mediating effect of employee's job satisfaction. The diagrammatic representation of the theoretical framework is duly presented in Figure 1: Theoretical Framework of Employee Retention

3.2. Research Hypothesis

The following hypotheses are drawn for the present study:

Hypothesis H_{1a}: Work Culture leads to Employee's Retention.
Hypothesis H_{1b}: Work Culture leads to Job Satisfaction.
Hypothesis H_{1c}: Work Culture leads to Employee's Retention mediating through Job Satisfaction.
Hypothesis H_{2a}: Training and Development leads to Employee's Retention.

Figure 1. Theoretical Framework of Employee Retention

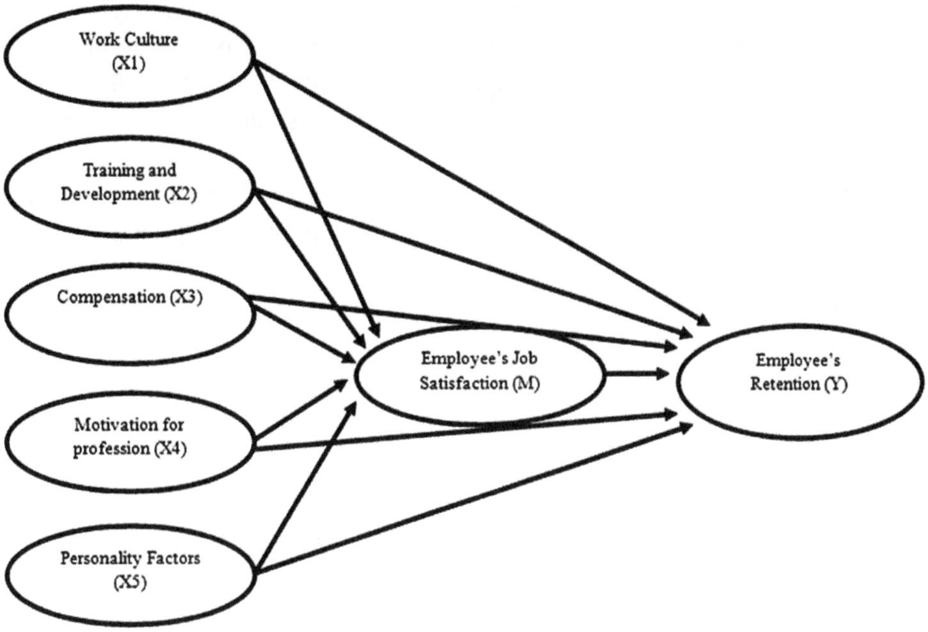

Hypothesis H_{2b}: Training and Development leads to Job Satisfaction.
Hypothesis H_{2c}: Training and Development leads to Employee's Retention mediating through Job Satisfaction.
Hypothesis H_{3a}: Compensation leads to Employee's Retention.
Hypothesis H_{3b}: Compensation leads to Job Satisfaction.
Hypothesis H_{3c}: Compensation leads to Employee's Retention mediating though Job Satisfaction.
Hypothesis H_{4a}: Motivation for Profession leads to Employee's Retention.
Hypothesis H_{4b}: Motivation for Profession leads to Job Satisfaction.
Hypothesis H_{4c}: Motivation for Profession leads to Employee's Retention mediating through Job Satisfaction.
Hypothesis H_{5a}: Personality Factors leads to Employee's Retention.
Hypothesis H_{5b}: Personality Factors leads to Job Satisfaction.
Hypothesis H_{5c}: Personality Factors leads to Employee's Retention mediating through Job Satisfaction.
Hypothesis H_6: Job Satisfaction leads to Employee's Retention.

3.3. Questionnaire Development

To develop measuring instrument for the survey, researchers has adopted three approaches:

1. **Exhaustive Literature Survey:** It includes work culture, training and development, compensation, motivation for profession, personality factors, employee job satisfaction and employee's retention.
2. **Experts Opinion Followed by Pretesting:** For the purpose of pretesting, 15 experts that include 4 top level employees from manufacturing organization, 5 director level professors with work experi-

ence of manufacturing organizations and 6 reputed academicians whose papers have been cited and proven authority. Study had 62 items originally which were reduced to 23 items based on pre-test results. The questionnaire was divided into two sections. First section was the demographic profile of the respondent and second section was related to the employee's retention study.

3. **Pilot Survey to Check the Reliability and Validity of Constructs:** It is important to ensure sufficient pilot work is carried out during development of measure. This will identify items that lack clarity or may not be appropriate for, or discriminate between respondents (Rattray and Jones, 2007). We have initially checked three dimensions during piloting of questionnaire:
    a. Items analysis
    b. Internal consistency
    c. Face or content validity

3.4. Data Collection

The questionnaire was sent to the executive level employees working in flexible laminates industry with work experience of three or more years. The questionnaire was designed with both positive and negative statements to avoid response biasness. Personal administering method was used and the respondents were assured regarding the confidentiality of their responses will be maintained.

After circulating 350 questionnaires, researcher could receive 295 responses where 206 were usable for the study. Questionnaire was prepared in both hard copies as well as on-line on google docs and respondents were asked to respond as per their convenience.

3.5. Measures

Measures were modified after adopting from extant research. The scales were not directly adopted as they were developed earlier in context to employee's turnover intentions. We used multi-item measures of constructs for our theoretical framework in order to improve reliability, reduce measurement error, ensure greater variability among survey individuals and improve validity (Churchill, 1979). Each construct was operationalised using at least three items for effective measurement and analysis, applying confirmatory factor analysis (Anderson & Gerbing, 1988). The under mentioned Table 2: Constructs of Theoretical Framework; presents the variables, their definitions, items drawn in each variable, and the corresponding references.

The questionnaire consisting of the above mentioned scale and items were captured using five-point Likert scales varying from strongly disagree to strongly agree.

4. DATA ANALYSIS

After collecting data which have been discussed in the preceding section, the following tests were performed for the study:

Table 2. Constructs of Theoretical Framework

S. No.	Variables	Definitions	Items Drawn for Study	References
Independent Variables				
1	Work Culture	Work culture can be defined as the organization's and its supervisor's support to employees, and encouragement to participate in organization's decision making.	• Supervisor's support • Organization's support • Safe environment • Employee's participation	Lambert et al., 2001; Iles, Chuai, and Preece, 2010
2	Training and Development	Training development refers to imparting valuable information to employees by which their skills, knowledge and abilities can be enhanced with a permanent change in their thoughts, behavior and actions.	• Induction program • Mentoring program • Staff development program	Hall, 1996; Prince, 2005
3	Compensation	Compensation means the adequate remuneration paid by an organization in terms of money or fringe benefits in lieu of services rendered by an employee.	• Satisfactory salary • Incentives	Ihsan and Naeem, 2009
4	Motivation for Profession	Motivation for profession refers to one's basic instinct to choose and carry forward their profession as their occupation and feel committed. It includes the employee's perception of the overall status of their profession.	• Social status of profession • Inspiration for profession • Views for profession	Gaan, 2008; Yanadoria and Katob, 2010
5	Personality Factors	Personality factors of employees are the inherent factors related to emotional and mental health including both positive and negative factors. Positive factors are reflected in the attitude of a person while negative factors lead to stress due to varied reasons.	• Positive Attitude • Anxiety** • Stress reducing confidence** • Stress through reporting head **	Makhijani et at., 2009; Kyriakidou and Ozbilgin, 2004
Dependent Variable				
6	Employee's Retention	Employee's retention is achieved when they decides to stay with their organization which is a result of certain factors like their commitment towards job and their organization, low absenteeism rate, and employees paying low or no attention to other job offers.	• Job commitment • Attention to job offers • Absenteeism **	Lambert et al., 2001
Mediating Variable				
7	Employee's Job Satisfaction	Job satisfaction is a positive emotional feeling as a result of one's evaluation towards own job or job experience due to job environment work relations, autonomy at work and better growth opportunities.	• Job satisfaction • Trust in organization's work • Satisfaction in job autonomy • Satisfaction in organization's improvement.	Irshad, 2012

** Negative statements drafted and reverse scored

4.1. Assessment of Psychometric Properties

It is pertinent to test the assumptions of psychometric properties for determining reliability and validity of measurement items before conducting further analysis on data. The following tests are conducted for existing study:

1. **Cronbach's Alpha:** It is a test conducted for construct reliability and is a reliability coefficient that assesses the consistency of entire scale (Patney, 2010; Cronbach, 1951) which is well within the range between 0.7 and 0.9 for existing constructs of the study.

Figure 2. Partial Regression Plot for Linearity Test

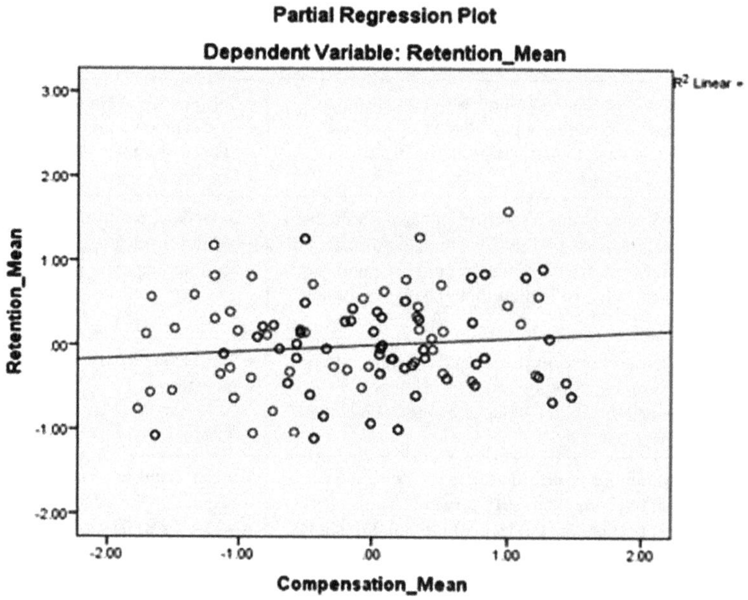

2. **Kaiser-Meyer-Olkin (KMO):** It is a measure of sampling adequacy and is used to quantify the degree of inter-correlations among the variables. KMO for existing study is >0.5 which is required for a satisfactory factor analysis to proceed (Kaiser, 1974) with significance level of 0.000 indicating appropriateness for factor analysis (Field, 2009).
3. **Normality Test:** It is a test to check the normal distribution of sample. The values of mean, median, and mode are nearly paced with each other and values of skewness and Kurtosis is less than 2 and 7 respectively thus meet the normality assumption.
4. **Linearity Test:** Linearity is a test to check the relationships between independent and dependent variables ensuring best representation in regression analysis (Hair et al., 2011) presented in the form of a graph duly presented in Figure 2: Partial Regression Plot for Linearity Test.
   The above mentioned partial regression plot in Figure 2 shows that the clutter around the line does not exhibit any non-linear pattern, thus indicating that the assumption of linearity is met.
5. **Homoedasticity:** The data is tested for homoedasticity assumption graphically through a scatter plot of ZEPRED plotted against ZRESID. It indicates the homogeneity of variance exhibited by the dependent variables across the range of independent variables. The evidence of homoedasticity is seen as an elliptical shaped random array of points, with absence of any patterns of increasing or decreasing residuals (Hair et al., 2011; Field, 2009). The scatter plot for homoedasticity test is duly presented in Figure 3: Scatter Plot for Homoedasticity Test.
   Since the above residual plot exhibits an elliptical shaped random array of points with a constant variance around the straight line, hence the assumption is justified.
6. **Multicollinearity:** Multicollinearity is the test of checking correlation errors through tolerance and Variation Inflation Factor (VIF). Multicollinearity exists when there is strong correlation between the independent variables. The presence of multicollinearity makes it difficult to assess the importance of individual independent variables (Fornell and Larcker, 1981). The assumption has

Figure 3. Scatter Plot for Homoedasticity Test

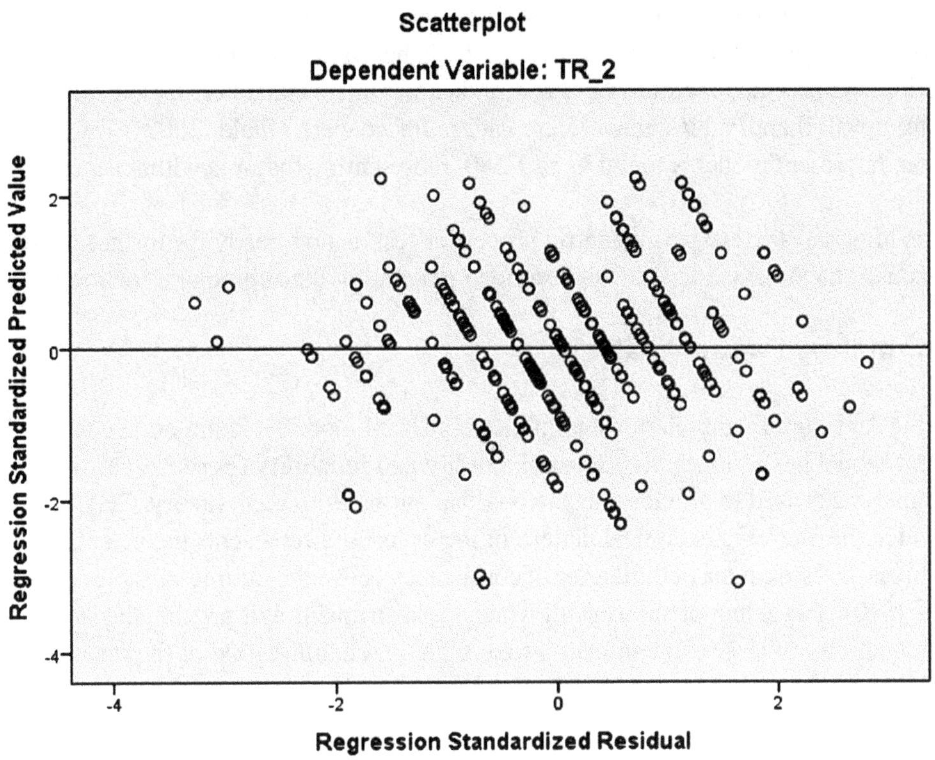

been tested via collinearity diagnostics of Variance Inflation Factor (VIF) and Tolerance, where, Tolerance is defined as the amount of variability of selected independent variable not explained by the other independent variables. The threshold value for assessing VIF is 10, whereas corresponding value for tolerance is 1.0 (Hair et. al. 2011; Field, 2009). The tolerance and VIF statistics have been duly presented in the Table 2: Tolerance and VIF for Multicollinearity test.

The above Table 3 shows acceptable values for VIF and Tolerance, indicating that the collinearity diagnostics meets the specifications. Hence it may be concluded that the data is free from multi-collinearity effect.

Table 3. Tolerance and VIF for Multicollinearity test

S. No.	Independent Variables	Collinearity Statistics	
		Tolerance	VIF
1	Work Culture	0.440	2.272
2	Training and Development	0.554	1.805
3	Compensation	0.681	1.468
4	Motivation for Profession	0.581	1.722
5	Personality Factors	.887	1.128

Source: Compiled from SPSS 20.0 Output Sheet

7. **Durbin-Watson:** Unlike multicollinearity, Durbin-Watson checks for uncorrelated errors. The assumption of Durbin-Watson states that for any two observations, the residual terms should be uncorrelated. Durbin Watson (1950, 1951) establishes upper and lower bounds for testing this assumption. While, the upper threshold is 4, indicating uncorrelated errors, the values below 1 and ranging up till 0 and values above 3 are causes for concern (Field, 2009). The value of Durbin Watson, in present model is found to be 1.580, representing that it is within acceptable limits.

The data, after satisfactory psychometric properties test, is now ready for further analysis. The statistical tools used on the existing data are described in detail in the subsequent section.

4.2. Confirmatory Factor Analysis

Before testing for a significant relationship in the significant model, it is imperative to confirm that the measurement model has a satisfactory level of validity and reliability (Fornell et al, 1981). Confirmatory factor analysis is used to provide a test to confirm our measurement theory. CFA is used to test the length to which the a-priori, theoretical pattern of factor loading represents the actual data.

Reliability is the assessment of the degree of consistency between multiple variables of a measurement (Hair et al., 2010). The extent of surety with which the instrument will provide the same result in case of re-administration of the instrument over a time scale is reliability, one of the measure of reliability is the reliability coefficient. This coefficient, Cronbach's Alpha, assesses the consistency of the entire scale. The value of Cronbach's Alpha increases as the correlations between the variables increases. A measurement scale, for which the calculated Cronbach's Alpha > 0.7, is generally accepted as consistent. The composite Cronbach's Alpha for the scale is 0.79.

Amongst many objectives of CFA, one of the major objectives is to assess the construct validity of the proposed measurement theory (Hair et al., 2010). Validity is the extent to how a scale accurately measures what it is expected to measure. Construct Validity comprises of Convergent Validity and Discriminant Validity. Converge Validity assures that the scale generated correlates with other similar scales while discriminant validity assures that the scale is distinct from other similar scales.

1. **Convergent Validity:** As the name suggests the items which are parts of the scale should converge i.e. share high proportion of their variance. The first measure of Convergent validity is that each factor loadings > 0.5. Higher the factor loadings are, greater the correlations will be resulting into higher convergent validity. Structural Composite Reliability (SCR) is also a measure of Convergent Validity. It is computed from the squared sums of factor loadings for each construct and the sum of the error variance terms for a construct. High SCR corresponds to higher internal consistency. The accepted levels for SCR are between 0.6 and 0.7. Though the reliability of a single measurement and the reliability of the construct (Composite Reliability) can be considered, neither one measure the amount of variance that is captures by the construct in relation to the amount of variance due to measurement error (Fornell and Larcker, 1981). The average variance extracted (AVE) captures this and is further used as a measure of convergent validity. AVE is calculated as the mean variance extracted for the items loading on a construct and is a summary indicator of convergence. Accordingly AVE > 0.5 is a good indicator of convergent validity (Hair et al., 2011). Table 4 shows the values of SCR and AVE. It is observed in Table 4: of SCR and AVE score; that the values of AVE for three of the factors are less than 0.5. Since for all those three factors theoretically and

Table 4. Details of SCR and AVE score

Variables	Items	Factor Loadings	SCR	AVE
Training and Development (TD)	Induction Program	.940	0.82	0.88
	Mentoring Program	.679		
	SDP	.691		
Compensation (C)	Satisfactory Salary	.979	0.80	0.67
	Incentives	.710		
Work Culture (WC)	Supervisor's support	.925	0.84	0.57
	Organization's support	.569		
	Safe Environment	.758		
	Employee's participation	.717		
Motivation for Profession (MFP)	Views for profession	.865	0.81	0.59
	Inspiration for profession	.617		
	Social status of profession	.790		
Personality Factors (PF)	* Anxiety	.944	0.83	0.56
	Positive Attitude	.686		
	* Stress reducing Confidence	.571		
	* Stress through reporting head	.735		
Employee Retention (ER)	Job Commitment	.940	0.96	0.89
	Attention to job offers	.979		
	* Absenteeism rate	.925		
Job Satisfaction (JS)	Job Satisfaction	.940	0.96	0.86
	Trust in organization's work	.979		
	Satisfaction in working independently	.925		
	Satisfaction in organization's improvement	.856		

* Reverse Scored

other tests of Reliability and Validity are positive, it is decided to consider the three factors to be included in the scale.

From the above Table 4, we can say that the constructs of our theoretical framework possesses convergent validity as we can see standardized factor loadings of the items are mostly greater than 0.7 and not less than 0.5. Secondly, the Scale Composite Reliability (SCR) is found to be greater than 0.7 and Average Variance Extracted (AVE) is found to be greater than 0.5. The subsequent section is including the Discriminant validity of the scale.

2. **Discriminant Validity:** The AVE calculated as above can be utilized for the evaluation of Discriminant Validity (Fornell and Larcker, 1981). The sum of the squares of the correlations of all the items in the scale should be less than the AVE for that factor. Discriminant validity reinforces the uniqueness of the scale developed and its distinctness from other similar scales. Table 5 shows the AVE in the leading diagonal while the squares of the correlations.

Table 5. Discriminant Analysis

Component	TD	C	WC	MFP	PF	ER	JS
TD	0.940						
C	.586	0.816					
WC	.528	.345	0.752				
MFP	.582	.338	.235	0.764			
PF	.132	.248	-.115	.443	0.746		
ER	.807	.664	.658	.607	.133	0.948	
JS	.748	.668	.564	.648	.337	.944	0.926

The leading bold shaded diagonal represents the Average Variance Extracted (AVE) and the lower half indicates the correlation coefficients between the constructs. According to Fornell and Larcker (1981), if the diagonal elements are greater than lower half, squared correlations coefficients as shown in Table 5, constructs possess discriminant validity. Thus, the constructs of our theoretical model possess adequate reliability, convergent validity and discriminant validity. With satisfactory reliability and validity of the measurement scale, we can now test the significant relationship of constructs as proposed in theoretical framework in subsequent section.

4.3. Hypothesis Tests through Mediating Regression Analysis

We tested our hypothesis using mediation effect of employee job satisfaction and was analyzed using regression analysis procedure suggested by Baron and Kenny (1986). As we have already checked our regression analysis assumptions under psychometric properties, thus presenting regression analysis output for 16 hypotheses in table 6. The path used for the mediating regression analysis is presented in figure 4.

Figure 4. represents path followed by mediating regression analysis. The variable is dependent if it has one or more unidirectional arrows pointing towards it (such as Y), using all variables that have unidirectional arrows that point toward Y as predictors/ independent variables. The first regression predicts Y from X (path C). The second regression predicts M from X (path A). The third regression predicts Y from both X and M (path C'). Thus, to estimate the strength of association that corresponds to each path in Figure 4, a series of three ordinary least squares (OLS) linear regression analyses can be run. Thus, the top panel of Figure 4 estimates the *total effect of X on Y and is denoted by path C.* The bottom panel of Figure 4 estimates the product of path A and B coefficients and the strength of the *mediated or indirect effect of X on Y,* that is, how much is the increase in Y that occurs as X due to M. The path C' coefficient estimates the strength of the *direct (also called partial) effect of X on Y,* any effect of X on Y that is not mediated by M.

There are two typical questions to be answered in mediation analysis. The first question is whether there is a statistically significant mediated path from X to Y via M (along with whether the part of the Y outcome variable score that is predictable from the path is large enough to be of practical importance). The unstandardized regression coefficient is used for this significance test. If this mediated path is judged to be non-significant, the mediation hypothesis is not supported, and the researcher would need to consider other explanations. If there is a significant mediated path, then the second question in mediation analysis is whether there is also a significant direct path from X to Y which is denoted C' in

Table 6. Mediating Regression Analysis Results of the present study with interpretations

Variables under Study	Hypothesis	Path connecting Variables	R Square	Adjusted R Square	Durbin Watson	Beta Coefficient	t-Statistics	Significance	Tolerance	VIF	Hypothesis Interpretation	z-Statistics	Sobel p-value	Mediation Interpretation
Work Culture (X1) to Employee's Retention (Y)	H_{1a}	Path C	.652	.651	1.580	.813	21.51	.000	1.000	1.000	Accepted	15.0	.000	Partial Mediation
	H_{1b}	Path A	.560	.558	1.760	.679	17.72	.000	1.000	1.000	Accepted			
	H_{1c}	Path C'	.914	.913	1.323	.232	8.16	.000	0.440	2.272	Accepted			
	---	Path B	.891	.890	1.471	.856	27.36	.000	0.440	2.272	----			
Training and Development (X2) to Employee's Retention (Y)	H_{2a}	Path C	.441	.438	1.601	.655	13.95	.000	1.000	1.000	Accepted	12.9	.000	Partial Mediation
	H_{2b}	Path A	.446	.444	1.891	.594	14.100	.000	1.000	1.000	Accepted			
	H_{2c}	Path C'	.893	.892	1.380	.060	2.16	.032	0.554	1.805	Accepted			
	---	Path B	.891	.890	1.471	1.002	32.18	.000	0.554	1.805	----			
Compensation (X3) to Employee's Retention (Y)	H_{3a}	Path C	.433	.430	1.421	.519	13.73	.000	1.000	1.000	Accepted	10.4	.000	Partial Mediation
	H_{3b}	Path A	.391	.316	1.422	.402	10.75	.000	1.000	1.000	Accepted			
	H_{3c}	Path C'	.914	.913	1.631	.145	8.09	.000	0.681	1.468	Accepted			
	---	Path B	.891	.890	1.471	.932	36.99	.000	0.681	1.468	----			
Motivation for Profession (X4) to Employee's Retention (Y)	H_{4a}	Path C	.368	.365	1.637	.599	11.99	.000	1.000	1.000	Accepted	12.4	.000	Complete Mediation
	H_{4b}	Path A	.419	.417	1.848	.577	13.357	.000	1.000	1.000	Accepted			
	H_{4c}	Path C'	.819	.890	1.479	-0.008	-0.287	.774*	0.581	1.722	Not Accepted			
	---	Path B	.891	.890	1.471	1.053	34.29	.000	0.581	1.722	----			
Personality Factors (X5) to Employee's Retention (Y)	H_{5a}	Path C	.018	.014	1.557	.132	2.114	.036	1.000	1.000	Accepted	5.5	.000	Partial Mediation
	H_{5b}	Path A	.113	.110	1.685	0.301	5.622	.000	1.000	1.000	Accepted			
	H_{5c}	Path C'	.929	.928	1.661	-0.206	-11.54	.000	0.887	1.128	Accepted			
	---	Path B	.891	.890	1.471	1.125	56.20	.000	0.887	1.128	----			
Job Satisfaction (M) to Employee's Retention (Y)	H_6	Path B	.891	.890	1.471	1.047	44.841	.000	1.000	1.000	Accepted	--	--	---

* Not Significant

Figure 4. Path used to analyze Mediating Regression Analysis

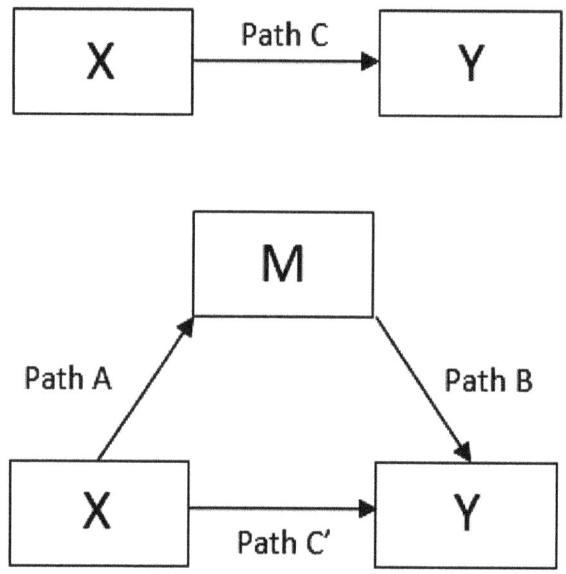

Figure 4. If C' is not statistically significant (or too small to be of any practical importance), a possible inference is that the effect of X on Y is completely mediated by M. If C' is statistically significant and large enough to be of practical importance, a possible inference is that the influence of X on Y is only partially mediated by M and that X has some additional effect on Y that is not mediated by M. Another inference can be, there could be additional mediators of X on Y which is unexplored.

After exploring the mediation analysis, next step is to evaluate the statistical significance of mediated model. This is done as a part of Baron and Kenny (1986) causal-step approach. The Sobel test can be performed by hand; or Preacher and Hayes (2008) provided an online calculator at http://people.ku.edu/~preacher/sobel/sobel.htm to compute the z test, standard errors or t ratios for the coefficient of path A and B. for the purpose of present study, the Sobel test results are drawn by online calculator. Sobel provides the following approximate estimate:

$$SE_{ab} \approx \sqrt{b^2 s_a^2 + a^2 s_b^2}$$

Where, a and b are the raw (unstandardized) regression coefficients that represents the effect of X on M and the effect of M on Y, respectively; s_a is standard error of the *a* regression coefficient; s_b is standard error of the *b* regression coefficient.

Using the standard error from the above equation as the divisor, the following *z* ratio for the Sobel (1982) test can be set up as follows:

$$z = ab/SE_{ab}$$

The *ab* product is judged to be statistically significant if *z* is greater than +1.96 or less than -1.96. Note that the *z* tests for the significance of *ab* assume that the values of this *ab* product are normally distributed across samples from the same population.

Table 6 is presenting the mediating regression analysis results of all 16 hypotheses drawn for the study including the corresponding path, R square statistics, adjusted R square statistics, Durbin-Watson, beta coefficient, *t* statistics, significance level, tolerance, variance inflation factor, *z* statistics, sobel test results and finally the mediation analysis interpretation in brief for each hypothesis. Note that the results to be significant and accepting the hypotheses drawn for the study, the value of Durbin-Watson is suppose to be between 1-3 (Durbin Watson, 1951), beta coefficient to be positive, *t* statistics to be positive, significance level to be less than 0.05, tolerance to be less than or equals to 1 (Hair et. al. 2011), VIF to be less than 4 (Hair et. al. 2011; Hair et al., 1998), *z* statistics to be greater than +1.96 or less than -1.96, and Sobel *p* value to be less than 0.05 (Sobel, 1982). The following interpretations are drawn after analyzing Table 6.

Hypothesis 1a, 1b, 1c, 2a, 2b, 2c, 3a, 3b, and *3c:* We can see from Table 6 that all hypotheses have positive beta coefficient, *t* statistics greater than 2, tolerance is less than 1, VIF is less than 4, and results are significant. Thus, all the nine hypotheses are accepted and the relationship among the constructs is partial mediation.

Hypothesis 4a, 4b, and 4c: We can see from Table 6 that the hypotheses *4a* and *4b* have positive beta coefficient, *t* statistics greater than 2, tolerance is less than 1, VIF is less than 4, and results are significant. Thus hypotheses *4a* and *4b* are accepted but the hypothesis *4c* has negative beta coefficient, *t* statistics less than 2, and not significant. Thus, the hypothesis *4c* is not accepted and the relationship among the constructs is complete mediation.

Hypothesis 5a, 5b, and 5c: We can see from Table 6 that the hypotheses *5a* and *5b* also have positive beta coefficient, *t* statistics greater than 2, tolerance is less than 1, VIF is less than 4, and results are significant. Thus hypotheses *5a* and *5b* are accepted but the hypothesis *5c* has negative beta coefficient and *t* statistics which is significant at the level of 0.000. Thus, the hypothesis *5c* is accepted and the relationship among the constructs is inversely proportional partial mediation.

Hypothesis 6: We can see in Table 6 that the hypothesis *6* also have positive beta coefficient, *t* statistics greater than 2, tolerance is less than 1, VIF is less than 4, and results are significant. Thus, hypothesis *6* is accepted.

The conclusions drawn on the basis of regression analysis are explained in subsequent section along with the conclusive model of the present study.

5. CONCLUSION

The mediating regression analysis output suggests that, in manufacturing organizations, the work culture, training & development, compensation, and personality factors has partial mediating effect of employee's job satisfaction on employee retention. The motivation for profession has complete mediating effect of employee's job satisfaction on employee retention whereas the personality factors have inversely proportional partial mediation effect on employee's retention.

The partial mediation effect means the relationship between the predicting variable and employee retention is direct as well as indirect mediating through job satisfaction as a mediator. The complete

Figure 5. Conclusive model for employee's retention in manufacturing organizations

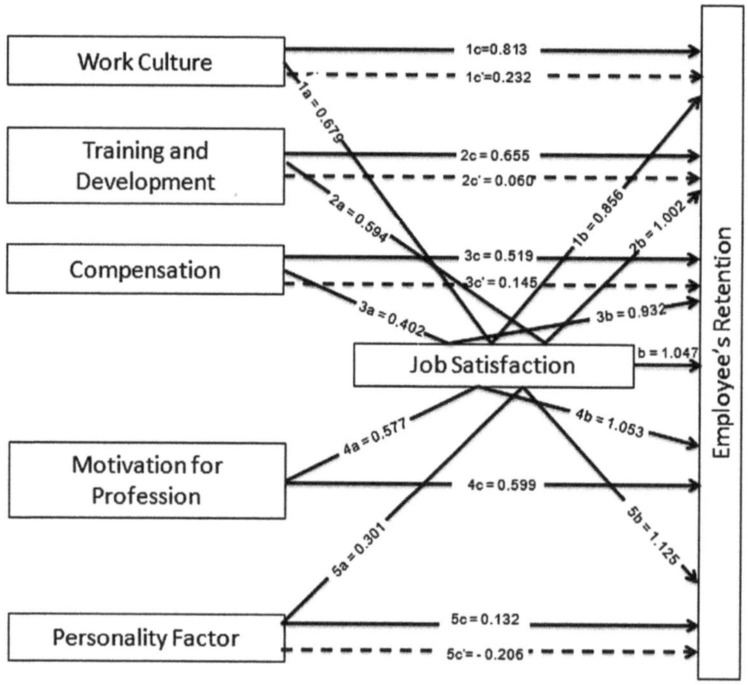

mediation refers to absence of direct relationship between predicting variable and employee retention but only through mediation effect of job satisfaction.

The highest beta coefficient for partial mediation of predictors and employee retention is path 5b = 1.125 of personality factors variable leading from job satisfaction to employee retention where impact of personality factors are studied under the mediation effect of job satisfaction. On contrary, the beta coefficient of path 5c' = -0.206 in controlling effect of job satisfaction indicating the inversely proportional relationship in the absence of mediator, job satisfaction.

The relationship between the variables is presented in conclusive framework in Figure 5: Conclusive Framework for Employee's Retention in Manufacturing organizations with mediating effect of Employee's Job Satisfaction.

5.1. Conclusive Framework

The conclusive framework concluded from the data analysis is presented in Figure 5: Conclusive Framework for Employee's Retention in Manufacturing organizations with mediating effect of Employee's Job Satisfaction; depicting the beta values of each corresponding path.

5.2. Unique Contribution

We started with systematic literature review and found the gaps in literature relating to employee's retention and job satisfaction. Previous researches helped us in understanding the concept of employee's retention in an organization which is a sub system of talent management. Literature helped us in locating issues and conceptualizes a theoretical framework to test empirically. We then developed a questionnaire with both positive and negative statements with expert's opinion and pre-testing. The responses received were empirically tested for goodness of fit using confirmatory factor analysis and hypotheses were tested using mediating regression analysis.

The present study is useful for retaining executive employees of manufacturing organizations in India. There is not sufficient research done on executive level of employees and thus study is relevant in understanding retention from employee's perspective. Moreover, talent retention strategies can be developed by organizations on the basis of results drawn from the present study.

Work culture can be improved by improving supervisor's support, promoting safe working environment, and promoting employee's to participate in yearly planning of the organization. Training and development exercises can be introduced as per the requirements of executive employees starting from their need based induction program, regular mentoring program under a experienced and old employee of organization, and regular staff development program can help executives in keeping them updated with competitive edge. An appropriate salary, and incentives planning for employees helps organizations in attracting and retaining executives for long term. Employee's personality factors are also important for their retention as the high levels of anxiety and stress will lead to employee turnover whereas high anxiety with job satisfaction will lead to employee retention. Thus, efforts should be made to increase job satisfaction of executive employees. It will contribute in increasing job satisfaction of these executives which in turn help them in retaining them with their organizations.

On contrary, there is motivation for profession of an executive employee which does not exhibit direct effect on their retention but when the employee's job characteristics are satisfied, only then they can be retained with their organizations.

Thus, employee's job satisfaction has an important role in their retention.

5.3. Further Research Directions

The present study can be extended as longitudinal study in context of retention of employees for turnover intentions. For the purpose of organization specific results and for strategy formulation, present study can be conducted on single firm and results can help top management to better manage executive employees. A comparative study of organizational factors and employee centric factors can also be conducted.

5.4. Limitations

The study duly recognizes the limitation of survey method and sampling errors, however, within the options available the most scientific and pragmatic approach has been duly adopted.

REFERENCES

The high performance work force: separating the digital economy's winners from losers. (2001). *Accenture: In the battle for retention accentures study.*

Allen, T. D. (2001). Family-Supportive Work Environments: The Role of Organizational Perceptions. *Journal of Vocational Behavior*, *58*(3), 414–435. doi:10.1006/jvbe.2000.1774

Amar, A. D. (2004). Motivating knowledge workers to innovate: A model integrating Motivation dynamics and antecedents. *European Journal of Innovation Management*, *7*(2), 89–101. doi:10.1108/14601060410534366

Anderson, J. C., & Gerbing, D. W. (1988). Structural equation modeling in practice: A review and recommended two-step approach. *Psychological Bulletin*, *103*(3), 411–423. doi:10.1037/0033-2909.103.3.411

Field, A. (2009). *Discovering statistics using SPSS*. Sage publications.

Arthur, J. (1994). Effects of human resource systems on manufacturing performance and turnover. *Academy of Management Journal*, *37*(3), 670–687. doi:10.2307/256705

Ashton, C., & Morton, L. (2005). Managing talent for competitive advantage. *Strategic Human Resources Review*, *4*(5), 28–31. doi:10.1108/14754390580000819

Axelrod, R., Riolo, R., & Cohen, D. (2002). Beyond Geography: Cooperation with Persistent Links in the Absence of Clustered Neighborhoods. *Personality and Social Psychology Review*, *6*(4), 341–346. doi:10.1207/S15327957PSPR0604_08

Baron, R. M., & Kenny, D. A. (1986). The Moderator-Mediator Variable Distinction in Social Psychological Research: Conceptual, Strategic, and Statistical Considerations. *Journal of Personality and Social Psychology*, *51*(6), 1173–1182. doi:10.1037/0022-3514.51.6.1173 PMID:3806354

Bassi, L. J., & Van Buren, M. E. (1999). Sharpening the leading edge. *Training & Development*, *53*(1), 23–32.

Bishop, J. W. (1998, February 14). The relationship between quality and support in the workplace. *Paper presented at the conference on Quality and Management: Quality now and Direction for the 21st Century*, Arizona State University, Tempe, Arizona, USA.

Black, S. E., & Lynch, L. M. (1996). Human capital investments and productivity. *The American Economic Review*, *86*(2), 68.

Bockerman, P., & Ilmakunnas, P. (2004). Job dimensions, job satisfaction, and on-the-job search: is there a nexus? *HECER Discussion Paper* No. 36, University of Helsinki.

Boudreau, J. W., & Ramstad, P. M. (2005). Talentship and the new paradigm for human resource management: From professional practice to strategic talent decision science. *Human Resource Planning*, *28*(2), 17–26.

Brannick, J. (1999). Employees; Job satisfaction; Labor turnover; Recruiting. *AMACOM (New York)*. Retrieved from http://64.17.184.140/wp-content/uploads/2012/12/V4I1-7.pdf

Buckingham, M., & Clifton, D. O. (2001). *Now, discover your strengths*. New York: Free Press.

Butler, T., & Waldroop, J. (2001). *Job sculpting: The art of retaining your best people. Harvard Business Review on finding and keeping the best people* (pp. 179–203). Boston: Harvard Business School Press.

Cable, D. M., & Judge, T. A. (1997). Interviewer's perception of person organization fit and organizational selection decisions. *The Journal of Applied Psychology*, *82*(4), 61. doi:10.1037/0021-9010.82.4.546 PMID:9378683

Cascio, W. F. (1987). *Applied psychology in personnel management*. Englewood Cliffs, NJ: Prentice-Hall.

Cheese, P., Thormas, R. J., & Craig, E. (2008). *The talent powered organization: strategies for globalization, talent management and high performance*. London, Philadelphia: Kogan.

Cheloha, R., & Swain, J. (2005). Talent management system key to effective succession planning. *Canadian HR Reporter*, *18*(17), 5–7.

Churchill, G. A. (1979). A paradigm for developing better measures for marketing constructs. *JMR, Journal of Marketing Research*, *16*(1), 64–73. doi:10.2307/3150876

Clarke, K. F. (2001). What businesses are doing to attract and retain employee- becoming an employer of choice. *Employee Benefits Journal*, 2001, 34–37. PMID:11272511

Coates, T. J., & Thoresen, C. E. (1976). Teacher anxiety: A review with recommendations. *Review of Educational Research*, *46*(2), 159–184. doi:10.3102/00346543046002159

Cranny, C. J., Smith, P. C., & Stone, E. F. (1992). *Job satisfaction: How people feel about their jobs and how it affects their performance. Lexington Books*. New York: Lexington Books.

Cronbach, L. J. (1951). Coefficient alpha and the internal structure of tests. *psychometrika*, 16(3), 297-334.

Davies, R. (2001). How to boost staff retention. *People management*, 7(8), 54-56.

Dobbs, R., Remes, J., Manyika, J., Roxburgh, C., Smit, S., & Schaer, F. (2012). *Urban world: Cities and the rise of the consuming class*. McKinsey Global Institute.

Edward, J. R. (1999). Person job fit: a conceptual integration, literature review, and methodological critique. In C.L. Cooper & I.T. Robertson (Ed.), International review of industrial and organizational psychology (Vol. 6, pp. 283-357). John Wiley and Sons Ltd.

Eisenberger, R., Fasolo, P., & Davis-LaMastro, V. (1990). Perceived organizational support and employee diligence, commitment, and innovation. *The Journal of Applied Psychology*, *75*(1), 51–59. doi:10.1037/0021-9010.75.1.51

Eskildesen, J., & Hahlgaard, J. (2000). A meta-analysis of antecedents and correlates of employee turnover: Update, Moderator test, and research implications for the next millennium. *Journal of Management*, *26*(3), 463–488. doi:10.1177/014920630002600305

Farber, B. (1991). *Crisis in education: Stress and burnout in the American teacher*. San Francisco, CA: Jossey-Bass.

Fornell, C., & Larcker, D. F. (1981). Evaluating Structural Equation Models with Unobservable Variables and Measurement Error. *JMR, Journal of Marketing Research*, *18*(1), 39–50. doi:10.2307/3151312

Frazis, H., Gittleman, M., Horrigan, M., & Joyce, M. (1998). Results from the 1995 survey of employer provided training. *Monthly Labor Review*, *21*(6), 3–14.

Freyermuth. (2007). Retaining Employees in a Tightening Labor Market. RSM McGladrey.

Friesen, D. (1986, April). Overall stress and job satisfaction as predictors of burnout. Paper presented at the annual meeting of the American Educational Research Association, San Francisco, CA, USA.

Gaan, N. S. (2008). Social Support, Job Attitudes and Job Outcome across Gender. *The Icfai University Journal of Organizational Behavior*, *52*, 34–44.

Gaiduk, R., Gaiduk, J., & Fields, D. (2009). Limiting the Brain Drain. Determinants of Employee Organizational attachment in Lithuania. *Baltic Journal of Management*, *4*(2), 149–168. doi:10.1108/17465260910958782

Garg, P., & Rastongi, R. (2006). New model of job design motivation employees Performance. *Journal of Management Development*, *25*(6), 572–587. doi:10.1108/02621710610670137

Garg, R. B. L. (1985). Management Education in India. *Management Accountant*, *20*(1), 19.

Gold, M. (2001). Breaking all the rules for recruitment and retention. *Journal of career planning and employment*, 61(3), 6-8.

Goldstein, I. (1991). Training in organizations. In M.D. Dunnette, & L.M. Hough (Ed.), Handbook of Industrial and Organizational Psychology (2nd ed.,Vol. 2, pp. 507-619).

Gomez-Mmejia, L. R., Balkin, D. B., & Cardy, R. L. (1995). *Managing human resources*. Englewood cliffs, NJ, USA: Prentice-Hall, Inc.

Graddick, M. N. (1988). *Corporate Philosophies of Employee Development, Career Growth and Human Resource Strategies: The Role of the Human Resource Professional in Employee Development* (M. London & E. M. Mone, Eds.). Westport, CT, USA: Quorum Books.

Greenhaus, J. H., & Callanan, G. A. (1994). *Career Management*. Fort Worth, Texas: The Dryden Press.

Hair, J. F. Jr, Black, W. C., Babin, B. J., Anderson, R. E., & Tantham, R. L. (2011). *Multivariate Data Analysis* (6th ed.). India: Person Education Inc.

Hall, D. (1996). *The Career Is Dead, Long Live the Career: A Relational Approach to Careers*. San Francisco, CA: Jossey-Bass Publishers.

Hall, D. T., & Moss, J. E. (1998). The new protean career contract: Helping organizations and employees adapt. *Organizational Dynamics*, *26*(3), 22–37. doi:10.1016/S0090-2616(98)90012-2

Highhouse, S., Stierwalt, S. L., Bachiochi, P., Elder, A. E., & Fisher, G. (1999). Effects of advertised human resource management practices on attraction of African American applicants. *Personnel Psychology*, *52*(2), 425–442. doi:10.1111/j.1744-6570.1999.tb00167.x

Hsu, M. K., Jiang, J. J., Klein, G., & Tang, Z. (2003). Perceived career incentives and intent to leave. *Information & Management*, *40*(5), 361–369. doi:10.1016/S0378-7206(02)00018-6

Huberman, M. (1989). The professional life cycle of teachers. *Teachers College Record*, *9*(1), 31–57.

Huselid, M. A. (1995). The impact of human resource management practices on turnover productivity, and corporate financial performance. *Academy of Management Journal*, *38*(3), 635–672. doi:10.2307/256741

Hytter, A. (2008). *Dark Side Leaders, Work Environment and Employee Health*. Retrieved from hvxu.se/ehv/forskning/hofreseminarier/2008/080514%20DarkSide%20Final%20version.pdf

Ihsan Malik & Basharat Naeem. (2009), Research methods in the social Sciences. Melbourne: Martin Press, Inc.

Iles, P., Chuai, X., & Preece, D. A. (2010). Talent management and HRM in multinational companies in Beijing: Definitions, differences and drivers. *Journal of World Business*, *45*(2), 179–189. doi:10.1016/j.jwb.2009.09.014

Irshad, M. (2012). Factors affecting Employee Retention Evidence from Literature Review. *Abasyn Journal of Social Sciences*, 4(1), 84-102. Retrieved from http://64.17.184.140/wp-content/uploads/2012/12/V4I1-7.pdf

Judge, T. A., Hanisch, K. A., & Drankoski, R. D. (1995). Human resource management and employee attitudes. In G. R. Ferris, S. D. Rosen, & D. T. Barnum (Eds.), *Handbook of human resources management*. United Kingdom: Blackwell Publishers, Ltd.

Kaiser, H. F. (1974). An index of factorial simplicity. *Psychometrika*, *39*(1), 31–36. doi:10.1007/BF02291575

Kim, J. K. (2000, February). Does attitude towards money moderate the relationship between intrinsic job satisfaction and voluntary turnover? *Human Relations*, *53*, 213–245.

Kirby, S., & Richard, O. C. (2000). Impact of marketing work-place diversity on employee job involvement and organizational commitment. *The Journal of Social Psychology*, *140*(3), 367–378. doi:10.1080/00224540009600477 PMID:10902377

Konovsky, M. A. (2000). Understanding procedural justice and its impact on business organizations. *Journal of Management*, *26*(3), 489–511. doi:10.1177/014920630002600306

Koustelios, A., & Bagiatis, K. (1997). The Employee Satisfaction Inventory (ESI): Development of a Scale to measure satisfaction of Greek employees. *Educational and Psychological Measurement*, *57*(3), 469–476. doi:10.1177/0013164497057003008

Kristensen, N., & Westergård-Nielsen, N. (2004). Does Low Job Satisfaction Lead to Job Mobility? *IZA Discussion paper no. 1026*.

Kumar, M. S., Udayasuriyan, G., & Vimala, B. (2008). Motivation among the employees of a public sector concern. *Journal of Community Guidance and Research*, *25*(3), 340–351.

Kyriakidou, O., & Ozbilgin, M. (2004). Individuals, organizations and careers: A relational perspective. *Career Development International*, *9*(1), 7–11. doi:10.1108/13620430410518101

Lachman, R., & Diamant, E. (1987). Withdrawal and restraining factors in teachers' turnover intentions. *Journal of Occupational Behavior*, *8*(3), 219–232. doi:10.1002/job.4030080304

Lambert, E. G., Hogan, N. L., & Barton, S. M. (2001). The impact of Job Satisfaction on Turnover Intent: A test of a structural model using a national sample of workers. *The Social Science Journal, Elsevier*, *38*(2), 233–250. doi:10.1016/S0362-3319(01)00110-0

Lauri, B., Benson, G., & Cheney, S. (1996). The top ten trends. *Training & Development*, *11*, 28–42.

Lawler, E. E. (1990). *Compensation management; Strategic planning; Pay-for-knowledge systems*. San Francisco: Jossey-Bass Publishers.

Lee, T., & Mowday, R. (1987). Voluntary leaving an organization: An empirical investigation of Steers and Mowday's model of turnover. *Academy of Management Journal*, *30*(4), 721–743. doi:10.2307/256157

Lewis, R. E., & Heckman, R. J. (2006). Talent management: A critical review. *Human Resource Management Review*, *16*(2), 139–154. doi:10.1016/j.hrmr.2006.03.001

Light, J. N. (2004). Relationships and effects of employee involvement, employee empowerment and employee satisfaction by job-type in a large manufacturing environment [Unpublished PhD. Dissertation].

Lum, L., Kervin, J., Clark, K., Reid, F., & Sirola, W. (1998). Explaining nursing turnover intent: Job satisfaction, pay satisfaction or organizational commitment? *Journal of Organizational Behavior*, *19*(3), 305–320. doi:10.1002/(SICI)1099-1379(199805)19:3<305::AID-JOB843>3.0.CO;2-N

Makhijani, N., Rajendran, K., & Creelman, J. (2009). *Best Practices in Aligning People with Strategic Goals*. Azkia Publishers.

Report, M. (2003). Mercer study raises red flags for employer pay and benefits plans (findings of the 2002 people at work survey). In Human resource department management report (pp. 8-15).

Messmer, M. (2000). Orientations programs can be key to employee retention. Strategic Finance, 81(8), 12-15.

Michaels, E., Handfield-Jones, H., & Axelrod, B. (2001). *The War for Talent*. Boston: Harvard Business School Press.

Netemeyer, R. G., Boles, J. S., McKee, D. O., & McMurrian, R. (1997). An investigation into the antecedents of organizational citizenship behaviors in a personal selling context. *Journal of Marketing*, *61*(3), 85–98. doi:10.2307/1251791

Noe, R. A. (1999). *Employee training and development* (pp. 212–218). New York: Irwin McGraw-Hill.

Olajide, A. (2000). *Getting the best out of the employees in a developing economy. A personal Psychology Guest Lecture Series*. Nigeria: Department of Guidance and Counseling, University of Ibadan.

Long-term care facility worker retention. (2004). Ontario Ministry of Health and Long-Term Care.

Parker, O., & Wright, L. (2001). Pay and employee commitment: the missing link. In Ivey Business Journal, 65(3), 70-79.

Pasewark, W. R., & Viator, R. E. (2006). Sources of Work- Family Conflict in the Accounting Profession. *Behavioral Research in Accounting*, *18*(1), 147–165. doi:10.2308/bria.2006.18.1.147

Patney, M.B. (2010). *Indian consumers and their mall patronage: Application of cultural-self and the theory of planned behavior to patronage intentions* [Thesis].

Pfeffer, J. (1998). Six myths about pay. *Harvard Business Review*, May-June, 109–119. PMID:10179647

Price, J., & Mueller, C. (1986). *Absenteeism and turnover among hospital employees*. Greenwich, CT: JAI Press.

Prince, B. J. (2005). Career-focused employee transfer processes. *Career Development International, 10*(4), 293–309. doi:10.1108/13620430510609136

Putti, J. M. (1989). Work Values and Organizational Commitment: A Study in Asian Context. *Human Relations, 42*(3), 275. doi:10.1177/001872678904200305

Ramlall, S. (2003). Organizational Application Managing Employee Retention as a Strategy for Increasing Organizational Competitiveness. *Applied H.R.M. Research, 8*(2), 63–72.

Rattray, J., & Jones, M. C. (2007). Essential Elements of Questionnaire Design and Development. *Journal of Clinical Nursing, 16*(2), 234–243. doi:10.1111/j.1365-2702.2006.01573.x PMID:17239058

OReilly, C. A., Chatman, J., & Caldwell, D. F. (1991). People and organizational culture: A profile comparison approach to assessing person organization fit. *Academy of Management Journal, 34*(3), 487–516. doi:10.2307/256404

OReilly, C. A., Chatman, J., & Caldwell, D. F. (1991). People and organizational culture: A profile comparison approach to assessing person organization fit. *Academy of Management Journal, 34*(3), 487–516. doi:10.2307/256404

Robertson, A., & Abbey, G. (2003). *Managing Talented People*. Harlow: Pearson Education Limited.

Roznowski, M., & Hulin, C. (1992). The scientific merit of valid measures of general constructs with special reference to job satisfaction and job withdrawal. In C. J. Cranny, P. C. Smith, & E. F. Stone (Eds.), *Job satisfaction: How people feel about their jobs and how it affects their performance*. New York: Lexington Books.

Schiemann, W. A. (2009). *Reinventing Talent Management*. Hoboken, NJ: John Wiley & Sons.

Shields, M. A., & Wheatley Price, S. W. (2002). Racial Harassment, Job Satisfaction and Intentions to Quit: Evidence from the British Nursing Profession. *Economica, 69*(274), 295–326. doi:10.1111/1468-0335.00284

Silbert, L. T. (2005). The effect of Tangible Rewards on Perceived Organizational Support. Management Sciences. Retrieved from http://etd.uwaterloo.ca/etd/lsilbert2005.pdf?origin=publication_detail

Sinha, D., & Sinha, S. (2008). Talent Retention in Academics: A Case Study on Management.

Sjoberg, A., & Sverke, M. (2000). The interactive effect of job involvement and organizational commitment on job turnover revisited: A note on the mediating role of turnover intention. *Scandinavian Journal of Psychology, 41*(3), 247–252. doi:10.1111/1467-9450.00194 PMID:11041307

Smilansky, J. (2006). *Developing Executive Talent*. Chichester: Jossey-Bass.

Smith, M. K. (2001) Young people, informal education and association *Informal Education*. Retrieved from www.infed.org/youthwork/ypandassoc.htm

Sobel, M. E. (1982). Asymptotic confidence intervals for indirect effects in structural equation models. *Sociological Methodology, 13*, 290–312. doi:10.2307/270723

Steel, R., & Ovalle, N. (1984). A Review and meta-analysis of research on the relationship between behavioral intentions and employee turnover. *The Journal of Applied Psychology*, *69*(4), 673–686. doi:10.1037/0021-9010.69.4.673

Steel, R. P., Griffeth, R. W., & Hom, P. W. (2002). Practical retention policy for the practical manager. *The Academy of Management Executive*, *18*(2), 149–169. doi:10.5465/AME.2002.7173588

Steers, R. M. (1997). Antecedents and outcomes of organizational commitment. *American Scientist*, *53*, 36–39.

Storey, J., & Sisson, K. (1993). *Managing Human Resources and industrial Relations* (pp. 22–31). Buckingham: Open University Press.

Thompson, C. A., & Prottas, D. J. (2005). Relationships among Organizational Family Support, Job Autonomy, Perceived Control, and Employee well-being. *Journal of Occupational Health Psychology*, *11*(1), 100–118. doi:10.1037/1076-8998.10.4.100 PMID:16551178

Tower, P. (2003). Rewards: The not-so-secret ingredient for managing talent Retention. *HRFocus*, *80*(1), 3–10.

Tranfield, D., Denyer, D., & Smart, P. (2003). Towards a methodology for developing evidence-informed management knowledge by means of systematic review. *British Journal of Management*, *14*(3), 207–222. doi:10.1111/1467-8551.00375

Verma, M. & Roy. (1988). Work Motivation and Job Satisfaction as Determinants of Organizational Commitment. *Indian Journal of Applied Psychology*, *25*(2), 21.

Verma, O. P. (1986). Organizational Commitment as a Function of Managerial Respect. Asian. *Journal of Psychology and Education*, *17*(1), 1–6.

Walker, J.W. (2001). Perspectives. *Human resource planning*, 24(1), 6-10.

Watson, W. (1999). *Work USA 2000: Employee commitment and the bottom line*. Bethesda, MD: Watson Wyatt.

Wetland, D. (2003). The strategic training of employee's model: balancing organizational constraints and training content. S.A.M. Advanced Management Journal, winter, 103-107.

Wildman, T. M., & Niles, J. A. (1987). Essentials of professional growth. *Educational Leadership*, *44*(5), 4–10.

Williams, L., & Hazer, J. (1986). Antecedents and consequences of satisfaction and commitment in turnover models: A re-analysis using latent variable structural equation models. *The Journal of Applied Psychology*, *71*(2), 219–231. doi:10.1037/0021-9010.71.2.219

Williams, M., & Dreher, G. (1992). Compensation systems attributes and applicant pool characteristics. *Academy of Management Journal*, *35*(3), 571–595. doi:10.2307/256487

Willis, C. (2000). Go for your goals. *Working Woman*, 2000, 6–7.

Woodruffe, C. (1999). Winning the talent war: A strategic approach to attracting, developing and retaining the best people. Chichester, UK: John Wiley & Sons. Retrieved from www.cpscccsp.ca/PDFS/CPSC%20 Final%20Report%20June28%20%207%20case%20studis2%20oct%207%2004.pdf

Yanadori, Y., & Kato, T. (2010). Work and Family Practices in Japanese Firms: Their Scope, Nature and Impact on Employee Turnover. *International Journal of Human Resource Management*, *20*(2), 439–456. doi:10.1080/09585190802673494

Zeytinoglu, I. U. & Denton, M. (2005). Satisfied workers, Retained Workers: Effects of work and work environment on Homecare worker's Job satisfaction, Stress, Physical Health, and Retention. *Canadian Health Services Research Foundation, Foundation canadienne de la recherché sur les services de sante*.

Chapter 3
Enablers of Sustainable Manufacturing Overview, Framework and Further Research Directions

Shibin K. T.
Symbiosis International University, India

ABSTRACT

The aim of this paper is to develop a conceptual sustainable manufacturing (SM) framework, which is the pressing need of highly competitive and natural resource scarce world. Systematic literature review has been adopted to identify enablers of SM. The enablers identified through systematic literature review includes product design, material selection & procurement, customer requirements, regulatory norms, social values & ethics, environmental priorities, energy costs, natural resource scarcity, supply chain & logistics systems, fuel price regulations, public awareness, technology and total quality management. Author has attempted to develop a contextual relationship among enablers of SM using interpretive structural modeling. The author has further classified enablers into four categories using MICMAC analysis. The author has also outlined the limitations of present study and identified further research directions.

1. INTRODUCTION

Sustainable manufacturing is the need of the current era as global warming, manmade pollution and natural resource scarcity are becoming like a sword of Damocles to the world. Insights from the literature on the topic show that, sustainability in manufacturing must be taken into consideration from the product design stage through production to the logistics stage. Sustainability has got three aspects in manufacturing. These aspects includes: 1) the manufacturing sustainability in the production process; 2) sustainability during the use of the product, 3) sustainability for the recycling and logistics point of view. We can say that product design, material selection, manufacturing process design and resource utilization as the factors of first aspect; environmental emission from the product, ease of maintenance

DOI: 10.4018/978-1-5225-0350-7.ch003

etc. as the factors of second aspect and ease of disassembly, recyclability, reusability of materials as the factors of third aspect. According to Garbie (2014); Carter and Easton (2011); Dyllick and Hockerts (2002), Environment, society and economy are the three widely accepted components of sustainability. Manufacturing can be defined as the process of converting raw materials into value added useful tangible finished products for a specific use through one or more operations. Hence, sustainable manufacturing can be defined as the innovative technology and organizational practices which ensures zero negative impact to the environment, community, employees with the minimum use of raw materials and energy and which produces high quality recyclable finished products. Different viewpoints and definitions on sustainability and sustainable development from different literature are listed in Table 1. Through systematic literature review, the authors have identified twelve key driving factors of sustainable manufacturing. Since these factors are much broad in their sense, sub-dimensions of these variables are also elaborated from literature.

Table 1. Definitions of Sustainability

Author(s), (Year)	Definition
IPCC, Copenhagen, Denmark (2014).	Sustainable Development is defined as the one which is based on three pillars such as economic, social and environmental and preserves the interests of future generations and ecosystem services and balance on which the existence of human kind is dependent on.
Institute for Sustainability (2011)	Sustainability can be defined as increasing the productivity by minimizing the resources used without compromising on service quality, profitability and competitiveness to help to preserve the environment.
Jayal et al. (2010)	Sustainability is the term which can be achieved only through an integrated approach incorporating closed loop supply chain, manufacturing process and systems, multiple product life-cycles etc. at product process and system levels with the objective of reduce, reuse, recover, redesign, remanufacture and recycle.
Ciceri et al. (2010)	Sustainability is a quality which helps to preserve, save and keep.
Hutton et al. (2007)	Sustainable development can be defined as the one which helps to fulfill the needs of the present without compromising the capability of the future to fulfill their own needs.
Barber (2007)	Sustainable production and consumption is defined as the system which helps to meet the human needs and improve the quality of life through social and economic improvement for all, including future generations, while maintaining the ecosystems on which human life is depended up on.
Asheim et al. (2007)	Sustainability is defined by using the framework of ethical social choice theory and explains that sustainable behavior is the ethically justifiable treatment of all future generations with equity and efficient consistent consumption of natural resources.
Seliger (2007)	Sustainability is defined as the direction in which the living standards of human being is improved at the same time the availability of natural resources and ecosystems are also preserved and improved for future generations.
Jenkins and Yakovleva (2006)	Sustainable development is defined as the progress in economic development, environmental protection and social consistency.
Szekely and Knirsch (2005)	Sustainability is the process by which organizations amalgamate their social, economic and environmental objectives with the organization strategic goals and balances all these three dimensions in an optimum way.
Stavins et al. (2003)	Sustainable growth is defined in an economic framework as the one which is both dynamically efficient and non-decreasing over time.
Deloitte and Touche (1992)	Sustainability can be defined as the process of adopting new business strategies and ideas which will fulfill today' needs of the organization and all other stakeholders, which will protect, sustain and enhance all natural and human resources for the requirement of the future.
UN - The World Commission on Environment and Development (1987)	Sustainable development can be defined as the usage of resources, investment guidelines, technological advancement and organizational changes are made to meet the need of the present as well as of future generations consistently.

2. LITERATURE REVIEW

As a part of systematic literature review, the authors started the review from analyzing the history of manufacturing itself, which helps to give insights into the evolution of sustainable manufacturing. According to Brundtland Commission (1987), the word sustainable implies the use of consistent practices which will ensure zero negative repercussions to the future generations due to the prevailing practices. Davidson et al. (2010) foresee the need to educate future engineers and new generation about sustainable development (SD) practices by including SD workshops and educational materials in their curriculum to meet increasing environmental, resource scarcity and social threats of 21st century. Miller et al. (2010) show how the theories of sustainable manufacturing help to attain competitive advantage for a small scale industry through waste reduction, supply chain network optimization and recycling of waste materials by using discrete event simulation and mathematical modeling technique. Pham and Thomas (2011) and Thomas et al. (2012) also explain the sustainability issues with small and medium scale industries and propose a fit manufacturing strategy starting from strategy development through financial planning to operationalization with one team one vision for better coordination. Joung et al. (2013) classifies the sustainability indicators into five major enablers of sustainable manufacturing such as environmental, social, economic, technology and performance but didn't go in detail about the interrelationship among the indicators.

2.1. Evolution of Sustainable Manufacturing

At the start of industrial revolution in the 17th century, the one and only one objective of manufacturers was to somehow explore the available natural resources and to convert them into to useful end products to meet the human need. As time passed, during the era of 18th century, manufactures tried to increase efficiency and productivity through automation and scientific industrial engineering techniques. As the living standard of people and the competition between manufactures increased, the major focus was on the product quality and price control during the 19th century. Pressure of competition forced manufactures to strive for excellence through various manufacturing concepts such as world class, agile, flexible, lean, and green manufacturing techniques. Jin-Hai et al. (2003) and Parkinson (1999) define agile manufacturing as a concept which is driven by competition and is designed to meet quickly changing customer tastes with a faster response rate and more flexibility. Slack (2005); Kaighobadi and Venkatesh (1994) explain the concept of flexible manufacturing systems as the one which helps manufactures to have wider variety and faster response to market changes and to get economic advantage of batch manufacturing through technology flexibility, labor flexibility and infrastructure flexibility. Hayes and Wheelwright (1985) classified manufacturing organizations into four stages, and used the term "world class manufacturing firms" for the first time, to refer to the fourth stage organizations. According to Hayes and Wheelwright, those organizations which are very progressive and having long range business plans through continuous improvement and newer technologies and strive to make manufacturing as a major competitive success factor of the organization can be called as world class organizations. The next challenge was to quantify the term world class, which resulted to the origin of a new term known as 'lean manufacturing'. Oliver et al. (1996); Pettersen (2009) and Womack et al. (1990) define lean as a management concept originated from Japan's Toyota Production System (TPS) which tries to minimize wastage or non-value adding processes and use less of everything including inventory through techniques such as Just in Time (JIT) to produce high quality products. According to National Oceanic and Atmospheric Administration report published in 2005, since industrial revolution there is a 35% increase in the level of carbon dioxide all over the world within a period of three centuries. Houe and Grabot (2009) show that the increased customer awareness on environmental issues forced companies for go for more natural friendly products and production processes. To get competitive advantage, Zhu et al. (2007) emphasizes that: with the new regulatory norms put by various governments to reduce carbon emissions and to control global warming also forced manufactures to shift their manufacturing strategy from merely output based to a more environmental friendly one. This resulted in the evolution of green manufacturing concept. Dubey and Bag (2014) define green manufacturing concept as the one which intends to achieve environmental sustainability and to minimize carbon emission by eliminating environmentally harmful wastages and using innovative technologies to change the existing processes. Deif (2011) elaborates the green manufacturing concept as the one which is eco-efficient, uses less non-toxic energy and material which produces no unwanted outputs and byproducts, and output to input conversion is possible. Sustainable manufacturing philosophy is evolved in twentieth century, which considers not only the environment like green manufacturing; but all other factors such as social and economic. Garbie (2014); Garetti and Taisch (2011) explicitly state that the three pillars of sustainability includes social, environment and economic. Molamohamadi and Ismail (2013) propose to add technology, education, ethic and accountability along with the main three pillars of social, environment and economic. Jayal et al. (2010) differentiates all these manufacturing concepts more clearly and in a very simple way as: traditional manufacturing is substitution-based; lean-manufacturing as waste reduction based; green manufacturing as environmentally begin and based on 3Rs such as Reduce, Reuse and Recycle; and finally sustainable manufacturing as based on 6R such as Reduce, Reuse, Recycle, Recover, Redesign, and Remanufacture with innovation. Kuik et al. (2011) further develop a clear framework by dividing the first 3Rs of Reuse, Recover and Redesign for product design and Reduce, Recycle and Remanufacture for process improvement towards the journey to sustainability. Stoughton and Ludema (2012) use a grounded theory approach to model and analyze perceptions of sustainability at senior leadership level, functional level and individual level of an organization and how to integrate all these views to achieve sustainable development within an organization. Literature reviewed on sustainable manufacturing and the brief description regarding the objective and findings of each paper is described in Table 2 given below.

2.2. Classification of Literature

All the literature are classified under four broad pillars of sustainable manufacturing such as economy, social, environment and technology based on ESET context proposed by Jovane et al. (2008). According to the global survey conducted by Kiron et al. (2013) from MIT Sloan Management Review and the Boston Consulting group, shows that the main three social issues in achieving sustainability as the health and well-being of employees, the community and customers; the top most environmental issues in achieving sustainability as energy efficiency, pollution and waste management and the top most economic concern in achieving sustainability as competitiveness. According to Kara et al. (2014), even though the sustainable product design strategy is different for different companies based on their size, regional location etc. some common driving factors are considered by every manufactures for sustainable product development, which include: customer awareness, legislation, economic benefits and competitive strategies and life cycle assessment. So, all the driving factors of sustainable manufacturing identified through systematic literature review are classified under these four different dimensions. Environment dimension includes Environmental priorities, Natural resource scarcity; Social dimension includes Regulatory norms, Fuel

Table 2. Research in Sustainable Manufacturing

Author(s), (Year)	Research Objective	Research Findings
Garbie (2014)	To model the components of sustainable manufacturing and to make a performance assessment framework up to sustainable development level from a micro-level point of view based on the three pillars of social, economic and environment.	Sustainable development index is developed by considering social, environmental and economic factors. Using analytical hierarchy Process weight of each sub-factors under each pillars are found out.
Garetti et al. (2012)	To identify the main aspects of sustainable manufacturing from literature review.	Review of selected papers is done and different viewpoints of sustainable manufacturing systems are explained.
Gunasekaran and Spalanzani (2012)	To build a framework for sustainable business development by using the methodology of literature review.	A classical review is made which helps to give the research areas and techniques used in sustainable business development in manufacturing and service. A framework is made for sustainable business development in service and manufacturing through literature review technique.
Garetti and Taisch (2011)	To make the roadmap and strategy for the manufacturing research by 2020 for achieving smart, sustainable and inclusive growth.	Sustainable manufacturing framework is made with technology and education as enablers and by using four broad research clusters for sustainable manufacturing.
Jayal et al. (2010)	To present an overview of recent trends in sustainable manufacturing product, process and system development.	Evolution of sustainable manufacturing is explained with the 6R concept for sustainable manufacturing product, process and systems development. Performance measurement for processes for achieving sustainability is also explored by the author.
Garbie (2013)	Development of enterprise level sustainability model for manufacturing firms, to become globalized sustainable enterprise in economic point of view.	Mathematically modeled the sustainability index by considering seven factors and suggests considering the sustainability from the starting of the design of an enterprise itself.
Davies and Kochar (2002)	To review the methodologies adopted so far to define the best practices in the manufacturing industry and to find out how these best practices can be used to improve the company performance.	A particular common best practice method can't be applied to all types of industries in general. Best practices must be tailor made to best suit to the particular company under study. Effectiveness of best practices must be evaluated by checking the overall performance improvement of the company.
Azzone and Noci (1998)	To define different green manufacturing strategies and to analyze, how to make an environmental performance measurement system framework for various green manufacturing strategies.	Environmental performance measurement system (PMS) design architecture is developed for each manufacturing strategies and concludes that it is more difficult to design PMS in pro-active companies rather than re-active companies.

price regulations, Public awareness, Social values & ethics; Economic dimension includes Material selection & procurement, Energy costs and Technology & Operations includes Customer requirements, Product design, Supply chain & logistics systems and Technology & total quality management.

2.2.1. Environment Dimension

Most of the literature reviewed on sustainability gives very high weightage for environment factor. The main factors discussed under environment dimension are carbon footprint concerns and the requirements for energy efficiency. Intergovernmental Panel on Climate Change Copenhagen, Denmark (2014) demands the full stoppage of fossil fuel usage by 2100, to control the world carbon footprint. According to Gotschol et al. (2014) environmental management is an economically sustainable business, because any investment in environmental management programs helps to gain economic benefits for companies in their long run. Flint and Golicic (2009) explain the success story of carbon zero certification program for wine companies in New Zealand, for achieving zero carbon footprint and environmental awareness and habitat rehabilitation programs to achieve sustainability. Halldorsson and Kovacs (2010) illustrate the importance of energy efficiency and minimizing carbon footprint in logistics and supply chain context. All 'green' concepts are fundamentally based on eco-friendliness concept and thus can be classified under the environment pillar of sustainability. According to Deif (2011); Houe and Grabot (2009), companies must consider the term 'green' as a tool for achieving competitive advantage to go global and to increase the market share as today's highly educated and highly aware customers prefer eco-friendly products. Unruh and Ettenson (2011) mention three ways to achieve green goals such as "accentuate" which means the re-positioning of the existing product portfolio by extending green attributes; "acquire", which means the complete acquirement of an existing green company and their product portfolios; "architect", which means the building of new green products from scratch. They also add that, going green doesn't mean that companies have to sacrifice their profit margin, but rather it improves the profitability through waste minimization and recycling of products and byproducts. In a case study from the wine industry in United States, Cholette and Venkat (2009) show the need to incorporate third party logistics providers in the supply chain network, the need to optimize the package, route and load of transport system and the need to encourage direct delivery of end products to consumers to minimize carbon emissions and maximize energy efficiency. According to Wu and Pagell (2011) environmental strategies adopted by organizations have a direct impact on the supply chain and competitiveness of the organization. They also adds that top level managers should benchmark and get insights from the practices and policies that are successfully implemented in top leading companies in achieving sustainability in their own organization. Law and Gunasekaran (2012) illustrates the need of clear strategic level policy and the need of coordination of the strategic level team with the tactic and operations levels of the organization for the introduction and implementation of sustainable development in an organization. According to Rimanoczy and Pearson (2010), change in organizational culture is required to achieve the sustainability goals of profit, people and planet and human resource department has a major role in institutionalization of sustainability. According to Dekker et al. (2012), in supply chain, transportation is the most important factor with respect to environment. Because transportation accounts fourteen percentages of total CO_2 emissions across the globe and the three factors such as mode of transportation, type and size of transportation unit and fuel choice must be optimized to minimize the CO_2 emissions. Edwards et al. (2010) indicates that e-commerce and online shopping practices has got the advantage of lesser carbon footprint than conventional shopping practices in logistics and supply chain aspect. According to Ramani et al. (2010), sustainability view point must be considered from the early stage of the design itself and along with Design for Assembly and Manufacturing (DFAM), Design for Environment (DFE) must also be carried out. Gudmundsson and Roheim (2000) explain the effective way of using the eco-labeling practice to maintain sustainable fisheries resource, which can be carried over for other resources as well to encourage customer to go for environmentally sustainable products.

Kaebernick et al. (2003), illustrates a four step methodology for sustainable product development which include: 1) making environmental awareness to consumers 2) assessing environmental performance as a design objective by modifying quality function deployment matrix into environmentally conscious quality function deployment matrix 3) conducting life cycle assessment during design process 4) evaluating the recycle, reuse potential of the product. Feng & Joung (2009) measure the performance of sustainable manufacturing only by considering energy used and Co_2 emission without considering other social and economic factors. Smith and Ball (2012); Despeisse et al. (2012) and Despeisse et al.

(2013) elaborate a process flow mapping methodology to quantify environmental performance of sustainability through material, waste and energy flow in a manufacturing firm, despite considering social and economic and technology factors in par with environment factor. Yuan et al. (2012) also consider only the environmental aspect of sustainable manufacturing by considering three major components of manufacturing such as technology, energy and material. Kim et al. (2010) emphasize for DFE method to reduce the environmental impact of manufacturing processes, but only consider the environmental aspect of sustainability. Paju et al. (2012) propose a process mapping method and quantify the sustainability indicators based on discrete event simulation, lifecycle assessment, and value stream mapping methodologies, but didn't consider economic and social indicators at all.

2.2.2. Social Dimension

According to Law and Gunasekaran (2012), external factors such as laws, regulation, social pressure, market trend and competition determines the internal factors of an organization such as performance, organization system etc. and strategic level policies must be made in line with the same to implement sustainable development in an organization. O'Brien (1999) emphasizes the need to have coordination between governments to make a common single worldwide regulation for the sustainable industrial development, regardless of the size and nature of the countries. He also adds the need to organize public awareness programs, to change the consumer buying behavior and to make the customers motivated to adopt green products. According to Hong et al. (2012), for achieving greater social reputation, organizations must implement sustainability practices not only at the strategic level, but at the employee level also. In most of the literature social dimension of sustainability is considered as corporate social responsibility (CSR) initiatives. Biggemann et al. (2014) show the importance of CSR initiatives in achieving sustainability through value creation by the integration of all stake holders. According to Adams et al. (2014) ecological and social factors are least considered for the performance measurement of sustainability. Very few organizations are voluntary going for social sustainability since they believe that social sustainability won't help out them much to gain any competitive advantage. According to Deif (2011), because of the rules and regulations by various governments and by United Nations, green manufacturing has become a mandatory thing for manufacturing firms to go globally. Strong rules and regulations by government are necessary to encourage organizations to go for social sustainability, through which sustainable development can be achieved. Labor sustainability and health and well-being of the society in which the organization is operating are coming under the social aspect of sustainability. Labor sustainability was considered by Tapiero and Kogan (2008). Health and well-being factor was explained by Labuschagne and Brent (2008), and socially responsible purchasing by Mont and Leire (2009). Socially responsible purchasing implies the selection and purchasing of materials which are not harmful and which will make zero negative effect on the customers and internal employees of the organization. McWilliams and Siegel (2001); Van De Ven and Jeurissen, (2005) add that social responsibility is closely related with the firm's business strategy. Labuschagne and Brent (2008) developed a social sustainability criteria frame work, which shows the importance of social assessment for the project and technology life cycle management.

2.2.3. Economic Dimension

Competition and profitability are the two major factors that must be taken into account while discussing the economic dimension of sustainability. According to Nidumolu et al. (2009) in Harvard Business

Review, companies are considering sustainability as a part of corporate social responsibility because they fear that they will be forced to go for major process and technology changes to achieve sustainability, for which customer won't be ready to pay for. But he further explains that, environmental friendly manufacturing and sustainable development cost must be included in mainstream business practices, as the one who is successful at the earliest in developing sustainability will definitely get a competitive advantage and economic advantage gradually. Garetti and Taisch (2011) explain that energy recovery by cross plant analysis, co-operation between industries to optimize the energy at various heat levels, re-manufacturing of products, optimization of the speed, load and nature of transport, use of fuels from waste materials instead of fossil fuels, etc. will significantly contribute to the profitability margin of the business organization. According to Clelland et al. (2000), waste and pollution minimizing practices will reduce the operational cost and indirectly create economic value to the firm. Dos Santos et al. (2014) show that Woolworths Company was able to save 9.3 million US Dollars after starting the triple bottom line sustainable business initiative, which include ecological, social and economic indices. According to Stavins et al. (2003) an economy is sustainable only if it is dynamically stable and efficient and total social welfare is non-decreasing over time. The path for sustainability through the above mentioned steps are economical not only through operating cost reduction perceptive, but also through the branding of the company as a green and environment friendly one among all stake holders. Heilala et al. (2008) demonstrate how energy usage and environmental impact can be quantified through a simulation technique.

2.2.4. Technology and Operations Dimension

Technology and innovation has an important role in achieving sustainability in manufacturing. Flexible technologies and new innovative ideas help to reduce cost and other efforts tremendously. According to Westkamper (2008), emerging technologies are the key enablers behind the efficient economic, social and environmental performance of a manufacturing organization. Lee et al. (2011) shows that digital manufacturing technique as a key technique to integrate product design with enterprise resource, facility, and process planning. Digital manufacturing helps to validate each and every design aspect of a new product in line with the existing processes in a virtual world before going for real life expensive implementations. Liu et al. (2011); Koren et al. (1999) and Liu & Liang (2008) assert the importance of reconfigurable manufacturing systems. These systems are another aspect of new technology which helps to minimize waste, operational cost, and improves responsive time and helps to produce a variety of products in comparatively short span of time. According to Gupta and Palsule-Desai (2011) remanufacturing can be considered as an effective tool to tackle the rapidly growing competitive pressure, and is the area, where the OEMs going to concentrate in future. Mani et al. (2013) emphasize that organizations must design sustainable processes to manufacture sustainable products and a clear well defined parameter must also be there to measure the sustainability level of the processes designed. Veleva et al. (2001) propose a frame work to measure the performance of sustainable production, with twenty two indicators and well defined parameters for the measurement of each indicator. Vinodh and Rajanayagam (2010), through a case study find that computed aided design (CAD) along with design for manufacturing (DFM) can significantly contribute in the sustainable product development process, and can ensure minimum environmental impact. According to Hong at al. (2012) lean production and restructuring of supply chain helps to achieve responsive product strategy, which in turn increases the enterprise performance. They also add that benchmarking is an effective practice to institutionalize sustainability practices. Pumar (2005) explains the way to operationalize the sustainable development ideas and theories through group interaction and social networking. Despeisse et al. (2012) argue that there are limited works which explain about the operationalization of sustainability theories and explains the sustainability practices followed in industry the importance of capturing such practices systematically. Beachy (2010) emphasizes the need to use science and technology effectively in achieving agricultural sustainability. Gonzalez-Torre et al. (2004); Dowlatshahi (2000) shows how reverse logistics helps to increase the utilization of resources for reuse and recycling of the product. These new innovative ideas in supply chain and manufacturing ensure minimum waste and high operational efficiency. According to Westkamper et al. (2001), the future challenge is to reduce the cost of manufacturing technology to a level suitable for the mass production of sustainable products by using sustainable technologies such as solar energy, hydrogen technologies and other renewable-energy systems. Ijomah et al. (2007) illustrate the key indicators that are to be taken care in product design stage as durable material selection, use of joining technologies that won't damage components during separation and discouraging the use of banned substances and processes based on the classification of design subassemblies into three categories such as repair, reconditioning, and remanufacturing to produce sustainable products. Marksberry and Jawahir (2008) develop methodology to measure tool wear and tool life with near dry machining concept to achieve process sustainability to avoid the usage of harmful metal working fluids. According to Seliger et al. (2008) modularity, easy diagnose and standardization are the three main steps to develop efficient remanufacturing products. Fang et al. (2011) explain the use of travelling sales man problem in productivity and environmental performance improvement of manufacturing processes by optimum job scheduling to improve sustainability performance. Gungor and Gupta (1999) elaborate the challenges in recycling and remanufacturing of the used products as collecting the used products to a single point, disassembly level fixing, and in the optimization of disassembly processes.

3. METHODOLOGY

Since the objective of the research in this paper is to find out the interrelationships among variables, Interpretive Structural Modeling (ISM) is adopted as the methodology. In ISM methodology, group expert judgment is used as a tool to predict the relationships among variables under study. Since there are limited literature which explains the interrelationships among the variables of Sustainable Manufacturing, this paper will contribute significantly by using ISM methodology. ISM was first introduced by Warfield (1974) and Malone (1975). The steps involved in ISM are:

- Literature collection on the topic
- Systematic literature review and identification of variables
- Approaching experts and explaining the guidelines of VAXO matrix formulation to them to make the structural self-interaction matrix.
- Experts were asked to fill the matrix by using V, A, X and O letters based on their expert knowledge in the area to define the relationship among two variables of the matrix.
- The structural self-interaction matrix was first converted to a binary matrix and then to a final reachability matrix by considering transitivity property.
- Identifying the level of variables depending on the dependence power and driving power of the variable from the final reachability matrix.

- Based on the levels of variables identified from the reachability matrix directed graph (DIGRAPH) is made.
- Converting the DIGRAPH into ISM Model which is self-explanatory about the relation amongst the variables.
- Reviewing the ISM model to validate the conceptual stability and make necessary changes in the model.

Opinion from five experts from TATA Motors having fifteen plus years of experience in manufacturing and three experts from TATA Technologies having twelve plus years of experience in manufacturing service industry are collected. The respondents of the survey include two experts from product design field, two from process design field, two from productivity improvement department and two from engineering research center vehicle integration and testing team.

Variables identified in sustainable manufacturing: V1 - Product design, V2 - Customer requirements, V3 - Supply Chain & Logistics systems, V4 - Energy costs, V5 - Technology & total quality management, V6 - Material selection & procurement, V7 - Natural resource scarcity, V8 - Environmental priorities, V9 - Regulatory norms, V10 - Fuel price regulations, V11 - Public awareness, V12 - Social values & Ethics

The variables of the models are identified through systematic literature review process and are listed out in Table 3 in the above section. The two parameters amongst which the relationship is checked is represented by say *i* and *j*. Four variables are used to represent the relationship among the variables in the survey, which includes V, A, X, O. The matrix is to be filled with:

V: If *i* leads to *j* but *j* doesn't lead to *i*
A: If *i* doesn't lead to *j* and *j* leads to *i*
X: If *i* and *j* lead to each other
O: If *i* and *j* are not related each other

Using the above criteria, we have developed Table 4. The next step is to convert the Structural Self Interaction Matrix (SSIM) into binary matrix for which the below rules are followed:

- If the *i,j* relationship in SSIM Matrix is V, the corresponding binary relationship is 1 for *i, j* and is 0 for *j,i* .
- If the *i,j* relationship in SSIM Matrix is A, the corresponding binary relationship is 0 for *i, j* and is 1 for *j,i* .
- If the *i,j* relationship in SSIM Matrix is X, the corresponding binary relationship is 1 for both *j,i* and *i,j* .
- If the *i,j* relationship in SSIM Matrix is O, the corresponding binary relationship is 0 for both *j,i* and *i,j* .

Transitivity Principle

Farris and Sage (1975); Sushil (2005a,b) say that to maintain the consistency of the model developed, transitivity property is always used in the ISM modeling. Transitivity principle can be explained with an illustrative example as: if a leads to b and b leads to c, the transitivity property implies that a leads

Table 3. Dimensions of Sustainable Manufacturing

Sustainable Manufacturing Drivers/Variables	Authors (Year)	Dimensions/Sub-Variables
Product design	Garetti and Taisch (2011)	• Mass customization • Eco-design: easiness of serviceability, disassembly. • Recyclability.
Customer requirements	Garetti and Taisch (2011); Garbie (2014)	• Involvement of customers in design, Innovative products
Supply chain & logistics systems	Dubey and Gunasekaran (2014); Jeffrey and Choi (2004)	• Green Logistics • Reverse Logistics • Supplier selection • Logistics network design
Energy costs	Garetti and Taisch (2011)	• Energy Conservation • Minimization fossil fuel energy use • Energy production from waste & garbage.
Technology & total quality management	Garetti & Taisch (2011); Garbie (2014); Garbie (2013); Law & Gunasekaran (2012)	• Advanced automation for high product quality • High-performance and quality-embedded manufacturing, • Minimization of customer complains • Reconfigurable & Flexible Manufacturing Systems • Strategic level management policies
Material selection & procurement	Kara et al. (2014)	• Environmental friendly material selection • Reusable & recyclable material usage • Material selection to minimize energy consumption & operating cost
Natural resource scarcity	Azzone and Noci (1998)	• Minimization of the impact on the state of natural resources
Environmental priorities	Mittal and Sangwan (2014a); Mittal and Sangwan (2014b)	Minimization of • Greenhouse gas emission • Degradation of soil and water
Regulatory norms	Zhu et al. (2007)	• Environmental rules & regulations by government
Fuel price regulations	Garetti and Taisch (2011)	• To promote bio-fuels & to discourage fossil fuels usage
Public awareness	Santos et al. (2014), Houe and Grabot (2009)	• Customers' awareness on environmental issues • Internal employee awareness
Social values & Ethics	Hoejmose et al. (2013)	• Social responsible supply chain

Table 4. Structural self-interaction Matrix

	V12	V11	V10	V9	V8	V7	V6	V5	V4	V3	V2	V1
V1	O	A	V	A	V	X	X	V	X	X	X	-
V2	A	X	A	O	O	V	V	V	V	V	-	
V3	O	O	A	A	A	O	V	X	O	-		
V4	O	O	X	A	O	A	O	A	-			
V5	O	O	V	V	V	O	X	-				
V6	A	A	O	A	V	V	-					
V7	A	A	X	O	A	-						
V8	A	A	O	V	-							
V9	O	O	O	-								
V10	O	O	-									
V11	O	-										
V12	-											

Table 6. Level matrix

V1,V3,V4,V6,V7,V10	Level 1
V5,V8,V9	Level 2
V2,V11	Level 3
V12	Level 4

to c. Transitivity property helps to remove the gaps among the variables if any. By adopting the above criteria and transitivity property, final reachability matrix is prepared and is shown in Table

Level Partitioning

The ranking of different variables into different levels is known as level partitioning. Reachability set and antecedent set are found out from the final reachability matrix (Warfield 1974). In any iteration, if the reachability set intersection antecedent set is the reachability set itself then those variables occupies the top levels of the hierarchy. The final output of level partitioning is shown in Table 6 below.

CMAC Analysis

After preparing the ISM model (Figure 1), MICMAC diagram of the variables is prepared based on their driving power and dependence power. Driving power and dependence power is calculated in the final reachability matrix and is shown in Table 5. Duperrin and Godet (1973) developed the MICMAC method for the first time. According to Warfield (1994) MICMAC Analysis is used to categorize variables in a complicated system. Mandal and Deshmukh (1994) explain that MICMAC will also help to analyze the driving and dependence power of variables of a complex system. According to Jharkharia and Shankar (2005), depending on the value of dependence power and driving power the variables can be classified into four categories such as autonomous, linkage, dependent and independent barriers. The first category

Figure 1. Model generated by interpretive structural modeling (ISM)

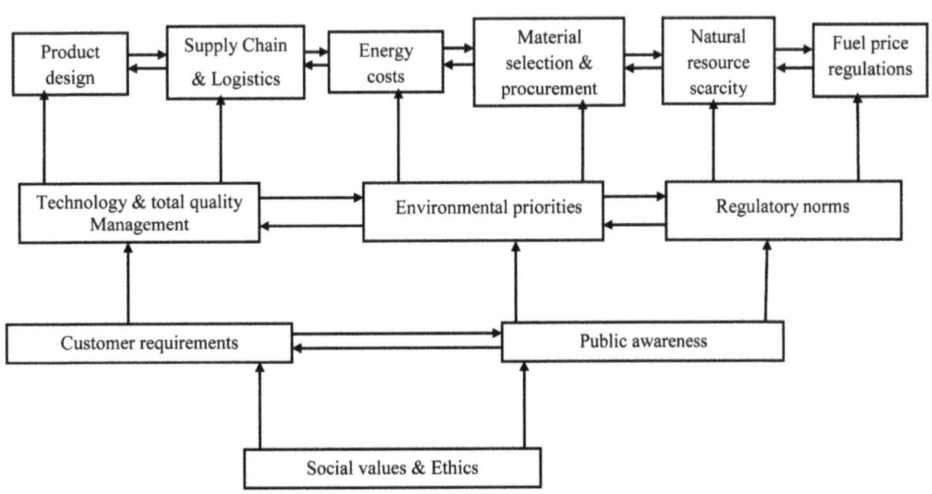

Table 5. Final reachability matrix

	V1	V2	V3	V4	V5	V6	V7	V8	V9	V10	V11	V12	Driving Power
V1	1	1	1	1	1	1	1	1	1*	1	1*	0	9
V2	1	1	1	1	1	1	1	1*	1*	1*	1	0	8
V3	1	1*	1	1*	1	1	1*	1*	1*	1*	0	0	4
V4	1	1*	1*	1	1*	1*	1*	1*	0	1	0	0	3
V5	1*	1*	1	1	1	1	1*	1	1	1	0	0	7
V6	1	1*	1*	1*	1	1	1	1	1*	1*	0	0	5
V7	1	1*	1*	1	1*	1*	1	1*	0	1	0	0	4
V8	1*	0	1	1*	1*	1*	1	1	1	1*	0	0	4
V9	1	1*	1	1	1*	1	1*	1*	1	1*	0	0	5
V10	1*	1	1	1	1*	1*	1	0	0	1	1*	0	5
V11	1	1	1*	1*	1*	1	1	1	1*	1*	1	0	6
V12	1*	1	1*	1*	1*	1	1	1	1*	1*	1*	1	5
Dependence Power	8	5	7	7	5	8	8	6	3	5	2	1	

Figure 2. MICMAC Diagram

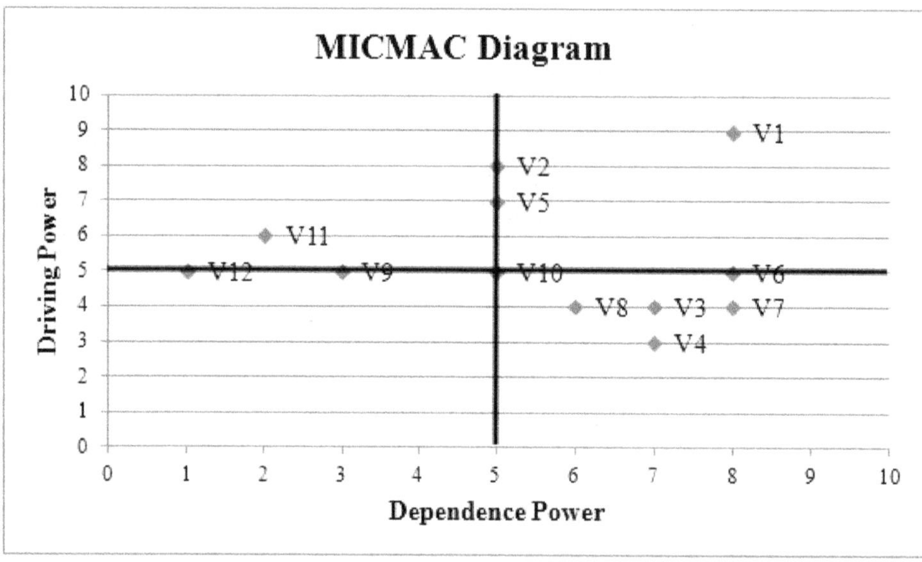

known as 'autonomous barriers' include the variables having weak driving as well as dependence power. MICMAC Diagram for the sustainable manufacturing variables under study is shown in Figure 2, and there is no variables coming in the first quarter which means that, there is no autonomous variable.

The variables coming in first quarter won't be having much connection with the system or with other variables. The variables V3, V4, V7 and V8 are coming in second quarter which is known as 'dependent barrier'. Dependent barrier variables are having weak driving power and strong dependence power. Since these variables depend heavily on other variables, any change on other variables will affect these variables. The variable V1 is coming in third quarter and is known as 'linkage variable'. These variables are characterized by their strong driving power and strong dependence power. These variables are highly sensitive. Any change in the system or other variables will affect linkage variables. The fourth quarter contains variable V11 which is known as 'independent barriers'. These variables have strong driving power and weak dependence power.

4. RESULTS AND CONCLUSION

The study helps to identify the driving factors of sustainable manufacturing system, and the frame work model built through ISM technique will help policy makers to get a clear overall picture of the system to develop and implement the sustainable manufacturing philosophy. From MICAMC analysis, we found that there are no autonomous variables, which implies that all the variables stated in the study will influence the development of sustainable manufacturing system. Product design is coming in the third quarter which implies that it is a linkage variable and is highly unstable. This implies that product design strategy must be changed in accordance with the other driving factors to achieve the manufacturing sustainability. The variable public awareness has high driving power and this factor will affect other enabling factors. Supply chain and logistics systems, Energy costs, Natural resource scarcity and Environmental priorities are coming in second cluster and are having high dependence power.

5. RESEARCH GAPS IDENTIFIED AND SCOPE OF FURTHER RESEARCH

From the global survey conducted by Kiron et al. (2013) from MIT Sloan Management Review and the Boston Consulting group, it can be seen that for thirty five percentages of the companies sustainability efforts are adding profits to business for consistently over the years. But, the percentage of companies that have successfully established a sustainability business case is only thirty seven percentages, and twenty percentages of companies are failing to establish a successful sustainability business case in their organization. This shows the requirement for a clear well defined framework for sustainable manufacturing, the relationship among driving variables and their relative importance in framing a sustainable manufacturing systems, which will help the organizations to achieve sustainability. We have found that many of the articles are focused only on environmental factor of sustainability rather than considering social, economic and technology factors in equal par to achieve better sustainability performance. So, future studies must focus on all the three factors of sustainability simultaneously to make better meaningful theoretical contributions. According to Nakano (2010), definition of sustainable manufacturing is not clear and technical approaches for tackling risks and sustainability issues in manufacturing must be explored further in detail. We acknowledge the same based on the review done and hence we have developed a comprehensive definition of sustainable manufacturing by reviewing the available definitions of sustainability, sustainable development, and sustainable manufacturing. We have found that at present there are limited literature on sustainable manufacturing which uses simulation and mathematical modeling techniques. Jimenez-Gonzalez et al. (2011) argue that there is a lack of total systems approach and emphasize the need to coordinate different engineering and science professionals to develop better sustainable products and processes. Throughout the literature we have identified the gap of total systems approach in developing the sustainable manufacturing models. In further research, the relative importance of all driving factors and the effect of one over the other variables are also to be explored. Design for Environment (DFE) method is getting much importance in the early product design stage along with Design for Assembly (DFA) and Design for Manufacturing. Future researches can be extended to make complete guidelines and check points for DFE methodology and to develop better ecofriendly products. There are limited works which consider the effect of soft dimensions such as top management strategic policies, organization human resource policies etc. on the sustainable manufacturing system development in an organization. From the literature review we understood that sustainable product and process design didn't get enough focus and attention as much as they were supposed to get. Our research aims to fill these gaps and to formulate a clear framework for sustainable manufacturing systems. ISM and MICMAC analysis is done only by taking opinion from eight experts, which is not sufficient to derive any statistical conclusion. So to overcome the same, Structured Equation Modeling (SEM) analysis must be performed by taking sufficient survey based on structured questionnaire.

6. REFERENCES

Adams, C. A., Muir, S., & Hoque, Z. (2014). Measurement of sustainability performance in the public sector. *Sustainability Accounting. Management and Policy Journal*, *5*(1), 46–67.

Asheim, G. B., Buchholz, W., & Tungodden, B. (2007). *Justifying sustainability* (pp. 33–51). Springer Netherlands.

Azzone, G., & Noci, G. (1998). Identifying effective PMSs for the deployment of "green" manufacturing strategies. *International Journal of Operations & Production Management*, *18*(4), 308–335. doi:10.1108/01443579810199711

Barber, J. (2007). Mapping the movement to achieve sustainable production and consumption in North America. *Journal of Cleaner Production*, *15*(6), 499–512. doi:10.1016/j.jclepro.2006.05.010

Beachy, R. N. (2010). Science and sustainability: The emerging consensus. *Bioscience*, *60*(6), 406–407. doi:10.1525/bio.2010.60.6.2

Biggemann, S., Williams, M., & Kro, G. (2014). Building in sustainability, social responsibility and value co-creation. *Journal of Business and Industrial Marketing*, *29*(4), 304–312. doi:10.1108/JBIM-08-2013-0161

Brundtland, G. H. (1987). *Report of the World Commission on environment and development:" our common future*. United Nations.

Business strategy for sustainable development: leadership and accountability for the'90s. (1992Deloitte & Touche, International Institute for Sustainable Development, and Business Council for Sustainable Development.

Carter, C. R., & Easton, P. L. (2011). Sustainable supply chain management: Evolution and future directions. *International Journal of Physical Distribution & Logistics Management*, *41*(1), 46–62. doi:10.1108/09600031111101420

Cholette, S., & Venkat, K. (2009). The energy and carbon intensity of wine distribution: A study of logistical options for delivering wine to consumers. *Journal of Cleaner Production*, *17*(16), 1401–1413. doi:10.1016/j.jclepro.2009.05.011

Ciceri, N. D., Garetti, M., & Sperandio, S. (2010). From product end-of-life sustainable considerations to design management. In Advances in Production Management Systems. New Challenges, New Approaches (pp. 152-159). Springer Berlin Heidelberg.

Clelland, I. J., Dean, T. J., & Douglas, T. J. (2000). Stepping towards sustainable business: An evaluation of waste minimization practices in US manufacturing. *Interfaces*, *30*(3), 107–124. doi:10.1287/inte.30.3.107.11661

Davidson, C. I., Hendrickson, C. T., Matthews, H. S., Bridges, M. W., Allen, D. T., Murphy, C. F., & Austin, S. et al. (2010). Preparing future engineers for challenges of the 21st century: Sustainable engineering. *Journal of Cleaner Production*, *18*(7), 698–701. doi:10.1016/j.jclepro.2009.12.021

Davies, A. J., & Kochar, A. K. (2002). Manufacturing Best Practice and performance studies: A critique. *International Journal of Operations & Production Management*, *22*(3), 289–305. doi:10.1108/01443570210417597

Deif, A. M. (2011). A system model for green manufacturing. *Journal of Cleaner Production*, *19*(14), 1553–1559. doi:10.1016/j.jclepro.2011.05.022

Dekker, R., Bloemhof, J., & Mallidis, I. (2012). Operations Research for green logistics–An overview of aspects, issues, contributions and challenges. *European Journal of Operational Research*, *219*(3), 671–679. doi:10.1016/j.ejor.2011.11.010

Despeisse, M., Ball, P. D., & Evans, S. (2012). Modelling and Tactics for Sustainable Manufacturing: An Improvement Methodology. In Sustainable Manufacturing (pp. 9-16). Berlin Heidelberg: Springer.

Despeisse, M., Mbaye, F., Ball, P. D., & Levers, A. (2012). The emergence of sustainable manufacturing practices. *Production Planning and Control*, *23*(5), 354–376. doi:10.1080/09537287.2011.555425

Despeisse, M., Oates, M. R., & Ball, P. D. (2013). Sustainable manufacturing tactics and cross-functional factory modeling. *Journal of Cleaner Production*, *42*, 31–41. doi:10.1016/j.jclepro.2012.11.008

Dos Santos, M. A., Svensson, G., & Padin, C. (2014). Implementation, monitoring and evaluation of sustainable business practices: Framework and empirical illustration. *Corporate Governance. The International Journal of Business in Society*, *14*(4).

Dowlatshahi, S. (2000). Developing a theory of reverse logistics. *Interfaces*, *30*(3), 143–155. doi:10.1287/inte.30.3.143.11670

Dubey, R., & Bag, S. (2014). Antecedents of Green Manufacturing Practices: A Journey towards Manufacturing. *Smart Manufacturing Innovation and Transformation: Interconnection and Intelligence*, 333.

Duperrin, J. C., & Godet, M. (1973). *The method for hierarchical system elements* (pp. 45–51). Paris: Rapp. Econ. De CEA.

Dyllick, T., & Hockerts, K. (2002). Beyond the business case for corporate sustainability. *Business Strategy and the Environment*, *11*(2), 130–141. doi:10.1002/bse.323

Edwards, J. B., McKinnon, A. C., & Cullinane, S. L. (2010). Comparative analysis of the carbon footprints of conventional and online retailing: A "last mile" perspective. *International Journal of Physical Distribution & Logistics Management*, *40*(1/2), 103–123. doi:10.1108/09600031011018055

Fang, K., Uhan, N., Zhao, F., & Sutherland, J. W. (2011). A new shop scheduling approach in support of sustainable manufacturing. In *Glocalized Solutions for Sustainability in Manufacturing* (pp. 305–310). Springer Berlin Heidelberg. doi:10.1007/978-3-642-19692-8_53

Farris, D. R., & Sage, A. P. (1975). On the use of interpretive structural modeling for worth assessment. *Computers & Electrical Engineering*, *2*(2), 149–174. doi:10.1016/0045-7906(75)90004-X

Feng, S. C., & Joung, C. B. (2009, December). An overview of a proposed measurement infrastructure for sustainable manufacturing. *Proceedings of the 7th Global Conference on Sustainable Manufacturing*. Retrieved from http://www.nist.gov/customcf/get_pdf.cfm

Flint, D. J., & Golicic, S. L. (2009). Searching for competitive advantage through sustainability: A qualitative study in the New Zealand wine industry. *International Journal of Physical Distribution & Logistics Management*, *39*(10), 841–860. doi:10.1108/09600030911011441

Garbie, I. H. (2013). DFSME: Design for sustainable manufacturing enterprises (an economic viewpoint). *International Journal of Production Research*, *51*(2), 479–503. doi:10.1080/00207543.2011.652746

Garbie, I. H. (2014). An analytical technique to model and assess sustainable development index in manufacturing enterprises. *International Journal of Production Research*.

Garetti, M., Mummolo, G., & Taisch, M. (2012). Special issue on "sustainable manufacturing". *Production Planning and Control*, *23*(2-3), 79–82. doi:10.1080/09537287.2011.591617

Garetti, M., & Taisch, M. (2011). Sustainable manufacturing: Trends and research challenges. *Production Planning and Control*, *23*(2-3), 83–104. doi:10.1080/09537287.2011.591619

Gonzalez-Torre, P. L., Adenso-Diaz, B., & Artiba, H. (2004). Environmental and reverse logistics policies in European bottling and packaging firms. *International Journal of Production Economics*, *88*(1), 95–104. doi:10.1016/S0925-5273(03)00181-6

Gotschol, A., De Giovanni, P., & Vinzi, V. E. (2014). Is environmental management an economically sustainable business? *Journal of Environmental Management*, *144*, 73–82. doi:10.1016/j.jenvman.2014.05.001 PMID:24921964

Gudmundsson, E., & Roheim, C. A. (2000). Ecolabeling seafood for sustainable production: Implications for fisheries management. *Marine Resource Economics*, *15*(2), 97–113. doi:10.1086/mre.15.2.42629294

Gunasekaran, A., & Spalanzani, A. (2012). Sustainability of manufacturing and services: Investigations for research and applications. *International Journal of Production Economics*, *140*(1), 35–47. doi:10.1016/j.ijpe.2011.05.011

Gungor, A., & Gupta, S. M. (1999). Issues in environmentally conscious manufacturing and product recovery: A survey. *Computers & Industrial Engineering*, *36*(4), 811–853. doi:10.1016/S0360-8352(99)00167-9

Gupta, S., & Palsule-Desai, O. D. (2011). Sustainable supply chain management: Review and research opportunities. *IIMB Management Review*, *23*(4), 234–245. doi:10.1016/j.iimb.2011.09.002

Halldorsson, A., & Kovacs, G. (2010). The sustainable agenda and energy efficiency: Logistics solutions and supply chains in times of climate change. *International Journal of Physical Distribution & Logistics Management*, *40*(1/2), 5–13. doi:10.1108/09600031011018019

Hayes, R. H., & Wheelwright, S. C. (1985), Restoring our Competitive Edge: Competing through Manufacturing, John Wiley & Sons, New York, NY.

Heilala, J., Vatanen, S., Tonteri, H., Montonen, J., Lind, S., Johansson, B., & Stahre, J. (2008, December). Simulation-based sustainable manufacturing system design. *Proceedings of theSimulation Conference WSC '08* (pp. 1922-1930). IEEE. doi:10.1109/WSC.2008.4736284

Hoejmose, S., Brammer, S., & Millington, A. (2013). An empirical examination of the relationship between business strategy and socially responsible supply chain management. *International Journal of Operations & Production Management*, *33*(5), 589–621. doi:10.1108/01443571311322733

Hong, P., Roh, J. J., & Rawski, G. (2012). Benchmarking sustainability practices: Evidence from manufacturing firms. *Benchmarking: An International Journal*, *19*(4/5), 634–648. doi:10.1108/14635771211258052

Hutton, R. B., Cox, D. B., Clouse, M. L., Gaensbauer, J., & Banks, B. D. (2007). The role of sustainable development in risk assessment and management for multinational corporations. *Multinational Business Review*, *15*(1), 89.

Ijomah, W. L., McMahon, C. A., Hammond, G. P., & Newman, S. T. (2007). Development of design for remanufacturing guidelines to support sustainable manufacturing. *Robotics and Computer-integrated Manufacturing*, *23*(6), 712–719. doi:10.1016/j.rcim.2007.02.017

Intergovernmental Panel on Climate Change, Copenhagen, Denmark. (2014, October 31). Retrieved from http://www.ipcc.ch/scripts/_session_template.php?page=_40ipcc.htm

Jayal, A. D., Badurdeen, F., Dillon, O. W. Jr, & Jawahir, I. S. (2010). Sustainable manufacturing: Modeling and optimization challenges at the product, process and system levels. *CIRP Journal of Manufacturing Science and Technology*, *2*(3), 144–152. doi:10.1016/j.cirpj.2010.03.006

Jenkins, H., & Yakovleva, N. (2006). Corporate social responsibility in the mining industry: Exploring trends in social and environmental disclosure. *Journal of Cleaner Production*, *14*(3), 271–284. doi:10.1016/j.jclepro.2004.10.004

Jharkharia, S., & Shankar, R. (2005). IT-enablement of supply chains: Understanding the barriers. *Journal of Enterprise Information Management*, *18*(1), 11–27. doi:10.1108/17410390510571466

Jimenez-Gonzalez, C., Poechlauer, P., Broxterman, Q. B., Yang, B. S., am Ende, D., Baird, J., & Manley, J. et al. (2011). Key green engineering research areas for sustainable manufacturing: A perspective from pharmaceutical and fine chemicals manufacturers. *Organic Process Research & Development*, *15*(4), 900–911. doi:10.1021/op100327d

Jin-Hai, L., Anderson, A. R., & Harrison, R. T. (2003). The evolution of agile manufacturing. *Business Process Management Journal*, *9*(2), 170–189. doi:10.1108/14637150310468380

Joung, C. B., Carrell, J., Sarkar, P., & Feng, S. C. (2013). Categorization of indicators for sustainable manufacturing. *Ecological Indicators*, *24*, 148–157. doi:10.1016/j.ecolind.2012.05.030

Jovane, F., Yoshikawa, H., Alting, L., Boër, C. R., Westkamper, E., Williams, D., & Paci, A. M. et al. (2008). The incoming global technological and industrial revolution towards competitive sustainable manufacturing. *CIRP Annals-Manufacturing Technology*, *57*(2), 641–659. doi:10.1016/j.cirp.2008.09.010

Kaebernick, H., Kara, S., & Sun, M. (2003). Sustainable product development and manufacturing by considering environmental requirements. *Robotics and Computer-integrated Manufacturing*, *19*(6), 461–468. doi:10.1016/S0736-5845(03)00056-5

Kaighobadi, M., & Venkatesh, K. (1994). Flexible manufacturing systems: An overview. *International Journal of Operations & Production Management*, *14*(4), 26–49. doi:10.1108/01443579410056029

Kara, S., Ibbotson, S., & Kayis, B. (2014). Sustainable product development in practice: An international survey. *Journal of Manufacturing Technology Management*, *25*(6), 848–872. doi:10.1108/JMTM-09-2012-0082

Kim, J., Park, K., Hwang, Y., & Park, I. (2010). Sustainable manufacturing: A case study of the forklift painting process. *International Journal of Production Research*, *48*(10), 3061–3078. doi:10.1080/00207540902791785

Kiron, D., Kruschwitz, N., Rubel, H., Reeves, M., & Fuisz-Kehrbach, S. (2013). Sustainability's next frontier: Walking the talk on the sustainability issues that matter most. *MIT Sloan Management Review Research Report*.

Koren, Y., Heisel, U., Jovane, F., Moriwaki, T., Pritschow, G., Ulsoy, G., & Van Brussel, H. (1999). Reconfigurable manufacturing systems. *CIRP Annals-Manufacturing Technology*, *48*(2), 527–540. doi:10.1016/S0007-8506(07)63232-6

Kuik, S. S., Nagalingam, S. V., & Amer, Y. (2011). Sustainable supply chain for collaborative manufacturing. *Journal of Manufacturing Technology Management*, *22*(8), 984–1001. doi:10.1108/17410381111177449

Labuschagne, C., & Brent, A. C. (2008). An industry perspective of the completeness and relevance of a social assessment framework for project and technology management in the manufacturing sector. *Journal of Cleaner Production*, *16*(3), 253–262. doi:10.1016/j.jclepro.2006.07.028

Law, K. M., & Gunasekaran, A. (2012). Sustainability development in high-tech manufacturing firms in Hong Kong: Motivators and readiness. *International Journal of Production Economics*, *137*(1), 116–125. doi:10.1016/j.ijpe.2012.01.022

Lee, C., Leem, C. S., & Hwang, I. (2011). PDM and ERP integration methodology using digital manufacturing to support global manufacturing. *International Journal of Advanced Manufacturing Technology*, *53*(1-4), 399–409. doi:10.1007/s00170-010-2833-x

Liu, S., Leat, M., & Smith, M. H. (2011). State-of-the-art sustainability analysis methodologies for efficient decision support in green production operations. *International Journal of Sustainable Engineering*, *4*(3), 236–250. doi:10.1080/19397038.2011.574744

Liu, W., & Liang, M. (2008). Multi-objective design optimization of reconfigurable machine tools: A modified fuzzy-Chebyshev programming approach. *International Journal of Production Research*, *46*(6), 1587–1618. doi:10.1080/00207540600943944

Malone, D. W. (1975). An introduction to the application of interpretive structural modeling. *Proceedings of the IEEE*, *63*(3), 397–404. doi:10.1109/PROC.1975.9765

Mandal, A., & Deshmukh, S. G. (1994). Vendor selection using interpretive structural modeling (ISM). *International Journal of Operations & Production Management*, *14*(6), 52–59. doi:10.1108/01443579410062086

Mani, M., Madan, J., Lee, J. H., Lyons, K., & Gupta, S. K. (2013). *Review on Sustainability Characterization for Manufacturing Processes*. US Department of Commerce, National Institute of Standards and Technology. doi:10.6028/NIST.IR.7913

Marksberry, P. W., & Jawahir, I. S. (2008). A comprehensive tool-wear/tool-life performance model in the evaluation of NDM (near dry machining) for sustainable manufacturing. *International Journal of Machine Tools & Manufacture*, *48*(7), 878–886. doi:10.1016/j.ijmachtools.2007.11.006

McWilliams, A., & Siegel, D. (2001). Corporate social responsibility: A theory of the firm perspective. *Academy of Management Review*, *26*(1), 117–127.

Miller, G., Pawloski, J., & Standridge, C. R. (2010). A case study of lean, sustainable manufacturing. *Journal of industrial engineering and management*, 3(1), 11-32.

Mittal, V. K., & Sangwan, K. S. (2014a). Development of a structural model of environmentally conscious manufacturing drivers. *Journal of Manufacturing Technology Management*, *25*(8).

Mittal, V. K., & Sangwan, K. S. (2014b). Modeling drivers for successful adoption of environmentally conscious manufacturing. *Journal of Modelling in Management*, *9*(2).

Molamohamadi, Z., & Ismail, N. (2013). Developing a New Scheme for Sustainable Manufacturing. *International Journal of Materials. Mechanics and Manufacturing*, *1*(1), 1–5. doi:10.7763/IJMMM.2013.V1.1

Mont, O., & Leire, C. (2009). Socially responsible purchasing in supply chains: Drivers and barriers in Sweden. *Social Responsibility Journal*, *5*(3), 388–407. doi:10.1108/17471110910977302

Nakano, M. (2010). A conceptual framework for sustainable manufacturing by focusing on risks in supply chains. In Advances in Production Management Systems. New Challenges, New Approaches (pp. 160-167). Springer Berlin Heidelberg. doi:10.1007/978-3-642-16358-6_21

After Two Large Gains, Rate of Atmospheric CO2 increase Returns to Average, NOAA Reports. (2005, March 31). *National Oceanic and Atmospheric Administration*. Retrieved from http://www.noaanews.noaa.gov/stories2005/s2412.htm

Nidumolu, R., Prahalad, C. K., & Rangaswami, M. R. (2009). Why sustainability is now the key driver of innovation. *Harvard Business Review*, *87*(9), 56–64.

O'Brien, C. (1999). Sustainable production–a new paradigm for a new millennium. *International Journal of Production Economics*, *60*, 1–7. doi:10.1016/S0925-5273(98)00126-1

Oliver, N., Delbridge, R., & Lowe, J. (1996). Lean Production Practices: International Comparisons in the Auto Components Industry. *British Journal of Management*, *7*(Suppl. 1), S29–S44. doi:10.1111/j.1467-8551.1996.tb00146.x

Paju, M., Heilala, J., Hentula, M., Heikkila, A., Johansson, B., Leong, S., & Lyons, K. (2010, December). Framework and indicators for a sustainable manufacturing mapping methodology. *Proceedings of the* Simulation Conference (WSC), *2010*(Winter), 3411–3422.

Parkinson, S. (1999). Agile manufacturing. *Work Study*, *48*(4), 134–137. doi:10.1108/00438029910272409

Pettersen, J. (2009). Defining lean production: Some conceptual and practical issues. *The TQM Journal*, *21*(2), 127–142. doi:10.1108/17542730910938137

Pham, D. T., & Thomas, A. J. (2011). Fit manufacturing: A framework for sustainability. *Journal of Manufacturing Technology Management*, *23*(1), 103–123. doi:10.1108/17410381211196311

Pumar, E. S. (2005). Social networks and the institutionalization of the idea of sustainable development. *The International Journal of Sociology and Social Policy*, *25*(1/2), 63–86. doi:10.1108/01443330510791298

Ramani, K., Ramanujan, D., Bernstein, W. Z., Zhao, F., Sutherland, J., Handwerker, C., & Thurston, D. et al. (2010). Integrated sustainable life cycle design: A review. *Journal of Mechanical Design*, *132*(9), 091004. doi:10.1115/1.4002308

Rimanoczy, I., & Pearson, T. (2010). Role of HR in the new world of sustainability. *Industrial and Commercial Training*, *42*(1), 11–17. doi:10.1108/00197851011013661

Seliger, G. (2007). *Sustainability in manufacturing: recovery of resources in product and material cycles*. Springer. doi:10.1007/978-3-540-49871-1

Seliger, G., Kim, H. J., Kernbaum, S., & Zettl, M. (2008). Approaches to sustainable manufacturing. *International Journal of Sustainable Manufacturing*, *1*(1), 58–77. doi:10.1504/IJSM.2008.019227

Slack, N. (2005). The flexibility of manufacturing systems. *International Journal of Operations & Production Management*, *25*(12), 1190–1200. doi:10.1108/01443570510633594

Smith, L., & Ball, P. (2012). Steps towards sustainable manufacturing through modelling material, energy and waste flows. *International Journal of Production Economics*, *140*(1), 227–238. doi:10.1016/j.ijpe.2012.01.036

Stavins, R. N., Wagner, A. F., & Wagner, G. (2003). Interpreting sustainability in economic terms: Dynamic efficiency plus intergenerational equity. *Economics Letters*, *79*(3), 339–343. doi:10.1016/S0165-1765(03)00036-3

Stoughton, A. M., & Ludema, J. (2012). The driving forces of sustainability. *Journal of Organizational Change Management*, *25*(4), 501–517. doi:10.1108/09534811211239191

Sushil, A. (2005a). Interpretive matrix: A tool to aid interpretation of management and social research. *Global Journal of Flexible Systems Management*, *6*(2), 27–30.

Sushil, A. (2005b). A flexible strategy framework for managing continuity and change. *International Journal of Global Business and Competitiveness*, *1*(1), 22–32.

Szekely, F., & Knirsch, M. (2005). Responsible leadership and corporate social responsibility: Metrics for sustainable performance. *European Management Journal*, *23*(6), 628–647. doi:10.1016/j.emj.2005.10.009

Tapiero, C. S., & Kogan, K. (2008). Sustainable infrastructure investment with labor-only production. *International Journal of Production Economics*, *113*(2), 876–886. doi:10.1016/j.ijpe.2007.12.004

The Institute for Sustainability. (2011). Retrieved from http://theinstituteforsustainability.com/

Thomas, A., Francis, M., John, E., & Davies, A. (2012). Identifying the characteristics for achieving sustainable manufacturing companies. *Journal of Manufacturing Technology Management*, *23*(4), 426–440. doi:10.1108/17410381211230376

Unruh, G., & Ettenson, R. (2011). Growing green: Three smart paths to developing sustainable products. *Harvard Business Review*, *87*(9), 23–39.

Van de Ven, B., & Jeurissen, R. (2005). Competing responsibly. *Business Ethics Quarterly*, *15*(2), 299–317. doi:10.5840/beq200515216

Veleva, V., Hart, M., Greiner, T., & Crumbley, C. (2001). Indicators of sustainable production. *Journal of Cleaner Production*, *9*(5), 447–452. doi:10.1016/S0959-6526(01)00004-X PMID:17208696

Vinodh, S., & Rajanayagam, D. (2010). CAD and DFM: Enablers of sustainable product design. *International Journal of Sustainable Engineering*, *3*(4), 292–298. doi:10.1080/19397038.2010.501870

Warfield, J. N. (1974). Toward interpretation of complex structural models. *IEEE Transactions on Systems, Man and Cybernetics*, 5, 405-417.

Warfield, J. N. (1994). *Science of Generic Design: Managing Complexity through Systems Design*. Iowa State Press.

Westkamper, E. (2008). Manufuture and sustainable manufacturing. In *Manufacturing Systems and Technologies for the New Frontier* (pp. 11–14). Springer London. doi:10.1007/978-1-84800-267-8_3

Westkamper, E., Alting, L., & Arndt, G. (2001). Life cycle management and assessment: Approaches and visions towards sustainable manufacturing. *Proceedings of the Institution of Mechanical Engineers. Part B, Journal of Engineering Manufacture*, *215*(5), 599–626. doi:10.1243/0954405011518557

Womack, J., Jones, D., & Roos, D. (1990). *The Machine that Changed the World*. New York, NY: Rawson Associates.

Wu, Z., & Pagell, M. (2011). Balancing priorities: Decision-making in sustainable supply chain management. *Journal of Operations Management*, *29*(6), 577–590. doi:10.1016/j.jom.2010.10.001

Yuan, C., Zhai, Q., & Dornfeld, D. (2012). A three dimensional system approach for environmentally sustainable manufacturing. *CIRP Annals-Manufacturing Technology*, *61*(1), 39–42. doi:10.1016/j.cirp.2012.03.105

Chapter 4
Environmental Accounting for Industrial Growth and Sustainable Development in India
An Overview and Theoretical Framework

Archana Singh
Symbiosis International University, India

Nehajoan Panackal
Symbiosis International University, India

Bhama Venkataramani
Symbiosis International University, India

ABSTRACT

Environmental accounting is an important function that provides industry with a means to incorporate information with business decision making and business operations. This chapter evaluates the extent to which the process of environmental accounting can be applied as a solid foundation for industrial growth and sustainability. Having an environmental accounting system in place allows the industry to better manage environmental costs and to discover new opportunities to minimize environmental costs through environmental thinking to the industrial sector. Researchers have adopted systematic literature review as a methodology to generate strong knowledge base for the study. In this paper, the theoretical foundation of environmental accounting and reporting is discussed with reference to India. The insights gained will shed light on positive outcomes of adopting environmental accounting for sustainability and manufacturing management in the industrial sector of India.

1. INTRODUCTION

There is a growing concern towards lack of adequate measures to ensure sustainability in our society. Environmental business management has focused its attention on how and why existing firms can

DOI: 10.4018/978-1-5225-0350-7.ch004

become greener. The field of green management has begun to broaden its arena and recently there has been a growing recognition of the importance of several factors impacting green management such as linking sustainability and innovation, the role of small and medium enterprises in innovating products and processes keeping sustainability in mind. The importance of sustainability in strategic business development and green practices in industries is must. For an industry to have a positive environmental influence, real improvement can be created only if the production processes, products and services are environment friendly. Environment is the aggregation of the eco system of living organisms and the non living components like air, water, soil, etc. around us which together affects the livelihood of men and animals on earth. Natural Capital Accounting (NCA) is the measurement and valuation of natures benefits in terms of ecosystems, goods and services-like fresh water, flood control and forest products-to be incorporated into a general standard format consistent with conventional national accounts. (http://www.conservation.org/projects/pages/valuing-and Accounting-for-Natural-Capital)

Natural capital accounting (NCA) gives insights on how economic development depends on natural resources both for production as well as for absorption of effluents and pollution. This information will help policy makers while charting the development plan.

A native American proverb states that, "Only when the last tree is cut, only when the last river is polluted, only when the last fish is caught only then they will realize that we cannot eat money." In this background, it is not only imperative that we are aware of the imporatnce of and need for sustainable development but it is also a requirement to bring on record and report to the stakeholders the extent of environment friendly measures or otherwise adopted and practiced by the organisation. Environmental accounting thus plays an important role in this regard and is now gaining importance in the reporting function of an organisation.

In the early 1970's, Norways ministry of environment initiated the first natural resource accounting system which was later followed by other countries. In India environmental accounting is optional. Since the 1990's, Europe has been focusing on sustainable accounting indicating that the accounting profession has a significant role to play through environment sustainability reporting. Accounting for social economic and environmental aspects and reporting the same for decision making of internal and external stakeholders is facilitated through financial statements A customized format for identifying and recording environment related data can be created for disclosure. Business is an integral part of society and such disclosure shows the responsibility and commitment of the entities. Many researchers have discussed environmental accounting with reference to India and it is noted that awareness and adoption of environmental accounting is not a choice but a necessity. Since the new economic policy 1991 was implemented in India, the intensity of global competition has increased. Eco auditing is essentially related to Green accounting (Jasch, 2001). Green accounting is concerned with the appropriate analysis and reporting of environmental issues of an organization.

The present civilization is involved in various economic and non-economic activities which generates waste with potential constituents. Responsibility towards the environment is one of the crucial areas of social responsibility. Environmental pollution is a worldwide phenomenon and we must not forget that the earth's environment has been handed to us by the previous generations and nature. Wealth is enough to satisfy everyone's need but not their greed. Therefore environment protection is a critical issue in the present day context. The indiscriminate disposal of such wastes contributes towards the environmental pollution and hazards. At present, the magnitude of environmental pollution is increasing at a very alarming magnitude in different parts of the globe.

Environmental issues are currently receiving due attention from the public and private sector. It has become crucial especially for manufacturing industries, to carefully monitor the effects of their activities on the environment. There exists a correlation between manufacturing industries and cost and challenges these industries need to face in the process of implementing environmental management accounting. Some of the companies adopt internally developed environment policies but they also face challenges in defining, separating, identifying, classifying, measuring and controlling the environmental protection cost. Namakonzi & Inanga. (2014).

The importance of environmental accounting information for management supports the prevalence of decision making usefulness theory in the implementation of environmental management systems. Nowadays members of the society appreciates the ill effects of exploitation of the environmental resources and are therefore concerned about the same. Environmental awareness is the new value emerging in our society and the same values need to be incorporated in the industrial sector. Every organisation needs to articulate its environmental management policy through an information system that helps to analyze the interaction between firms and environment. Being a corporate citizen, an entity is evaluated by its financial as well as environmental performance. The Indian corporate sector needs to improve as far as environmental performance. Laws have been laid down and amended for them to fulfill corporate social responsibility for the economic development of the Indian Economy. In the global and Indian context, environmental responsibility is of vital importance and a very crucial area. India being a developing country whose focus is on growth and development must ascertain the benefits and costs of environmental damages and degradation. The gross domestic product is highly impacted due to the environmental cost. Accounting is a language of business and it is very important that through accounting, the government and entities should fulfill the responsibilities towards the environment by measuring and reporting the costs of used or misuse of natural resources. Thus the need of environmental accounting has emerged. At the global level, the environment movement has grown giving birth to the United Nations Environment Programme (UNEP). When environment issues became a worldwide problem in the 1980's the world commission on environment and development (WCED) known as Bruntland commission was established by the United Nations. It published a report on sustainability called "Our common future."

Green economy has become the language at the global level to develop the environment. There are different social dimensions of sustainability and many concerns in the emerging framework for sustainable development. It is very important for the industry to raise general questions about measuring sustainability and to focus on sustainable economic renewal. Davies & Mullin (2011).

1.1 Environment Accounting: A Conceptual Analysis

Industries utilize natural resources for their production process. Consequently, it leads to environment pollution and degradation. Environmental consciousness is increasing day by day at the global level. The role of sustainable development is major and central as an organizing principle in industry as human activity and its impact on environment has become more evident. The two major concerns emerging out of it is to address innovation in our mode of thinking and policy in practice settings to explore knowledge. One outcome is development of different disciplines progressing sustainable development like sustainability science. Bebbington & Larrinaga (2014).

Corporate stakeholders need to pay attention on environmental accounting for reducing and avoiding costs and to improve the quality of the environment. Environmental accounting can be supported at the micro and macro level in relation to financial accounting, cost accounting and managerial accounting.

Industrial, commercial and service industry at different levels must measure and evaluate the actual environmental impacts of various projects for development decisions which are economically and environmentally sound. Issuance of environmental regulations for new solutions needs to be updated for which accounting can play a vital role.

The consequences of environmental accounting are on the economic, social and legal status of a nation. Undoubtedly, the industry enjoys the competitive advantage by following proper environmental management. Government policies have not come out with great outcomes in this regard. It is very important for a developing nation like India to come up with incentives and rewards for quality environmental reporting. Stakeholders expect the business entities to meet environmental standards. Smith (2002) and companies need to provide environmental management system to support the same. Fortis (2002).Accountants that are responsible for the communication of business need to open their eyes to the impact of environmental issues on accounting, auditing and disclosure. Globally accountants are expected to be proactive and show their concern for the environment protection process. Liberalization and globalization has made environment accounting of vital importance today.

Environmental accounting is a growing field. Industries are taking serious attempts to identify the environment issues by the business houses. There is no single definition encompassing the scope of Environmental accounting. Data that have been included in the environmental accounts of different countries includes pollutant eminents in physical terms by economic sector, actual environmental protection expenditures by economic sector and hypothetical expenditures that would be required to further reduce pollution(sometimes called "maintenance cost"), by economic sector.

It assesses the costs and benefits to an enterprise due to environment preservation activities that are undertaken. Environmental accounting involves the identification, measurement and allocation of environmental costs, the integration of these costs into business, identifying environmental liabilities if any and finally communicating this information to the company's stakeholders.

The three main aspects of environmental accounting are—economic, social, and environment (Ball 2005; Milne 1996). Studies have been conducted for evaluating the financial and sustainable impact on financial management (Lamberton, 2005; Schaltegger & Wagner, 2006; Taplin, Bent & Aeron-Thomas, 2006). Gray (2006) was concerned that sustainability accounting must include revealing the right information to stakeholders for a sustainable organization. As a consequence, sustainability accounting needs to take care of environmental and social accounting issues. Environmental accounting and its disclosure is not mandatory. In the current era environmental problems are vital issues which makes environment accounting a strong branch of accounting. Legal authorities and standard setting bodies have made a conceptual framework of environmental accounting and its disclosures. Pramanik et al (2007).

Major functions of environmental accounting are to minimize the negative environmental effects of old accounting practices, to know environment related costs and to take active steps to set up initiatives in order to ameliorate the existing environmental effects of conventional accounting practices. Published sources of ACCA acknowledge full environmental costing to be adopted by business. Bebbington, Gray, Hibbit & Kirk (2001).

The forms of environmental accounting are environmental management accounting (EMA), environmental financial accounting (EFA) and environmental national accounting (ENA).EMA is an internal environmental accounting tool to support sustainable environmental management activities and environmental reporting. It includes segment environmental accounting, eco balance environmental accounting and corporate environmental accounting. EMA emphasis on energy and material flow information and environmental cost information. Economic performance is evaluated through environment management accounting theoretically, the concern remains rare. Christ & Burritt.(2013). It's an internal environmental accounting tool to select a project or investment activity. It consists of segment environmental accounting, eco balance environmental accounting and corporate environmental accounting. Segment environmental accounting includes environmental accounting tool to select an investment activity, eco balance environmental accounting talks about sustainable environmental management activities and corporate environmental accounting is concerned with informing proper financial reporting in accordance with environmental accounting. EFA focuses on environmental liability cost and other environmental costs. EFA is a method of reporting environmental liability cost and other significant environmental cost through financial accounting. ENA focuses on natural resources and environment cost at the national level. The specific focus of ENA is natural resources, stocks and flows, environmental cost and externality costs, etc.

The need for environmental accounting includes the factor show in Figure 1.

Environmental accounting is required to fulfill the responsibility of different stakeholders. The objectives include increasing environmental transparency, to determine positive relation between the company and society. The environmental accounting consumerism movement launched by the environmental lobby groups encourages the consumers to purchase environmental friendly products. Industries producing green products enjoy marketing advantage by doing so. Through this method of accounting, industries can show commitments by making environmental disclosures. Industries that are not focused on environment friendliness may use green accounting to overcome negative public opinions.

Due to environmental accounting, environmental awareness is enhanced and environmental commitment becomes a shared responsibility amongst the stakeholders. Environmental costs can be offset by generating higher revenues through sale of by-products or recycling. It helps in deciding cost of the product and helps in focusing perpetually on natural resource asset. Sustainability at the ecological level enhances national wealth and national economic accounts to find an economy's total wealth. It is important to know the physical flow accounts to determine the right policies by taking into consideration data on pollution and waste along with designing and evaluating environmental taxes. To asses economic progress there is a need to meet regulatory requirements of the industrial sector, to operate in an environmental friendly manner with an environmental attitude controlling operational efficiency and optimize cost and waste. The important aspects of environmental accounting involves both internal and external factors. Internally they are the business and the investment made by the industry for minimizing losses to environment. This includes environment saving equipment and measurement in monetary terms is possible.

Figure 1. Need for environmental accounting

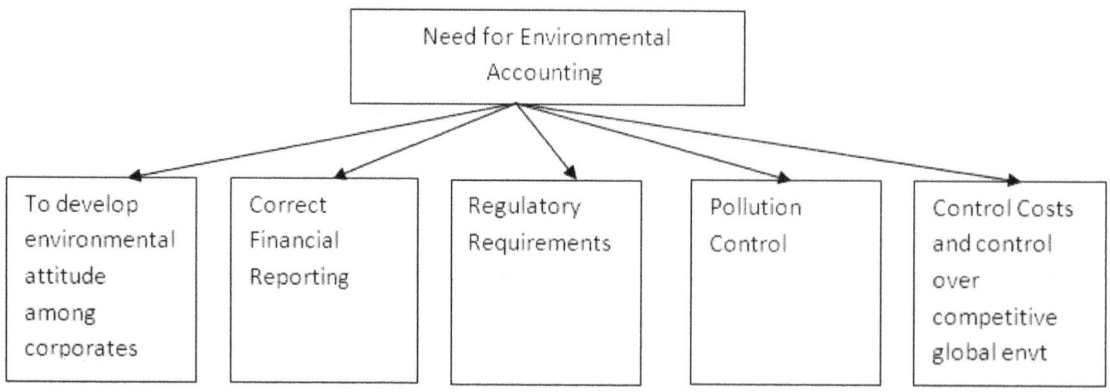

Externally it constitutes costs of de gradation and destruction of soil, coastal and marine pollution and exploitation of non renewable sources like minerals, gas etc along with deforestation and land use. This step of accounting is very challenging and complicated. It is difficult to measure the loss accurately.

2. ENVIRONMENTAL ACCOUNTING IN INDIA: AN OVERVIEW

The developing countries like India are currently facing the challenge of promoting economic development without causing any negligence to the environment. A tradeoff between environmental protection and development is a major concern today. The concern of environmental responsibility and the sustainable industrial development has given rise to a novel branch of accounting known as environmental accounting.

Accounting is defined by the AICPA as "The art of recording, classifying, and summarizing in a significant manner and in terms of money, transactions and events which are, in part at least, of financial character, and interpreting the results thereof."

Following the end of the British rule, India plunged into serious troubles of communal riots. Initially India did not pay much attention towards the environment. The five year plans were implemented with a view to the economic development of the country. The second five year plan focused on industrial development. But the plans remained unconcerned about protecting nature. The public awareness on environmental issues grew after the beginning of the seventies when a law for protection of environment was passed. As there was requirement of environment related information from business on a periodic basis, the ministry of environment and forests has proposed that every company shall in the report, in the report of its board of directors disclose briefly the measures taken towards the adoption of clean technologies for the prevention of pollution, waste minimization, waste recycling and utilization, pollution control measures, investment on environmental protection and its impact.

Countries like India have to focus on environmental problems as well as make sure that the pace of economic development is high. Therefore it is must to have a tradeoff between the two. Proper evaluation of cost and damages to the environment should be studied to know the impact of major problems in terms of monetary costs. Environmental accounting will work as a tool to enumerate economic efficiency and environmental efficiency as a whole. In India disclosures related to environment in the corporate annual reports both financial and non-financial is not an encouraging level. Neither company law nor the accounting standards issued by ICAI prescribes disclosing norms for environment related matter in the corporate financial statements. There is a lack of environment awareness, poor enforcement of the of the environment protection acts is also partly responsible for freeing the companies from disclosure of such information. In the absence of standardized environmental accounting practices and disclosure techniques at both national and global level along with legal enforcement to advocate environmental accounting practices from a global perspective.

In the past, issues related to the environment were ignored by, large companies and small medium enterprises to a great extent. However, due to increasing concerns and awareness environmental accounting is receiving increased attention in the recent times. Many companies now consider it important to be green and follow green practices, as many investors place a high value on environmental responsibility. Environmental costs can be significantly reduced and eliminated due to business decisions, ranging from operational to investment in greener process technology to redesign processes or products. Better natural resource and green accounts would provide valuable insights into the interaction between the environment and the economy. Industrialization is the main cause for environment degradation. Though it is vital for economic growth, but it drastically damages the environment. Optimum utilization of resources is the prime responsibility of every industry which has given birth to a new branch of accounting called environmental accounting. It is a must for organizations to implement the concept of sustainable development for balancing ecological activities and to measure the organization in an economic manner.

In the early 1990's Indian companies faced cut throat international competition wherein International firms disclosed non-financial information including environmental information to meet the expectation from Indian companies and to improvise their corporate image showing socially responsible behavior. Thus it was expected that Indian companies should report their environmental performance. But the disappointing part is that in the developing countries very little attention has been given to the state of environmental accounting. (Ahmed & Sulaiman, 2004; Thompson & Zakaria, 2004; Nafez & Kamal, 2000). Pahuja (2009) concluded that companies which have better environmental performance provide more environmental information in the annual reports as compared to poor performers. Her study also indicated the existence of statistically significant ''size'' and ''profitability'' effect. Hence, larger and more profitable companies disclose more information on environment as compared to smaller and less profitable companies. Environmental reporting by the public sector is higher than the private sector. Jaffar et al (2002) measured environmental performance on the basis of legal action taken on companies for noncompliance of environmental regulations by the department of environment. Companies having better environmental performance disclose more on the environment.

Rapid industrialization growth has great effect on the environment and is of major concern to the economy. The government has been supporting small enterprise sectors and they are tended to be the worst polluters. It is very important to pay attention to environmental issues as a part of their operations. Although existing environmental legislations serve some purpose of protecting the environment, the small scale and large enterprises have significant differences in the way they approach environmental issues (D'Souza & Peretiatko, 2002)

2.1 Environmental Accounting in Indian Industry: The Road Ahead

Eresi (1996) in his study disclosed that companies gave only positive information about the practices adopted by them in the context of environment conservation. Kumar (2002), examined the Indian Scenario and stated that accounting and reporting aspects of environmental protection is a must for policy statements in an annual report and emphasized that the environmental reporting in India was limited to little more than a sentence or two each in the annual report. Vasal (1998) administered a study on 144 companies found that the practice of preparing stand-alone environmental reports was yet to rake roots in India. Roy (2000) observed that companies giving environment disclosure in the public sector was more than in the private sector. Pradhan & Bal (2002) agreed that a company must reveal information environmental policy, environmental audit report, future targets. Shah analyzed 50 annual reports of large and medium sized companies and found that hardly any company report on failure regarding environmental protection or damage caused by the company to the environment and society. Environmental issues are of low priority for some companies who consider environmental protection as a trustee duty. Industries who consider environmental goals are supplementary to the business attempt to pilot and control environmental impacts in an efficient manner, industries located in the upper right hand treat environmental issues as central to their core business

Environmental accounting can be implemented by industries of large, medium and small scale sizes. Whether administered by a global corporation or a small business, elements need to be in place for success. The business entities management must support these practices. These leaders need to focus in setting a positive tone when communicating the benefits of environmental accounting practices to the employee population and implementing it. Consisting of employees across all business lines, including finance, sales, manufacturing and procurement, these teams ensure that all environmental accounting policies and procedures are communicated and followed. The green entrepreneurs activities have an overall positive impact on the natural environment and moves towards a more sustainable future. All their business is structured and operated in such a way that every component has a neutral or positive impact on the environment. Environmental and social entrepreneurs, or ecopreneurs, lead socially committed, break-through ventures that are driven by environmental, social and economic goals. There exists a strong link between youth entrepreneurialism and environmentalism. A model called as Green-Works business model has been developed for attaining sustainability of the environment and social objectives of the organization. Sarah E.A. Dixon et al. Entrepreneurship can be driven by moral attitudes. Anderson (1998).

It has become important to give ecopreneurship a serious consideration as we are rapidly losing our natural resources which are the true capital of any nation. Unlike past decades where economic gains were the only measure of a nation's progress, now a day's sustainability is given high importance. Joshi, TOI. Environmental awareness and reporting in India is a wide road ahead to understand the reasons for environmental disclosure by taking into consideration the global scenario. Due to the depletion rate there is a contradiction in the growth of renewable resources and growing demand. Industries need to realize and make a genuine attempt on the optimum use of resources and curtail environmental damage. Initially GNP was considered growing with the growth of industrialization and technological advancement, but in reality industrialization has resulted in heavily polluted environments having a negative impact on the quality of life. Therefore industries must realize the prime responsibility for clean technologies which would contribute to the nation in the long run. Cost volume profit (CVP) analysis, standard costing and variance analysis cannot be adequately used as a measure of control in the present manufacturing system. (Bromwich and Bhimani, 1994; Kaplan, 1983; Lucas, 1997).Production process and less labor usage consequently may lead to non-existence of above tools (Drury et al, 1993; Hilton, 2002).

Economic, social and ecological systems of environment integrate to sustainability. Thus companies need to acquire sustainable development. Dixon & Fallon (1989).A sustainability management accounting system (SMAS) refers to sustainability and management accounting concepts and practices that deal with environmental and social issues. Development areas are related to human, social, economic and environmental aspects. Goodland (2002).

Environmental accounting seeks to better measure sustainability by expanding gross measures of national welfare (product, investment, etc.) to include non-market values, in particular ones associated with environmental goods and services. In addition, it also seeks to incorporate costs and benefits of environmental protection and depletion of natural capital – two measurements not typically included in national accounting systems such as gross domestic product. While opinions vary on how to perform green accounting, the technique is used worldwide and is well-established in the US.

Indian industry needs to have accounting practices which are conservative and innovative. It is very important to develop future accountants who can measure environmental degradation in quantitative terms. It is a challenge and duty for India to understand the importance of the implementation of environmental accounting in the curriculum. Business should take care of social environmental liabilities.

Since 1991, the liberalization of the Indian economy has increased the pressure for international competition and changed the information needs of Indian companies. The future emphasis in India is on traditional practices and less on the new techniques, because higher benefits were derived from such techniques. Size (total assets) has an influential factor in the adoption of the newly developed practices. Most of the differences could be attributed to the differences in cultural values. Indian management generally avoids risk, is quite conservative, and less innovative in adopting new management accounting techniques. Since, Indians have a long history of their heritage; it takes them longer time to change their societal values and practices, which also seems true in the case of adopting new management accounting practices. Due to cost pressures, customer awareness, supply chain relations and activities of environmental campaigners encourage the companies to go for environmental initiatives Perry & Sheng(1999). Developed countries are highly concerned with the quality of their lives due to the arising alarm of pollution. The same concern is shared by developing countries hence; many laws pertaining to environmental protection are enacted. Industries need to integrate environmental issues into their strategic planning process. Banerjee (2001).

3. ENVIRONMENTAL ACCOUNTING: THE REGULATORY FRAMEWORK COST AND BENEFITS

Environmental accounting and reporting practices are in the evolution stage in India. Even though Indian industries comply with the rules and regulations with regard to environmental protection, till now no clear cut policies are formulated at the National, State or even at the industry level, for ensuring the level of compliance to environmental norms. Indian practice, voluntary environmental reporting with regard to the environmental parameters identified.

There has been an increased trend towards environment management and disclosure in annual reports, but meaningful disclosure has not been achieved in socially related company activities such as pollution control. So, reporting is yet to attain the desired shape in terms of both the quantity and the quality. This shortcoming may be because of the absence of any standard specifying the kind of information required to be disclosed, and the absence of any compulsion on the firms regarding such disclosures. Regular attempts have been made in this regard in the different countries to streamline the environmental reporting practices for industry. They are yet to come up with form of legislations; a comprehensive accounting standard. In India, a few such legislations require submission of different environment statements to be respective regulatory bodies. But nothing exists by now, which requires disclosure of environmental information through annual reports. In such situations it is quite encouraging to note that industries are voluntarily disclosing environmental (howsoever inadequate that may be) through their annual reports. Data on environmental accounting plays an important role to different users at different levels for different reasons. Now a days there is a huge demand on financial and economic data on environmental and natural resources for civilizing environmental performance and for the own self interest of the industry. Singh & Joshi. (2009).

Industry environment reporting is an important tool for developed countries driven by legal compliance. Green economic development aims at reducing environmental damage and government regulations along with voluntary self regulatory measures to curtail the impact on local environment is a must. Such environmental management practices should be synchronized in different sectors of the society to promote green economic development. Gibson &Collins.

In the constitution under the directive principle of state policy" it is expressly mentioned that it is the duty of the state to protect and improve the environment and to safeguard the forest and wildlife of the country." It is the fundamental duty of every citizen of India to protect and improve the natural environment including forest, lakes, rivers and wildlife." Government of India enacted numbers of laws which can directly and indirectly help the protection of environment and natural resources. In addition to that union ministry of environment and forest and corresponding state governments have made a provision for environmental clearance before any industrial unit actually set up. However, there is no any specific guideline regarding accounting and disclosure of environmental cost and liabilities except provisions of company about disclosure of conservation of energy, technology absorption in director's report under sec 217(1). Recently ministry of corporate affairs laid more emphasis on corporate social responsibility and accordingly issued national voluntary guidelines on social environmental and economic responsibility of the business in 2011. In such guidelines, it also mentions that business should provide goods and service that are safe and contribute to sustainability throughout their lifecycle. It is explicitly mentioned that business should respect, protect and make efforts to restore environment. In pursuance of that a business should report:

Percentage of materials used that are recycled input materials, total energy consumed by the business entity for its operations, statement on use of energy saving processes and the total energy saved due to use of such processes, use of renewable energy as percentage of total energy consumption, Total water consumed and the percentage of water that is recycled and reused, statement on quantum of emissions of greenhouse gases and efforts made to reduce the same, statement on discharge of water and effluents indicating the treatment done before discharge and the destination of disposal, details of efforts made for reconstruction of bio-diversity.

The disclosures of above mentioned information are voluntary and there are no specific guidelines on format of reporting. On the other hand accounting standard board and Ministry of Corporate Affairs is till silent on environmental accounting and disclosure issue in corporate accounting and reporting. Environmental conservation benefit is measured in physical units and is the benefit obtained from the prevention, reduction, and/or avoidance of environmental impact, removal of such impact, restoration following the occurrence of a disaster, and other activities. By assessing the economic value of environmental conservation benefits which are measured in physical units, results are described in monetary values by some cases. Currently, in the field of environmental economics, a number of valuation methods are being developed for the purpose of determining the environment's economic value.

At the global level, environmental protection concern has increased; countries like India also had to set up a union ministry of environment along with essential legislations for coordination among the states and various industries. Although licensing for industry is abolished in India but environmental clearance is of vital importance.

The various laws relevant to environmental protection are as under: (a) Directly related to environment protection: i. Water (Prevention and Control of Pollution) Act, 1974. ii Water (Prevention and Control of Pollution) Cess Act, 1977. iii. The Air (Prevention and Control of Pollution) Act, 1981. iv. The Forest (Conservation) Act, 1980. v. The Environment (Protection) Act, 1986. (b) Indirectly related to environment protection: vi. Constitutional provision (Article 51A). vii. The Factories Act, 1948. viii. Hazardous Waste (Management & Handling) Rules, 1989. ix. Public Liability Insurance Act, 1991. x. Motor Vehicle Act, 1991. xi. Indian Fisheries Act, 1987. xii. Merchant of shipping Act, 1958. xiii. Indian Port Act. xiv. Indian Penal Code. xv. The National Environment Tribunal Act, 1995. It is important to note that all new projects require environment clearance. This clearance concerns both the Union Ministry of Environment and Forests and the corresponding State Govt. department of environment. Guidelines have been issued and all such projects are expected to obtain environmental and antipollution clearance before they are actually set up. A Central Pollution Control Board (CPCB) has also been set up. Wherever cases of violating of standards of water or air pollution have been detected, show cause notices have been issued to industrial units and all such units are being kept under regular inspection. (A report on environmental accounting: a general overview and Indian perspective social statistics division central statistics office ministry of statistics & programme implementation New Delhi).

System of integrated environmental and economic accounting (SEEA) is concerned with integrating different methodologies into an organised framework. It comprises of pure physical data relating to flow of material and energy, it takes those elements that are relevant to good management of the environmental and has environment related transactions and accounts of environmental assets measured in physical and monetary terms. It emphasis on depletion, defensive expenditures relating to environmental degradation.(Green National Accounts in India A Framework A Report by an Expert Group Convened by the National Statistical Organization Ministry of Statistics and Programme Implementation Government of India March 2013). It is vital to implement environmental accounting system for accurate information on environmental analysis, Mein-chin (2002), as it will ensure environment friendly management decisions. The usage of environmental accounting has changed over the past 2 decades. Now it is an important source of information for a company's manager. Schaltegger et al (1996).International accounting standards address the general principles for the measurement and disclosure of environmental matters in the financial report. (IAS 37). Any industry failing to comply with legal compliance requirements concerning environmental matters may face fines and penalties. It is important that some entities must recognize environmental obligations as liabilities in the financial statement. For example: obligations related to solid waste. The industry in which the entity operates must focus on environmental issues and should have accounting treatment adopted for environmental cost and environmental restoration liabilities (Pinckard & Wendy, 2000).

Environmental accounting techniques and the incomplete efforts of accountants and their allies to overcome environmental issues makes it mandatory to explore cost–benefit analysis. The identification and allocation of environmental costs requires accounting and book keeping knowledge. There is no universally accepted definition of environmental cost. It depends on the industry and its entity how the information will be used. Bouma (1998). According to the United Nations division for sustainable development, "the main problem of environmental management accounting is that we lack a standard definition of environmental costs". Environment cost that has been incurred by an entity with no return or benefit like damages done which cannot be recovered is environmental loss. The Canadian Institute of Chartered Accountants (CICA) (1993)Environmental cost can be listed as atmospheric emissions waste, energy consumption, material usage, oil & gas consumption, packaging, transportation costs, waste and waste resource use. By integrating social and ecological costs and benefits resulting from the natural environment into traditional economic accounting systems, green accounting aims to capture the interdependency and dynamic interactions between the three pillars of sustainability (economy, society, and environment). More accurately valuing natural resources costs and benefits may contribute to the development of more appropriate and sustainable economic, trade, and development policies.

Product related cost and research and development expenditure incurred can be categorized as capital investment, operating costs, research and development cost, environment administration and planning, expenditures for remedial measures and recovery measures (Pramanik, 2002). From the time of acquisition to the cost incurred for capitalization of environment expenditure needs to be justifiable to mitigate environmental contamination. Contingent environment liability may or may not occur but to compensate a third party suffered is an obligation to pay future expenditure for environmental damage. Quantifiable environmental liabilities are shown with the accurate balance in the balance sheet and unquantifiable one are shown as foot notes explaining the nature. All natural assets are environmental assets but not vice versa (Mohamed, 2003).

4. HOW CAN ENVIRONMENTAL ACCOUNTING CONTRIBUTE TO THE SUSTAINABLE INDUSTRY GROWTH?

Sustainability is recognised the world over as the litmus test for any new project to be launched. The process of sustainability is uninterrupted and has forms diversity Rizkallah & Bone (1998). Mebratu (1998).The World Commission on Environment and Development was asked to formulate a global agenda for change. The General Assembly of the United Nations was called upon to propose long-term environmental strategies in order to achieve sustainable development. The term sustainable development was discussed by the World commission on environment and development in 1987 in their report which was titled as "Our Common Future". Sustainable development that "meets the needs of the present without compromising the ability of future generations to meet their own needs."(World Commission on Environment and Development (1987). Our Common Future. Oxford: Oxford University Press, p.43.) is supposed to be the global mission, necessity and hopefully, reality. Awareness of this problem1 leads to the fact that "all the world is in the same boat", that everyone should somehow contribute to the sustainable development and to communicate about problems, activities and accomplishments in this threatening area. Sustainability needs to pay attention to the poor of the present and future generation as they are the wealth of the nation. Sen (2013). Kordestani et al. (2015) have summarized the area of research towards developing sustainability. Manners in 2014 proposed sustainability economics as subservient to society. The concept of sustainability is important for all stakeholders, who take and implement the concept of sustainability in different stakes. Felix et al (2015). HR policies and interventions are pivotal to the success of sustainability (Dubey & Gunasekaran, 2015).

The heart of sustainable development is accountability of socio-economic policies for their environmental impacts. Development analysis comprises of non-economic objectives that do not lend themselves to monetary valuation. Integrating environmental concerns into economic policies is accounting for economic performance. Social evaluation of these objectives by means of norms, standards, and targets is required for integrated development.

Long term sustainability can be achieved through environmental instruments to decrease environmental footprint and for the betterment of relationship with the stakeholders. It is very important for an industry to comply all their operations with the principles of sustainable development as far as environmental responsibilities are concerned. They must inculcate more and greener environment practices in their business strategies for which developing and implementing environmentally sustainable practices and assessment of the same through environmental accounting and reporting need to be emphasized. Sandra & Dubravka (2014).Environment problem impacts every area globally and it is very important to study and explore the roots of environmental accounting (Moorthy, Yacob) 'Green procurement' practices and conceptualizes a strategic framework is required to further strengthen the linkage between the business entity and industry. Dubey, Bag, Ali & Venkatesh (2013). The need for consistent, decisive environmental accounting principles is pervasive. A loss of nature's carrying capacity entails damage from an accounting perspective. Generally accepted accounting principles illustrate the need for environmental accounting. Environmental accounting entails a subset of GAAP and gives a holistic picture of both financial activity and associated risks. It would require additional disclosure to cover environmental contingencies; thus creating better awareness of a firm's overall financial position, and better educating management.

Accounting is a product of environment and it is shaped by the environment. (Choi & Mueller, 1992; Perera, 1989; Radebaugh & Gray, 1993). Core sectoral company's disclosure covering informational needs of stakeholders supports suitable framework for environmental disclosure. Environmental accounting is an emerging issue required to support the Indian industry. Incorporating environmental accounting into national economic accounts could provide a measure of sustainability; however, considerable advances in methods of measurement and valuation are needed. From a purely accounting perspective, particular forms of capital could be diminished or, in an extreme case, wholly eliminated without decreasing overall welfare if other forms can be substituted for it. There are, of course, no substitutes for the life-sustaining services of nature and the question of when and how to account for this fact is the source of many ongoing debates in green accounting.

To progress from an economic and social angle without endangering the natural equilibrium of the planet is a primary condition for sustainable development. Citizens as well as industries as their true stake of the future must avoid environment degradation. Environmental accounting is the expression of a new ethics that answers to the human worries for planetary development and progress. Letiţia & Dănuleţiu (2010).The progress and welfare of an economy is measured by the per capita income. Per capita income is calculated by dividing population with the national income. National income accounting is a tool for economic analysis reflecting changes in investment, savings, structure of industry, etc. Environmental degradation and scarcity of natural resources has a negative impact on development. Deterioration of environment and scarce natural resources add to the cost of development. Increase in the cost can hamper the future development of an economy. Some of the overarching methods currently in use for green accounting include natural resource accounts; emissions accounting; disaggregation of conventional national accounts; value of non-marketed environmental goods and services; and, green gross domestic product.

Identifying such costs promotes better decision making and better utilization of resources. In addition, it gives an incentive to management to save on environmental protection costs such as pollution reduction, fines, waste management, legal fees, monitoring, insurance, and regulatory reporting and consequently improve the environment. This calls for accountants to adopt and understand better environment management and accounting systems that can aid in identification, classification, allocation and control of environmental costs for better decision making and better environmental management among other benefits. Roger Adams, (2002) affirms that much of the interest surrounding environmental issues has been directed at external reporting and financial accounting and less in EMA which is intended for internal decision making. Environment accounting might also avoid huge, but predictable swings in revenue and expenses. Environmental economists are, in fact, already attempting to place a value on nature's services, something that a generation or two ago might have been considered unquantifiable, or worse, limitless. Many critical environmental valuations and estimates, accountants are the natural leaders for a team of professionals necessary to incorporate all of this valuable environmental data into a financial reporting package.

Industrial sustainability is a great challenge and accounting as a source of information can meet the criteria of sustainability. The 'going concern' principle outlines the accounting point of view of sustainability. Pollution control, equitable distribution of wealth poses challenges to industry. Sustainability is a concept understood at the global level and applied at the organizational level (Gray & Milne, 2002). Industrial sustainability is a situation of achieving economic, social and environmental performance.

Raising awareness of environmental issues and developing guidelines to assist identification of environmental issues and evaluation and reporting is required to provide education programs across disciplines focused on environmental issues and their accounting treatment. Indian civilization has a long tradition of taking care of the environment.

It includes Environmental conservation benefit associated with the inputs of resources into business operations, Environmental conservation benefit associated with environmental impact and waste emissions from business operations,

Environmental conservation benefit associated with the goods and services produced by business operations and Environmental conservation benefit associated with transports and other operations.

New methodologies and concepts for the successful implementation of environmentally sound and sustainable socio-economic development has increased its attention globally and has been emphasized in the report of world commission on environment and development.

5. LIMITATIONS OF ENVIRONMENT ACCOUNTING

Environmental accounting has various limitations which include difficulty in measuring cost and also measuring the benefits of environmental accounting. It is very critical to evaluate the right environmental cost and benefits in a systematic quantitative manner. Another paradox is consideration of internal cost and exclusion of cost to society. As of now, there are no standard accounting methods for environmental accounting. It is very difficult to do an intra firm comparison analysis between two firms and two countries. It is almost impossible to track the use of energy and material.

The limits of the accounting regulations to distinguish the environmental cost are an obstacle to the internalization of environmental externalities and do not support the main issue of a proactive environmental system. Bracci & Maran.(2013)

Valuation techniques for environmental goods and services are imperfect and shadow prices depict only partial valuations. The change in values for environmental goods and services are uncertain. Economic values are marginal and incremental, not absolute and total reliable industry data are not readily available. Environmental accounting is an ongoing and a long process, it needs to be integrated with financial accounting which in turn makes it complicated. Environmental accounting is a longitudinal process. Therefore, to draw a conclusion with help of it is not easy. Integration of environmental accounting with financial accounting, cost accounting and management accounting is a difficult task. User of environmental accounting needs to have adequate knowledge about rules and regulations and environmental aspects. In India, environmental accounting is still in the developing phase. It is a must for the corporate to formulate firm environmental policies for sustainable development.

Costs and benefits relevant to the environment are not easily available and measurable. Proper tracking of the use of energy and material or the cost of inefficient materials use, waste management and related issue is not done by many business and the Government organizations and they underestimate the environmental cost. It mainly considers the cost internal to the company and excludes cost to society. The user of information contained in the environmental accounting needs adequate knowledge of the process of environmental accounting as well as rules and regulations prevailing in that country either directly or indirectly related to environmental aspects.

6. CONCLUSION

The primary challenge of environmental accounting is the absence of standards in identifying all that needs to be measured and deciding how it is to be measured. A number of initiatives have been undertaken to address the issues including the one undertaken by the United Nations. Experimental ecosystems accounting (EEA). An accounting framework for ecosystem is attempted to be developed by using a consistent set of principles to document information on ecosystem benefits as well changes to ecosystems over time. Wealth Accounting and Valuation of Ecosystems Services(WAVES), a World Bank let global partnership is also working in this area to promote sustainable development through implementation of NCA in several countries. However, the sheer intangibility of many of the resources does not lend itself for quantification and it is here that the concept of environmental accounting as a whole has not received the seriousness and commitment that it deserves.

One of the challenges of the reporting community at large in India is to improve comparability among environmental reports. Reviewed reports currently do not address issues on environmental reporting. It can be interpreted in any manner. Environmental accounting is a must for understanding the role played by natural environment in the development of an economy. It consists of data that contains the contribution of natural resources to the economic well-being and emphasizes on the costs imposed by environmental pollution and resource degradation. Reporters must give careful consideration to, how they identify issues for reporting. Even if researchers have propounded theories of environmental accounting, there is a need to investigate the feasibility and dependability of such studies by studying the longitudinal data. Involvement and commitment of corporate accountants in environmental management appears to be limited due to lack of regional reporting guidelines. Inviting inputs from stakeholders, while formulating guidelines, will be a valuable means of engaging stakeholders and enhancing mutual interests and priorities (Sustainability Ltd and UNEP, 1999). Such a bold and participative approach would ensure benefits of enduring value both to the industry and its stakeholders. Environmental awareness provides a transparent and dynamic way of reporting environmental performance for the business. It provides the internal and external users with reports which helps in capital decisions and pricing decisions. Environmental management accounting is more for the internal use of the business. Bartolomeo et al (2000).

Efficiency of the Indian economy is dependent to a great extent on business environmental efficiency. Environmental accounting reporting is concerned with cost and benefits arising due to the process and product of firms which also have environmental impact. It is very difficult to state whether corporate enterprises are environmentally sustainable ones. It is a must for the corporate enterprise to consider the impact on environment and then determine their profit level to assure that they leave the planet same at the start and end of the accounting period. Environmental accounting is vital and mandatory for a green economy. Developing countries like India may have to choose various possible approaches as per their needs, level of development and environmental management problems. Hence, a country wise analysis should be done based on a simple accounting framework with practical recommendations to comprehensively cover the needs of an economy for a successful planning process.

Moreover, the relationship between the business entity and the environment is receiving high attention. Environmental reporting is yet to find a permanent position in the financial sections of the annual report. Positive relationship exists between the profitability, size of the company. It indicates that the companies in India are becoming more enlightened as to the corporate environmental disclosures and social responsibility in this regard. At present, a debate on global climatic change and biodiversity is becoming more commonplace in corporate world, together with concern over water, air and soil pollution. Although the sustainability of economic development is a shared responsibility of organization and stakeholders, but business entities can play a vital role in slowing down the environmental degradation. A business entity has all the resources, technological knowledge, institutional capability and long-term vision to find out various environmental solutions. Effectiveness, efficiency and innovation can improve and streamline the manufacturing process, products with optimum use of raw materials along with decrease in the waste generated from production. Waste disposal methods and activities to keep pollution under control should be designed for quality environment management. Consequently environmental performance, customers' satisfaction and company performance will improve.

Environmental disclosure practices by Indian companies are of casual nature quantitative information and proper accounting pronouncements from regulatory authorities needs to be mentioned in the annual reports. The annual reports must disclose all information related to the environment. The lack of awareness and commitment on the part of company management about the social responsibility of the firm also keeps the firm away from reporting environmental costs and benefits. Thus, it can be concluded that the absence of a standardized environmental accounting practice and disclosure norms at national as well as international levels spur the corporate to be away from the environmental accounting practices and to shut their eyes towards the deterioration in the environment. In India, currently environmental accounting and reporting is at a preliminary stage and limited to demonstrating compliance to regulations and rules in the act to some extent.

Environment related disclosures in the corporate annual reports require due consideration. There are no latest company law and accounting standards; even ICAI does not prescribe the disclosure norms for environmental related aspects in the corporate financial reports. For few industries environmental accounting is mandatory such as oil and petroleum, natural gas, cement, steel, etc. For the rest environmental disclosures are voluntary in nature. The poor environmental performance of the company may also bind them to no-disclosure.

ACKNOWLEDGMENT

We are extremely grateful to Dr Rameshwar Dubey and his editorial team members for their high cooperation during the publication process.

REFERENCES

Abu-Baker, N., & Naser, K. (2000). Empirical evidence on corporate social disclosure (CSD) practices in Jordan. *International Journal of Commerce and Management*, *10*(3/4), 18–34. doi:10.1108/eb047406

Almeida, C. M. V. B., Santos, A. P. Z., Bonilla, S. H., Giannetti, B. F., & Huisingh, D. (2013). The roles, perspectives and limitations of environmental accounting in higher educational institutions: An energy synthesis study of the engineering programme at the Paulista University in Brazil. *Journal of Cleaner Production*, *52*, 380–391. doi:10.1016/j.jclepro.2013.03.008

Amran, A., & Devi, S. S. (2008). The Impact of Government and Foreign Affiliate Influence on Corporate Social Reporting: The Case of Malaysia. *Managerial Auditing Journal*, *23*(4), 386–404. doi:10.1108/02686900810864327

Anant, T. C. A., Basu, K., Chopra, K., Dasgupta, P., Desai, N., Gundimeda, H. P., ... & Sundaram, K. (2013). Green National Accounts in India: A Framework.

Anderson, A. R. (1998). Cultivating the Garden of Eden: Environmental entrepreneuring. *Journal of Organizational Change Management*, *11*(2), 135–144. doi:10.1108/09534819810212124

Ball, A. (2005). Environmental accounting and change in UK local government. *Accounting, Auditing & Accountability Journal*, *18*(3), 346–373. doi:10.1108/09513570510600738

Bansal, P. (2002). The corporate challenges of sustainable development. *The Academy of Management Executive*, *16*(2), 122–131. doi:10.5465/AME.2002.7173572

Barbera, A. J., & McConnell, V. D. (1990). The impact of environmental regulations on industry productivity: Direct and indirect effects. *Journal of Environmental Economics and Management*, *18*(1), 50–65. doi:10.1016/0095-0696(90)90051-Y

Bebbington, J. (2001, June). Sustainable development: A review of the international development, business and accounting literature. *Accounting Forum*, *25*(2), 128–157. doi:10.1111/1467-6303.00059

Bennett, M., & James, P. (Eds.). (1998). *The Green bottom line: environmental accounting for management: current practice and future trends*. Greenleaf Publishing.

Blacconiere, W. G., & Patten, D. M. (1994). Environmental disclosures, regulatory costs, and changes in firm value. *Journal of Accounting and Economics*, *18*(3), 357–377. doi:10.1016/0165-4101(94)90026-4

Bouma, J. J., & van der Veen, M. (2002). Wanted: A theory for environmental management accounting. In Environmental management accounting: Informational and institutional developments (pp. 279-290). Springer Netherlands. doi:10.1007/0-306-48022-0_22

Bracci, E., & Maran, L. (2013). Environmental management and regulation: Pitfalls of environmental accounting? *Management of Environmental Quality: An International Journal*, *24*(4), 538–554. doi:10.1108/MEQ-04-2012-0027

Buckley, R. (2002). Tourism ecolabels. *Annals of Tourism Research*, *29*(1), 183–208. doi:10.1016/S0160-7383(01)00035-4

Burritt, R., & Saka, C. (2006). *Quality of physical environmental management accounting information. Lessons from pollutant release and transfer registers* (pp. 373–407). Springer Netherlands; doi:10.1007/978-1-4020-4974-3_17

Burritt, R. L., & Saka, C. (2006). Environmental management accounting applications and eco- efficiency: Case studies from Japan. *Journal of Cleaner Production*, *14*(14), 1262–1275. doi:10.1016/j.jclepro.2005.08.012

Cairns, R. D. (2000). Sustainability accounting and green accounting. *Environment and Development Economics*, *5*(1), 49–54. doi:10.1017/S1355770X00000048

Chatterjee, B., & Zaman Mir, M. (2008). The current status of environmental reporting by Indian companies. *Managerial Auditing Journal*, *23*(6), 609–629. doi:10.1108/02686900810882138

D'Souza, C., & Peretiatko, R. (2002). The nexus between industrialization and environment: A case study of Indian enterprises. *Environmental Management and Health*, *13*(1), 80–97. doi:10.1108/09566160210417859

Davies, A. R., & Mullin, S. J. (2011). Greening the economy: Interrogating sustainability innovations beyond the mainstream. *Journal of Economic Geography*, *11*(5), 793–816. doi:10.1093/jeg/lbq050

Dixon, S. E. A., & Clifford, A. (2007). Ecopreneurship – a new approach to managing the triple bottom line. *Journal of Organizational Change Management*. doi:10.1108/09534810710740164

Dorweiler, V. P., & Yakhou, M. (2004). Environmental Accounting: An Essential Component of Business Strategy. *Asian Institute of Technology Thailand Online*, *13*(2), 65–77.

Dubey, R., Bag, S., Ali, S. S., & Venkatesh, V. G. (2013). Green purchasing is key to superior performance: An empirical study. *International Journal of Procurement Management*, *6*(2), 187–210. doi:10.1504/IJPM.2013.052469

Dubey, R., & Gunasekaran, A. (2015). Shortage of sustainable supply chain talent: An industrial training framework. *Industrial and Commercial Training*, *47*(2), 86–94.

El Serafy, S. (1997). Green accounting and economic policy. *Ecological Economics*, *21*(3), 217–229. doi:10.1016/S0921-8009(96)00107-3

Eresi, K. (1996). Environment-Information Disclosure in Annual Reports. *Chartered Accountant-New Delhi*, *44*, 45–48.

Foster, J. (2001). Education as sustainability. *Environmental Education Research*, *2001*, 153–165. doi:10.1108/ICT-08-2014-0052

Frederick, W. C. (1994). From CSR1 to CSR2 The Maturing of Business-and-Society Thought. *Business & Society*, *33*(2), 150–164. doi:10.1177/000765039403300202

Frederick, W. C. (1994). From CSR1 to CSR2 The Maturing of Business-and Society Thought. *Business & Society*, *33*(2), 150–164. doi:10.1177/000765039403300202

Godschalk, S. K. (2009). Does Corporate environmental accounting make business sense? In *Environmental management accounting for cleaner production* (pp. 249–265). Springer Netherlands. Doi:10.1007/978-1-4020-8913-8_13

Gray, R. (2001). Thirty years of social accounting, reporting and auditing: What (if anything) have we learnt? *Business Ethics (Oxford, England)*, *10*(1), 9–15. doi:10.1111/1467-8608.00207

Gray, R. (2010). Is accounting for sustainability actually accounting for sustainability… and how would we know? An exploration of narratives of organisations and the planet. *Accounting, Organizations and Society*, *35*(1), 47–62. doi:10.1016/j.aos.2009.04.006

Gray, R., & Bebbington, J. (2007). Corporate sustainability, accountability and the pursuit of the impossible dream. Handbook of sustainable development, 376-394.

Gray, R., Owen, D., & Maunders, K. (1988). Corporate social reporting: Emerging trends in accountability and the social contract. *Accounting, Auditing & Accountability Journal*, *1*(1), 6–20. doi:10.1108/EUM0000000004617

Harte, G., & Owen, D. (1991). Environmental disclosure in the annual reports of British companies: A research note. *Accounting, Auditing & Accountability Journal*, *4*(3). doi:10.1108/09513579110144442

Herzig, C., Viere, T., Schaltegger, S., & Burritt, R. L. (2012). *Environmental Management Accounting: Case Studies of South-East Asian Companies*. Routledge.

Janković, S., & Krivačić, D. (2014). Environmental accounting as perspective for hotel sustainability: Literature review. *Tourism and Hospitality Management*, *20*(1), 103–120.

Jasch, C. (2001). Environmental Management Accounting Metrics, Procedures and Principles. *UN Division for Sustainable Development*. DOI:10.1007/0-306-48022-0_3

Johnson, S. (2004). *Environmental management accounting*. Association of Chartered Accountants.

Kopnina, H. (2011). Consumption in environmental education: Developing curriculum that addresses cradle to cradle principles. *Factis Pax*, *5*(3), 374–388.

Kordestani, A., Peighambari, K., & Foster, T. (2015). Emerging trends in sustainability research: A look back as we begin to look forward. *International Journal of Environment and Sustainable Development*, *14*(2), 154–169. doi:10.1504/IJESD.2015.068602

Lamberton, G. (2005, March). Sustainability accounting—a brief history and conceptual framework. *Accounting Forum*, *29*(1), 7–26. doi:10.1016/j.accfor.2004.11.001

Lanen, W., & Anderson, S.W. (2002). Economic Transition, Strategy and the Evolution of Management Accounting Practices: The Case of India.

Lintott, J. (1999). The Environmental Accounting: Useful to Whom and for What? *Ecological Economics*, *16*(3), 179–190. doi:10.1016/0921-8009(95)00097-6

Lohmann, L. (2009). Toward a different debate in environmental accounting: The cases of carbon and cost–benefit. *Accounting, Organizations and Society*, *34*(3), 499–534. doi:10.1016/j.aos.2008.03.002

Malarvizhi, P., & Yadav, S. (2008, May). Corporate Environmental Reporting On The Internet: An Insight Into Indian Practices. *Proceedings of the11th Annual Convention of the Strategic Management Forum, May* (pp. 8-10).

Mathews, M. R. (1997). Twenty-five years of social and environmental accounting research: Is there a silver jubilee to celebrate? *Accounting, Auditing & Accountability Journal*, *10*(4), 481–531. doi:10.1108/EUM0000000004417

Milne, M. J., Kearins, K., & Walton, S. (2006). Creating adventures in wonderland: The journey metaphor and environmental sustainability. *Organization*, *13*(6), 801–839. doi:10.1177/1350508406068506

Milne, M. J., & Patten, D. M. (2002). Securing organizational legitimacy: An experimental decision case examining the impact of environmental disclosures. *Accounting, Auditing & Accountability Journal*, *15*(3), 372–405. doi:10.1108/09513570210435889

Minimol, M.C., & Makesh, K.G. (2013). Green accounting and reporting practices among indian corporates. *Asia Pacific Journal of Research*, 1(14).

Mohamed, E. K., & Lashine, S. H. (2003). Accounting knowledge and skills and the challenges of a global business environment. *Managerial Finance*, *29*(7), 3–16. doi:10.1108/03074350310768319

Moorthy, K., & Yacob, P. (2013). Green accounting: Cost measures. *Open Journal of Accounting*, *2*(01), 4–7. doi:10.4236/ojacct.2013.21002

Namakonzi, R., & Inanga, E. L. (2014). Environmental management accounting and environmental management in manufacturing industries in Uganda. *African Journal of Economic and Sustainable Development*, *3*(4), 288–329. doi:10.1504/AJESD.2014.065578

Nazli Nik Ahmad, N., & Sulaiman, M. (2004). Environment disclosure in Malaysia annual reports: A legitimacy theory perspective. *International Journal of Commerce and Management*, *14*(1), 44–58. doi:10.1108/10569210480000173

Onishi, Y., Kokubu, K., & Nakajima, M. (2009). Implementing material flow cost accounting in a pharmaceutical company. In *Environmental Management Accounting for Cleaner Production* (pp. 395–409). Springer Netherlands.

Ouellette, J. (1996). Coagulants and Flocculants Rise. *Chemical Marketing Reporter*, 250(15).

Pahuja, S. (2007). Environmental Reporting Verification: A Critical Evaluation of Accountants' Views and Corporate Practices in India. *Social Responsibility Journal*, *3*(2), 22–31. doi:10.1108/17471110710829696

Pahuja, S. (2009). Relationship between environmental disclosures and corporate characteristics: A study of large manufacturing companies in India. *Social Responsibility Journal*, *5*(2), 227–244. doi:10.1108/17471110910964504

Panackal, N., & Singh, A. (2015). Using Interpretive Structural Modeling to Determine the Relation between Youth and Sustainable Rural Development. *IBMRD's Journal of Management & Research*, *4*(1), 58–74. doi:10.17697/ibmrd/2015/v4i1/60352

Perry, M., & Tse Sheng, T. (1999). An overview of trends related to environmental reporting in Singapore. *Environmental Management and Health*, *10*(5), 310–320. doi:10.1108/09566169910289667

Pinckard, M., & Wendy, V. (2000, October). Environmental Audit–Working within a Restricted Mandate (Draft paper 20).

Pittman, K. (1991). Promoting Youth Development: Strengthening the Role of Youth Serving and Community Organizations. *Paper prepared for USDA, Extension Service, National Youth at Risk Initiative Task Force.*

Pradhan, B. B., & Bal, R. K. (2002). Corporate Environmental Reporting: Perceptions of Corporate Managers. *Environmental Accounting and Reporting.*

Pramanik, A. K., Shil, N. C., & Das, B. (2007). Environmental accounting and reporting with special reference to India.

Ramanathan, K. V. (1976). Toward a Theory of Corporate Social Accounting. *Accounting Review*, *51*(3), 516–528.

Rof, L. M., & Danuletiu, D. C. (2010). Environmental Accounting and Corporate Social Responsibility. *Annals of Faculty of Economics*, *1*(2), 885–891.

Sahay, A. (2004). Environmental reporting by Indian corporations. *Corporate Social Responsibility and Environmental Management*, *11*(1), 12–22. doi:10.1002/csr.51

Schaltegger, S., Bennett, M., & Burritt, R. (Eds.). (2006). *Sustainability accounting and reporting* (Vol. 21). Springer. doi:10.1007/978-1-4020-4974-3

Sen, A. (2013). The ends and means of sustainability. *Journal of Human Development and Capabilities*, *14*(1), 6–20. doi:10.1080/19452829.2012.747492

Shukla, A., & Vyas, N. (2000). Environmental Accounting & Reporting in India. *Bharat Petroleum Company Limited & Oil & Natural Gas Company Limited.*

Singh, G., & Joshi, M. (2009). Environment management and disclosure practices of Indian companies. *International Journal of Business Research*, *9*(2), 116–128.

Sushil. (2001). Framework. *Global Journal of Flexible Systems Management.*

Taplin, J.R.D., Bent, D., & Aeron-Thomas, D. (2006). Developing a Sustainability Accounting Framework to Inform Strategic Business Decisions: a Case Study from the Chemicals.

The Handbook of National Accounting, Studies in Methods Series F, No. 78 Integrated Environmental and Economic Accounting: An Operational Manual. (2000). United Nations, New York.

Thompson, P., & Zakaria, Z. (2004). Corporate social responsibility reporting in Malaysia. *Journal of Corporate Citizenship*, *2004*(13), 125–136. doi:10.9774/GLEAF.4700.2004.sp.000014

Upadhyay, D. K. ((2009). Promoting Sustainable Development from the Grassroots Level: Indian Perspectives. *Madhya Pradesh Journal of Social Sciences.*

Yakhou, M., & Dorweiler, V. P. (2004). Environmental accounting: An essential component of business strategy. *Business Strategy and the Environment*, *13*(2), 65–77. doi:10.1002/bse.395

Yee, L. M., Devi, S., & Khin, E. W. S. (2011). Kyoto Protocol and Social Accounting Implication on Global-Warming in Malaysia. *African Journal of Agricultural Research*, *6*(6), 1489–1499.

Zadeh, F. O., & Eskandari, A. (2012). Firm Size as Company's Characteristic and Level of Risk Disclosure: Review on Theories and Literatures. *International Journal of Business and Social Science*, *3*(17), 9–17.

Zadeh, F. O., & Eskandari, A. (2012). Firm Size as Company's Characteristic and Level of Risk Disclosure: Review on Theories and Literatures. *International Journal of Business and Social Science*, *3*(17), 9–17.

Zeldin, S., Camino, L., & Calvert, M. (2007). Toward an understanding of youth in community governance: Policy priorities and research directions. *Análise Psicológica*, 2007, 77–95.

Chapter 5
Identifying Green Supply Chain Management Enablers in South African Mining Industry using Ecological Modernization Theory Approach

Surajit Bag
Tega Industries South Africa Pty Ltd, South Africa

ABSTRACT

Green supply chain management is a popular practice in any world class organization. The traditional supply chain was managed with the objectives to reduce cost without focusing on environmental and social dimensions. However with passage of time institutional pressures has directed firms to design supply chain network which takes care of environmental and social dimensions as well. Green supply chain management concept has evolved gradually over a period of time. The main reason behind evolution of green supply chain management is conservation of natural resources which human beings ignored for long. Published reports show that major portion of the energy is supplied by fossil fuels and thereby depletion of natural resources is occurring at an accelerating rate with time. Industrial manufacturing report also suggest that demand for manufacturing products is likely to double by the year 2050. This will result in high generation of pollutants and gases. If the greenhouse gases emissions remain unchecked, then it will create a catastrophic effect by 2050. Globally manufacturers and industry associations are now acting proactively to conserve the natural resources and curb greenhouse gases emissions. The level of popularity of green supply chain management can be estimated from the high research output in leading operations journals such as IJPE, IJPR, Resources, Conservation and Recycling Journals.

Green supply chain management is a concept that is also gaining popularity in the South African region. For many organizations in this region it is a process to exhibit their sincere commitment to sustainability. However, if green supply chain management practices are to be fully adopted by all mining and minerals industry in South Africa, a demonstrable link between such measures and understanding the role of associated variables is necessary. This paper endeavors to develop a green supply chain management

DOI: 10.4018/978-1-5225-0350-7.ch005

model for mining and minerals industry in the South African region. The approach extends the domain of green supply chain research and produce theories that have greater explanatory power than the current practices.

For this purpose a conceptual model was developed from literature review and data collected using a structured questionnaire mailed to a sample of 174 firms which are leading edge ISO14001 certified mining companies in South Africa. Data analysis is further performed through exploratory factor analysis and regression analysis. This paper presents the first empirical evaluation of the link between green supply chain management practices amongst a sample of mining and minerals companies in South Africa.

1. INTRODUCTION

"We are living as if we have an extra planet at our disposal. We are using 50 per cent more resources than the Earth can sustainably produce and unless we change course, the number will grow fast by 2030. Even two planets will not be enough." The statement was given by Dr. Morne Du Plessis, CEO of World Wildlife Fund South Africa.

South Africa has an ecological footprint of 2.59 per hectare. Ecological footprint refers to a measure of how much biologically productive land, water and individual population or activity requires producing all the resources it consumes, and to absorb the waste generates using prevailing technology and resource management practices (WWF 2012).

Supply chain practitioners in mining industry confront with various social and environmental issues which do not directly impact daily operation but ultimately has a profound impact on organization performance. These issues are threat to the organizational sustainability and must be understood and managed in a proactive manner. South Africa is fast becoming savvy at demanding social and environmental friendly business practices, ensuring that the mineral resources are exploited in a sustainable and eco-friendly manner.

In recent years green supply chain management (GSCM) has gained popularity in mining companies in South Africa. There has been a strong move towards mining companies "being green" and focusing on sustainability. To accelerate GSCM movement the Chartered Institute of Logistics and Transport: South Africa, the Consumer Goods Council of South Africa and Supply Chain Today launched the GSCM awards program in 2009. The objective of such an award is to recognize and reward companies making true efforts to reduce environmental impact of their supply chain activities.

The continued academic growth of GSCM and its development requires further exploration and more insights be generated. Here author present ecological modernization theory for explanatory purpose with a strong focus on GSCM practices and performance outcome.

The study starts with a brief overview defining the ecological modernization theory which is used as a theoretical foundation for further analysis. The identification of GSCM key dimensions is done based on systematic literature review approach. The final section presents the conclusions drawn from data analysis, recommendations, managerial implications and future research directions.

1.1 Background

Ecological modernization theory (EMT) has its root in the sociological theory which was further developed and applied in the field of management science. The key objective is to achieve industrial development and protecting environment through innovation and technological development (Murphy & Gouldson, 2000). This theory was earlier used to explain environmental planning by Governments (Murphy, 2000). The first dimension which influences GSCM practices is related to evolution of regulations and policies that impact environmental innovation. The second dimension concentrates on technological up gradation and innovation (Murphy & Gouldson, 2000). EMT suggests that regulations and policies can motivate companies to adopt GSCM practices. A legal framework by Government strongly drives GSCM practices and such policy is necessary for GSCM program development. GSCM practices are supported by the concept of environmental innovation from the EMT perspective. Companies generally adopt GSCM practices by considering the hard technological innovation and soft technological innovation. The hard technological innovation involves innovative technology, eco-friendly equipment for cleaner production. The soft technological innovations involve supplier collaboration for green design and increased supplier relationship management for innovation. GSCM research based on Ecological modernization theory clearly demonstrates that company can gain from social, economic and environmental perspective which is in line with the triple bottom line approach of achieving sustainability.

2. LITERATURE REVIEW

The main purpose of this chapter is to explore green supply chain management (GSCM) in depth and understand how GSCM practices within a mining company can be a source of competitive advantage based upon review of past articles published in the journals. Literature review is considered heart of any research study. It is found that past researchers have mostly used traditional review; however systematic literature review is a recent trend adopted by researchers for conducting literature review to organize findings (e.g. Pittaway et al., 2004; Van Aken., 2005; Lightfoot et al., 2013). Therefore present review is done based on systematic literature review (SLR) approach. Here SLR has been followed as per steps recommended by Tranfield et al., 2003. This eradicates the issues related to the application of correct methodology and easily helps to develop the later sections of the study. In this process researcher has adhered to the principles that are an integral part of systematic literature review. For literature review scholarly works from databases such as Science Direct, Compendex, Ebsco, Emerald and Scopus has been utilized.

2.1 Evolution of GSCM

The evolution of supply chain management was a gradual process. Over the last century, there were major revolutions in the field of supply chain management which were examined in the context of the broader evolution in the economic and technological environment.

Going green or sustainable supply chain practices started when organizations witnessed the ill effects of the traditional supply chain process. The results were terrible with marked impact on environment which caused irreversible damage to the environment. Apart from human beings it even posed threat to the animals and plant life on this planet. This created a concern among the policy makers, government and manufacturing associations globally and thus the concept of green supply chain emerged.

Companies such as Dell now allow customers to return any Dell-branded product back to the company – for free through its "no computer should go to waste" recycling program. The company has even moved a step ahead and established programs that accept computers, monitors, or printers from other companies for safe disposal, as well. The main objective of the program is to take care of the safe disposal of the electronics components after its useful life. This is why green supply chain management is also called closed loop supply chain. Here care is taken by the manufacturer to minimize the environmental impact of the product during the entire life cycle.

The traditional supply chain was managed with the objectives to reduce cost without any focus towards environmental and social dimensions (Simpson et al., 2005; Sarkis et al., 2011). However with the passage of time institutional pressures directed firms to design supply chain network which takes care of environmental and social dimensions as well (Srivastava, 2007; Gavronski et al., 2008; Guide & Van Wassenhove, 2003; Gunasekaran & Spalanzani, 2012).

GSCM dimensions are green design, green purchasing, green manufacturing, and green distribution and transportation (Hervani et al., 2005). The environmental awareness which initiated in the 1960s, in the western part of the globe gradually spread all over the world (Sarkis, 2011; Nelson et al., 2012). The GSCM principle was guided by a single objective, i.e, environmental performance. However in this dynamic business scenario firms understood that only focusing on environmental dimension does not help to achieve sustainability. Therefore, firms initiated integrating the three important dimensions and adapting a more comprehensive performance framework, i.e., Triple Bottom Line (e.g. Awaysheh & Klassen, 2010, Paulraj & De Jong, 2011; Gimenez et al., 2012; Giovanni, 2012).

Firms are exposed to many elements of risk and these elements are classified into economic, environmental and social element. Sustainable development encompasses the simultaneous adoption of Environmental, Economic and Social principles. This is called "Triple Bottom Line" approach.

Globally manufacturers are now focusing on safe operations keeping in line with the three basic elements of sustainability. This necessitates the adoption of sustainable supply chain management. World leading firms including IBM, Dell, HP, Sony, Toyota, and Nokia adopted sustainable supply chain management (SSCM) as a strategic initiative. Various respected work exists in the area of sustainable supply chains (Linton et al., 2007; Longo & Mirabelli., 2008; Pereira., 2009; Azevedo et al., 2011; Gupta & Desai., 2011; Wu. & Pagell., 2011; Wang. et al., 2011; Kang et al., 2012; Barari et al., 2012; Hassisni et al., 2012; Gimenez et al., 2012; Gunasekaran & Spalanzani 2012; Dubey & Bag., 2013; Chaabane et al., 2012; Ageron et al., 2012; Liu et al., 2012; Reefke & Trocchi., 2013; Mirhedayatian et al., 2014; Brandenburg et al., 2014; Bag & Anand 2015). It can be concluded that in the last two decades, supply chain management has gradually transformed into sustainable supply chain management.

2.2 Definition of GSCM

In this section, effort has been made to define GSCM based on review of past scholarly works.

- Schrettle et al., (2014) defined GSCM as a "tool which helps to position a company from strategic viewpoint."
- Gunasekaran and Spalanzani, (2012) defined GSCM as "an organizational philosophy which provides competitive edge to an organization."
- Ali and Govindan (2011) defined GSCM as "an organizational philosophy to reduce environmental risks."

- Johnny et al., (2009) defined GSCM as a "process that added the green element to existing supply chain management and portrayed how reverse supply chain, an organization and an innovative activity reconstruct a system."
- Zhu et al., (2008a) defined GSCM as "adoption of ecofriendly practices considering internal environmental management, green purchasing, cooperation with customers, and eco design for developing corporate and operational strategies for the firm's environmental sustainability."
- Seuring and Muller's (2008) defined GSCM as " the management of material, information, and capital flows as well as cooperation among companies along the supply chain while taking goals from all three dimensions of sustainable development" which are derived from customer and stakeholder requirements."
- Carter and Rogers (2008) defined GSCM as an "integration of environmental dimensions with traditional supply chain network."
- Srivastava (2007) defined GSCM as "integrating environmental thinking into supply chain management, including product design, material sourcing and selection, manufacturing processes, delivery of the final products to the consumers, and end-of-life management of the product after its useful life."
- Vachon and Klassen (2006) defined GSCM as a "strategy, which assists to reduce wastages in supply chain network."
- Kogg (2003) defined GSCM "the set of supply chain management policies held, actions taken and relationships formed in response to concerns related to the natural environment with regard to the design, acquisition, production, distribution, use, re-use and disposal of the firm's goods and services."
- Sarkis (2003) defined GSCM as a "combination of the activities of an environmental dimension and reverse logistics."
- Gilbert (2000) defined GSCM as an "integration of environmental criteria with traditional supply chain network by redesigning purchasing policies and involving suppliers in the entire procurement process."
- Beamon (1999) defined GSCM as "utilizing the supply chain between a central company and a cooperating company, to support the company with knowhow of eco management practices and development of clean manufacturing techniques."
- Narshiman and Carter (1998) defined GSCM as "involving the procurement strategies that reduced the use of materials in addition to recycling and reuse."
- Godfrey (1998) defined GSCM as "organizational practices that continuously monitor the environmental impact of a supply chain and improve the performance."

Based on the above definitions author proposes GSCM definition as "integrating the green dimensions into the traditional supply chain management so that the product will impose reduced environmental impact during its entire life cycle. It is the set of clean and hygienic practices which starts from the product design stage and flows through all activities of the supply chain such as procurement, manufacturing, packaging, transportation and end of life management. Firm's internal resources mediate the relationship to external forces and its adoption. It improves the organizational performance and is the source of competitive advantage to sustain in the turbulent business environment."

2.3. Identification of Enablers for Successful GSCM Practices

On the basis of systematic literature review the enablers are classified into nine broad categories. Effort is made to explain each of these enablers/variables based on the understanding from review of past research studies.

Top Management Commitment

Top management commitment for implementation of GSCM is considered as a very important variable. Top management commitment helps in proper allocation of resources; organizational adaptability; proper flow of information and success of any project.

Regulatory Pressures

South African government has come to the realization that mineral is a national asset and key to the future success of its people. For this reason, South African government have framed stringent legislation to ensure that the benefit of mineral assets accrue to its countryman. The Relevant South African Laws and policies that impact Sustainable practices are Constitution of the Republic of South Africa, Act 108 of 1996; National Environmental Management Act, 107 of 1998 (NEMA); National Environmental Management: Waste Bill; Hazardous Substances Act, 15 of 1973; Consumer Protection Bill, 2007; Energy Efficiency Strategy of the Republic of South Africa, 2005; National Climate Change Response Strategy, 2004; White Paper on the Energy Policy of the Republic of South Africa, 1998; White Paper on Integrated Pollution and Waste Management for South Africa; Draft Green Paper on the Consumer Policy Framework, 2004; White Paper on Environmental Management Policy, 1998; Mineral and Petroleum Resources Development Act ("MPRDA"); Black economic empowerment ("BEE") legislation and Mine Health and Safety legislation. These laws and policies aims at preventing pollution and ecological degradation; promote conservation; and secure ecologically sustainable development and use of natural resources while promoting justifiable economic and social development.

Market Pressures

Market pressures influence GSCM adoption in mining industry. Market pressure dimensions include pressures from customers and sensitivity of customers towards adopting green practices. Modern customer relationship theory refers to the idea of 'integrating the customer'. This new way of looking at the business involves integrating the customer (more precisely the customer's relevant people and processes) into all aspects of the supplier's business, and vice versa. This implies a relationship that is deeper and wider than the traditional 'arms-length' supplier-customer relationship.

Collaborative Relationships

Organizations are generally interested in environmental collaboration with suppliers to extract positive benefits from successful adoption of GSCM. Various case studies from European and Australian organizations show that green procurement practices are used as a tool to gain competitive advantage which indirectly enhances operational and financial performance.

Without collaborative relationship with suppliers, underlying benefits may never be realized, the business might consider procurement "missing in action" when something goes wrong, and suppliers may begin to see procurement as just negotiating, i.e. not involved in the real business. As a result, they might not make their best effort going forward in green practices or offering innovative suggestions.

Green Technology Adoption

Green technology is another important variable that must be considered by mining industry. Green technology deals with improving energy governance and requires companies to improve its ability to address long term issues such as the development of renewable energy resources to achieve a more sustainable energy mix. Green technology is an important dimension of hard technology innovation having its foundation in the Ecological modernization theory. Mining companies adopt such green technology to save energy, natural resources and focus on efficient mining equipment and machineries.

Reduction in Carbon Emission

The dawning of the 21st century bought with it a tendency towards greater investor awareness of sustainability matters. With the establishment of indices such as the Dow Jones Sustainability Index, FTSE for Good, and the Johannesburg Stock Exchange (JSE Limited) Socially Responsible Investment (SRI) Index, and with trillions of dollars invested in green, social and/or sustainability funds, much more attention is being afforded to matters still commonly referred to as 'non-financial'. Low carbon emission (Hitchock, 2012) and measuring the carbon footprints across the supply chain (Sundarakani et al., 2010) provides new direction to the present GSCM literature. Technologies for cleaner production positively influenced carbon emission reduction. (e.g., Sikdar & Howell, 1998; Xu et al., 2006; Baas, 2007; Montalvo, 2008; Zhang et al., 1997; Hoof & Lyon, 2013).

Compliance with Product/Services

This must be considered by mining companies while adopting GSCM. The offerings must comply with the regulations. Life cycle analysis, FEA analysis is commonly used to achieve the desired result and achieve sustainability in the long term.

Profit

A continuous growth in operating, safety and environmental performance is found to stabilise the production of mine operations resulting in enhanced profitability. Increased profitability is supported by real improvements in EBITDA.

Market Share

Mining companies who have blended GSCM into their corporate strategy enjoys manifold success in terms of superior environmental performance, strong image, and high market share.

Based on literature review, the GSCM enablers and its dimensions are presented in Table 1.

Table 1. Dimensions of GSCM enablers

GSCM Enablers/ Variables	Sub Variables	Authors (year)
Top Management Commitment	Well defined environmental policy	Diabat, & Govindan, (2011); Pagell, & Wu (2009); Hsu, & Hu (2008); Zhu et al.,(2008a); Zhu et al., (2008b); Zhu, & Sarkis, (2006); Hervani et al., (2005); Handfield et al., (2005); Zhu et al., (2005); Zhu & Sarkis (2004); Barratt, (2004); Lamming, & Hampson, (1996)
	Employee awareness on firm's environmental policy	
	Support environmental programs	
	Approve special fund for investment in cleaner technologies	
	Positive attitude towards green practices	
	Motivate and support new ideas received from junior executives	
	Employees recognition for innovative ideas and reward	
Regulatory Pressures	Pollution control board (PCB) pressures to adopt green practices	Diabat, & Govindan, (2011); Sarkis et al., (2011); Lee, (2008); Zhu et al., (2008a); Zhu et al., (2008b); Walker et al., (2008); Darnall et al., (2008); Zhu et al., (2007); Chien et al., (2007); Zhu & Sarkis. (2007); Vachon, & Klassen, (2006); Zhu & Sarkis, (2006); Hervani et al., (2005); Zhu et al., (2005); Zhu, & Sarkis, (2004); Zhu & Sarkis, (2004); Christmann & Taylor (2001); Hall, (2000)
	Compliance with government regulations and guidelines in controlling pollution level	
	PCB monitoring the pollution level of firm	
	Penalty fee charged by pollution control board	
Market Pressures	Export oriented sales	Sarkis et al., (2011) ; Walker et al., (2008); Lee, (2008); Zhu et al.,(2008a); Srivastava,(2007); Zhu et al., (2007); Zhu, & Sarkis, (2007); Zhu et al., (2005); Hall, (2000); Beamon, (1999)Sarkis et al., (2011); Zhu et al.,(2008a); Srivastava,(2007); Beamon, (1999)
	Sensitivity of foreign customers towards green practices	
Collaborative relationships	Consideration of environmental criteria while selecting suppliers	Sarkis, et al., (2011) ; Carter & Rogers, (2008); Zhu et al., (2008c); Vachon, & Klassen, (2008); Darnall et al., (2008); Zhu et al., (2008); Chien et al., (2007); Vachon, & Klassen, (2006); Zhu, & Sarkis, (2006); Rao, & Holt, (2005); Simpson, & Power, (2005); Hervani et al., (2005); Barratt, (2004); Zhu, & Sarkis, (2004); Sarkis, (2003); Stock, & Lambert, (2001); Croom et al., (2000); Nagel, (2000); Beamon, (1999); Walton et al., (1998); Choi, & Hartley, (1996)
	Environment collaboration with suppliers	
	Technological integration with suppliers	
	Educating suppliers in implementing ISO 14001	
	Environmental audit of suppliers	
Green Technology Adoption	Green design of products	Sarkis, et al., (2011) ; Darnall et al., (2008); Vachon, (2007); Linton, et al.,(2007); Rao & Holt (2005); Hervani et al., (2005); Sarkis, (2003); Florida (1996)
	Wastage reduction	
	Availability of real time information through Information technology infrastructure SAP/ERP	
	Alternate source of energy	
	Process optimization	
	Eco friendly materials for packaging	
	Reduction of emission of Green House Gases	

continued on following page

Table 1. Continued

GSCM Enablers/ Variables	Sub Variables	Authors (year)
Reduction in Carbon emission	Reduction of solid waste generation	Benjaafar et al., (2013); Hsu et al., (2013); Shi et al., (2012); Chaabane et al., (2012); Wang et al., (2011); Diabat & Govindan, (2011); Sarkis et al., (2011); Walker et al., (2008); Carter & Rogers, (2008); Matthews et al., (2008); Nagel, (2000)
	Effluent meets CPCB norms	
	Environmental accidents and health hazards	
	Eco friendly raw materials costs	
	Reduction in inventory levels	
	Reduction in cost of energy consumption	
Compliance in product/ services	Improvement in customer satisfaction	Zhu et al., (2008) ; Seuring, & Müller, (2008); Chien, & Shih, (2007); Zhu et al., (2007); Darnall et al., (2006); Zhu & Sarkis, (2006); Zhu et al., (2005); Hervani et al., (2005); Zhu et al., (2005); Zhu & Sarkis (2004); Barratt, (2004); Zhu & Cote, (2004); Sarkis, (2003); Croom et al., (2000); Van Hoek, (1999)
	Recovery of end of life products from customers	
	Customers appreciate ecofriendly products	
Profit	Green practices improve firm's profit	Carter & Rogers, (2008); Chien, & Shih, (2007); Zhu et al., (2007); Rao & Holt (2005); Zhu et al., (2005); Sheu et al., (2005); Guide et al., (2003); Carter et al., (2000)
Market Share	Green practices improve firm's market share	Testa, & Iraldo, (2010); Zhu et al., (2008d); Green Jr, et al., (2008); Zhu, & Sarkis, (2007); Chien, & Shih, (2007); Zhu et al., (2005); Rao, & Holt, (2005); Zhu, & Sarkis, (2004); Zhu, & Cote (2004);

2.7. Research Gaps

The review reveals various insights and gaps in the existing GSCM literature.

1. Existing GSCM literature could not capture the key enablers influencing GSCM practices in context to mining industry in South Africa.
2. Existing GSCM literature is not capable to explain the underlying relationships among key enablers influencing GSCM practices in mining industry of South Africa.

2.8. Research Objectives

Literature review has helped to identify the research gaps and further develop the below two specific research objectives for the present study:-

1. To identify the key enablers influencing GSCM practices in mining sector mining industry of South Africa.
2. To develop a GSCM model for mining industry of South Africa.

2.9. Research Questions

1. What are the key enablers influencing GSCM practices in mining industry in South Africa?
2. What is the GSCM model for mining industry in South Africa?

3. RESEARCH METHODOLOGY

This section discusses on theoretical model development, the constructs, designing questionnaire, pre-testing of the questionnaire, population size, sampling design and data collection technique applied.

The purpose of research methodology is to present the research design that will be used to conduct the empirical research for this study. The research design connects the broader assumptions of a study to its detailed methods of data collection, analysis, and interpretation. Research design enable researchers to achieve the goal of answering research questions as validly, objectively, accurately, and economically as possible.

Research methodology specifies way to systematically solve the research problem. It may be understood as a science of studying the way research is conducted scientifically. In it researcher define the various steps that are required in studying the research problem along with the logic behind them. It is necessary for the researcher to know not only the research methods/techniques but also the methodology. Researcher also needs to understand the assumptions underlying various techniques and need to know the criteria by which researcher can determine the appropriateness of the techniques and procedures for specific problems. The scope of research methodology is wider than that of research methods.

3.1. Research Process

Before moving to the details of research methodology and techniques applied, it seems appropriate to present a brief overview of the research process. Research process consists of series of actions or steps necessary to effectively carry out research and the desired sequencing of these steps. The research process flow chart is presented in Figure 1.

- Based on the synthesis of literature review and experts' opinion from mining industry; the key enablers influencing GSCM practices were identified.
- Exploratory factor analysis was applied to reduce the number of items.
- The functional relationship of the proposed GSCM model was tested by applying Regression analysis.

3.2. Hypothesis Development

Hypothesis 1: There is a positive relationship between GSCM factors and firm performance;

3.3. Assumptions of Study

To test hypothesis, the researcher has made certain assumptions are;

1. The variables are assumed to be orthogonal in nature;
2. Macro variables are assumed to constant;
3. To measure firm performance researcher has assumed perception of senior mining engineer and plant head of the mining company;

Figure 1. Research process flow chart

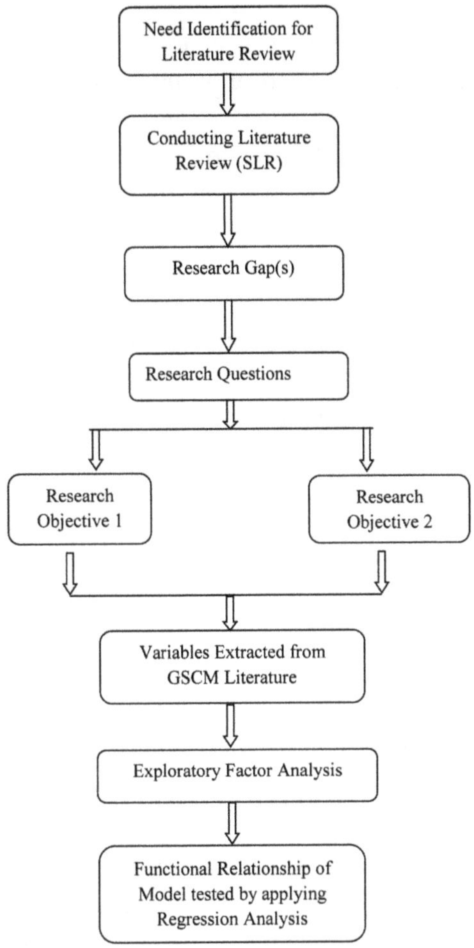

4. There will be response and non-response error. However, researcher has taken utmost care to minimize these sampling errors by designing questionnaire scientifically and adopting random sampling;

3.4. Questionnaire Design

To design questionnaire review was done on latest published work. The items of questionnaire were selected from these studies and some items were also included based upon opinion of practitioners/experts. The questionnaire was divided into three parts, i.e, *Part A* which consists of open ended questionnaire and dichotomous scale.

The questionnaire, *Part B* consists of close ended questionnaire, on five point Likert scale. Initially this part consists, 45 items.

3.5. Pretesting

Before questionnaire is finalized for survey, it was pretested with 15 experts from mining companies who are working in respective mining sector for over twenty years and professional members of "mining society of South Africa" and "association of mine managers South Africa." On the basis of pretesting wordings of the questionnaire were changed. In response to experts' comment, nine items from Part B was dropped to avoid any confusion among respondents.

3.6. Sampling

In our study we focused on mining industry in South African region. In this case mail survey method is employed. Initially we targeted randomly to 250 mining companies to avoid any response bias. In response, we received 174 usable questionnaires, after four months of regular follow up with targeted firms, which represent (70%) response rate.

4. DATA ANALYSIS

This section deals with exploratory factor analysis on collected data using SPSS 19.0, reliability testing on construct and items of the framework and regression analysis for testing research hypothesis.

4.1 Exploratory Factor Analysis

Factor analysis is used to know the contribution of each variable. It is used for the identification of the key GSCM factors that lead to improvement in firm performance.

The method used to get the extraction is the principal component analysis using the varimax rotation, with Kaiser Normalization for Factor rotation. Table 2 presents the KMO and Bartlett's test for the variables. The KMO value is 0.938. The approximate chi-square value is significant at the .000 level of significance.

Table 3 shows the summary of the result of factor analysis for the GSCM factors.

In total 11 variables have been loaded, factor analysis was done using the technique of principal component analysis with varimax and Kaiser Normalization. A total of five factors were generated. The total variance extracted is also presented in the Table 3. The results of the principal components factor analysis using a varimax rotation yielded five factors with Eigen values higher than 1.00 that explained 79.63% of the variance.

Table 2. KMO and Bartlett Test

Kaiser-Meyer-Olkin Measure of Sampling Adequacy.		.938
Bartlett's Test of Sphericity	Approx. Chi-Square	8.48403
	df	630
	Sig.	.000

Table 3. Rotated Matrix

Variables Included in the Factor	Component				
	1	2	3	4	5
Approval of special fund for investment in cleaner technologies	0.715				
Positive attitude towards green practices		0.703			
Motivate and support new ideas received from junior executives		0.788			
PCB pressures to adopt green practices				0.768	
Compliance with government regulations and guidelines in controlling pollution level				0.762	
PCB monitoring the pollution level of firm				0.727	
Environment collaboration with suppliers	0.788				
Educating suppliers in implementing ISO 14001	0.813				
Environmental audit of suppliers	0.779				
Eco friendly raw materials costs					0.732
Reduction in energy consumption			0.701		
Eigen Value	22.84	1.658	1.571	1.365	1.235
% of variance explained	21.3	20.33	14.49	12.539	10.971

Dependent Variables

Factor analysis was conducted also for the dependent variables i.e. firm performance. The data adequacy test as shown in Table 4 shows that KMO value is 0.50 shows that extracted factors through exploratory factor analysis represent the constructs well and further suitable for regression analysis and statistically significant at 0.000.

The summary of factor analysis for firm performance is shown in Table 5. The PCA output indicates that two measures of firm performance are loaded on single factor. Hence for simplicity for further regression analysis it will be treated as a single measure. The Cronbach alpha is 0.982 and indicates that the constructs chosen for study are reliable.

In order to obtain these stable structure researcher has assumed two criteria as:

- The researcher has eliminated variables having weak factor loadings(less than 0.7);
- Those variables having cross loadings on two factors or more;

Table 4. KMO and Bartlett Test

Kaiser-Meyer-Olkin Measure of Sampling Adequacy.		.500
Bartlett's Test of Sphericity	Approx. Chi-Square	88.973
	df	1
	Sig.	.000

Table 5. Dependent variable factor analysis output

Factor Interpretation	Eigen Value	% of Variance Explained	Factor Loadings	Variables Included in the Factor	Chronbach's Alpha
Firm Performance	1.634	81.81	0.818	Green practices improve profitability	0.982
			0.818	Green practices improve market share	

Table 6. Loadings on factors

Components	Items	Factor Loadings
Collaborative relationships (F1)	Approved special fund for investment in cleaner technologies	0.869
	Environment collaboration with suppliers	0.882
	Educating suppliers in implementing ISO 14001	0.874
	Environmental audit of suppliers	0.862
Top Management Commitment (F2)	Positive attitude towards green practices	0.794
	Motivate and support new ideas received from junior executives	0.846
Reduction in Carbon Emission (F3)	Reduction in cost of energy consumption	0.872
Regulatory Pressures (F4)	PCB pressures to adopt green practices	0.823
	Compliance with government regulations and guidelines in controlling pollution level	0.695
	PCB monitoring the pollution level of firm	0.782
Green Procurement (F5)	Eco friendly raw materials costs	0.633

Factor loadings represent the correlation between an original variable and its factor. In determining a significance level for the interpretations of loadings, an approach similar to determining the statistical significance of correlation coefficients could be used. However, research studies have demonstrated that factor loadings have substantially larger standard errors than typical correlations. Thus factor loadings should be evaluated at considerably stricter levels. In a sample of 174 respondents factor loadings of 0.45 and above are significant (Table 6) (Hair et al., Multivariate data analysis, 7th edition).

4.1.1 Reliability Analysis

The reliability of the data is checked for errors that cause inconsistency. The final Cronbach's Alpha score for all the 11 listed items is 0.930, which shows that data has satisfactory internal consistency reliability.

Table 7. Regression analysis

R	R Square	Adjusted R Square	Std. Error of Estimate	Change Statistics					Durbin Watson
				R Square Change	F Change	df1	df2	Sig. F Change	
0.905	0.820	0.814	0.324	0.82	152.73	5	168	0.00	1.746

4.2 Regression Analysis

Exploratory factor analysis (EFA) output was used as an input for regression analysis. The regression model to be tested is proposed in Figure 2.

The above regression model can be represented in the form of linear regression equation as

$$F6= \beta0+\beta1F1+\beta2F2+\beta3F3+\beta4F4+ \beta5F5+\varepsilon$$

where;

F1= Collaborative relationships

F2= Top Management Commitment

F3= Reduction in Carbon Emission

F4= Regulatory Pressures

Figure 2. Proposed model

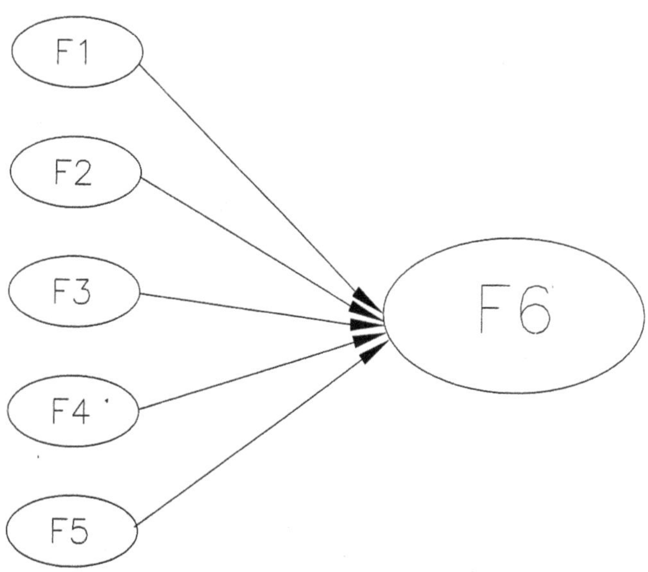

F5= Green Procurement

F6= Firm Performance

β0= Firm performance intercept

β1,….., β5= slopes of F1,F2,F4,F5

The regression analysis output is discussed in Table 7.

The above Table 7 shows that regression model is statistically significant. The value of adjusted R^2=0.814 which indicates that the regression model suggest that the predictors of the regression model explain over 81.4% of the total variance of firm performance. The adjusted- R^2 is close to R^2, shows that size of the data is sufficient to perform multiple regression analysis.

The F- statistics= (SSR/df1)/ (SSRes/df2) where df1=5 and df2=168.

At 95% significance level, the critical value of F for df1=5 and df2=168 is calculated from statistical tables as 2.21 which indicates that for given model F statistics is much above than critical value of F suggest that model explain good. In simple words it can be said that the assumption that five-parsimonious factors are strong determinants of the firm performance and the proposed model explain nearly 82% of the total variance of firm performance is a good model.

The regression analysis output (Table 8) suggests F4 (Regulatory Pressures) whose t-statistics is 1.268 which is between -2 and +2, is not statistically significant. Except F2 all variables F1, F2, F3 and F5 are statistically significant. Hence above table can be represented mathematically as:

Firm performance (F6) = -2.92 +0.115F1+0.355F2+0.386F3+0.268F5+ ε;

Table 8. Regression analysis output

	Unstandardized Coefficients		Standardized Coefficients			Collinearity Statistics	
	B	**Std. Error**	**Beta**	**t**	**Sig**	**Tolerance**	**VIF**
Constant	-2.92	0.104		-2.795	0.006		
Collaborative relationships (F1)	0.820	0.400	0.115	2.069	0.400	0.345	2.902
Top Management Commitment (F2)	0.337	0.450	0.355	7.460	0.000	0.474	2.111
Reduction in Carbon Emission (F3)	0.336	0.410	0.386	8.145	0.000	0.478	2.093
Regulatory Pressures (F4)	0.620	0.490	0.600	1.268	0.207	0.475	2.103
Green Procurement (F5)	0.216	0.290	0.268	7.380	0.000	0.815	1.227

5. CONCLUSION

Conclusions are drawn from the data analysis. Here researcher has adopted exploratory factor analysis to develop scales which are suitable for the study. Most of the time raw input is used for regression analysis which doesn't produce strong result. In order to take care of this limitation researcher has carried out exploratory factor analysis to identify orthogonal factors which was used as explanatory variables.

The exploratory factor analysis reduces eleven items into five factors which are parsimonious and orthogonal. The factor analysis output is further used for constructing regression model for regression analysis. The regression analysis output suggests that the regression model constructed using factors explain nearly 81.4% of the total variance of firm performance. Out of five factors one factor (F4) is not statistically significant; this may be due to response error. Remaining four factors (F1), (F2), (F3) and (F5) are positively supporting mining company performance.

In the present study Ecological modernization theory has been used as a theoretical lens to explain GSCM model in South African mining industry which can be considered as a unique contribution of the study.

It is found that "Top management support" is essential for success of green supply chain program in mining industry. Top management support is key dimension of soft technological innovation having its foundation in the Ecological modernization theory. Top management need to drive innovation that will lead to successful GSCM adoption. The success of future generations and mining company rests on the shoulders of top management who are the leaders and role models. Leaders drive GSCM sustainability project. They must understand the diverse nature of business stakeholders and their expectations. Focus is also required on non-financial parameters in ensuring long term sustainability of mining companies.

Another important factor emerged from the study is "Green procurement" which has gained huge popularity in South African mining companies. It involves collaboration with suppliers that further drive green supply chain management project. Collaboration with suppliers is another dimension of soft technological innovation having foundation in Ecological modernization theory. Supplier relationship management (SRM) with partners played a pivotal role in innovation and reducing carbon emissions. This is supported by previous studies (e.g., Bierma & Waterstraat, 1999; Vachon & Klassen, 2006; Hsu & Hu, 2009; Bai & Sarkis, 2010; Testa & Iraldo, 2010; Hoof & Lyon,2013). Various case studies from European and Australian organizations also supported that fact that green procurement practices are a tool to achieve competitive advantage and enhance operational and financial performance.

Another finding from the analysis is that mining companies in South Africa are focusing mainly in reducing carbon emissions through various strategic initiatives. This finding is also supported by past studies such as low carbon emission (Hitchock, 2012) and measuring the carbon footprints across the supply chain (Sundarakani et al., 2010) which provides new direction to the present green supply chain practices. Clean technology is a hard dimension of technology innovation having foundation in Ecological modernization theory. Technologies for cleaner production positively influence carbon emission reduction which is supported by past studies (e.g., Sikdar & Howell, 1998; Xu et al., 2006; Baas,2007; Montalvo,2008; Hoof & Lyon,2013).

Therefore, the findings from the data analysis present insights which fully support Ecological modernization theory. South African mining ecology is in the verge of modernization through adoption of GSCM. This study can be extended in future to conduct comparative studies considering another developing country.

5.1 Recommendation

- Sustainable development requires that the development, use and exploitation of renewable resources do not exceed the level beyond which their integrity is jeopardised. Negative impacts on the environment must be anticipated and prevented, and where they cannot be altogether prevented, must be minimised and remedied.
- Responsibility for the environmental health and safety consequences of a policy, programme, project, product, process, service or activity exists throughout its life cycle. There must be inter-organisational coordination and harmonisation of policies, legislation and actions relating to the environment.
- Top management should encourage the design, development and use of products and services that are safe and energy and resource efficient, considering their full life-cycle impacts.

5.2 Managerial Implications

The outcome of the empirical survey provides an insight into GSCM and its impact on South African mining company performance. This study will benefit supply chain practitioners interested in deriving superior performance. They must focus on Top Management Commitment, Reduction in Carbon Emission, collaborative relationships, and Green Procurement practices which are critical GSCM enablers.

5.3 Future Research Directions

This study further motivates to conduct study from different sector and to understand impact of moderating variables on GSCM performance.

REFERENCES

Ageron, B., Gunasekaran, A., & Spalanzani, A. (2012). Sustainable supply management: An empirical study. *International Journal of Production Economics*, *140*(1), 168–182. doi:10.1016/j.ijpe.2011.04.007

Awaysheh, A., & Klassen, R. D. (2010). The impact of supply chain structure on the use of supplier socially responsible practices. *International Journal of Operations & Production Management*, *30*(12), 1246–1268. doi:10.1108/01443571011094253

Azevedo, S. G., Carvalho, H., & Machado, V. C. (2011). The influence of green practices on supply chain performance: A case study approach. *Transportation Research Part E, Logistics and Transportation Review*, *47*(6), 850–871. doi:10.1016/j.tre.2011.05.017

Baas, L. (2007). To make zero emissions technologies and strategies become a reality, the lessons learned of cleaner production dissemination have to be known. *Journal of Cleaner Production*, *15*(13), 1205–1216. doi:10.1016/j.jclepro.2006.07.017

Bag, S., & Anand, N. (2015). Modelling barriers of sustainable supply chain network design using interpretive structural modelling: An insight from food processing sector in India. *International Journal of Automation and Logistics*, *1*(3), 234–255. doi:10.1504/IJAL.2015.071722

Bai, C., & Sarkis, J. (2010). Green supplier development: Analytical evaluation using rough set theory. *Journal of Cleaner Production*, *18*(12), 1200–1210. doi:10.1016/j.jclepro.2010.01.016

Barari, S., Agarwal, G., Zhang, W. C., Mahanty, B., & Tiwari, M. K. (2012). A decision framework for the analysis of green supply chain contracts: An evolutionary game approach. *Expert Systems with Applications*, *39*(3), 2965–2976. doi:10.1016/j.eswa.2011.08.158

Barratt, M. (2004). Understanding the meaning of collaboration in the supply chain. *Supply Chain Management: an international journal, 9*(1), 30-42.

Beamon, B. M. (1999). Designing the green supply chain. *Logistics information management, 12*(4), 332-342.

Benjaafar, S., Li, Y., & Daskin, M. (2013). Carbon footprint and the management of supply chains: Insights from simple models. *IEEE Transactions on* Automation Science and Engineering, *10*(1), 99–116.

Bierma, T. J., & Waterstraat, F. L. (1999). Cleaner Production for Chemical Suppliers: Understanding shared saving contracts'. *Journal of Cleaner Production*, *7*(2), 145–158. doi:10.1016/S0959-6526(98)00073-0

Brandenburg, M., Govindan, K., Sarkis, J., & Seuring, S. (2014). Quantitative models for sustainable supply chain management: Developments and directions. *European Journal of Operational Research*, *233*(2), 299–312. doi:10.1016/j.ejor.2013.09.032

Carter, C. R., Kale, R., & Grimm, C. M. (2000). Environmental purchasing and firm performance: An empirical investigation. *Transportation Research Part E, Logistics and Transportation Review*, *36*(3), 219–228. doi:10.1016/S1366-5545(99)00034-4

Carter, C. R., & Rogers, D. S. (2008). A framework of sustainable supply chain management: Moving toward new theory. *International Journal of Physical Distribution & Logistics Management*, *38*(5), 360–387. doi:10.1108/09600030810882816

Chaabane, A., Ramudhin, A., & Paquet, M. (2012). Design of sustainable supply chains under the emission trading scheme. *International Journal of Production Economics*, *135*(1), 37–49. doi:10.1016/j.ijpe.2010.10.025

Chien, M. K., & Shih, L. H. (2007). An empirical study of the implementation of green supply chain management practices in the electrical and electronic industry and their relation to organizational performances.

Choi, T. Y., & Hartley, J. L. (1996). An exploration of supplier selection practices across the supply chain. *Journal of Operations Management*, *14*(4), 333–343. doi:10.1016/S0272-6963(96)00091-5

Christmann, P., & Taylor, G. (2001). Globalization and the environment: Determinants of firm self-regulation in China. *Journal of international business studies*, 439-458. Retrieved from http://www.gurn.info

Croom, S., Romano, P., & Giannakis, M. (2000). Supply chain management: an analytical framework for critical literature review. *European journal of purchasing & supply management, 6*(1), 67-83.

Darnall, N., Jolley, G. J., & Handfield, R. (2008). Environmental management systems and green supply chain management: Complements for sustainability? *Business Strategy and the Environment, 17*(1), 30–45. doi:10.1002/bse.557

De Giovanni, P. (2012). Do internal and external environmental management contribute to the triple bottom line? *International Journal of Operations & Production Management, 32*(3), 265–290. doi:10.1108/01443571211212574

Deloitte report (Energy Resources). (n. d.). Retrieved from www.deloitte.com

Diabat, A., & Govindan, K. (2011). An analysis of the drivers affecting the implementation of green supply chain management. *Resources, Conservation and Recycling, 55*(6), 659–667. doi:10.1016/j.resconrec.2010.12.002

Dubey, R., & Bag, S. (2013). Exploring the dimensions of sustainable practices: An empirical study on Indian manufacturing firms. *International Journal of Operations and Quantitative Management, 19*(2), 123–146.

Florida, R. (1996). Lean and green: The move to environmentally conscious manufacturing. *California Management Review, 39*(1), 80–105. doi:10.2307/41165877

Gavronski, I., Klassen, R. D., Vachon, S., & do Nascimento, L. F. M. (2011). A resource-based view of green supply management. *Transportation Research Part E, Logistics and Transportation Review, 47*(6), 872–885. doi:10.1016/j.tre.2011.05.018

Gilbert, S. (2000). Integrated summary of the conference on "Greening Supply Chain".

Gimenez, C., Sierra, V., & Rodon, J. (2012). Sustainable operations: Their impact on the triple bottom line. *International Journal of Production Economics, 140*(1), 149–159. doi:10.1016/j.ijpe.2012.01.035

Godfrey, R. (1998). Ethical purchasing: developing the supply chain beyond the environment. In *Greener purchasing: Opportunities and innovations* (pp. 244-251).

Green, K. W. Jr, Whitten, D., & Inman, R. A. (2008). The impact of logistics performance on organizational performance in a supply chain context. *Supply Chain Management: An International Journal, 13*(4), 317–327. doi:10.1108/13598540810882206

Guang Shi, V., Lenny Koh, S. C., Baldwin, J., & Cucchiella, F. (2012). Natural resource based green supply chain management. *Supply Chain Management: An International Journal, 17*(1), 54–67. doi:10.1108/13598541211212203

Guide, V. D. R., Harrison, T. P., & Van Wassenhove, L. N. (2003). The challenge of closed-loop supply chains. *Interfaces, 33*(6), 3–6. doi:10.1287/inte.33.6.3.25182

Gunasekaran, A., & Spalanzani, A. (2012). Sustainability of manufacturing and services: Investigations for research and applications. *International Journal of Production Economics, 140*(1), 35–47. doi:10.1016/j.ijpe.2011.05.011

Gupta, S., & Palsule-Desai, O. D. (2011). Sustainable supply chain management: Review and research opportunities. *IIMB Management Review, 23*(4), 234–245. doi:10.1016/j.iimb.2011.09.002

Hall, J. (2000). Environmental supply chain dynamics. *Journal of Cleaner Production*, *8*(6), 455–471. doi:10.1016/S0959-6526(00)00013-5

Handfield, R., Sroufe, R., & Walton, S. (2005). Integrating environmental management and supply chain strategies. *Business Strategy and the Environment*, *14*(1), 1–19. doi:10.1002/bse.422

Hassini, E., Surti, C., & Searcy, C. (2012). A literature review and a case study of sustainable supply chains with a focus on metrics. *International Journal of Production Economics*, *140*(1), 69–82. doi:10.1016/j.ijpe.2012.01.042

Hervani, A. A., Helms, M. M., & Sarkis, J. (2005). Performance measurement for green supply chain management. *Benchmarking International Journal (Toronto, Ont.)*, *12*(4), 330–353.

Hitchcock, T. (2012). Low carbon and green supply chains: The legal drivers and commercial pressures. *Supply Chain Management: An International Journal*, *17*(1), 98–101. doi:10.1108/13598541211212249

Ho, J. C., Shalishali, M. K., Tseng, T., & Ang, D. S. (2009). Opportunities in green supply chain management. *The Coastal Business Journal*, *8*(1), 18–31.

Hsu, C. W., & Hu, A. H. (2008). Green supply chain management in the electronic industry. *International Journal of Environmental Science and Technology*, *5*(2), 205–216. doi:10.1007/BF03326014

Hsu, C. W., & Hu, A. H. (2009). Applying hazardous substance management to supplier selection using analytic network process. *Journal of Cleaner Production*, *17*(2), 255–264. doi:10.1016/j.jclepro.2008.05.004

Hsu, C. W., Kuo, T. C., Chen, S. H., & Hu, A. H. (2013). Using DEMATEL to develop a carbon management model of supplier selection in green supply chain management. *Journal of Cleaner Production*, *56*, 164–172. doi:10.1016/j.jclepro.2011.09.012

Kang, J., Bennett, M. W., Carbado, D. W., Casey, P., Dasgupta, N., Faigman, D. L., & Mnookin, J. (2012). Implicit bias in the courtroom. *UCLA Law Review*, *59*(5).

Kogg, B. (2003). Greening a cotton-textile supply chain. *Greener Management International, 2003*(43), 52-64.

Lamming, R., & Hampson, J. (1996). The environment as a supply chain management issue. *British Journal of Management*, *7*(Suppl. 1), S45–S62. doi:10.1111/j.1467-8551.1996.tb00147.x

Lee, S. Y. (2008). Drivers for the participation of small and medium-sized suppliers in green supply chain initiatives. *Supply Chain Management: An International Journal*, *13*(3), 185–198. doi:10.1108/13598540810871235

Lightfoot, H., Baines, T., & Smart, P. (2013). The servitization of manufacturing: A systematic literature review of interdependent trends. *International Journal of Operations & Production Management*, *33*(11/12), 1408–1434. doi:10.1108/IJOPM-07-2010-0196

Linton, J. D., Klassen, R., & Jayaraman, V. (2007). Sustainable supply chains: An introduction. *Journal of Operations Management*, *25*(6), 1075–1082. doi:10.1016/j.jom.2007.01.012

Liu, S., Kasturiratne, D., & Moizer, J. (2012). A hub-and-spoke model for multi-dimensional integration of green marketing and sustainable supply chain management. *Industrial Marketing Management*, *41*(4), 581–588. doi:10.1016/j.indmarman.2012.04.005

Longo, F., & Mirabelli, G. (2008). An advanced supply chain management tool based on modeling and simulation. *Computers & Industrial Engineering*, *54*(3), 570–588. doi:10.1016/j.cie.2007.09.008

Matthews, H. S., Hendrickson, C. T., & Weber, C. L. (2008). The importance of carbon footprint estimation boundaries. *Environmental Science & Technology*, *42*(16), 5839–5842. doi:10.1021/es703112w PMID:18767634

Mirhedayatian, S. M., Azadi, M., & Saen, R. F. (2014). A novel network data envelopment analysis model for evaluating green supply chain management. *International Journal of Production Economics*, *147*, 544–554. doi:10.1016/j.ijpe.2013.02.009

Montalvo, C., & Kemp, R. (2008). Cleaner technology diffusion: Case studies, modeling and policy. *Journal of Cleaner Production*, *16*(1), S1–S6. doi:10.1016/j.jclepro.2007.10.014

More than two planets needed by 2030. (2012). WWF Living planet report. Retrieved from www.wwf.org.za

Murphy, J. (2000). Ecological modernisation. *Geoforum*, *31*(1), 1–8. doi:10.1016/S0016-7185(99)00039-1

Murphy, J., & Gouldson, A. (2000). Environmental policy and industrial innovation: Integrating environment and economy through ecological modernisation. *Geoforum*, *31*(1), 33–44. doi:10.1016/S0016-7185(99)00042-1

Nagel, M. H. (2000). Environmental supply-chain management versus green procurement in the scope of a business and leadership perspective. *Proceedings of the 2000 IEEE International Symposium onElectronics and the Environment ISEE '00* (pp. 219-224). IEEE. doi:10.1109/ISEE.2000.857652

Narasimhan, R., & Carter, J. R. (1998). *Environmental supply chain management*. Center for Advanced Purchasing Studies.

Nelson, D., Marsillac, E., & Rao, S. (2012). Antecedents and evolution of the green supply chain. *Journal of Operations and Supply Chain Management*, *1*(1), 29–43.

Pagell, M., & Wu, Z. (2009). Building a more complete theory of sustainable supply chain management using case studies of 10 exemplars. *Journal of supply chain management, 45*(2), 37-56.

Paulraj, A., & de Jong, P. (2011). The effect of ISO 14001 certification announcements on stock performance. *International Journal of Operations & Production Management*, *31*(7), 765–788. doi:10.1108/01443571111144841

Pereira, J. V. (2009). The new supply chain's frontier: Information management. *International Journal of Information Management*, *29*(5), 372–379. doi:10.1016/j.ijinfomgt.2009.02.001

Pittaway, L., Robertson, M., Munir, K., Denyer, D., & Neely, A. (2004). Networking and innovation: A systematic review of the evidence. *International Journal of Management Reviews*, *5*(3-4), 137–168. doi:10.1111/j.1460-8545.2004.00101.x

Rao, P., & Holt, D. (2005). Do green supply chains lead to competitiveness and economic performance? *International Journal of Operations & Production Management, 25*(9), 898–916. doi:10.1108/01443570510613956

Reefke, H., & Trocchi, M. (2013). Balanced scorecard for sustainable supply chains: Design and development guidelines. *International Journal of Productivity and Performance Management, 62*(8), 805–826. doi:10.1108/IJPPM-02-2013-0029

UNEP Report. (2011). Retrieved from http://www.unep.org/greeneconomy/greeneconomyreport/tabid/29846/default.aspx

SAMDA report. (n. d.). Retrieved from www.samda.co.za

Sarkis, J. (2003). A strategic decision framework for green supply chain management. *Journal of Cleaner Production, 11*(4), 397–409. doi:10.1016/S0959-6526(02)00062-8

Sarkis, J., Gonzalez-Torre, P., & Adenso-Diaz, B. (2010). Stakeholder pressure and the adoption of environmental practices: The mediating effect of training. *Journal of Operations Management, 28*(2), 163–176. doi:10.1016/j.jom.2009.10.001

Sarkis, J., Zhu, Q., & Lai, K. H. (2011). An organizational theoretic review of green supply chain management literature. *International Journal of Production Economics, 130*(1), 1–15. doi:10.1016/j.ijpe.2010.11.010

Sarkis, J., Zhu, Q., & Lai, K. H. (2011). An organizational theoretic review of green supply chain management literature. *International Journal of Production Economics, 130*(1), 1–15. doi:10.1016/j.ijpe.2010.11.010

Schrettle, S., Hinz, A., Scherrer-Rathje, M., & Friedli, T. (2014). Turning sustainability into action: Explaining firms' sustainability efforts and their impact on firm performance. *International Journal of Production Economics, 147*, 73–84. doi:10.1016/j.ijpe.2013.02.030

Seuring, S., & Müller, M. (2008). From a literature review to a conceptual framework for sustainable supply chain management. *Journal of Cleaner Production, 16*(15), 1699–1710. doi:10.1016/j.jclepro.2008.04.020

Sheu, J. B., & Chen, Y. J. (2012). Impact of government financial intervention on competition among green supply chains. *International Journal of Production Economics, 138*(1), 201–213. doi:10.1016/j.ijpe.2012.03.024

Sikdar, S. K., & Howell, S. G. (1998). On developing cleaner organic unit processes. *Journal of Cleaner Production, 6*(3-4), 253–259. doi:10.1016/S0959-6526(98)00026-2

Simpson, D. F., & Power, D. J. (2005). Use the supply relationship to develop lean and green suppliers. *Supply chain management:International Journal (Toronto, Ont.), 10*(1), 60–68.

Srivastava, S. K. (2007). Green supply-chain management: A state-of-the-art literature review. *International Journal of Management Reviews, 9*(1), 53–80. doi:10.1111/j.1468-2370.2007.00202.x

Stock, J. R., & Lambert, D. M. (2001). *Strategic logistics management* (Vol. 4). Boston, MA: McGraw-Hill/Irwin.

Sundarakani, B., De Souza, R., Goh, M., Wagner, S. M., & Manikandan, S. (2010). Modeling carbon footprints across the supply chain. *International Journal of Production Economics*, *128*(1), 43–50. doi:10.1016/j.ijpe.2010.01.018

Testa, F., & Iraldo, F. (2010). Shadows and lights of GSCM (Green Supply Chain Management): Determinants and effects of these practices based on a multi-national study. *Journal of Cleaner Production*, *18*(10), 953–962. doi:10.1016/j.jclepro.2010.03.005

Tough choices facing the South African mining industry. (n. d.). *Deloitte Blog*. Retrieved http://deloitteblog.co.za/wp-content/uploads/2013/08

Tranfield, D. R., Denyer, D., & Smart, P. (2003). Towards a methodology for developing evidence-informed management knowledge by means of systematic review. *British Journal of Management*, *14*(3), 207–222. doi:10.1111/1467-8551.00375

Vachon, S. (2007). Green supply chain practices and the selection of environmental technologies. *International Journal of Production Research*, *45*(18-19), 4357–4379. doi:10.1080/00207540701440303

Vachon, S., & Klassen, R. D. (2006). Extending green practices across the supply chain: The impact of upstream and downstream integration. *International Journal of Operations & Production Management*, *26*(7), 795–821. doi:10.1108/01443570610672248

Vachon, S., & Klassen, R. D. (2008). Environmental management and manufacturing performance: The role of collaboration in the supply chain. *International Journal of Production Economics*, *111*(2), 299–315. doi:10.1016/j.ijpe.2006.11.030

Van Aken, J. E. (2005). Management research as a design science: Articulating the research products of mode 2 knowledge production in management. *British Journal of Management*, *16*(1), 19–36. doi:10.1111/j.1467-8551.2005.00437.x

Van Hoek, R. I. (1999). From reversed logistics to green supply chains. *Supply Chain Management: An International Journal*, *4*(3), 129–135. doi:10.1108/13598549910279576

van Hoof, B., & Lyon, T. P. (2013). Cleaner production in small firms taking part in Mexico's Sustainable Supplier Program. *Journal of Cleaner Production*, *41*, 270–282. doi:10.1016/j.jclepro.2012.09.023

Walker, H., Di Sisto, L., & McBain, D. (2008). Drivers and barriers to environmental supply chain management practices: Lessons from the public and private sectors. *Journal of Purchasing and Supply Management*, *14*(1), 69–85. doi:10.1016/j.pursup.2008.01.007

Walton, S. V., Handfield, R. B., & Melnyk, S. A. (1998). The green supply chain: Integrating suppliers into environmental management processes. *Journal of Supply Chain Management*, *34*(2), 2.

Wang, F., Lai, X., & Shi, N. (2011). A multi-objective optimization for green supply chain network design. *Decision Support Systems*, *51*(2), 262–269. doi:10.1016/j.dss.2010.11.020

Wu, Z., & Pagell, M. (2011). Balancing priorities: Decision-making in sustainable supply chain management. *Journal of Operations Management*, *29*(6), 577–590. doi:10.1016/j.jom.2010.10.001

Xu, H. B., Zhang, Y., Li, Z. H., Zheng, S. L., Wang, Z. K., Qi, T., & Li, H. Q. (2006). Development of a new cleaner production process for producing chromic oxide from chromite ore. *Journal of Cleaner Production, 14*(2), 211–219. doi:10.1016/j.jclepro.2004.09.001

Zhang, H. C., Kuo, T. C., Lu, H., & Huang, S. H. (1997). Environmentally conscious design and manufacturing: A state-of-the-art survey. *Journal of Manufacturing Systems, 16*(5), 352–371. doi:10.1016/S0278-6125(97)88465-8

Zhu, Q., & Cote, R. P. (2004). Integrating green supply chain management into an embryonic eco-industrial development: A case study of the Guitang Group. *Journal of Cleaner Production, 12*(8-10), 1025–1035. doi:10.1016/j.jclepro.2004.02.030

Zhu, Q., & Sarkis, J. (2004). Relationships between operational practices and performance among early adopters of green supply chain management practices in Chinese manufacturing enterprises. *Journal of Operations Management, 22*(3), 265–289. doi:10.1016/j.jom.2004.01.005

Zhu, Q., & Sarkis, J. (2004). Relationships between operational practices and performance among early adopters of green supply chain management practices in Chinese manufacturing enterprises. *Journal of Operations Management, 22*(3), 265–289. doi:10.1016/j.jom.2004.01.005

Zhu, Q., & Sarkis, J. (2006). An inter-sectoral comparison of green supply chain management in China: Drivers and practices. *Journal of Cleaner Production, 14*(5), 472–486. doi:10.1016/j.jclepro.2005.01.003

Zhu, Q., Sarkis, J., Cordeiro, J. J., & Lai, K. H. (2008). Firm-level correlates of emergent green supply chain management practices in the Chinese context. *Omega, 36*(4), 577–591. doi:10.1016/j.omega.2006.11.009

Zhu, Q., Sarkis, J., & Geng, Y. (2005). Green supply chain management in China: Pressures, practices and performance. *International Journal of Operations & Production Management, 25*(5), 449–468. doi:10.1108/01443570510593148

Zhu, Q., Sarkis, J., & Lai, K. H. (2007). Green supply chain management: Pressures, practices and performance within the Chinese automobile industry. *Journal of Cleaner Production, 15*(11), 1041–1052. doi:10.1016/j.jclepro.2006.05.021

Zhu, Q., Sarkis, J., & Lai, K. H. (2008). Green supply chain management implications for "closing the loop". *Transportation Research Part E, Logistics and Transportation Review, 44*(1), 1–18. doi:10.1016/j.tre.2006.06.003

Zhu, Q., Sarkis, J., & Lai, K. H. (2008). Confirmation of a measurement model for green supply chain management practices implementation. *International Journal of Production Economics, 111*(2), 261–273. doi:10.1016/j.ijpe.2006.11.029

Zhu, Q., Sarkis, J., Lai, K. H., & Geng, Y. (2008). The role of organizational size in the adoption of green supply chain management practices in China. *Corporate Social Responsibility and Environmental Management, 15*(6), 322–337. doi:10.1002/csr.173

Chapter 6
Modeling the Soft Skills of Green Procurement Professionals using Interpretive Structural Modeling Approach

Surajit Bag
Tega Industries South Africa Pty Ltd, South Africa

ABSTRACT

Green procurement demands a creative attitude and innovation oriented culture which is still missing in many organizations. It is noted that most of the procurement professionals lack essential soft skills which are more important than hard skills. Human resource managers need to develop appropriate teaching methodologies for procurement professionals to enhance the essential soft skills. The purpose of this study is to identify the essential set of soft skills required for green procurement practices. Interpretive structural modeling approach was applied to find the interrelationships. This study also provides deep insight to human resource managers and will be helpful in developing training programs for procurement professionals.

1. INTRODUCTION

The world class purchaser is an individual who visualizes and approaches his or her job from a strategic perspective in dealing with the supplier firm-purchaser and firm-customer linkage. This individual continually embraces and leverages his or her skills and knowledge of critical supply chain activities to provide value in meeting corporate and customer objectives. - Giunipero (2000).

Green procurement consists of complex activities which necessitate procurement processionals to be equipped with hard and soft skills.

Organizational approaches to green procurement range from strategic, organization-wide commitments, to small-scale single purchases. The range of approaches includes: strategic commitments such

DOI: 10.4018/978-1-5225-0350-7.ch006

as green procurement policies; targets and public reporting of progress; supplier strategies and strategic supply chain management; green contracts and tenders; strategic assessment of procurement and priority spend areas highlighted; green procurement action teams and action plans; communications to support and promote green purchasing measurement and review mechanisms and specific projects or initiatives (Green Purchasing Australia report, 2009).

Green procurement demands a creative attitude and innovation oriented culture which is still missing in many organizations. It is noted that most of the procurement professionals lack essential soft skills which are more important than hard skills. Human resource managers need to develop appropriate teaching methodologies for procurement professionals to enhance the essential soft skills.

Though the concept of soft skills is of recent origin, it was Dale Carnegie who spearheaded its crucial significance in his 1936 book titled "How to win friends and influence people" and set the motion for soft skill training and development. Soft skills not only empower hard skills and create new opportunities, but also help to advance personal development and ethics in professionalism.

According to Simpson (2006) the term "soft skill" has two origins- first originated from employers identifying the need for non-traditional skills which relate to the individual and how they interact with others. Owing to the changing nature of business and industry where soft skills play a dominant role and when unemployed people have to be inducted into employment, the target groups have to be identified with specific skills which are different from hard skills (Dash, 2001; Gorman, 2000).

Soft skills development is intended to enable and enhance personal development, participation in learning and success in employment but there is little in the way of research or evidence on how well this assessment is done (Gibb 2014).

According to Simpson (2006), soft skill is an ability or competence, either inherent or acquired which can be repeatedly performed. It is a skill which can be verified and assessed through its performance only and can be demonstrated, learnt, taught, trained or coached but acquired only by performing them and can be improved through reputation and practice.

Moss and Tilly (2001) view soft skills "as skills, abilities and traits that pertain to personality, attitude and behaviour rather than to formal or technical knowledge".

Soft skills taught in the office technology soft skills course are by vast majority congruent with the soft skills employers require in today's entry-level office work (Ellis et al., 2014).

There are few seminal papers on purchasing skill development available in the vast purchasing& supply management literature.

Zheng et al. (2007) considers a broad and rather fragmented body of empirical evidence and analyses 42 relevant empirical studies on the future of purchasing and supply management. Study concluded that less strategic purchases can be automated or outsourced while more strategic task to be tackled by smaller number of professionals with highly developed broader skill sets. More emphasis to be given on teamwork in the skill sets for purchasing professionals.

Tassabehji and Moorhouse (2008) explored the current status of purchasing professionals role and identify the skills they need to develop in order to effectively manage the dynamic situation; and presented a procurement skills effectiveness framework to enable managers to assess the likely sophistication level of procurement and its impact.

Feisel et al., (2011) conducted research to better understand the development of the human aspect in the purchasing and supply management function and found organizations apply a broad variety of HRD instruments to improve performance at different levels and also to improve internal support and integration.

Table 1. Research questions and research tools

Research Questions		Research Tool
RQ1	What is meant by human capital?	Through literature review
RQ2	What type of soft skills is essential in procurement professionals doing green procurement practices?	Through literature review and expert opinion
RQ3	What is the relationship between soft skill sets and green procurement performance? Can a research model be developed?	ISM modeling

The foundation of present research study has been developed based on the Human Capital theory approach. Human capital is the stock of knowledge, skills and abilities, which can help employees increase productivity and performance at work. There have been many studies on this aspect (dealing with hard skills) but present study focuses on documenting the presence and need for soft skills in an emerging economy like India. This study contributes to the need for soft skills at the workplace for successful green procurement practices.

While human capital itself is an approach to economic development, approaches to human capital development focus on training to improve business, technical and soft skills of employees (OECD, 2000). The present study is concerned more in human capital development at the workplace.

The research questions need to be explored and solved. Each of the identified questions and the research strategies for each is presented below in a tabulated form in Table 1 as:

2. LITERATURE REVIEW

The objective of this section is to identify essential soft skills of procurement professionals involved in green procurement practices. Here author reviewed refereed journals like Journal of Purchasing and Supply Management, International Journal of Procurement Management, European Journal of Purchasing and Supply Management, International Journal of Operations & Production Management and International Journal of Production Economics to identify the variables.

Keeping the research objectives in mind, the focus is to identify essential soft skills in procurement professionals doing green procurement practices. Here researcher based on literature review prepared a non-exhaustive list of soft skills are presented in Table 2.

In this step, the green purchasing literature was thoroughly reviewed. As a result of the review, sixteen variables were found to be effective in green procurement practices. An expert panel was formed comprising eight supply chain practitioners' who are experts in green procurement. The group was asked to analyze the listed variables, take out the repeated ones and rearrange the list including the ones thought to be valid in Indian context. The first round resulted with reducing each of the antecedents to 12. Through the second round 10 and the third round nine total variables were determined.

The nine variables are as follows:

1. Environmental Knowledge
2. Leadership
3. Commitment
4. Motivation

Table 2. Soft skills of procurement professionals

Soft Skills	References
Creativity	Giunipero, (2000)
Flexibility	Giunipero, (2000)
Leadership	Hamner, (2006); Lee, (2008); Green Purchasing Australia report (2009); Yen & Yen, (2012)
Information sharing	Humphreys et al., (2001); Chen(2005); Rao (2002)
Delegation &accountability	Giunipero, (2000)
Motivation	Giunipero, (2000); Lynes, & Dredge, (2006).
Commitment	Min & Galle,(2001); Zhu et al., (2005); Handfield et al., (2002)
Environmental concern	Kim, & Choi, (2005); Mainieri et al., (1997)
Knowledge	Giunipero, (2000)
Relationship Management	Giunipero, (2000); Green Purchasing Australia report (2009); Lintukangas (2011)
Initiative	Giunipero, (2000); Green Purchasing Australia report (2009)
Negotiation	Giunipero, (2000)
Integrity	Giunipero, (2000)
Decision making	Giunipero, (2000)
Tenacity	Giunipero, (2000)
Innovation	Green et al., (1998); Rao and Holt (2005); Green Purchasing Australia report (2009)

5. Innovation
6. Creativity
7. Tenacity
8. Relationship Management
9. Flexibility

3. METHODOLOGY

The research was carried out using the ISM method, a method that is used to identify and clarify the enablers leading to successful green practices and their interrelations in terms of their power value. It is an interactive planning methodology whereby a set of directly and indirectly related factors are structured into a comprehensive systematic model.

3.1. ISM based Model

ISM has been used in the past by several researchers due to multiple benefits. It guides and records the results of group response on complex issues in an efficient and systematic manner, (Source: Attri, Dev & Sharma, 2013; Dubey& Ali, 2014; Sushil, 2012; Sushil, 2009; Sushil, 2005a; Sushil, 2005b; Warfield 1999; Warfield 1994; Warfield 1974).

The ISM steps are presented in fig 1 below.

Figure 1. Flowchart for the ISM Methodology

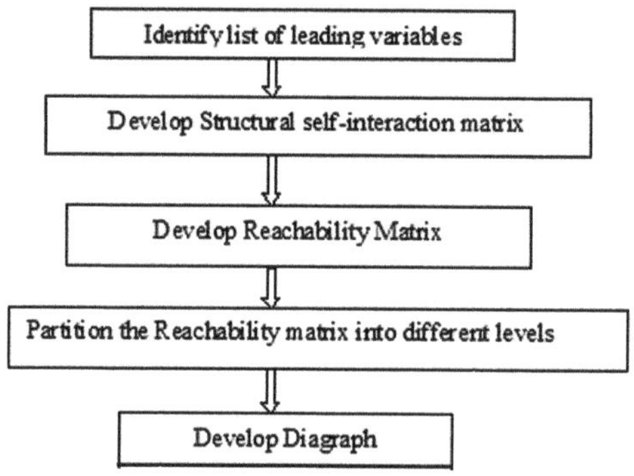

Table 3. Structural self-interaction matrix

	9	8	7	6	5	4	3	2	1
1	V	V	O	O	O	V	V	O	
2	V	V	O	V	V	V	V		
3	V	V	A	O	V	A			
4	A	V	A	A	A				
5	A	A	A	A					
6	O	A	A						
7	O	V							
8	V								
9									

4.1. Structural Self-Interaction Matrix

Developing a structural self-interaction matrix (SSIM) of elements, indicates pair-wise relationships between elements of the system (Table 3). To analyze the soft skills, a contextual relation of 'achieve' was selected. This means that one soft skill will achieve another soft skill; the latter will be achieved by another soft skill; the two soft skills will help achieve each other or the soft skills will be unrelated. For developing SSIM, the following four symbols have been used to denote the direction of relationships between soft skills (i and j):

V: i lead to j but j does not lead to i
A: i do not lead to j but j lead to i
X: i lead to j and j lead to i
O: i and j are unrelated to each other

Table 4. Initial RM

	1	2	3	4	5	6	7	8	9	Driving power (Y)
1	1	0	1	1	0	0	0	1	1	5
2	0	1	1	1	1	1	0	1	1	7
3	0	0	1	0	1	0	0	1	1	4
4	0	0	1	1	0	0	0	1	0	3
5	0	0	0	1	1	0	0	0	0	2
6	0	0	0	1	1	1	0	0	0	3
7	0	0	1	1	1	1	1	1	0	6
8	0	0	0	0	1	1	0	1	1	4
9	0	0	0	1	1	0	0	0	1	3
Dependence Power (X)	1	1	5	7	7	4	1	6	5	

4.2. Reachability Matrix

The SSIM has been converted into a binary matrix, called the RM (Table 4) by substituting V, A X and O by 1 and 0. The substitution of '1' and '0' is done as below:

1. If the (i, j) entry in the SSIM is V, then the (i,j) entry in the rechability matrix becomes '1' and (j,i) entry becomes '0'
2. If the (i, j) entry in the SSIM is A, then the (i,j) entry in the rechability matrix becomes '0' and (j,i) entry becomes '1'
3. If the (i, j) entry in the SSIM is X, then the (i,j) entry in the rechability matrix becomes '1' and (j,i) entry also becomes '1'
4. If the (i, j) entry in the SSIM is O, then the (i,j) entry in the rechability matrix becomes '0' and (j,i) entry also becomes '0'

4.3. Transitivity Principle

In this step the concept of transitivity is introduced so that some of the cells of the initial reachability matrix are filled by inference. Transitivity can be explained with the following example if element "i" relates to element "j" and element "j" relates to element "k", then transitivity implies element "i" relates to element "k". Transitivity is the basic assumption in ISM and is always used in this modelling approach (Watson 1978, Sushil, 2005a; Sushil, 2005b). It also helps in maintaining the conceptual consistency. The final reachability matrix will then consist of some entries from the pair-wise comparisons and some inferred entries. The transitivity concept is used to fill the gap, if any. Following the above rules, the initial reachability matrix is prepared. After incorporating the transitivity concept as described above, the final reachability matrix is obtained Table 5.

Table 5. Final RM

	1	2	3	4	5	6	7	8	9	Driving Power (Y)
1	1	0	1	1	1*	0	0	1	1	6
2	0	1	1	1	1	1	0	1	1	7
3	0	0	1	1*	1	0	0	1	1	5
4	0	0	1	1	1*	1*	0	1	1*	6
5	0	0	0	1	1	0	0	0	0	2
6	0	0	0	1	1	1	0	0	0	3
7	0	0	1	1	1	1	1	1	1*	7
8	0	0	0	1*	1	1	0	1	1	5
9	0	0	0	1	1	0	0	0	1	3
Dependence Power(X)	1	1	5	9	9	5	1	6	7	

Note: *Indicate transivity.

4.4. Level Partitioning

The final reachability matrix obtained above in Table 5 is now partitioned into different levels. After the first iteration, the variables classified to level 1 are discarded and the partitioning procedure is repeated on the remaining variables to determine the level 2. These iterations are continued until the level of each variable has been determined. The results for iterations 1 to 5 are presented from Table 6 to Table 10

Table 6. Level matrix (Iteration1)

Variables	RS	AS	IS Level	Level
1	1,3,4,5,8,9	1	1	
2	2,3,4,5,6,8,9	2	2	
3	3,4,5,8,9	1,2,3,4,7	3,4	
4	3,4,5,6,8,9	1,2,3,4,5,6,7,8,9	3,4,5,6,8,9	1
5	4,5	1,2,3,4,5,6,7,8,9	4,5	1
6	4,5,6	2,4,6,7,8	4,6	
7	3,4,5,6,7,8,9	7	7	
8	4,5,6,8,9	1,2,3,4,7,8	4,8	
9	4,5,9	1,2,3,4,7,8,9	4,9	

Table 10. Level matrix (Iteration 5)

Variables	RS	AS	IS Level	Level
1	1	1	1	5
2	2	2	2	5
7	7	7	7	5

Table 7. Level matrix (Iteration2)

Variables	RS	AS	IS Level	Level
1	1,3,8,9	1	1	
2	2,3,6,8,9	2	2	
3	3,8,9	1,2,3,7	3	
6	6	2,6,7,8	6	2
7	3,6,7,8,9	7	7	
8	6,8,9	1,2,3,7,8	8	
9	9	1,2,3,7,8,9	9	2

Table 8. Level matrix (Iteration3)

Variables	RS	AS	IS Level	Level
1	1,3,8	1	1	
2	2,3,8	2	2	
3	3,8	1,2,3,7	3	
7	3,7,8	7	7	
8	8	1,2,3,7,8	8	3

Table 9. Level matrix (Iteration 4)

Variables	RS	AS	IS Level	Level
1	1,3	1	1	
2	2,3	2	2	
3	3	1,2,3,7	3	4
7	3,7	7	7	

4.5. Diagraph

From the final Reachability matrix and level partitions, the structural model is generated by means of vertices or nodes and lines of edges. If there is a relationship between the variables j and i, this is shown by an arrow which points to from i to j. This graph is called graph or diagraph (Figure 2).

Figure 2. Diagraph

Here, 1 to 9 represents variables as: 1) Environmental Knowledge; 2) Leadership; 3) Commitment; 4) Motivation; 5) Innovation; 6) Creativity; 7) Tenacity; 8) Relationship Management; 9) Flexibility

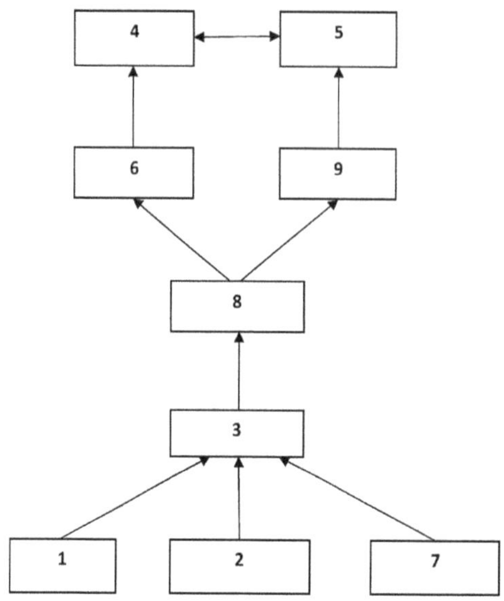

5. CONCLUSION

The diagraph indicates that Environmental Knowledge, Leadership and Tenacity leads to Motivation and Innovation skill sets among procurement professionals involved in green procurement practices within an organization. The diagraph shows that Environmental Knowledge, Leadership and Tenacity plays a key role in developing commitment which together leads to relationship management. Intervention programmes like training, coaching and mentoring influence soft skill acquisition. At the work place, training, experience, ideastorming, business reading and mentoring may enhance the intensity of soft skill acquisition. The impact can be assessed by comparing before and after intervention scenarios.

The findings can further be empirically tested by conducting a survey among procurement professionals in India.

5.1. Theoretical and Managerial Implications

With changes in business scenario worldwide purchasing function moves toward more leading edge and eco-friendly practices. This requires distinct set of skills which are more sophisticated and the knowledge required is more advanced. With this in mind, the research findings will allow for a faster and smoother move toward developing the procurement professional in green procurement practices. Human resource managers should focus in creating a comprehensive training model to trigger the soft dimensions among procurement professionals. The assessment metrics must be quantifiable and should be converted into measurable goals keeping in view the internal and external environment of organization.

5.2. Limitations and Future Directions of Research

The present study has employed ISM analysis. Like every methodology, ISM has its own limitations because it is purely based on expert opinion and need to be validated statistically. In future, effort will be taken to conduct an empirical study to test the relationship among variables.

REFERENCES

Attri, R., Dev, N., & Sharma, V. (2013). Interpretive Structural Modeling (ISM) approach: An overview. *Research Journal of Management Sciences*, *2*(2), 3–8.

Chen, C. C. (2005). Incorporating green purchasing into the frame of ISO 14000. *Journal of Cleaner Production*, *13*(9), 927–933. doi:10.1016/j.jclepro.2004.04.005

Dash, J. (2001). Schools push soft skills for info security majors. *Computerworld*, *35*(6), 24.

Dubey, R., & Ali, S. S. (2014). Identification of flexible manufacturing system dimensions and their interrelationship using total interpretive structural modeling and fuzzy MICMAC analysis. *Global Journal of Flexible Systems Management*, *15*(2), 131–143. doi:10.1007/s40171-014-0058-9

Ellis, M., Kisling, E., & Hackworth, R. G. (2014). Teaching Soft Skills Employers Need. *Community College Journal of Research and Practice*, *38*(5), 433–453. doi:10.1080/10668926.2011.567143

Feisel, H., Hartmann, E., & Giunipero, L. (2011). The importance of human aspect in the supply function: Strategies for developing PSM proficiency. *Journal of Purchasing and Supply Management*, *17*(2), 54–67. doi:10.1016/j.pursup.2010.07.001

Gibb, S. (2014). Soft skills assessment: Theory development and the research agenda. *International Journal of Lifelong Education*, *33*(4), 455–471. doi:10.1080/02601370.2013.867546

Giunipero, L. C. (2000). *A skills-based analysis of the world class purchaser*. Center for Advanced Purchasing Studies.

Gorman, H. (2000). Winning hearts and minds? - Emotional labour and learning for care management. *Journal of Social Work Practice*, *14*(2), 149–158. doi:10.1080/02650530020020546

Green, K., Morton, B., & New, S. (1998). Green purchasing and supply policies: Do they improve companies' environmental performance? *Supply Chain Management: An International Journal*, *3*(2), 89–95. doi:10.1108/13598549810215405

Green Purchasing Australia report. (2009). Retrieved from http://www.ecobuy.org.au

Hamner, B. (2006). Effects of green purchasing strategies on supplier behaviour. In *Greening the supply chain* (pp. 25–37). Springer London. doi:10.1007/1-84628-299-3_2

Handfield, R., Walton, S. V., Sroufe, R., & Melnyk, S. A. (2002). Applying environmental criteria to supplier assessment: A study in the application of the Analytical Hierarchy Process. *European Journal of Operational Research*, *141*(1), 70–87. doi:10.1016/S0377-2217(01)00261-2

Humphreys, P. K., Shiu, W. K., & Chan, F. T. S. (2001). Collaborative buyer-supplier relationships in Hong Kong manufacturing firms. *Supply Chain Management: an international journal, 6*(4), 152-162.

Investing in Education: Analysis of the 1999 World Education Indicators. (2000). OECD, Paris.

Kim, Y., & Choi, S. M. (2005). Antecedents of green purchase behavior: An examination of collectivism, environmental concern, and PCE. *Advances in Consumer Research. Association for Consumer Research (U. S.), 32*, 592.

Lee, S. Y. (2008). Drivers for the participation of small and medium-sized suppliers in green supply chain initiatives. *Supply Chain Management: An International Journal, 13*(3), 185–198. doi:10.1108/13598540810871235

Lintukangas, K. (2011). Supplier relationship management capability in global supply management. *Int. J. of Procurement Management, 4*(1), 1–19. doi:10.1504/IJPM.2011.037382

Lynes, J. K., & Dredge, D. (2006). Going green: Motivations for environmental commitment in the airline industry. A case study of Scandinavian Airlines. *Journal of Sustainable Tourism, 14*(2), 116–138. doi:10.1080/09669580608669048

Mainieri, T., Barnett, E. G., Valdero, T. R., Unipan, J. B., & Oskamp, S. (1997). Green buying: The influence of environmental concern on consumer behavior. *The Journal of Social Psychology, 137*(2), 189–204. doi:10.1080/00224549709595430

Mandal, A., & Deshmukh, S. G. (1994). Vendor selection using interpretive structural modeling (ism)'. *International Journal of Operations & Production Management, 14*(6), 52–59. doi:10.1108/01443579410062086

Min, H., & Galle, W. P. (2001). Green purchasing practices of US firms. *International Journal of Operations & Production Management, 21*(9), 1222–1238. doi:10.1108/EUM0000000005923

Moss, P., & Tilly, C. (2001). *Stories Employers Tell. Race, Skill and Hiring in America*. New York: Russell Sage Foundation.

Rao, P. (2002). Greening the supply chain: A new initiative in South East Asia. *International Journal of Operations & Production Management, 22*(6), 632–655. doi:10.1108/01443570210427668

Rao, P., & Holt, D. (2005). Do green supply chains lead to competitiveness and economic performance? *International Journal of Operations & Production Management, 25*(9), 898–916. doi:10.1108/01443570510613956

Simpson, J. A. (2006). *The measurement and recognition of soft skills-Developing a Common Standard?* Brussels: European Union.

Sushil. (2005a). Interpretive matrix: A tool to aid interpretation of management and social research. *Global Journal of Flexible Systems Management*, 6(2), 27-30

Sushil. (2005b). A flexible strategy framework for managing community and change. International Journal of Global Business and Competitiveness, 1(1), 22-32

Sushil. (2009). Interpretive ranking process. *Global Journal of Flexible Systems Management*, 10(4), pp.1-10

Sushil. (2012). Interpreting the Interpretive Structural Model. *Global Journal of Flexible Systems Management*, 13(2), 87-106

Tassabehji, R., & Moorhouse, A. (2008). The changing role of procurement: Developing professional effectiveness. *Journal of Purchasing and Supply Management*, *14*(1), 55–68. doi:10.1016/j.pursup.2008.01.005

Warfield, J. N. (1974). Structuring complex systems. Columbus, O.H: Battele Memorial Ins. 4.

Warfield, J. N. (1994). *A science of generic design: Managing complexity through systems design*. Iowa: Iowa State University press.

Warfield, J. N. (1999). Twenty laws of complexity: Science applicability in organizations. *Systems Research and Behavioral Science*, *16*(1), 3–40. doi:10.1002/(SICI)1099-1743(199901/02)16:1<3::AID-SRES241>3.0.CO;2-F

Watson, R. H. (1978). Interpretive structural modeling—A useful tool for technology assessment? *Technological Forecasting and Social Change*, *11*(2), 165–185. doi:10.1016/0040-1625(78)90028-8

Yen, Y. X., & Yen, S. Y. (2012). Top-management's role in adopting green purchasing standards in high-tech industrial firms. *Journal of Business Research*, *65*(7), 951–959. doi:10.1016/j.jbusres.2011.05.002

Zheng, J., Knight, L., James, K., Harland, C., & Humby, S. (2007). An analysis of research into the future of purchasing and supply management. *Journal of Purchasing and Supply Management*, *13*(1), 69–83. doi:10.1016/j.pursup.2007.03.004

Zhu, Q., Sarkis, J., & Geng, Y. (2005). Green supply chain management in China: Pressures, practices and performance. *International Journal of Operations & Production Management*, *25*(5), 449–468. doi:10.1108/01443570510593148

Chapter 7
Role of Sustainable Procurement in Sustainable Manufacturing Operations
An Indian Insight

V.G. Venkatesh
Symbiosis Institute of Business Management, India

Sunil Luthra
Government Polytechnic, Jhajjar, India

ABSTRACT

Sustainable Procurement is an emerging concept in India in the recent times. It is imperative for the production firms to look the procurement activity as one of the strategic enabler for sustaining the business in the competitive global environment. Moreover, sustainability in India has been understood in multiple meanings and it is essential to bring the awareness on sustainable procurement and its impact on manufacturing. This chapter elucidates the key drivers of sustainable procurement along with a complete discussion on the importance of those practices in the Indian context. It also lists some of the key trends in the current business environment. The study endorses that manufacturing industry dominated by SMEs are highly reluctant to adopt sustainability as they lack awareness and perceive them as a cost escalation activity. It also reveals challenges faced by the firms and gives a conclusion on the future directions for sustainable procurement research in Indian context.

INTRODUCTION

In the emerging markets like India, sustainable manufacturing has taken a new momentum as there is a focus beyond Green operations. Industry professionals also started practicing the principles in multiple operational domains with diversified objectives. The concept is to be supported by sustainable procurement in the upstream and sustainable transport/Logistics in the downstream directions. Both of them are having an effective role on the core operations and look for including the partners inclusively and to

DOI: 10.4018/978-1-5225-0350-7.ch007

work in the long term focus. In recent decade, governments across the world have become increasingly conscious about the adverse and undesirable impact of their purchases on society and environment, and are adopting sustainable procurement practices facilitated by legal reforms and policy guidelines (Terrin report, 2012). At present, Indian government is looking for a momentum in manufacturing domain through its new policy program called "*Make in India*" and for that Sustainability is becoming an essential element of Supply Chain (SC) systems. Due to this, many of the foreign players are trying to establish in Indian market with advanced and inclusive concepts in manufacturing, which have to be inducted to Indian system diligently. These firms are looking at India as an emerging market as a potential one because of abundant and versatility in resources' availability. Naturally, when they try to invest in manufacturing domains, they bring in the latest trends and to be adopted in the localized condition to sustain in the market. They also give the directions to their Indian counterparts to be ready to implement and go further on their current level of understanding on sustainability. Also, it would help Indian players to compete globally by adopting stringent global norms followed in the operations which are considered to be sustainable.

As earlier said, Sustainable procurement is becoming an integrated component in the supply chain systems with many players are becoming aware of the regulations and needs of the customers closely. Moreover, the research in the said domains is in nascent stage and any work in sustainable manufacturing, procurement and logistics would help industry policy makers to look upon and devise their strategy accordingly. The domain is explored in the public procurement domain at the global level with very few research reports on private firm orientation. Further, Indian industries face lots of challenges in implementing the same. This chapter will identify and discuss those challenges along with a brief report on the current status of sustainable procurement in India.

Sustainable Procurement: An Insight

Supply chain domain is currently dominated by the buzz word "Sustainability" in its processes and practices. Of late, it has been used as a misnomer with "Green approach," while Chartered Institute of Purchasing and Supply defines it as a process beyond "Green". Welsh Procurement Initiative (2004) defines Supply Chain Management (SCM) as a process that considers the negative ecological and societal impacts of the procurement while distributing or receiving the services/products. The firms are expected to draft a holistic procurement practices with socially, ethically and environmental responsible procurement activities. This is to minimize environmental impact through the supply chain and to deliver economically accepted solutions as well. So, these practices would create the values beyond the usage and not only to the purchasing organization, but also express the tangible benefits to the society and economy are generally named as Sustainable procurement practices. In today's competitive environment, organizations are also in need of continuous and swift improvements to get new competitive advantages. For that purchasing plays a major role in sustainable development. Therefore, these practices, strategies and policies must be extended across all levels of their SCs irrespective of firms' boundaries (Luthra et al., 2014a). That includes involvement of the suppliers in all the levels to create a value in the transformation process. In the similar lines, Walker et al. (2012) suggested that the Sustainable Procurement (SP) is the planning of long term development plans/objectives to include environmental, social and economic aspects in the purchasing and supply processes. Sustainable procurement practices may be a good solution to achieve that as well environment and social benefits in the SC activities. A sustainable purchasing means consideration of economic, technical, and legal purchasing requirement to achieve community and ecological advantages in conjunction with the conventional performance gains which the business seeks (Spence & Bourlakis, 2009). Guiding principles on the sustainability motivate procurement function to make assessment that incorporates the ecological, financial and societal elements of the Triple Bottom Line (Meehan & Bryde, 2011). Sustainable procurement helps to promote good governance and ensures environmental as well as societal benefits; and is consistent with the philosophy of sustainable development, (Walker & Brammer, 2009). The domain is an emerging area, but appears to be undergoing a rapid development phase.

Several research gaps have been reported in this emerging area that provides many future research directions and one vital gap is to identify the factors affecting to sustainable procurement with its effect on organizational performance (Walker et al., 2012). Further, Most of the research in this emerging area (purchasing and SC research) have been made at SC level and focused on ecological/green concerns. Negligible or very modest research exists in the literature that examined social sustainability (Miemczyk et al., 2012). The function of purchasing and SCM is gaining due importance in an organization's environmental management extending beyond regulations, compliance and cost savings (Green et al., 1998). Nevertheless, the field definition is growing and it is not limited by any such single definition. The sustainability can be extended to manufacturing operations as well and it is impacted by the procurement operations. Sustainability in production processes and supporting activities have drawn increasing attention over the past few years (Luthra et al., 2014b). Predominantly, the domain has been identified as the Eco-friendly or Green manufacturing. Further, it can be broadened further with an inclusion of reverse supply chain processes, buy- back policies and arrangements in the entire supply chain systems. But many of the firms are not having a clear policy in their operations for managing the day to day operations and methodologies or procedures to concentrate. Currently, organizations are trying to enter all possible domains with sustainable developments for the long term objectives. According to recent Ernst and Young report, it is stressed that the purchasing is playing a vital function in integrating the sustainable development principles across all the firm's business functions. While many organizations are acknowledging the importance of sustainability, implementation of an effective strategy to achieve a sustainable organization remains challenge (Galphin et al., 2015). One such area is procurement which contributes to sustainability in terms of resource consumption and through the negative impacts of production activities, transport, distribution and disposal of products after its useful life (Walker & Brammer, 2012).

Research has not comprehensively addressed what enables sustainable procurement in Indian context. Therefore, it has been observed important to explore this emerging area in India context to identify role of sustainable procurement in sustainable manufacturing operations. So, it is imperative to discuss the role of sustainable procurement in sustaining the manufacturing operations too. In the last decade, it is becoming as hot topic of discussions for management thought leaders and scientists (Corbett & Kleindorfer, 2003; Corbett & Klassen, 2006; Pagell & Wu, 2009). With that background, this chapter helps in achieving the following objectives:

- To identify key drivers to achieve sustainable procurement.
- To know current status of sustainable procurement in India.
- To know role of sustainable procurement in India.
- To identify challenges to achieve sustainable procurement in India.

The study is done by Systematic literature Review proposed by Tranfield et al. (2003) by organizing as follows: It starts with the key drivers for Sustainable procurement practices at the International level, current status of sustainable procurement in India, Challenges and opportunities for sustainable procurement, how does it impact the manufacturing operations with a discussion on key trends .

Key Drivers to Promote Sustainable Procurement Practices

There are several key drivers that lead to sustainable procurement practice in an organization. An attempt has been made to list some of the key driverspromoting Sustainable Procurement practices in the International context.Following are the key drivers on which the procurement practices are directed.

Government Authority and Legalizations

Government bodies and legislations play has been recognized as a key rationale for business organizations to take ecological action, and the regulatory sector has acknowledged the maximum concentration in the literature (Zhu & Sarkis, 2006; Björklund, 2011). Government authority and legalizations normally play a significant role in persuading business organizations to focus on the sustainability principles and practices. Government authorities may adopt a 'command and control' perception e.g. carbon tax, fines and penalties to non-maintenance of environmental standards and subsidizing certain industries as well as technologies (Gupta & Desai, 2011).

Societal Pressure

Walker et al. (2008) identified that pressure from a variety of societies such as NGOs and green promotion groups' forces business organizations to critically consider about their environmental and social sustainability plans. The broader societal considerations may include occupational health and safety practices, local society issues, employability practices, and stakeholder dimensions etc. (Bai & Sarkis 2010).

Top/Middle Management

When the leadership has complete knowledge on the business organizations' sustainability objectives, they make decision accordingly (funds allocation, priority setting, technology & infrastructure advancement etc. Top/middle management leadership is very influential in motivating the business organizations to assess their role in society and are accountable for the business organizations' sustainability development programs (Giunipero et al., 2012). This is applicable to procurement domain also.

Reward, Measurement and Benchmarking Systems

It has been reported by previous researchers that the reward, performance measurement and benchmarking systems have significant positive impacts on the purchase activities (Preuss, 2007; Björklund, 2011).

Employees

Employees of a firm should understand that what they do and how it affects the environment and society. This will led to new environmental and social friendly programs towards sustainable development initiated by employees and supported by management (Rondinelli & Berry, 2000; Szekely & Knirsch, 2005). Employee knowledge, awareness, interest and confidence may play very important role in purchasing decisions (Van Weele, 2005; Björklund, 2011).

Suppliers' Involvement

Suppliers' involvement in environmental issues has got significant importance in previous literatures. Supplier involvement in environmental design is necessary to improve environmental and social performances (Tate et al., 2012). Collaboration with suppliers is another way to achieve sustainability (Chabbane et al., 2012). Effective communication with suppliers affects sustainable procurement practices in the organization (Walker & Brammer, 2012).

Competitors

Competitors are very important to proactively tackle environmental and societal issues in purchasing (Zsidisin & Siferd, 2001; Björklund, 2011). In today's competitive environment, organizations have need of continuous and swift improvements to get new competitive advantage (Spence & Bourlakis, 2009).

Education and Training of Supply Chain Partners

Customers have started to comprehend that their purchasing activities may cause a huge impact on environment. Customers may put pressure on business organizations to produce environment friendly and socially responsible products (Wahid et al., 2011). An organization cannot expect to monitor their supplier's environmental awareness and benchmarking when the buying organization does not actively participate in such activities (Palisi, 2012). Business organizations should recognize education and training needs of their SC members and the education training must include: environmental policy and green issues, job specific environmental impacts and benefits of improved performance etc. SC members' education and training will help to attain the sustainable consumption and production practices in the society (Sambasivan & Fei, 2008; Ageron et al., 2012).

Current Status of Sustainable Procurement in India

Sustainable procurement concept is relatively a new concept in developing countries as compared developed countries. India has boarded on a route of economic reform and progressive integration with the world economy that intends to place it on a route of fast and sustained growth. The average Gross Domestic Product (GDP) growth rate for the Indian economy since 2001 has been reported as 7.5 percent and India now ranks as fourth largest economy in the world, considered on purchasing power parity (Center for Strategic & International Studies, 2013). Indian government has been spending approximately 21 percent of the GDP or 12 to 15 lakh crores rupees/annum on public procurement in India. Given the massive size of public spending, public sector in India can be a prime driver towards sustainable production and consumption and can create environmental and economic benefits (Terrin Report, 2012). The government has been taking many policy initiatives (such as mitigation of greenhouse gases other air/water pollutants) since post-liberalization in the country in 1991. In current wave of growing global competition and increasing importance on environmental concerns, business organizations under heavy pressure to produce high quality and innovative products at minimum cost, but also to develop SCs that are sustainable in the long term (Gupta & Palsule-Desai, 2011).

Few public sectors in India and government departments have started internalizing environmental and energy efficiency criteria in their procurement decisions. For instance, Indian Railways, National Thermal Power Corporation (NTPC), Bharat Heavy Electricals Limited (BHEL), and Indian Oil Corporation (IOC) are promoting sustainable public procurement in a decentralized way in project specific sites with a major focus on procurement of energy conserving equipment (TERI, 2007) in the procurement process of small items. However, such efforts have been undertaken in isolation and have not been replicated or scaled up across organizations, sectors, and levels of governments. There are few reports available in the sustainable procurement practices in the public sector enterprises and no standard practices area available in disclosing the practices (Mansi, 2015).

Need of Sustainable Procurement in India

In general, sustainability is an underestimated concept in India with the misconception that it is an additional overhead by practicing green as well as sustainable concepts. Further, professionals are also under the negative mindset that it is not going to add a value to the products or systems. Since many decades, procurement professionals had not paid enough attention to environmental impacts in manufacturing including their cost of services as well (Ho et al., 2010). Nevertheless, practices from western countries show a different direction with innovation and new product development. Corporates which are eyeing on the sustainable procurement principles would automatically sail on the new product development route. India as a country, with a status of *"country to be watched"* in global sustainable index rank has to understand the reasons behind implementation. Organizations need to prioritize sustainable procurement practices for three main reasons, Financial, Environmental and Social aspects.

In the financial reasons, the main objective is to reduce the life cycle cost of the procured goods, increased utilization of the product, to reduce the costs of waste, emissions and the disposal waste, to sail through the regulatory norms with less cost of operations, to improve productivity and competitive advantage. Sustainable procurement is also essentially making organizations to respond environmentally also by reducing the impacts of ecological risks and reducing the liability. It also focuses on the reduction of factors that cause negative health, well-being and safety impacts of procuring and using goods and services. There is always a growing demand for using the organic products to be used in any system, for which sustainable procurement is providing the complete help and roadmap. For the social reasons, the activity helps to promote the reputation of the firm, to create sustainable consumption and markets for its products/ services, and alignment with the society and needs and to attract and to retrain the talented and qualified staff and maintain the interest of the stakeholders by helping them to stay in the business for the long term.

Sustainable procurement considers products as well as the suppliers. This includes key issues such as: resource extraction and consumption; manufacturing and production; transport and logistics; product and asset design; use and maintenance; recycling and disposal options; employee rights and conditions, corruption, unfair competition and ethical behavior (Walker & Brammer, 2012). With growing awareness on sustainable issues, Indian government has established few regulatory norms to reduce the emissions and waste. But, very less emphasis has been given by Indian government to develop product standards. Till date, there is not a single product oriented standards. At the moment, green/sustainable procurement issues are far more relevant for exporting business organizations. Ultimately, these issues are anticipated to influence the entire Indian industry in big way. Currently the firms involved in sustainable procurement practices are not having the good disclosure practices covering the economic, social benefits. And, there is no standard disclosure practice followed by many organizations including public sector enterprises (Mansi, 2015). This lead to no standardized scales or Indices on which sustainable practices are developed and assessed by sector wise similar to other developed nations. The design, development and implementation of product oriented standards will bring about a change in Indian market e.g. availability of green/sustainable products. It will also help to promote sustainable production and consumption patterns in India (GPNI, 2015).

In last few years, few steps have been taken by Indian government to promote sustainable public procurement in India. First, introduction of Public Procurement Bill-2012 in parliament stating that environmental criteria of a product may be adopted as one of the criteria for evaluation of tender. Second, a committee nominated by Ministry of Environment and Forests (MoEF) has recommended for introducing legislation to encourage shift in demand towards greener products and services. Although, legal framework is necessary first step, merely enacting a law would not lead automatically to sustainable procurement unless a close observation has to be given on the ways it enables stakeholders to effectively collaborate and build sustainability across the entire supply chain. Incorporation of concept like eco-label, Life cycle cost and Life cycle accounting in procurement may in fact make decision making more difficult for procurers (Kumar, 2013).

Role of Sustainable Procurement in Sustainable Manufacturing Operations

It is expected that Indian economy needs to grow at 8-10% per year for next 2-3 decades to pay attention to its human development deficit. Economic growth is needed to offer the necessary public services such as education, infrastructure and health as well as to generate employment opportunities (Parikh, 2012). Manufacturing sector in India has a key contribution to make towards a more sustainable society. The importance of the manufacturing sector can be assessed from the following:

- Contributing 17 percent of India's GDP.
- Contributing 75 percent of India's total exports.
- Employment opportunities for 11 percent of India' workforce.
- Force multiplier as every rupee of investment increases GDP by four rupees.

In today's competitive business has also made organizations to consider sustainability in their manufacturing and service activities (Ageron et al. 2012). The growth in Indian manufacturing sector has been a quest unwritten and unavoidable for our growth process. Manufacturers are becoming more proactive towards cleaner and more ethical products/practices in improving their environmental and social performances linked to cost reduction (Despeisse et al., 2012).

More and more manufacturing organizations in India have started to voluntarily adopt ISO 14001 as a method for continuous improvements to meet up the goals of sustainability. The significance of harmonizing ecological, societal and financial objectives in business organizations' development has culture and increasing consciousness on the development of SSCM practices and related activities (Nikolopoulou & Ierapetritou 2012). India government's dream project *'Make in India'* programme implementation will help to inspire the large domestic market through credit provision, land and labor reforms, infrastructure development, stimulating domestic enterprise development, and enhancing ease of doing business to stimulate domestic investment and FDI. The government is also looking for alternative policies and measures to make informal employment more productive and inclusive; focusing on employment of women, particularly in the unorganized sector; and placing greater emphasis on creating decent jobs (UNIC, New Delhi).

Moreover, while practicing sustainable procurement practices in the manufacturing domain, in few areas where the manufacturing systems need to be modified to adopt procured materials also. To be specific, the product nature of the material which has been supplied through the sustainable procurement system may lead to the change in the production routing, systems and machinery requirements. Eventually, there is an impact on the lead time as well. The processing time for the products also may vary in terms of achieving the right quality by adjusting the entire production parameters. This innovation in the sustainability system also leads to the modification in the lead time and overall order/service execution. There could be extended lead time to achieve quality or there is a possibility to get reduction in the lead time, as it may increase production efficiency due to the changes in the procurement system. Indian manufacturing firms may not enjoy that kind of high level of awareness on the procurement practices or standards which lead to sustainability in the overall, as they are completely dominated by the participation of Small Medium Enterprises (SMEs) on one side. These SMEs are not having any orientation towards the sustainability concept as they are completely dominated by the OEMs (Original Equipment Manufacturers) to produce products at the lower cost.

In some cases, these firms are tied up and directed by the clients to participate in the overall sustainability exercise, but unfortunately, it leads to the big question of sustaining those operations in the overall system, even when the OEMs leaves from the business. So, at the given situation, it is imperative for the industries to promote the sustainable concepts in a collective effort to get the attractive business to the Indian manufacturing domain. Moreover, several small players are highly encouraged to go for the third party certifications and accreditations for sustainability promotions to make them as a notable sustainable partner in the supply chains. However, these numbers are very low and highly disturbed by the larger cost structure with the inspections and varied standards required by the Inspection authorities. The vendors sustaining their operations according to these requirements are labeled as the preferable partners in the manufacturing or service supply chains. As a result, currently there is a huge role to be played by the Sustainable procurement in India in both government and private procurement systems. Its role in manufacturing operations will help to ensure the growth of Indian industries towards a sustainable future.

Challenges in Achieving Sustainable Procurement in India

Sustainable procurement has increasingly been an important perspective for business organizations. There are various challenges to achieve Sustainable procurement practices have been identified after review of available literature. Though the list is exhaustive, few of the important challenges have been discussed by relating their international trends with Indian context.

LACK OF CONSENT ABOUT SUSTAINABILITY

Business organizations have acknowledged the importance of sustainability as a key strategic aspiration to attain ecological, financial and societal benefits. But, there is no single universally accepted definition of sustainability (Giunipero et al., 2012). Due to non-availability of a universal definition or language for talk about sustainability, some business organizations describe it more broadly, whereas others originations describe it more narrowly (Berns et al., 2009). Indian companies still depend on foreign companies to establish the standards. They do not have their own developed standards especially on the meaning of Sustainability. In India, the major industrial production is based on SMEs and these SMEs are not having the uniform understanding on Sustainability.

Economic Uncertainty and Cost Issues

The status of supply management sustainability initiatives are motivated by the regulations/legalizations and top management programs, but are normally constrained by economic uncertainty and cost issues (Walker et al., 2008; Giunipero et al., 2012). Indian firms still look at sustainability as a cost escalation activity not a business development activity. It is a general feeling of insecurity that investment is not going to give returns resulting in the less interest towards the concept itself. The economic volatility in Indian conditions is also a big challenge for the organization to adopt certain sustainable concepts in their operations. Apart from financial and cost barriers, corruption and bribery also impede sustainable procurement practices in India (Sheth, 2008).

Absence of Strong Regulatory Framework

Sustainable procurement is on governmental agenda throughout the world, and trying to buy commodities, products and services in most economic ways by taking consideration of environment and public concerns (Walker et al., 2012). Regulatory liabilities means legally enforceable restrictions relating to use of the materials, products or services (Zsidisin & Ellram, 1999; Zsidisin, 2003; Lavastre et al., 2012). Regulatory boundaries may also play a key role in the economic boundaries for organizations (Sarkis, 2012). Complicated regulatory requirements may disturb the suppliers and negatively impacts on the business performances i.e. penalties and fines due to non-confirming of the regulatory requirements (Ageron et al., 2012). India does not have strong regulatory structure compared to their western counter parts and developed nations and few ministries along with some government regulatory bodies are taking care of green procurement legislation. This is with a considerable overlap in their accountability and oversight.

Conflict between Short Term and Long-Term Strategic Goals

Giunipero et al. (2012) suggested that a most of business organizations did not have a strong business case for sustainability. Therefore, organizations face trouble in planning beyond the short term strategic goals e.g. uncertainty to meet requirements of regulatory bodies and customers. In the Indian context, companies are having the profit as their prime orientation than growth strategy. This priority stops the organization to look the industry horizontally for the sustainable growth.

Resistance to Change in SC Network

The purchasing and supply management practices have been affected by sustainable development and therefore foster a change in the business organization's supply network (Crespin-Mazet & Dontenwill, 2012). This could be a big challenge in the Indian environment as well, as many of them are used to their outmoded systems and they do not want to adopt themselves to the newer systems.

Green Products Perceived to be More Expensive

Going green could be an expensive undertaking in short term basis. A Business organization that decides to commence green or sustainable practices will have to front the cost for a broad array of upgrades from more energy efficient machines to recycled printer paper (Giunipero et al., 2012). Up gradation in technologies and machinery will increase initial investments of business organizations. Still, there is a perception in Indian consumers that the green products and processes are high cost ones.

Lack of Managerial Support

Many organizations fail at gaining top management support for sustainable purchasing practices. Without support from top management to change the mindset of the organization, the purchasing department will never become aware of how big of an impact it can make on the bottom line (Palisi, 2012). Many of the SMEs are family run business and work on only the profit motive not on the long term expansion/ sustainable mode. Managers or Process owners are not completely aware of the benefits of the sustainable practices in the long term and they challenge these initiatives vehemently which is the big obstacle of sustainable practices.

Lack of Environmental Awareness among Society

There is a lack of strong environmental conservation or citizen groups in India (Muduli et al., 2013). Low demand of green/sustainable products is due to lack of environmental awareness among society (Mudgal et al., 2010; Luthra et al., 2011; Govindan et al., 2014). Due to lack of guidelines and lower environmental awareness in society, it is very difficult to promote green/sustainable procurement.

Lack of Tools and Information

Firsthand experience and specific implementation tools and information sources to make sustainable purchasing more effective are missing (Miemczyk et al., 2012). These tools are less customized to Indian situation constraints and there is a good scope for the development for the Index pertaining to the same. Poor Information & technological and resources capabilities of suppliers to transfer timely and relevant information to OEMs/buyers may be real problem to manage supply chain (Zsidisin, 2003; Handfield et al., 2006). Infrastructure and communication issues with suppliers may one of major risk to manage the supply chain (Lavastre et al., 2012). This is also one of the important challenges for Sustainable manufacturing in India, as they are not keen in developing their own tools and techniques/ systems for operations. Outsourcing the activity is very common in Indian industry which created gaps of communication.

Absence of a Single Uniform Approach for Defining Criteria/Norms

Berns et al. (2009) suggested that there is no a single consistent approach for defining criteria/norms about sustainability means to a company, which creates misalignment between more profit oriented and sustainability goals of business organizations. Sustainability is multi-dimensional approach with lacking of consensus among business organizations hinders sustainable purchasing efforts (Giunipero et al., 2012). It is a general perception that they fail to look beyond greener approach.

Lack of Information, Training and Accountability

Lack of information, training and accountability are major hurdles to integrate sustainable procurement practices. Inconsistent and incomplete information makes buyer's confusion, what to purchase and what not. Lack of information, training and accountability reduces the motivation of procurement managers to buy green/sustainable products (Chari & Chiriseri, 2014).

Fluctuation in Volume and Mix Requirement

Supplier unable to meet demand fluctuations in quantity and type for a component or service may be real risk to manage SC (Zsidisin, 2003). Problems with the supplier when the raw material has to be changed and his contract tenure left. Searching for a new supplier or old one being transformed takes time. This produces discontinuities in supply chain. A SC cannot be responsive to varying market trends and consumer preferences if the correct market signals cannot be predictable (Christopher & Lee, 2004). This situation is very prominent in the Indian conditions. The lead time and volume management are not followed strictly according to the world class standards available across all the industries.

Poor Management Policies

Poor management policies (supplier's management attitude, lack of vision and ability to foresee market changes) are main supply risk to manage the supply chain (Zsidisin, 2003). The challenges associated with changes of managerial decision or crashes in operational schedule planning or failure of control tools/methods comes under this (Giunipero & Eltantawy, 2004).

Discussions

Although the literature urges the need to investigate SP practices globally (Walker & Brammer, 2012), sustainable procurement studies within the Indian context has been neglected. In response, in this paper we have explored the literature on sustainable procurement practice by taking Indian perspective. Key drivers to promote sustainable procurement practices in India have been recognized through a review. This has helped us to identify current status and the need of sustainable procurement in India. The review endorses that sustainable procurement is a new and emerging area in India. Its role in sustainable manufacturing operations has also to be explored. Although, the literatures on theory and practice of sustainable procurement have been increasing rapidly but still many firms are exploring for the best strategies to incorporate sustainability principles in their supply chain activities. These practices are in very initial stage especially developing countries due to existence of many challenges. Therefore, in this chapter, an attempt has been made to identify major challenges to achieve sustainable procurement practices in India. Understanding of the challenges, problems, or hurdles in implementing sustainable procurement practices will help addressing them. Finally, possible future research directions from the gaps in the literature have been recognized.

Possible Future Research Directions

Though the sustainable procurement domain is very nascent research domain in India, some of the industries have understood the importance of the same. Nevertheless, in general sustainability is a misunderstood concept in India due to low awareness about their related benefits. Researchers need to focus on this direction. At the moment, the literature particularly associated with sustainable procurement practices in the initial phase of development, with related articles dealing mainly with theoretical discussions and anecdotal evidence. There is an awareness level on the benefits of using sustainable products to be increased among various sectors of society, which would help to achieve sustainable consumption and production practices. In the context of global warming, climate change problems and resources shortage problems are emerging at a very high rate in India; therefore, an urgent need is required to diversification towards sustainability in the supply chain i.e. design, procurement, production, marketing & distribution and consumption. It has reported that Indian public procurement system is plagued lack of transparency and accountability, which impede its effective and efficient functioning. India currently has no clear rules for regulating public private partnership project (Mansi, 2015). Therefore, defined regulations and frameworks for regulating public private partnership are required. In course of reviewing the literature, it has been found that very less work has been reported in the area of sustainable procurement practices implementation and performance evaluation. Sustainable procurement in public as well as private sector continues to be an important research agenda among the researchers. Also, a need has been identified to design central procurement mechanisms and disclosure practices. In course of reviewing the literature, it has been reported that strong regulatory framework and supportive policies to promote sustainable procurement practices are missing in Indian context. Therefore, strong legalizations and supportive government policies and should be designed such as incentives and subsidies etc. to the investors to encourage their investments in sustainable practices. In the future research, possible solutions to address those listed challenges, which would help to increase effectiveness of sustainable procurement practices in India, can also be studied in depth. Also, they may be validated and analyzed through empirical study/case studies. Sustainable procurement is not a simple procedure as its implementation raise new risks, practical paradox and challenges (Carter & Rogers, 2008). Future research may be focused on identifying and analyzing risks in achieving sustainable procurement practices. Multi Criteria Decision Making (MCDM) methods and techniques may be used to design decision making framework to understand the Critical Success Factors (CSFs) affecting sustainable procurement practices. Efficient sustainable procurement initiatives may play useful role in achieving the environmental, economic and social, performances towards sustainable development. Therefore, need arises to explore an emerging area. Role of consumer in sustainable procurement also needs to be explored.

It is suggested that in further research opportunities should not overlook the above mentioned points, only the purpose to increase effectiveness of sustainable procurement practices in public as well as in private sectors. Research must draw new theories/prepositions/efforts to increase the effectiveness of sustainable procurement practices in India.

Concluding Remarks

An increasing trend of research literature has been observed on green/sustainable procurement in recent years. It suggests need for maturation of this field with great scope for further research. It has been reported that at the moment, sustainable procurement are very initial stage of development in developing countries especially in India. The main aim of the paper is to increase awareness on the benefits of sustainable procurement practices in India. It gives a holistic view of sustainable purchasing with its role in sustainable manufacturing operations towards theory building. Research suggested that sustainable procurement in public as well as private sector in India will help to achieve resources saving and energy, reducing risks, enhance transparency and accountability in manufacturing operations to ensure the growth of Indian industries towards a sustainable future. This chapter represents an up to date knowledgebase researchers/SC practitioners/Procurement managers examining the concept of sustainable procurement and help managers/practitioners to boost their understanding towards sustainable procurement practices and benefits. It is believed that this chapter may act as a catalyst in compressing the learning curve with respect to effective implementation of sustainable procurement practices in India and may provide as a good base for broadening research in area of sustainable procurement in Indian perspective.

REFERENCES

Ageron, B., Gunasekaran, A., & Spalanzani, A. (2012). Sustainable supply management: An empirical study. *International Journal of Production Economics*, *140*(1), 168–182. doi:10.1016/j.ijpe.2011.04.007

Bai, C., & Sarkis, J. (2010). Integrating sustainability into supplier selection with grey system and rough set methodologies. *International Journal of Production Economics*, *124*(1), 252–264. doi:10.1016/j.ijpe.2009.11.023

Berns, M., Townend, A., Khayat, Z., Balagopal, B., Reeves, M., Hopkins, M. S., & Kruschwitz, N. (2009). The business of sustainability: What it means to managers now. *MIT Sloan Management Review*, *51*(1), 20–26.

Björklund, M. (2011). Influence from the business environment on environmental purchasing—Drivers and hinders of purchasing green transportation services. *Journal of Purchasing and Supply Management*, *17*(1), 11–22. doi:10.1016/j.pursup.2010.04.002

The emerging Indian economy. (2013). *Center for Strategic & International Studies*. Retrieved from http://csis.org/files/publication/130206_Khambatta _IndiaEmergEcon_Web.pdf

Chaabane, A., Ramudhin, A., & Paquet, M. (2012). Design of sustainable supply chains under the emission trading scheme. *International Journal of Production Economics*, *135*(1), 37–49. doi:10.1016/j.ijpe.2010.10.025

Chari, F., & Chiriseri, L. (2014). Barriers to sustainable procurement in Zimbabwe. *Greener Journal of Business and Management Studies*, *4*(1), 14–18.

Christopher, M., & Lee, H. (2004). Mitigating supply chain risk through improved confidence. *International Journal of Physical Distribution & Logistics Management*, *34*(5), 388–396. doi:10.1108/09600030410545436

Corbett, C. J., & Klassen, R. D. (2006). Extending the horizons: Environmental excellence as key to improving operations. *Manufacturing & Service Operations Management*, *8*(1), 5–22. doi:10.1287/msom.1060.0095

Corbett, C. J., & Kleindorfer, P. R. (2003). Environmental management and operations management: Introduction to the third special issue. *Production and Operations Management*, *12*(3), 287–289. doi:10.1111/j.1937-5956.2003.tb00203.x

Crespin-Mazet, F., & Dontenwill, E. (2012). Sustainable procurement: Building legitimacy in the supply network. *Journal of Purchasing and Supply Management*, *18*(4), 207–217. doi:10.1016/j.pursup.2012.01.002

Despeisse, M., Mbaye, F., Ball, P. D., & Levers, A. (2012). The emergence of sustainable manufacturing practices. *Production Planning and Control*, *23*(5), 354–376. doi:10.1080/09537287.2011.555425

Galpin, T., Whitttington, J. L., & Bell, G. (2015). Is your sustainability strategy sustainable? Creating a culture of sustainability. *Corporate Governance: The International Journal of Business in Society*, *15*(1), 1–17. doi:10.1108/CG-01-2013-0004

Giunipero, L. C., & Eltantawy, R. A. (2004). Securing the upstream supply chain: A risk management approach. *International Journal of Physical Distribution & Logistics Management*, *34*(9), 698–713. doi:10.1108/09600030410567478

Giunipero, L. C., Hooker, R. E., & Denslow, D. (2012). Purchasing and supply management sustainability: Drivers and barriers. *Journal of Purchasing and Supply Management*, *18*(4), 258–269. doi:10.1016/j.pursup.2012.06.003

Govindan, K., Kaliyan, M., Kannan, D., & Haq, A. N. (2014). Barriers analysis for green supply chain management implementation in Indian industries using analytic hierarchy process. *International Journal of Production Economics*, *147*, 555–568. doi:10.1016/j.ijpe.2013.08.018

Green, K., Morton, B., & New, S. (1998). Green purchasing and supply policies: Do they improve companies' environmental performance? *Supply Chain Management: An International Journal*, *3*(2), 89–95. doi:10.1108/13598549810215405

Consumer perceptions on green products. (2015). *Green Purchasing Network India-GPNI*. Retrieved from http://gpnindia.org/green-products/

Gupta, S., & Palsule-Desai, O. D. (2011). Sustainable supply chain management: Review and research opportunities. *IIMB Management Review*, *23*(4), 234–245. doi:10.1016/j.iimb.2011.09.002

Handfield, R. B., Krause, D. R., Scannell, T. V., & Monczka, R. M. (2000). Avoid the pitfalls in supplier development. *Mit Sloan Management Review*, 41(2).

Ho, L. W., Dickinson, N. M., & Chan, G. (2010, February). Green procurement in the Asian public sector and the Hong Kong private sector. In *Natural resources forum*. UK: Blackwell Publishing Ltd. doi:10.1111/j.1477-8947.2010.01274.x

Kumar, S. (2013).Strategy to implement sustainable public procurement in India. *Encore BCSD India, 7*(1). Retrieved from http://sustainableprocurement.blogspot.in/2013/02/strategy-to-implement-sustainable.html

Lavastre, O., Gunasekaran, A., & Spalanzani, A. (2012). Supply chain risk management in French companies. *Decision Support Systems, 52*(4), 828–838. doi:10.1016/j.dss.2011.11.017

Luthra, S., Garg, D., & Haleem, A. (2014a). Sustainable purchasing in Indian automobile industry: An ISM approach. In *K. Jain, L. Ganapathy, R. Gokhale, & R. Mukundan (Eds.), Manufacturing Excellence: Imperatives for Emerging Economies* (pp. 95–108). Mumbai: Excel India Publishers.

Luthra, S., Garg, D., & Haleem, A. (2014b). Critical success factors of green supply chain management for achieving sustainability in Indian automobile industry. *Production Planning & Control.*

Luthra, S., Kumar, V., Kumar, S., & Haleem, A. (2011). Barriers to implement green supply chain management in automobile industry using interpretive structural modeling technique: An Indian perspective. *Journal of Industrial Engineering and Management, 4*(2), 231–257. doi:10.3926/jiem.2011.v4n2.p231-257

Mansi, M. (2015). Sustainable procurement disclosure practices in central public sector enterprises: Evidence from India. *Journal of Purchasing and Supply Management, 21*(2), 125–137. doi:10.1016/j.pursup.2014.12.002

Meehan, J., & Bryde, D. (2011). Sustainable procurement practice. *Business Strategy and the Environment, 20*(2), 94–106. doi:10.1002/bse.678

Miemczyk, J., Johnsen, T. E., & Macquet, M. (2012). Sustainable purchasing and supply management: A structured literature review of definitions and measures at the dyad, chain and network levels. *Supply Chain Management: An International Journal, 17*(5), 478–496. doi:10.1108/13598541211258564

Mudgal, R. K., Shankar, R., Talib, P., & Raj, T. (2010). Modelling the barriers of green supply chain practices: An Indian perspective. *International Journal of Logistics Systems and Management, 7*(1), 81–107. doi:10.1504/IJLSM.2010.033891

Muduli, K., Govindan, K., Barve, A., Kannan, D., & Geng, Y. (2013b). Role of behavioral factors in green supply chain management implementation in Indian mining industries. *Resources, Conservation and Recycling, 76*, 50–60. doi:10.1016/j.resconrec.2013.03.006

Nikolopoulou, A., & Ierapetritou, M. G. (2012). Optimal design of sustainable chemical processes and supply chains: A review. *Computers & Chemical Engineering, 44*, 94–103. doi:10.1016/j.compchemeng.2012.05.006

Pagell, M., & Wu, Z. (2009). Building a more complete theory of sustainable supply chain management using case studies of 10 exemplars. *Journal of Supply Chain Management, 45*(2), 37–56. doi:10.1111/j.1745-493X.2009.03162.x

Palisi, F. X. (2012). Sustainable purchasing practices: An investigation into current industry awareness and practice [Master's Thesis College of Technology Paper 65]. Retrieved from http://docs.lib.purdue.edu/techmasters/65

Parikh, K. (2012). Sustainable development and low carbon growth strategy for India. *Energy*, *40*(1), 31–38. doi:10.1016/j.energy.2012.01.013

Preuss, L. (2007). Buying into our future: Sustainability initiatives in local government procurement. *Business Strategy and the Environment*, *16*(5), 354–365. doi:10.1002/bse.578

Rahbar, E., Shwu Shyan, T., & Abdul Wahi, N. (2011). Factors influencing the green purchase behavior of Penang environmental volunteers. *International Business Management*, *5*(1), 38–49. doi:10.3923/ibm.2011.38.49

Rondinelli, D. A., & Berry, M. A. (2000). Environmental citizenship in multinational corporations: Social responsibility and sustainable development. *European Management Journal*, *18*(1), 70–84. doi:10.1016/S0263-2373(99)00070-5

Sambasivan, M., & Fei, N. Y. (2008). Evaluation of critical success factors of implementation of ISO 14001 using analytic hierarchy process (AHP): A case study from Malaysia. *Journal of Cleaner Production*, *16*(13), 1424–1433. doi:10.1016/j.jclepro.2007.08.003

Sarkis, J. (2012). A boundaries and flows perspective of green supply chain management. *Supply Chain Management: An International Journal*, *17*(2), 202–216. doi:10.1108/13598541211212924

Seuring, S., & Müller, M. (2008). From a literature review to a conceptual framework for sustainable supply chain management. *Journal of Cleaner Production*, *16*(15), 1699–1710. doi:10.1016/j.jclepro.2008.04.020

Sheth, J. N. (2008). *Chindia rising: How India and China will benefit your business*. New Delhi: Tata McGraw Hill.

Spence, L., & Bourlakis, M. (2009). The evolution from corporate social responsibility to supply chain responsibility: The case of Waitrose. *Supply Chain Management: An International Journal*, *14*(4), 291–302. doi:10.1108/13598540910970126

Szekely, F., & Knirsch, M. (2005). Responsible leadership and corporate social responsibility: Metrics for sustainable performance. *European Management Journal*, *23*(6), 628–647. doi:10.1016/j.emj.2005.10.009

Tate, W. L., Ellram, L. M., & Dooley, K. J. (2012). Environmental purchasing and supplier management (EPSM): Theory and practice. *Journal of Purchasing and Supply Management*, *18*(3), 173–188. doi:10.1016/j.pursup.2012.07.001

Sustainable Public Procurement: Towards a low carbon economy (2007GL01). (2007). TERI, New Delhi.

Terrin Report. (2012). Retrieved from http://www.teriin.org/policybrief/docs/spp2013.pdf

Tranfield, D., Denyer, D., & Smart, P. (2003). Towards a methodology for developing evidence-informed management knowledge by means of systematic review. *British Journal of Management*, *14*(3), 207–222. doi:10.1111/1467-8551.00375

United Nations Information Centre (UNIC). (n. d.). New Delhi. Retrieved from http://www.unic.org.in/items/India_and_the_MDGs_small_web.pdf

Van Weele, A. J. (2005). *Purchasing & supply chain management: analysis, strategy, planning and practice*. London: Thomson learning.

Walker, H., & Brammer, S. (2009). Sustainable procurement in the United Kingdom public sector. *Supply Chain Management: An International Journal*, *14*(2), 128–137. doi:10.1108/13598540910941993

Walker, H., & Brammer, S. (2012). The relationship between sustainable procurement and e-procurement in the public sector. *International Journal of Production Economics*, *140*(1), 256–268. doi:10.1016/j.ijpe.2012.01.008

Walker, H., Di Sisto, L., & McBain, D. (2008). Drivers and barriers to environmental supply chain management practices: Lessons from the public and private sectors. *Journal of Purchasing and Supply Management*, *14*(1), 69–85. doi:10.1016/j.pursup.2008.01.007

Walker, H., Mayo, J., Brammer, S., Touboulic, A., & Lynch, J. (2012). Sustainable procurement: An international policy analysis of 30 OECD countries. *Proceedings of the5th International Public Procurement Conference,* August, Seattle, USA (p. 19).

Buy now don't pay later. (2004). *Welsh Procurement Initiative*. Retrieved from http://prp.wales.gov.uk/docs/prp/generalgoodsservices/130612buynowdontpaylater.pdf

Zhu, Q., & Sarkis, J. (2006). An inter-sectoral comparison of green supply chain management in China: Drivers and practices. *Journal of Cleaner Production*, *14*(5), 472–486. doi:10.1016/j.jclepro.2005.01.003

Zsidisin, G. A. (2003). Managerial perceptions of supply risk. *Journal of supply chain management, 39*(4), 14-26.

Zsidisin, G. A., & Ellram, L. (1999). Supply risk assessment analysis. *PRACTIX. Best Practices in Purchasing & Supply Chain Management*, *2*(4), 9–12.

Zsidisin, G. A., & Siferd, S. P. (2001). Environmental purchasing: A framework for theory development. *European Journal of Purchasing & Supply Management*, *7*(1), 61–73. doi:10.1016/S0969-7012(00)00007-1

Chapter 8
Supply Chain Social Sustainability and Manufacturing

Mani V
Indian Institute of Technology, India

Vinay Sharma
Indian Institute of Technology, India

Rajat Agrawal
Indian Institute of Technology, India

Kavitha T.N
Government Polytechnic, India

ABSTRACT

This chapter gives the overview of social sustainability in the supply chain of manufacturing industry. The author through literature survey identifies various definitions to social sustainability and suggest the approach to social sustainability problem in the supply chain. The readers are exposed with different social sustainability problems and what are the enablers and barriers to address such problem. In addition, the author identifies various social dimensions related to manufacturing supply chain and suggest as to how it can be addressed by various practices. Finally, the author identifies the research gaps in developing countries such as India, and suggests directions for future research. This is a vital contribution to the development of theory related to social sustainability in manufacturing Industries.

1. INTRODUCTION

The importance of sustainability in manufacturing has been talked over many decades. Formerly, the corporations were focusing predominately on fast paced production and innovative technologies. Now the focus has changed to sustainable manufacturing, because of increased awareness on limited resources, coupled with strict regulations, and corporate voluntary actions to meet the stakeholder's requirements. Sustainability is defined as "the development that meets the needs of today without compromising the ability to serve the future generations" WCED (1987) cited by Watkinson (2009). Due to climatic change and global warming, sustainability though it encompassed three dimensions such as environment, economic, and social, much emphasis was given on environmental dimension for over a decade. However, increasing stakeholder's awareness on safety, health, equity issues, and living conditions put back the

DOI: 10.4018/978-1-5225-0350-7.ch008

focus on socially sustainable manufacturing and business practices. As more companies commit to sustainability and CSR policies, there is increasing pressure to consider social impacts throughout the supply chain. United Nations Human development Index (HDI) measures the countries based on income disparity, education and mortality, and the countries are ranked accordingly. This acts as a motivating factor for these countries to improve upon social sustainability activities. Majority of the developing countries were rated poorly because of social issues.

There were measures such as Sarbanes-Oxley act, Dow Jone's economic index (1896) and financial reporting making mandatory requirements for the corporations to be economically sustainable. Similary, there has been tremendous research done on environmental aspects of sustainability in the supply chain. However, very little has been done in terms of social sustainability in the supply chain because of very complex human issues involved in it. On the other hand, the research to study the inter relationship among the economic, environment, and society are integral to the concept of sustainability. It is very essential to characterize their interactions to understand overall impact on future generations. Although, social sustainability was included and considered as an integral part of sustainability, the role played by social is rarely equal to the economic and environmental concerns. The Global Reporting Initiative (GRI, established in 1997) has reported that social performance occurs infrequently and inconsistently across organizations. This was again described by the recent major study by Western Australian Council of Social Services (WACOSS), who noted that, "while there has been tremendous work done relate to environment and economic aspects, the social has tended to fall off the sustainability agenda.

Further, social sustainability studies were so far, discussed and emphasized on legislative, health and safety rather on comprehensive way of focusing all the relevant issues. Especially, in manufacturing supply chain, studies on social issues are bit scant. In manufacturing, the sustainability is measured based on the entire value chain of the corporate. When it comes to sustainability in the supply chain, the supply chain comprises of many individual companies. The sustainability of the chain is depending on many standalone companies who are part of the corporate value chain. The impact of sustainability is not only restricted to the manufacturing companies but also its suppliers (Ashby et al., 2012). More recently many manufacturing companies develop their supplier capabilities in developing countries for low cost advantage. Invariably, the actions and behavior of these suppliers to the societies where they operate from, impacts the global companies in their own locations. We can notice these incidents, for example the recent development of unethical standards followed by Chinese manufacturing companies, the usage of melamine tainted milk and lead-tainted toys and toxic tooth paste, defective tires and fake medicines led to 24 million customer settlements by menu foods and many companies in China (Tybout & Roehm, 2009). These acts not only tarnished the image of the buyer, it also impacted on their financial performance. Yet another example of unethical actions such as "expired meat" supplied by one of the Mc Donald's largest meat supplier damaged the image of the company. This was resulted in suspension of burger products in China and US (CBS News, 2014). There were several other incidents of social issues related to supply chain. More recently, Wal-Mart's most preferred "shrimp" was procured from Thailand based supplier, whose facilities lack the basic living conditions, health and hygiene, and poor wages, etc. This was highlighted by many social organizations.

Especially, in developing countries, the issues related to social sustainability, still needs to be addressed. The recent noble price winner Khailash Satyarthi and his social efforts recovered and rehabilitated many child and bonded labourers from manufacturing set up in India. Further, TATA, a leading corporate brand was questioned by the World Bank, for the employee living conditions in their plantations. These are many social sustainability issues faced by the employees in the supply chain of manufacturing industries

(World Bank report, 2014). Most of the developing countries are still plagued with child labor, gender inequality, and poverty, health problems because of lack of awareness on social sustainability measures.

To understand the importance of social issues better, a practitioner from leading corporate house says [...] "In operations, people factor involve 85% of work and decisions, and the involvement of machines and automation is only 15%" [...] It means, the people play a vital role in the operations. This reinforces the need for the study, which would focus on social aspects in the supply chain which can lead to sustainability of the manufacturing.

In this chapter therefore, we attempt to bring together the available literature on social sustainability, emphasizing more on manufacturing. What enables the social sustainability? And the various impediments to social sustainability adoption were explored. In addition, the relevant social issues and how these issues should be approached and addressed were examined. So that social sustainability becomes reality in manufacturing.

2. SOCIAL SUSTAINABILITY

Many of the scholars referring to 'social sustainability' draw mainly upon the definition of sustainable development provided by the Brundtland Commision Report, *Our Common Future*: "Development that meets the needs of the present without compromising the ability of meeting future generation needs" (Bruntland, 1987). This definition appeals enormously because it holds the possibility of reconciling people needs with bio-physical environmental management goals through economic development. It captures the essence of a much larger construct that attempts to address both tangible and less tangible necessities for life which, in turn, was seen to depend on reviving growth; meeting essential needs for jobs, changing the quality of growth, food, energy, water and sanitation; conserving and enhancing the resource base; ensuring a sustainable level of population; reorienting technology and managing risk; merging the environment and economics in decision-making.

The Bruntland Commission report on sustainability much emphasized in terms of management of natural resources, but not in terms of society, people and culture (Chatterjee & Finger, 1994). Later research, advocated three important aspects of sustainability: environment, long-time focus on environment and equity (Pearce et al., 1989). Similar study conducted by a few researchers argued that sustainable development also 'embraces the need for equity' (Elkin et al., 1991). In urban planning circles sometimes sustainability means three E's: Environment, economy, and equity without much thought of whether equity alone can lead to social sustainability or not.

Many authors have also concurred with the view that for the enterprise to be sustainable, it needs to internalize social costs, grow and maintain capital stocks, foster democracy and enlarge peoples' choices and distribute property rights fairly (Gladwin et al., 1995). The United Nations Conference on Sustainable Development (UNCSD, 1998) has specified in Section 1 of Agenda 21(Economic Dimensions), promotion of economic growth; in Section 2(Social Dimensions),creation of productive employment and achieving equality; in Section3(Environmental Dimensions), reduction in use of natural resources and protection of the natural environment. United Nations conference on sustainable development set an agenda for social development. It also discussed about the inter relationship between environment and social requirements. Many researchers have even proved that if the basic human needs are not addressed, it will be very difficult to address the environment sustainability goals because, in developing economies people tend to spoil the natural resources for their own personal benefits. Later, many scholars did extensive research on social sustainability; notably, Sachs (1999) and Godschalk (2004), through their research titled "Social Sustainability and Whole Development" identified a number of essential elements including equitable income and access to goods, social homogeneity and services and employment. Sachs (1999) also pointed out the importance of 'cultural sustainability' which required balancing externally imposed change with continuity and development from within, and of 'political sustainability' based on human rights, democracy and effective institutional control.

Magis and Shinn (Cited in Dillard, 2008) review the literature and found the four emergent principles of social sustainability: human well-being, equality, democratic government, democratic society. Further they argue that these four principles create self-reinforcing virtuous cycle that facilitates progress towards even environmental sustainability.

Bansal (2005) described social equity as an important component of social sustainability and asserted that all members of society had equal access to resources and opportunities. In organisations, social equity term was extended to give the meaning of "fair and equitable treatment of employees" (Krause et al., 2009).

Later, Bramley and Power (2009) asserted that social sustainability was often equated with social cohesion, social capital and social inclusion. This also underlines the importance of focusing on higher needs such as access to necessary goods and services, and basic societal development. Further, research on social sustainability have emphasised providing intra and inter-generational equity and distribution of power, employment, resources, education, freedom, provision of basic infrastructure and services, justice, and access to influential decision-making fora (Bramley & Power, 2009; Mitlin & Satterthwaite, 1996; Källström & Ljung, 2005; Redclift, 2005).

More recently, Vallance (2011) in his research titled "What is Social Sustainability" described a three-fold literature schema comprising: (1) development sustainability addressing basic needs and creation of social capital, (2) bridge sustainability concerning behavioural changes to achieve bio and physical environmental goals, and (3) maintenance sustainability, which refers to social and cultural characteristics such as ways in which people actively resist or embrace change. On the basis of the studies mentioned above and their outcomes, it is clear that social sustainability is all about social issues and the ways in which they can be addressed which may lead to long term well being of the people in the societies. In the next section, an attempt is made to identify various definitions of social sustainability.

2.1 Defining Social Sustainability

In this section, we examine the issues involved in defining social sustainability for the purpose of creating common research agenda for society and organization usage. A working definition of social sustainability is provided below. However, it was much problematic in this task of arriving at single useful definition and instead, this section suggests various useful approaches that could be adopted. As we aware that social sustainability is related much with the people and human of the particular society. Though, the social sustainability can be an universal bigger agenda, it may bound to differ in terms of "where" and "what" because of different social issues that evolve over a period in societies. These issues are highly contextual. For examples, the social issues between western countries and other developing nations vary. Further to elaborate, the issues such as juvenile arrests, living wages, and recreations are not relevant in developing nations. In developing countries issues like poverty, health, safety and sanitation, hygiene and bonded and child labour becomes highly relevant. Though, all the issues irrespective of the society looks relevant under bigger umbrella, the time and priority of the issues makes it more contextual in nature.

Generally, there has been a strong focus on defining sustainability as a condition, and measuring it with a series of indicators. Social sustainability is known to be life enhancement feature within communities, and has the process that can achieve life enhancement feature. The following features are possible indicators to social sustainability:- equity in accessing the key services including health, education, transport, housing and recreation, and equity between generations: meaning that future generations should not be disadvantaged by the activities of current generations. Further, we will elaborate in detail about social dimensions and indicators under "social dimensions section". Many scholars through various researches have defined social sustainability.

To name a few, Lafferty and Langhelle (1999), Sharma and Ruud (2003) defined "social sustainability as a human code of conduct which needs to be achieved in an equitable, inclusive and prudent manner".

"Development that improves the quality of human life while living within the carrying capacity of supporting ecosystems" (WWF, UNEP, IUCN, 1991).

"The preservation of the planet and its ecosystems, society and its communities, for finest, equitable environmental and human health and wellbeing" (Hill, 1998).

Social Responsibility is defined as ''the continuing obligation by business to behave ethically and contribute to economic development while improving the quality of life of the workforce and their families, local community and society at large (WBCSD, 2000).

Polése and Stren (1999) define the social sustainability of a city as: 'Development and/or growth that is compatible with the harmonious evolution of civil society, fostering an environment conducive to the compatible cohabitation of culturally and socially diverse groups while at the same time encouraging social integration, with improvements in the quality of life for all segments of the population."

Western Australian Council of Social Services (WACOSS, 2000)

Social sustainability occurs when the formal and informal processes, systems, structures and relationships actively support the capacity of current and future generations to create healthy and liveable communities. Socially sustainable communities are equitable, diverse, connected and democratic and provide a good quality of life.

Robert Putnam (2001) describes "social capital" is an important aspect of social sustainability, just as environmentalists describe "natural capital" to environmental sustainability, and amenities.

According to Dillard et al. (2008) a working definition of social sustainability developed through seminars and workshops over a period: that social sustainability should be understood as

1. The process that generates social health and well being now and in the future and
2. Those social institutions that facilitate economic and environmental sustainability now and for future

A most appropriate definition is provided by Harris and Goodman (2001): A socially sustainable system must achieve fairness in distribution and opportunity and adequate provision of social services including health, education and gender equity and political accountability and participation

Wackernagel (2001) pointed out the importance of social sustainability by saying: "human health aspects are essential for the well-being of a society, but they should not be confused with environmental sustainability"

Sustainability is essentially an anthropocentric concept of inter- and intra-generational justice (Grunwald 2001),

Littig and Griessler (2005) defined as "claiming the right to a dignified life to humans" The Organization for Economic Cooperation and Development (OECD) views social sustainability in two diverse dimensions: human and social. The human dimension refers to man's skills, abilities knowledge and talents, which determines an individual's performance in the labor market and is also relevant to the process of economic growth (OECD Insights, 2008) cited in Tracey & Anne (2008). Social capital facilitates collective actions that are beneficial to the society (Hobbs, 2000; Milward, 2003).

With the extensive review of literature and understanding, social sustainability was defined and touched upon various social issues that falls under the ambit of "humanness". Hence, for the purpose of understanding, a working definition can be written:

"Social sustainability is all about addressing social issues of the people of today in the societies that can create a sustainable future for future generations"

Since this study focuses on supply chain related social issues, we further limit ourselves to various social aspects that are specifically related to supply chain that are discussed in the next section.

3. SOCIAL SUSTAINABILITY IN THE SUPPLY CHAIN

Social sustainability in the supply chain of manufacturing industry is of paramount importance because of increased stake holders awareness on not only "where" the product is made but also "how" and "at what" conditions the products are produced. Though, this creates somewhat undesirable situations for the corporate, due to involvement of cost, the benefits of human capital and its importance for strategic advantage of the corporations was well researched and explored. Hence, the corporate in the modern times, understand social issues and attempt to find the different ways of integrating the human and social aspects into the supply chain. Therefore, social sustainability in the supply chain talks about how people can be protected from the products and processes that affects the human being safety, health and well being. Before we define social sustainability in the supply chain, we need to understand "Who", "What" and "How", it means who are the people being targeted? What issues should be addressed? And how should we address these issues? (Wood, 1991).

The answer to the first question: stakeholder theory that addresses all the stakeholders in the supply chain including suppliers, employees, society, NGO's, and customers (Freeman, 1984, 2004; Stieb, 2009; Campbell, 2007). Yet another research described the similar view: social responsibility was seen as an organization's ability to manage stakeholders (Waddock & Bodwell, 2004; Clarkson, 1995). Further, resource based view supports that human resources, which can't be easily imitated or copied by the competitors, gives the company a strategic advantage in the market (Barney, 1981). Second, many scholars have identified various issues related to people in the manufacturing supply chain. For instance, human rights and labour working conditions were identified by Emmelhainz and Adams (1999). The study in US emphasised on health and safety, diversity, philanthropy, human rights and ethics in the supply chain (Carter & Jennings (2002a, 2002b, 2004). This research also proved that ethics can't be used as social sustainability dimension in the supply chain.

Dyllick and Hockerts (2002) provided one representative definition of corporate sustainability: "meeting the needs of a firm's direct and indirect stakeholders (such as shareholders, employees, clients, pressure groups, communities, etc.), without compromising its ability to meet the needs of future stakeholders as well". Further, they emphasised the importance of human capital and societal capital in social sustainability of the corporations. Similar research by many scholars insisted on various human issues such as safety, health, diversity, working conditions, child and bonded labor, and poverty in the supply chain (Whooley, 2004; Carter, 2005; Maloni & Brown, 2006; Kortelainen, 2008; Vachon & Mao, 2008; Andersen & Skjoett-Larsen, 2009). Further, some scholars have identified criteria of fair and equitable treatment, human rights, child and forced labour, training, employment and wages as issues in the supply chain (Pagell & Wu, 2009). There was another research that linked social issues of supply chain and country's economic performance through life cycle analysis (Hutchins & Sutherland, 2008).

Labuschagne et al. (2005) revealed practices related to poverty alleviation, administering justice, human rights, and welfare of all employees in the supply chain. Leire and Mont (2010) explained social issues such as reduction of unemployment, protecting employee health and safety, ensuring equal treatment and preventing social exclusion and how these dimensions were linked and incorporated in the supply chain

Finally, many authors described through various studies, in which how social issues can be addressed in the supply chain. One way to address is, incorporating these parameters in corporate's buying process known as socially responsible buying (SRB). Socially responsible buying (SRB) refers to the incorporation of social issues in purchase decisions advocated by organizational stakeholders (McWilliams & Siegel, 2001).

Further, Carter and Jennings(2002a, 2002b, 2004), Ciliberti et al. (2008) described the role of purchasing managers in the area of social responsibility within the supply chain and how the enactment of these roles could lead to improved trust and supplier commitment through purchasing social responsibility (PSR) and logistics social responsibility (LSR). Later research by Leire and Mont (2010) indicated how social criteria could be used to monitor suppliers and ensure their compliance. Bai and Sarkis (2010), Mani et al. (2014) have demonstrated usage of social parameters, in supplier selection. It means, incorporating social parameters in evaluation and selection of suppliers. Goworek (2011) in his research emphasized on socially responsible and environmentally sustainable sourcing practices and how it could be applied to clothing industries. Socially responsible supplier development (SRSD) and its importance in measuring the buyer's sustainability, and the relationship between social sustainability and supplier development efforts were established in China (see figure 1) (Lu et al., 2012).

With above analysis, we possibly can form the working definition of social sustainability in the supply chain as "how the corporate manage and addresses the various social issues such as equity, health, safety, wages, working conditions, philanthropy and labour rights across the supply chain that will help the corporate to maintain, and sustain the human resources that in turn leads to long time survival of the corporation"

4. ANTECEDENTS AND IMPEDIMENTS TO SOCIAL SUSTAINABILITY

Another aspect of social sustainability in the supply chain deals with what enables social sustainability adoption. Many renowned scholars have identified ways and means of integrating social parameters into the supply chain. For example, one researcher asserted that socially responsible organizational buying fully depended on two factors: first, the skillful policy entrepreneurs who institute new policies with zeal for social wellness in their policy decisions, and second, the organizational context in which decisions are made (Drumwright, 1994). There have been many other studies focusing on how social sustainability can be achieved if corporates enact their roles responsibly. Drumwright (1994), Clarkson (1995), Strong (1997), McWilliams (2001), Campbell (2007) along with Ehrgott et al. (2011) argue that various parameters such as customer requirements, stakeholder requirements, employee requirements,

Figure 1. Supply chain stages in manufacturing and relevant social issues
Source: Mani et al. (2015a)

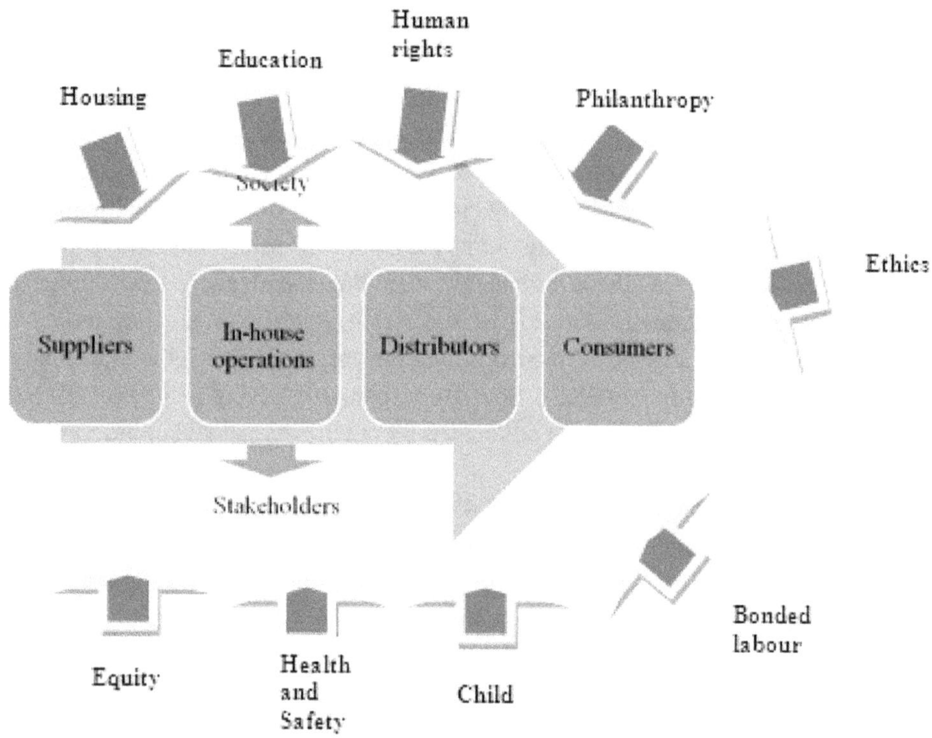

skilled policy entrepreneurs, economic status of corporates, and public and private regulation influence social sustainability adoption(Table-1). Mani et al (2015b) have established many new enablers for social sustainability adoption in emerging economies, which include awareness on social sustainability, social organisations pressure, pressure from employee union, investor's pressure, and financial status of the company.

There are many enablers and barriers which can lead to social sustainability adoption in the manufacturing supply chain and these are explained in the preceding section with the help of table-1.

4.1 Antecedents to Social Sustainability

a) Awareness of Social Sustainability

Awareness of social activities can act as a key parameter to accelerate social sustainability (Haugh & Talwar, 2010; Maloutas, 2003). If the manufacturing companies are aware of the social sustainability parameters and their relevance to social capital development, this may help in adoption of social sustainability in the supply chain.

Table 1. Antecedents to social sustainability adoption in the supply chain

S. NO	Antecedents	References
1	People oriented organisation culture	Carter and Jennings(2000, 2002)
2	Top Management leadership or skillful policy entrepreneur	Drumwright (1994),Carter and Jennings(2000, 2002), Ehrgott(2011), Mani et al. (2015b)
3	Government regulation	Carter and Jennings(2000, 2002), Aguilera et al. (2007), Campbell (2007), Ehrgott(2011)Mani et al. (2015b)
4	Customer pressures	Carter and Jennings(2000, 2002), Ehrgott(2011), Mani et al. (2015b)
5	Individual values	Carter and Jennings(2000, 2002)
6	Pressure from employee union	Aguilera et al. (2007), Campbell (2007), Mani et al. (2015b)
7	Direct incentives	Mani et al. (2015b)
8	Awareness on social sustainability	Mani et al. (2015b)
9	Stakeholders pressure	Aguilera et al. (2007), Campbell (2007), Mani et al. (2015b)
10	Competitive pressures	Campbell (2007), Mani et al. (2015b)
11	Social concerns	Mani et al. (2015b)
12	International certifications	Mani et al. (2015b)
13	Shareholders pressure	Mani et al. (2015b)
14	Ability to spend(Financial status)	Campbell (2007), Mani et al. (2015b)
15	Social organisations pressure	Campbell (2007),Mani et al. (2015b)

b) Competitive Pressure

In modern economy, the companies are always very much conscious about the competitor's actions that may potentially strengthens the corporate capabilities. The social measures adopted by competitors may force the other corporations to adopt similar measures for sustainability (Lamming & Hampson, 1996; Sarkis, 2003; Zhu & Sarkis, 2006; Carter *et al.*, 2000; Rao & Holt, 2005). For example, corporations with inimitable resources and capabilities by the competitors, are necessary for gaining sustainable advantage. Further, Rao and Holt (2005) have established relationship between sustainable measures and competitive advantages.

c) People Oriented Organizational Culture

Organization culture can be defined as set of values, believes, assumptions, and ways of thinking and shared by organizations members, and taught to the new members of the organization (Barney, 1986; Chapman & Jehn, 1994; Weiner, 1988). This means that organizational culture influence the work behavior. Similar research has established, relationship between culture and organizations success. Further it was used by Carter and Jennings and proved that it helps in social sustainability adoption (Carter & Jennings, 2000).

d) Customer Requirements

In manufacturing, many of the companies are expected to comply with customer requirements because of dependence and the economic relationship between these organizations. Many studies have proved that these requirements and their necessity in developing socially sustainable suppliers and their impact on buyers image and financial performance (Christmann, 2004; Drumwright, 1994; Sen & Bhattacharya, 2001; Wilkes, 2008; Lu & Lee, 2012; Carter & Dresner, 2001; Klassen & Vachon, 2003; Sarkis, 2003; Zhu & Sarkis, 2006).

e) Direct Incentives

Government direct and indirect initiatives in the form of tax benefits and other subsidies may drive the social sustainability (Cerin & Karlson, 2002; Hilson & Nayee, 2002; Ostrom & Wynne, 1993). For examples extending property rights acted as incentives for corporations to practice sustainable measures (Cerin & Karlson, 2002). Governments other incentives such as permitting individual harvest, territorial rights, and price ecosystem services, coupled with public research and monitoring promote sustainability in fishing industry (Campbell, 2007).

f) Corporations Financial Ability

The company's economic health and the level of competition determine the probability of adopting socially responsive behaviors. Further, small companies first tend to focus on economic sustainability and as they move on in the ladder, they extend their focus on social sustainable measures (Campbell, 2007; AlKhidir & Zailani, 2009; Luthra *et al.*, 2011).

g) International Certifications

Many international certifications that include 26000, ILO, OHSAS 18001, AA 1000, ISEA 1999 for social standards exerts pressure on the corporations for social compliance. This will in turn impact on corporate image and operational performance (Viscusi, 1979; Handfield et al., 2002; Gonzalez & Sarkis, 2008; Zuckerman, 2000). Some voluntary certifications and labeling have helped in achieving social sustainability in coffee industry (Reynolds *et al.*, 2007).

h) Investors Pressure

Shareholders and investor's actions influence the adoption of social sustainability measures in the organization (Tagesson *et al.*, 2009; Buchholz & Rosenthal, 2005; Laplume *et al.*, 2008). The investors positive role in bringing sustainability reporting was identified by many case studies (Tagesson *et al.*, 2009).

i) Easy to Implement without Resistance

Top executives 'resistance to change' attitude acts as barriers to social sustainability adoption. On the contrary, attitude to accept the change may speed up social responsiveness in the organizations (Shrivastava, 1995).

j) Pressure from Employee Unions

Employee unions, in the bigger organizations act as catalyst in negotiation and demanding for social sustainability measures. With bargaining power, these unions exerts pressures on the top management to address social issues and adopt corrective measures on top priority. (Sen & bhattacharya, 2001; Hansson et al., 2003; Nortan, 2014).

k) Regulatory Compliance (Government Regulation)

Regulatory authorities or Government regulations as part of external stakeholder group certify the corporations to streamline and regulate for social measures. Government –protector of the society bring in many policies to regulate and orient the companies in their socially responsive activities(Green et al., 1996; Gonzalez & Sarkis (2008); Beamon, 1999; Raynolds et al., 2007) .Further, There is a positive relationship between government regulation and social sustainability adoption in corporations(Carter & Jennings, 2000; Ehrgott et al., 2011) .

l) Skillful Policy Entrepreneur

The top executive's role in adopting social sustainability was established by many researchers. Drumwright (1994) in his research advocated the policy entrepreneur who invests to institute his policies can drive the social sustainability (Banerjee *et al.*, 2003; Buckholz, 1991). Yet In another research, Carter *et al.* (2007) mapped the social network within an organization to demonstrate how a low level manager effectively championed and drove a safety related supplier management project.

m) Social Concern

The organization's social concern can have a positive role in social sustainability adoption. Many organizations, through several reporting standards display their social concern (Laufer, 2003; Walker & Preuss, 2008).

n) Social Organizations Pressure

Social groups encourage corporations to implement social sustainability practices. Walker and Preuss (2008) identified the importance of social groups for corporate performance. For example, non-governmental organizations (NGO's) that works on child and bonded labor, living conditions, gender equity, health and hygiene. In India, a few of these popular social organizations are Samhita, Ruchika, Sulabh International, Sambhav, and the Parikrma foundation.

o) Stakeholder's Pressure (Others)

Other than internal stakeholders, the external stakeholders, i.e., Media influence the organizations in social sustainability adoption (Jones, 1995; Post et al., 2002; Maignan, 2002; Sharma & Vredenburg, 1998; Sarkis et al., 2010). A research conducted by Jones (2005) on the 18 top corporations showed how

Table 2.

S. No	Impediments	References
1	Lack of awareness	Haugh & Talwar (2010), Maloutas (2003)
2	Lack of competitive pressure	Lamming and Hampson, (1996), Carter *et al.,* (2000), Rao and Holt, (2005), Campbell(2007), Rao and Holt (2005)
3	Lack of customers' requirements	Ehrgott *et al.* (2010), Christmann (2004), Drumwright (1994), Sen and Bhattacharya (2001), Wilkes(2008), Carter and Dresner (2001), Klassen and Vachon, (2003), Sarkis(2003), Zhu and Sarkis (2006).
4	Lack of investor pressure	Tagesson *et al.(*2009)
5	Lack of pressure from employee unions	Bhattacharya et al.(2012), Hansson *et al.(*2003), Haugh and Talwar (2010)Campbell (2007), Aguilera et al. (2007)
6	Lack of pressure from regulatory authorities (Government regulation	Campbell (2007), Green *et al.* (1996), Gonzalez *et al.* (2008), Beamon(1999), Raynolds *et al (*2007).
7	Lack of interest from skillful policy entrepreneur	Banerjee, 2003; Buckholz, 1991). Ehrgott *et al.* (2010)
8	Lack of social concern	Laufer (2003) Walker and Preuss *(*2008)
9	Lack of pressure from social organizations	Walker and Preuss (2008) Campbell (2007), Bartley (2003)
10	Lack of stakeholder's pressure	Jones (1995), Post *et al.,(*2002), Maignan and Ralston (2002), Sharma and Vredenburg (1998), Laplume *et al.(*2008)

companies practiced social sustainability measures due to stakeholders pressure to feature their ratings in Dow Jones Sustainability Index(DJSI)

4.2 Impediments to Social Sustainability

Many scholars while discussing the barriers in the supply chain of manufacturing Industries identified various aspects leads non-adoption and act as bottle neck for the social sustainability. For instance, Bebbington and Dillard (2008) described that there are four factors contribute relative neglect of social sustainability by the organizations. First, the corporate's economic aims, commonly known as increasing the wealth of the shareholders. Second, the issues related to social sustainability has its modern origin in environmental sustainability issues. Thus social sustainability sometimes seen as related to developing world issues, such as access to water, education and health care etc. The third aspect of social sustainability is the social issues are often seen to be the responsibility of the state or society. Four, social sustainability appears to bring in severe challenges in understanding and communication than the other forms of sustainability. Further, research on social sustainability established many other barriers as described in Table 2.

a) Lack of Awareness

Lack of awareness of social sustainability measures may hinder sustainable development in the supply chain. On the other hand, awareness of relevant social activities can accelerate social sustainability (Haugh & Talwar, 2010). Maloutas (2003) emphasises the need for creating awareness of social sustainability as the lack of cognizance in adopting social measures is impeding the movement in the direction of social sustainability.

b) Lack of Competitive Pressure

A moderate level of competition among the corporations tends to encourage them to behave in socially responsible ways. On the other hand, high or low competitions tend to make them socially less responsible (Campbell, 2007). As some organisations utilize their socially sustainable image for overall strategic advantage, other organisations are compelled to adopt socially sustainable practices. In a global environment, competitive pressure plays a vital role in the supply chain. In social sustainability, the companies will have the resources and capabilities which cannot be easily imitated by their competitors who, therefore, will be forced to own and maintain such resources of their own, helping the cause of long-term sustainability (Lamming & Hampson, 1996; Carter *et al.,* 2000; Rao & Holt, 2005). Rao and Holt (2005) identify the linkages between sustainability measures and competitive advantage and also establish the relationship between sustainability and financial performance of a firm.

c) Lack of Customer Requirements

The customers or consumers can exert a strong pressure on the firms to adopt social sustainability measures through CSR if their requirements necessitate such measures (Aguilera *et al.,* 2007). In fact, the customers' requirements with regard to social dimensions may go a long way in achieving social sustainability. These demands not only mandate the suppliers to adopt sustainable practices in the short-term but also guarantee long-term sustainability in the supply chain. The growing requirement of the development of a socially responsible supplier base and its impact on the company image and financial operations has been affirmed by several researchers (Ehrgott *et al.,* 2010; Christmann, 2004; Drumwright, 1994; Sen & Bhattacharya, 2001; Wilkes, 2008; Carter & Dresner, 2001; Klassen & Vachon, 2003; Sarkis, 2003; Zhu & Sarkis, 2006).

d) Lack of Investor Pressure

The role of investor pressure in social sustainability reporting has been brought out by many case studies. As shown by these studies, privately owned companies do not disclose as much social information as the state owned organisations (Tagesson *et al.,* 2009).

e) Lack of Pressure from Employee Unions

Employee unions are the members of the internal stakeholder group, involved in pursuing the corporates to behave in a socially responsible way through negotiations and dialogue (Campbell, 2007). Aguilera *et al.* (2007) affirm that employee unions have an important role in compelling the corporates to be more socially sustainable. They play a role of catalyst in successfully bargaining for social measures (Bhattacharya et al., 2012; Hansson *et al,* 2003; Haugh & Talwar, 2010).

f) Lack of Pressure from Regulatory Authorities (Government Regulation)

Government regulations through law enactment and law enforcement can bring in social change in the corporations (Aguilera *et al.,* 2007). The government, as part of the external stakeholder group in the developing countries, monitors the corporations in their social sustainability measures (Campbell, 2007;

Green *et al.*, 1996; Gonzalez *et al.*, 2008; Beamon, 1999; Raynolds *et al.*, 2007). There is a positive relationship between government pressure and implementation of social sustainability measures (Ehrgott *et al.*, 2010).

g) Lack of Interest of Skillful Policy Entrepreneurs

As Drumwright (1994) affirms, socially responsible buying happens because of two key understandings: skillful policy entrepreneur who uses his resources to institute policies; and the organisational setting in which these policies operate (Banerjee, 2003; Buckholz, 1991). Ehrgott *et al.* (2010), in their research, mapped the social network within an organisation to show how a handler at the lower ranks effectively championed and drove to a safety related supplier management project.

h) Lack of Social Concern

Organisations' social concern can have a positive impact on the social sustainability measures (Laufer, 2003; Walker & Preuss, 2008). Organisations increasingly display their social measures through many reporting standards carried out to show the corporate's social concern to the stakeholders and shareholders of the organisation. Lack of social concern may hinder social sustainability in the corporations.

i) Lack of Pressure from Social Organisations

The NGOs and other social organisations that monitor the corporations for socially responsible behaviour may tend to increase social sustainability (Campbell, 2007; Bartley, 2003). Moreover, socially responsible corporate behaviour tends to overlap with the extent to which the firms belonging to the industrial and corporate systems are amenable to dialogue. The NGOs, through boycotts, campaigns and multi-party dialogues, can influence the corporates to adopt socially responsible practices (Aguilera *et al.*, 2007). Walker and Preuss (2008), in their research on sustainability, emphasise the importance of the social groups' pressure on corporate performance. In fact, the presence of active social groups warrants the adoption of social sustainability practices by the corporates.

j) Lack of Stakeholder Pressure

Internal stakeholders through their direct strategic decisions, and external stakeholders exercising their collective voice, can influence the acceptance of socially sustainable practices in the corporations (Aguilera *et al.*, 2007). On the other hand, the media stakeholders can also influence the organisations in their adoption of social sustainability criteria (Jones, 1995; Post *et al.*, 2002; Maignan & Ralston, 2002; Sharma & Vredenburg, 1998; Laplume *et al.*, 2008). An inquiry conducted by Jones (2005) in 18 top corporations revealed how companies were adopting social sustainability measures pushed primarily by the stakeholder pressure, in addition to their ratings in DJSI to be specifically recognised for their name and report.

5. Social Sustainability Dimensions

One of the fundamental questions that must be answered if we are to move forward with social sustainability in manufacturing: What are social sustainability dimensions? How do we measure these aspects in the supply chain? A consensus has not been reached on the measurements of social sustainability because it is highly contextual in nature. Further, sustainability indicators were largely developed by consultancy firms (such as Elkington's own company) who extend services to large companies by helping them to arrive at indicator system for their corporate sustainable reporting. Social sustainability is very difficult to quantify than economic and environmental impact and it is the most neglected part of triple bottom line reporting. All the social sustainability indicators are too general to be useful and specific indicators needs to be developed for particular state and industry, further down to company level. This means that the usefulness of these indicators to academic discourse in particular context of social sustainability is questionable (table 3) (Mckenzie, 2004).

Social sustainability dimensions were first proposed by United Nations Division for Sustainable Development (UNDSD, 2001) and widely accepted by the researchers, academicians and practitioners. This framework describes themes and sub themes and under sub themes-indicators etc. The first of these framework proposed by United Nations Division for Sustainable Development measures progress of social sustainability indicators that are tied to Millennium Development Goals. According to this framework, indicators first classified by the primary dimensions (Social, environmental, and economic) then by their theme (for example: education), and finally by sub theme(for example., literacy). The broad themes under social dimensions are: equity, health, education, housing, security, and population. Under each sub

Table 3. Comprehensive list of various social sustainability indicators adopted by many scholars across the globe.

Social Sustainability Indicators	
Human Rights Income disparity Justice to all Juvenile arrests Life companion Living Wage Objective basic Needs Objective equal opportunities Objective social resources Abuse and disciplinary practices Cultural Sustainability Employment Employment gender ratio Environmental reports Equity Ethics Gender Ratio (Employment) Harassment Housing Security Social homogeneity Subjective basic needs Subjective equal opportunities Subjective social resources	Philanthropy Population Poverty Quality Education Recreation Reliable and sufficient social system Safety Fair distribution of burden between generations Financial position Freedom of association Gender Discrimination Wages Working Hours Access to goods and services Affordable housing Bonded labour Child abuse or Neglect Child labour Creation of social capital Cultural diversity Health Health Insurance

Source: Mani et al. (2014a)

theme, at least one or as many as three quantifiable indicators associated with it. These indicators very useful and are shared to many countries by Human Development Report Office of UNEP.

Further, there is another frame work called "The sustainability Reporting Guidelines" developed in conjunction with many stakeholders, including business houses, non-profit organizations, investor organization, and trade unions. This was primarily the effort of United Nations Environment Programme (UNEP) in association with Coalition for Environmentally Responsible Economics (CERES). Though this framework is very similar to that of UNDSD, it is different in many areas. These framework differentiates "core" and "additional indicators –core indicators are more relevant to reporting organizations and stakeholders, others are named as additional indicators. However, majority of these indicators are subjective and qualitative. Within 'sustainable reporting guidelines' many of these indicators are informative however, it is very difficult to incorporate such qualitative guidelines in decision making tools required by business. Carter and Jennings (2000, 2002, 2004) through his research in US has proposed many social dimensions, including diversity, human rights, philanthropy, safety, in manufacturing Industries.

Labuschangne et al. (2005) has contributed to the improvement of sustainability framework by proposing additional indicators that incorporates criteria from number of key frameworks, namely: United Nations Commission on sustainable development, Global Reporting Initiative, Sustainability metrics of institution of chemical engineers, and Wuppertal sustainability indicators. They have concluded that

Table 4. List of criteria and sub criteria and Indicators proposed by UNDSD and adopted by Mani et al. (2014) in India.

Criteria	Sub Criteria	Indicators
Equity	• Poverty • Gender equality • Gender discrimination	• Percentage of population living below poverty level • Unemployment rate • Gini index of income inequality • Percentage ratio of average female wage vs male wage • The practice of granting or denying rights of privileges based on gender
Health & Safety	• Nutritional Status • Mortality • Sanitation • Drinking Water • Healthcare Delivery • Safety measures provided by corporates	• Children nutritional status • Mortality rate under 5 years old • Population who have adequate sewage facilities (%) • Access to safe drinking water • Access to primary health care facilities • Safety measures undertaken by the company to protect the employees
Wages	• Wage standard	• Standard of wages against man hours spent
Education	• Literacy Level • Education Level	• Adult literacy ratio • Access to primary education up to 5th STD
Philanthropy	• Donation to Temples & MUTT • Donations to NGO's • Donations to Schools & Colleges	• Money donated to religious organizations • Money given to Non Governmental Organization (NGO's) • Money given to schools and colleges
Human Rights	• Human Rights- Environment	• How human rights are protected (right to associate, speak)
Child and Bonded Labour	• Child labour • Bonded labour	• The percentage of child labour employed • Percentage of bonded labour employed
Housing	• Living Conditions	• Human and floor ratio • Hygienic conditions
Ethics	• Supplier Ethics • Ethical Environment	• Ethical values adopted by the supplier • Ethical environment

Source: Mani et al. (2014a)

the indicators within these frameworks are not "adequate" as a result eight supplement guidelines were reviewed to identify additional indicators. Similarly, Supply chain counsel has proposed, "SCOR people" framework through Supply Chain Operation Reference (SCOR) model for manufacturing supply chain. However, sustainability reporting framework, proposed by Global Reporting Initiative (GRI), A non-governmental organization is widely accepted among corporate circles (Table 4).

Though, there were many adequate dimensions and measures for social sustainability in manufacturing industries as we discussed earlier. The dimensions and indicators cannot be generalized because "social-ness" evolves over a period of time in the society and social issues are highly contextual. For example the social issues in developing countries are different from that of developed economies.

Various researches conducted by many renowned scholars have identified specific social sustainability dimensions that can be used as a tool for measuring social sustainability in manufacturing industries.

For examples Sachs (1999) identified dimensions such as access to goods, employment, cultural sustainability, human rights in New York City. Further, The research done in US manufacturing companies brought many social dimensions such as, diversity philanthropy, safety, and human rights (Carter &

Table 5. List of various social sustainability dimensions used by the researchers in different countries.

S.No	Authors(References)	Social Dimensions used in Various Researches
1	Poist(1989)	Housing, diversity, Philanthropy, safety, health and hygiene, hunger, education
2	Carter et al. (1999)	Procurement from minority supplier
3	Emmelhainz and Adams(1999)	Human rights, labour practices.
4	Carter & Jennings (2000)	Diversity, philanthropy, safety, human rights, ethics
5	UNDSD(2001)	Equity, health and hygiene
6	Carter & Jennings (2002)	Diversity, philanthropy, safety, human rights
7	Whooley(2004)	Diversity, safety, equity, health and hygiene.
8	Carter (2005)	Diversity, philanthropy, safety, equity.
9	Hens (2005)	Equity, poverty, health and hygiene
10	Maloni and Brown (2006)	Safety, human rights, labour practices, health and hygiene
11	Kortelainen(2008)	Human rights, child and bonded labour, poverty.
12	Hutchins and Sutherland (2008)	Philanthropy, health, safety, equity.
13	Vachon and Mao (2008)	Equity, labour, poverty
14	Andersen (2009)	Safety, human rights, labour practices, child and bonded labour, procurement from minority suppliers, education.
15	Pagell and Wu(2009)	Safety, child and bonded labour.
16	Wong et al. (2010)	Safety, human rights, labour practices, child and bonded labour.
17	Leire and Mont (2010)	Safety, equity, employment, health and hygiene
18	Yakovleva et al. (2012)	Equity, employment, wages.
19	Gopalakrishnan (2012)	Diversity, safety, human rights, health, hygiene, ethics
20	Lu et al. (2012)	Ethics.
21	Husgafvel et al. (2014)	Health and safety, lobour practices, human rights, wages, education and training.

Source: Mani et al. (2015b)

Jennings, 2000, 2002, 2005). In this research, however the authors could not establish direct relationship between social sustainability and financial performance. Instead, they could find trust and learning played a vital role between suppliers and buyers, invariably that led to cost reduction for the buying organization. Similar study, by Gupta (2007) uncovered human rights and labour standards in Indian manufacturing companies. Yet another research in German, the social dimensions such as objective basic needs, subjective basic needs, objective social resources, subjective social resources, cultural diversity, and social security system were established and ranked based on priority (Omann & Spangenberg, 2002).

There were other social indicators including financial position, life companion, and recreation have been used by Kallstorm and Ljung (2005). In Canada, another empirical study done by Vachon and Mao (2008) suggested the dimensions such as labour, employment, and gender. Hutchins and Sutherland (2008) in their research in US established relationship between social sustainability indicators such as wages, health, safety, equity, quality education, and philanthropy and the financial position of the country and through life cycle analysis. In various other studies conducted by many researchers in countries such as UK, Sweden, and London, China established many social performance indicators including ethics, environment, employee compensation, philanthropy, health and safety, child labour, bonded labour, education and housing facilities, etc. as listed in Table-1 (Yakovleva et al., 2012; Kogg & Mont, 2012; Schlossberg & Zimmerman, 2010; Andersen et al., 2009; Lu & Lee, 2012; Husgafvel et al., 2014).

However, caution needs to be applied while selecting the social indicators in a particular country. This might look different view than that of conventional way of thinking that all the societies are getting merged because of globalization and liberalized economy. And we get an illusion that all the social issues may be same. However, it is not true; this can be well noticed by the research done by Mani et al. (2015c) in two different global supply chains and their social sustainability practices in India and other continents. In which, the author emphasized that the social issues and indicators to measure such issues are different from one continent to other (Table 5).

CONCLUSION

A sincere attempt was made to expose the concepts and literature pertaining to social sustainability and how it was practiced in manufacturing supply chain under various domains. Based on the above literature and analysis, there were many dimensions and indicators, related to social sustainability in the supply chain, emerged. Social sustainability as a subject, it was very difficult to define due to the fact that it is highly contextual in nature further, it is a growing phenomenon, as the researcher keeps dwelling around "socialness" under various domains in different geographies. When people apply social sustainability in any domain, bit caution, and care need to be maintained as the indicators tend to vary from one domain to another. We conclude that social sustainability parameters can't be generalized and still there is researchable gaps exist in different geographies. For example, research on social sustainability in developing countries can bring in some visible and actionable results that could help in decision making and policies.

REFERENCES

Aguilera, R. V., Rupp, D. E., Williams, C. A., & Ganapathi, J. (2007). Putting the S back in corporate social responsibility: A multilevel theory of social change in organizations. *Academy of Management Review*, *32*(3), 836–863. doi:10.5465/AMR.2007.25275678

Al Khidir, T., & Zailani, S. (2009). Going green in supply chain towards environmental sustainability. *Global Journal of Environmental Research*, *3*(3), 246–251.

Andersen, M., & Skjoett-Larsen, T. (2009). Corporate social responsibility in global supply chains. *Supply Chain Management: An International Journal*, *142*, 75–86.

Ashby, A., Leat, M., & Hudson-Smith, M. (2012). Making connections: A review of supply chain management and sustainability literature. *Supply Chain Management: An International Journal*, *175*, 497–516.

Bai, C., & Sarkis, J. (2010). Integrating sustainability into supplier selection with grey system and rough set methodologies. *International Journal of Production Economics*, *124*(1), 252–264. doi:10.1016/j.ijpe.2009.11.023

Banerjee, S. B. (2003). Who sustains whose development? Sustainable development and the reinvention of nature. *Organization Studies*, *24*(1), 143–180. doi:10.1177/0170840603024001341

Bansal, P. (2005). Evolving sustainably: A longitudinal study of corporate sustainable development. *Strategic Management Journal*, *26*(3), 197–218. doi:10.1002/smj.441

Barney, J. B. (1986). Organizational culture: Can it be a source of sustained competitive advantage? *Academy of Management Review*, *11*(3), 656–665. doi:10.5465/AMR.1986.4306261

Bartley, T. (2003). Certifying forests and factories: States, social movements, and the rise of private regulation in the apparel and forest products fields. *Politics & Society*, *31*(3), 433–464. doi:10.1177/0032329203254863

Beamon, B. M. (1999). Measuring supply chain performance. *International Journal of Operations & Production Management*, *19*(3), 275–292. doi:10.1108/01443579910249714

Bhattacharya, C. B., Sen, S., & Korschun, D. (2012). Using corporate social responsibility to win the war for talent. *MIT. Sloan Management Review*, 49(2).

Bramley, G., & Power, S. (2009). Urban form and social sustainability: The role of density and housing type. *Environment and Planning. B, Planning & Design*, *361*(1), 30–48. doi:10.1068/b33129

Bruntland, G. 1987. Our common future: The world commission on environment and development.

Buchholz, R. A. (1991). Corporate responsibility and the good society: From economics to ecology. *Business Horizons*, *34*(4), 19–31. doi:10.1016/0007-6813(91)90003-E

Buchholz, R. A., & Rosenthal, S. B. (2005). Toward a contemporary conceptual framework for stakeholder theory. *Journal of Business Ethics*, *58*(1-3), 137–148. doi:10.1007/s10551-005-1393-8

Campbell, J. L. (2007). Why would corporations behave in socially responsible ways? An institutional theory of corporate social responsibility. *Academy of Management Review*, *32*(3), 946–967. doi:10.5465/AMR.2007.25275684

Carter, C.R., Jennings, M.M. (2002a). Logistics social responsibility: an integrative framework. *Journal of business logistics*, 23(1), 145-180.

Carter, C. R., & Jennings, M. M. (2002b). Social responsibility and supply chain relationships. *Transportation Research Part E, Logistics and Transportation Review*, *38*(1), 37–52. doi:10.1016/S1366-5545(01)00008-4

Carter, C. R. (2005). Purchasing social responsibility and firm performance: The key mediating roles of organizational learning and supplier performance. *International Journal of Physical Distribution & Logistics Management*, *35*(3), 177–194. doi:10.1108/09600030510594567

Carter, C. R., & Dresner, M. (2001). Purchasing's Role in Environmental Management: Cross-Functional Development of Grounded Theory. *Journal of Supply Chain Management*, *37*(3), 12–27. doi:10.1111/j.1745-493X.2001.tb00102.x

Carter, C. R., & Jennings, M. M. (2004). The role of purchasing in corporate social responsibility: a structural equation analysis. *Journal of business Logistics, 25*(1), 145-186.

Carter, C. R., Auskalnis, R., & Ketchum, C. (1999). 'Purchasing from minority business Enterprises: A cross-industry comparison of best practices. *Journal of Supply Chain Management*, *35*(1), 28–32. doi:10.1111/j.1745-493X.1999.tb00053.x

Carter, P. L., Carter, J. R., Monczka, R. M., Slaight, T. H., & Swan, A. J. (2000). The Future of Purchasing and Supply: A Ten-Year Forecast1. *Journal of Supply Chain Management*, *36*(1), 14–26. doi:10.1111/j.1745-493X.2000.tb00066.x

Shrimp sold at Walmart Costco tied to slave labour. (2014, June 11). CBS News report. Retrieved from http://www.cbsnews.com/news/shrimpsold-at-walmart-costco-tied-to-slave-labor/

Cerin, P., & Karlson, L. (2002). Business incentives for sustainability: A property rights approach. *Ecological Economics*, *40*(1), 13–22. doi:10.1016/S0921-8009(01)00275-0

Chapman, J. A., & Jehn, K. A. (1994). Assessing the relationship between industry characteristics and organizational culture: How different can you be? *Academy of Management Journal*, *37*(3), 522–553. doi:10.2307/256699

Chatterjee, P., & Finger, M. (1994). *The earth brokers: Power, politics, and world development*. US: Taylor & Francis.

Christmann, P. (2004). Multinational companies and the natural environment: Determinants of global environmental policy. *Academy of Management Journal*, *47*(5), 747–760. doi:10.2307/20159616

Ciliberti, F., Pontrandolfo, P., & Scozzi, B. (2008). Logistics social responsibility: Standard adoption and practices in Italian companies. *International Journal of Production Economics*, *113*(1), 88–106. doi:10.1016/j.ijpe.2007.02.049

Clarkson, M. E. (1995). A stakeholder framework for analyzing and evaluating corporate social performance. *Academy of Management Review*, *20*(1), 92–117.

Dillard, J., Dujon, V., & King, M. C. (Eds.). (2008). *Understanding the social dimension of sustainability*. Routledge.

Drumwright, M. E. (1994). Socially responsible organizational buying: Environmental concern as a noneconomic buying criterion. *Journal of Marketing*, *58*(3), 1–19. doi:10.2307/1252307

Dyllick, T., & Hockerts, K. (2002). Beyond the business case for corporate sustainability. *Business Strategy and the Environment*, *11*(2), 130–141. doi:10.1002/bse.323

Ehrgott, M., Reimann, F., Kaufmann, L., & Carter, C. R. (2011). Social sustainability in selecting emerging economy suppliers. *Journal of Business Ethics*, *98*(1), 99–119. doi:10.1007/s10551-010-0537-7

Elkin, T., McLaren, D., & Hillman, M. (1991). *Reviving the City*. London: Friends of the Earth.

Emmelhainz, M. A., & Adams, R. J. (1999). The apparel industry response to “sweatshop” concerns: A review and analysis of codes of conduct. *Journal of Supply Chain Management*, *35*(2), 51–57. doi:10.1111/j.1745-493X.1999.tb00062.x

Freeman, R. E. (1984). *Strategic Management: A Stakeholder Approach*. Englewood Cliffs, NJ: Prentice-Hall.

Gladwin, T. N., Krause, T. S., & Kennelly, J. J. (1995). Beyond eco efficiency: Towards socially sustainable business. *Sustainable Development*, *31*(1), 35–43. doi:10.1002/sd.3460030105

Sustainability reporting guidelines. (2000Global Reporting Initiative. Amsterdam: Global Reporting Initiative.

Sustainability reporting guidelines. (2002Global Reporting Initiative. Boston, MA: Global Reporting Initiative.

Godschalk, D. R. (2004). Land use planning challenges: Coping with conflicts in visions of sustainable development and livable communities. *Journal of the American Planning Association*, *70*(1), 5–13. doi:10.1080/01944360408976334

Gonzalez, P., Sarkis, J., & Adenso-Diaz, B. (2008). Environmental management system certification and its influence on corporate practices: Evidence from the automotive industry. *International Journal of Operations & Production Management*, *28*(11), 1021–1041. doi:10.1108/01443570810910179

Gopalakrishnan, K., Yusuf, Y. Y., Musa, A., Abubakar, T., & Ambursa, H. M. (2012). Sustainable supply chain management: A case study of British Aerospace (BAe) Systems. *International Journal of Production Economics*, *140*(1), 193–203. doi:10.1016/j.ijpe.2012.01.003

Goworek, H. (2011). Social and environmental sustainability in the clothing industry: A case study of a fair trade retailer. *Social Responsibility Journal*, *7*(1), 74–86. doi:10.1108/17471111111114558

Green, K., Morton, B., & New, S. (1996). Purchasing and environmental management: Interactions, policies and opportunities. *Business Strategy and the Environment*, *5*(3), 188–197. doi:10.1002/(SICI)1099-0836(199609)5:3<188::AID-BSE60>3.0.CO;2-P

Handfield, R., Walton, S. V., Sroufe, R., & Melnyk, S. A. (2002). Applying environmental criteria to supplier assessment: A study in the application of the Analytical Hierarchy Process. *European Journal of Operational Research, 141*(1), 70–87. doi:10.1016/S0377-2217(01)00261-2

Hansson, J., Backlund, F., & Lycke, L. (2003). Managing commitment: Increasing the odds for successful implementation of TQM, TPM or RCM. *International Journal of Quality & Reliability Management, 20*(9), 993–1008. doi:10.1108/02656710310500815

Haugh, H. M., & Talwar, A. (2010). How do corporations embed sustainability across the organization? *Academy of Management Learning & Education, 9*(3), 384–396. doi:10.5465/AMLE.2010.53791822

Hens, L., & Nath, B. (2005). *'The world summit on sustainable development*. Dordrecht: Springer. doi:10.1007/1-4020-3653-1

Hilson, G., & Nayee, V. (2002). Environmental management system implementation in the mining industry: A key to achieving cleaner production. *International Journal of Mineral Processing, 64*(1), 19–41. doi:10.1016/S0301-7516(01)00071-0

Husgafvel, R., Pajunen, N., Virtanen, K., Paavola, I. L., Päällysaho, M., Inkinen, V., ... & Ekroos, A. (2014). Social sustainability performance indicators–experiences from process industry. *International Journal of Sustainable Engineering*.

Hutchins, M. J., & Sutherland, J. W. (2008). An exploration of measures of social sustainability and their application to supply chain decisions. *Journal of Cleaner Production, 16*(15), 1688–1698. doi:10.1016/j.jclepro.2008.06.001

World conservation strategy: living resource conservation for sustainable development. (1980). IUCN/UNEP/WWF, Gland, Switzerland.

J., Elkington, (1999). Triple bottom line revolution: reporting for the third millennium. *Australian CPA*, l6(9), 75.

Jones, R. (2005). Finding sources of brand value: Developing a stakeholder model of brand equity. *The Journal of Brand Management, 13*(1), 10–32. doi:10.1057/palgrave.bm.2540243

Jones, T. (1995). Instrumental stakeholder theory: A synthesis of ethics and economics. *Academy of Management Review, 20*(2), 404–437.

Källström, H.N., Ljung, M. (2005). Social sustainability and collaborative learning. *AMBIO: A Journal of the Human Environment*; 34(4), 376-382.

Klassen, R. D., & Vachon, S. (2003). Collaboration and evaluation in the supply chain: The impact on plant-level environmental investment. *Production and Operations Management, 12*(3), 336–352. doi:10.1111/j.1937-5956.2003.tb00207.x

Kogg, B., & Mont, O. (2012). Environmental and social responsibility in supply chains: The practice of choice and inter-organisational management. *Ecological Economics, 83*, 154–163. doi:10.1016/j.ecolecon.2011.08.023

Kortelainen, M. (2008a). Dynamic environmental performance analysis: A Malmquist index approach. *Ecological Economics*, *64*(4), 701–715. doi:10.1016/j.ecolecon.2007.08.001

Kortelainen, K. (2008b). 'Global supply chains and social requirements: Case studies of labour condition auditing in the People's Republic of China'. *Business Strategy and the Environment*, *17*(7), 431–443. doi:10.1002/bse.634

Krause, D. R., Vachon, S., & Klassen, R. D. (2009). Special topic forum on sustainable supply chain management: introduction and reflections on the role of purchasing management. *Journal of Supply Chain Management*, *45*(4), 18–25. doi:10.1111/j.1745-493X.2009.03173.x

Labuschagne, C., Brent, A. C., & van Erck, R. P. G. (2005). Assessing the sustainability performances of industries. *Journal of Cleaner Production*, *13*(3), 73–85.

Lafferty, W. M., & Langhelle, O. (Eds.), (1999). Towards sustainable development: on the goals of development-and the conditions of sustainability. Macmillan.

Lamming, R., & Hampson, J. (1996). The environment as a supply chain management issue. *British Journal of Management*, *7*(Suppl. 1), S45–S62. doi:10.1111/j.1467-8551.1996.tb00147.x

Laplume, A. O., Sonpar, K., & Litz, R. A. (2008). Stakeholder theory: Reviewing a theory that moves us. *Journal of Management*, *34*(6), 1152–1189. doi:10.1177/0149206308324322

Laufer, W. S. (2003). Social accountability and corporate greenwashing. *Journal of Business Ethics*, *43*(3), 253–261. doi:10.1023/A:1022962719299

Barron, L., & Gauntlet, E. (2002, March 3–6). WACOSS 'Housing and Sustainable Communities Indicators Project.' Paper presented at 'Sustaining our Communities' International Local Agenda 21 Conference, Adelaide (p 4). Retrieved from http://www.adelaide.sa.gov.au/soc/pdf/barron_gauntlett.pdf

Leire, C., & Mont, O. (2010). The implementation of socially responsible purchasing. *Corporate Social Responsibility and Environmental Management*, *17*(1), 27–39.

Littig, B., & Griessler, E. (2005). Social sustainability: A catchword between political pragmatism and social theory. *International Journal of Sustainable Development*, *8*(1/2), 65–79. doi:10.1504/IJSD.2005.007375

Lu, R. X., Lee, P. K., & Cheng, T. C. E. (2012). Socially responsible supplier development: Construct development and measurement validation. *International Journal of Production Economics*, *140*(1), 160–167. doi:10.1016/j.ijpe.2012.01.032

Luthra, S., Kumar, V., Kumar, S., & Haleem, A. (2011). Barriers to implement green supply chain management in automobile industry using interpretive structural modeling technique: An Indian perspective. *Journal of Industrial Engineering and Management*, *4*(2), 231–257. doi:10.3926/jiem.2011.v4n2.p231-257

Polése, M., & Stren, R. (1999). *The social sustainability of cities: diversity and the management of change*. Toronto: University of Toronto Press.

Maignan, I., & Ralston, D. A. (2002). Corporate social responsibility in Europe and the US: Insights from businesses self-presentations. *Journal of International Business Studies*, *33*(3), 497–514. doi:10.1057/palgrave.jibs.8491028

Maloni, M. J., & Brown, M. E. (2006). Corporate social responsibility in the supply chain: An application in the food industry. *Journal of Business Ethics*, *68*(1), 35–52. doi:10.1007/s10551-006-9038-0

Maloutas, T. (2003). Promoting social sustainability: The case of Athens. *City*, *7*(2), 167–181. doi:10.1080/1360481032000136732

Mani, V., Agrawal, R., & Sharma, V. (2014a). Supplier selection using social sustainability: AHP based approach in India. *International Strategic Management Review*, *2*(2), 98–112. doi:10.1016/j.ism.2014.10.003

Mani, V., Agrawal, R., Sharma, V. (2015d). Supply Chain Social Sustainability: A Comparative Case Analysis in Indian Manufacturing Industries. *Procedia-Social and behavioural science.*

Mani, V., Agrawal, R., Sharma, V. (2015c). Social sustainability in the supply chain: Analysis of enablers. *Management Research Review*, 38(9).

Mani, V., Agrawal, R., Sharma, V. (2015b). Social Sustainability Practices in the Supply Chain of Indian Manufacturing Industries. *International Journal of Automation and Logistics*, 1(2).

McKenzie, S. (2004). *Social sustainability: towards some definitions. Magill.* Hawke Research Institute, University of South Australia.

McWilliams, A., & Siegel, D. (2001). Corporate social responsibility: A theory of the firm perspective. *Academy of Management Review*, *26*(1), 117–127.

Mitlin, D., & Satterthwaite, D. (1996). Chapter One Sustainable Development and Cities. *Sustainability. Environment and Urbanization.*

Linking economy, society and environment. (2008). *OECDInsights*. Retrieved from http://www.oecd.org/insights/sustainabledevelopmentlinkingeconomysocietyenvironment.htm

Ostrom, E., Schroeder, L., & Wynne, S. (1993). *Institutional Incentives and Sustainable Development: Infrastructure Policies in Perspective*. Westview Press.

Pagell M, Wu Z *(*2009). Building a more complete theory of sustainable supply chain management using case studies of 10 exemplars. *Journal of supply chain management*, 45(2), 37-56.

Pearce, D. W., Markandya, A., & Barbier, E. (1989). *Blueprint for a green economy*. Earthscan.

Poist, R. F. (1989). Evolution of conceptual approaches to the design of logistics systems: A sequel. *Transportation Journal*, *28*(3), 35–39.

Post, J. E., Preston, L. E., & Sachs, S. (2002a). Managing the extended enterprise: The new stakeholder view. *California Management Review*, *45*(1), 6–28. doi:10.2307/41166151

Post, J. E., Preston, L. E., & Sachs, S. (2002b). *Redefining the corporation: Stakeholder management and organizational wealth*. Stanford University Press.

Putnam, R. (2001). Social capital: Measurement and consequences. *Canadian Journal of Policy Research, 2*(1), 41–51.

Rao, P., & Holt, D. (2005). Do green supply chains lead to competitiveness and economic performance*? International Journal of Operations & Production Management, 25*(9), 898–916. doi:10.1108/01443570510613956

Raynolds, L. T., Murray, D., & Heller, A. (2007). Regulating sustainability in the coffee sector: A comparative analysis of third-party environmental and social certification initiatives. *Agriculture and Human Values, 24*(2), 147–163. doi:10.1007/s10460-006-9047-8

Redclift, M. (2005). Sustainable development (1987–2005): An oxymoron comes of age. *Sustainable Development, 13*(4), 212–227. doi:10.1002/sd.281

Sachs, I. (1999). Social sustainability and whole development: exploring the dimensions of sustainable development. *Sustainability and the social sciences: a cross-disciplinary approach to integrating environmental considerations into theoretical reorientation* (pp. 25-36).

Sarkis, J. (2003). A strategic decision framework for green supply chain management. *Journal of Cleaner Production, 11*(4), 397–409. doi:10.1016/S0959-6526(02)00062-8

Sarkis, J., Gonzalez-Torre, P., & Adenso-Diaz, B. (2010). Stakeholder pressure and the adoption of environmental practices: The mediating effect of training. *Journal of Operations Management, 28*(2), 163–176. doi:10.1016/j.jom.2009.10.001

Schlossberg, M., & Zimmerman, A. (2003). Developing state wide indices of environmental, economic, and social sustainability: a look at Oregon and the Oregon Benchmarks. *Local Environment, 86*, 641–660.

Sen, S., & Bhattacharya, C. B. (2001). Does doing good always lead to doing better? Consumer reactions to corporate social responsibility. *JMR, Journal of Marketing Research, 38*(2), 225–243. doi:10.1509/jmkr.38.2.225.18838

Sharma, S., & Ruud, A. (2003). On the path to sustainability: Integrating social dimensions into the research and practice of environmental management. *Business Strategy and the Environment, 12*(4), 205–214. doi:10.1002/bse.366

Sharma, S., & Vredenburg, H. (1998). Proactive corporate environmental strategy and the development of competitively valuable organizational capabilities. *Strategic Management Journal, 19*(8), 729–753. doi:10.1002/(SICI)1097-0266(199808)19:8<729::AID-SMJ967>3.0.CO;2-4

Shrivastava, P. (1995). Environmental technologies and competitive advantage. *Strategic Management Journal, 16*(Suppl. 1), 183–200. doi:10.1002/smj.4250160923

Spangenberg, J. H., Omann, I., & Hinterberger, F. (2002). Sustainable growth criteria: Minimum benchmarks and scenarios for employment and the environment. *Ecological Economics, 42*(3), 429–443. doi:10.1016/S0921-8009(02)00125-8

Stieb, J. A. (2009). Assessing Freeman's stakeholder theory. *Journal of Business Ethics, 87*(3), 401–414. doi:10.1007/s10551-008-9928-4

Strong, C. (1997). The problems of translating fair trade principles into consumer purchase behaviour. *Marketing Intelligence & Planning*, *15*(1), 32–37. doi:10.1108/02634509710155642

Tagesson, T., Blank, V., Broberg, P., & Collin, S. O. (2009). What explains the extent and content of social and environmental disclosures on corporate websites: A study of social and environmental reporting in Swedish listed corporations. *Corporate Social Responsibility and Environmental Management*, *16*(6), 352–364. doi:10.1002/csr.194

Tracey, S., & Anne, B. (2008). *OECD Insights Sustainable Development Linking Economy, Society, Environment: Linking Economy, Society, Environment*. OECD Publishing.

Tybout, A. M., & Roehm, M. (2009). Let the response fit the scandal. *Harvard Business Review*, *87*(12), 82–88.

United Nations Commission on Sustainable Development. Measuring Changes in Consumption and Production Patterns. (1998UNCSD, Division for Sustainable Development, Department of Economic and Social Affairs, New York.

Indicators of sustainable development: guidelines and methodologies. (2001). UNDSD. Retrieved from http://www.un.org/esa/sustdev

Vachon, S., & Mao, Z. (2008). Linking supply chain strength to sustainable development: A country-level analysis. *Journal of Cleaner Production*, *16*(15), 1552–1560. doi:10.1016/j.jclepro.2008.04.012

Vallance, S., Perkins, H. C., & Dixon, J. E. (2011). What is social sustainability? A clarification of concepts. *Geoforum*, *42*(3), 342–348. doi:10.1016/j.geoforum.2011.01.002

Viscusi, W. (1986). Kip (1979).The Impact of Occupational Safety and Health Regulation. *The Bell Journal of Economics*, *10*(1), 117–140. doi:10.2307/3003322

Wackernagel, M. (2001). Shortcomings of the Environmental Sustainability Index. Retrieved from http://www.anti-Iomborg.com/ESI%20critique.rtf

Waddock, S., & Bodwell, C. (2004). Managing responsibility: What can be learned from the quality movement? *California Management Review*, *471*(1), 25–37. doi:10.2307/41166285

Walker, H., & Preuss, L. (2008). Fostering sustainability through sourcing from small businesses: Public sector perspectives. *Journal of Cleaner Production*, *16*(15), 1600–1609. doi:10.1016/j.jclepro.2008.04.014

Watkinson, J. (2009). WCED (1987)–Copenhagen (2009) Will We Ever Take the Environment Seriously. Retrieve from myliberaldemocratpoliticalramblings.wordpress.com/2009/08/27lwced-1987-copenhagen-2009-wiIl-we-ever-take-theenvironment-seriously

Model of social sustainability, (n. d.). Western Australian Council of Social Services (WACSS). Retrieved from http://www.wacoss.org.au/downloads/socialsustainable.pdf

Whooley, N. (2004). Social Responsibility in Europe. *PWC.com*. Retrieved from www.pwc.com/extweb/newcolth.nsf/0/503508DDA107A61885256F35005C1E35S

Wiener, Y. (1988). Forms of value system: A focus on organizational effectiveness and cultural change and maintenance. *Academy of Management Review*, *13*(4), 534–545.

Wilkes, R. S. (2008). Low linolenic soybeans and beyond. *Lipid Technology*, *20*(12), 277–279. doi:10.1002/lite.200800072

Wong, S., Liu, J., & Culpan, T. (2010, June 3). Why Apple and others are nervous about Foxconn. *Bloomberg Business Week*.

Wood, D. J. (1991). Corporate social performance revisited. *Academy of Management Review*, *16*(4), 691–718.

The more things change: The World Bank, TATA and enduring abuses on India's tea plantations. (2014). *World Bank report*. Retrieved from https://web.law.columbia.edu/sites/default/files/microsites/human-rights-institute/files/tea_report_final_draft-smallpdf.pdf

Our common future. (1987). World Commission on Environment and Development. Oxford University Press, Oxford.

Yakovleva, N., Sarkis, J., & Sloan, T. (2012). Sustainable benchmarking of supply chains: The case of the food industry. *International Journal of Production Research*, *50*(5), 1297–1317. doi:10.1080/00207543.2011.571926

Zhu, Q., & Sarkis, J. (2006). An inter-sectoral comparison of green supply chain management in China: Drivers and practices. *Journal of Cleaner Production*, *14*(5), 472–486. doi:10.1016/j.jclepro.2005.01.003

Zuckerman, A. (2000). Quality Assurance through ISO 9000. *School Administrator*, *57*(6), 12–16.

Chapter 9
Sustainable Business Development by Responding to Climate Change:
A Case of the Tata Group

Sujata Mukherjee
SVKM's NMIMS, India

Arunavo Mukerjee
Tata Clean Tech Capital Ltd., India

ABSTRACT

The UN Intergovernmental Panel on Climate Change Report (2007) concluded that the globally averaged net effect of human activities since 1750 has been one of the sources of global warming. The Stern Report (2007) corroborates this statement and states that anthropogenic CO2 influences the climate and stresses that the cost of mitigating against climate change is significantly lower than the cost of climate change. The Tata group companies have been actively seeking out experiences of other global companies to develop an effective action plan against climate change. The present paper seeks to review the role of the Tata group in addressing and abating the climate change. It further looks at the various Tata group companies like Tata Chemicals Limited, Tata Steel Europe, Tata Communications and Tata Motors and their ways to stay the course towards sustainable development

INTRODUCTION

The environment, despite its vast spread and bounty, has only limited cache of resources at its disposal. Climate has been changing naturally at its own pace since the beginning of evolution but presently, it has gained momentum due to inadvertent anthropogenic (manmade) disturbances (Khan et.al., 2009). The UN Intergovernmental Panel on Climate Change (IPCC) Report (2007) concluded that the globally averaged net effect of human activities since 1750 has been one of the sources of global warming. Emissions of greenhouse gases due to human activities like extensive use of fossil fuels, deforestations

DOI: 10.4018/978-1-5225-0350-7.ch009

etc. have altered GHG in the atmosphere in ways that are expected to affect the climate. The equally well-known Stern Report (2007) corroborates this statement by declaring that we are now beyond the discussion on whether anthropogenic CO_2 influences the climate and stresses that the cost of mitigating against climate change is significantly lower than the cost of climate change.

India is concerned about climate change since this phenomenon might have substantial adverse impact on agriculture, submergence of coastal areas and increased frequency of extreme events. As a developing economy, India cannot afford the risks and economic backlashes which industrialised nation can. With 27.5 per cent of the population still living below the poverty line, reducing vulnerability to the impacts of climate change is essential (India Climate Change Portal, 2011). The IPCC report (2007) also suggests that India will experience the greatest increase in energy and GHG emissions in the world if it has to sustain the growth rate. To add, the energy needs (nearly 70% of the electricity supply comes from coal fuelled power plants) of the country are met by burning fossil fuels. The International Energy Agency predicts that India will become the third largest emitter of GHGs by as early as 2015.

Climate change presents a real and tangible threat to the world today. It has impacted not just public policy and legislations but also changing market mechanics and behaviour patterns of consumers in key market. It has created a challenging business environment where it leads companies to relook at their businesses. According to Ernst and Young (2008) companies are addressing the issues of climate change on account of one or more of the following four reasons:

- **Cost Reduction Opportunities:** Reducing carbon emission cuts costs. Some people, particularly in an economic downturn, are using it as an opportunity purely to bring down costs, but in a way that it is also perceived as doing the right thing.
- **New Business Opportunities:** Companies are finding opportunities for new services, new products and new ways of doing things in new marketplaces.
- **Regulatory Demands:** Depending on where a company stands in the use of carbon and other greenhouse gases, there will either be a cost or opportunity.
- **Stakeholder Response:** Companies are looking at their competitors and the benefits they are reaping in terms of new products or stakeholder endorsements, or the costs they have saved.

Organizations seem to believe that focusing on climate change will not only lead to sustainability but will lead to emergence of many new opportunities (Mukherjee, 2012; Connell, 2008). A few examples (Mukherjee, 2012) are as follows:

- IKEA sawmill waste, mixed and recycled plastic, produces a strong material that minimizes the use of resources.
- Statoil Hydro built one of the world's earliest carbon capture and sequestration projects in Sleipner in 1996.
- Siemens has developed a portfolio of "recession proof" green businesses.
- Yara has built significant new business around their application technology that removes more NOx globally than the total emission in Norway.
- Wilhelmsen had developed a model of low emission ship which will be powered by solar, wind and natural gas.
- Standard Chartered and Deutsche Bank buttress their position as "lighthouse organizations", given their vital role in providing finance to green business ventures.

The present study seeks to review the role of the Tata group in addressing and abating the climate change. It further looks at the various Tata group companies like Tata Chemicals Limited, Tata Steel Europe, Tata Communications and Tata Motors and their ways to stay the course towards sustainable development.

BACKGROUND

The concept of sustainability is complex (Faber et.al, 2005). Issues of sustainability cut across many boundaries and are trans-functional and trans-organizational in nature. Muthuri and Huang (2010) posited that sustainability had become a mantra for the 21st century. It addresses the social, economic and environmental concerns of actors. It is garnering ever –greater public attention and ranks high on legislative agendas of most governments, boardroom agenda of businesses and the media coverage of the topic has proliferated (Berns, 2009).

The most common and accepted definition on sustainability is the one given by the World Commission for Economic Development in 1987: "Sustainable development means meeting the needs of the present without compromising the ability of future generations to meet their own needs"(Loucks et.al, 2010; Hult, 2011; Sisaye, 2011; Mukherjee-Saha, 2011; Elliot, 2012). In 2008, the World Wildlife Fund in their report 'The Living Planet' published that the earth as a planet has limited capacity "to support a thriving diversity of species, humans included" (pg.2) and that it was essential to engage in more socially and environmentally sustainable practices. The concept of sustainability or sustainable development has been adopted by business as an overarching concept designed to describe and encompass a broad range of practices (Loucks, et.al, 2010). There is a growing acceptance that well-embedded sustainability practices can be a key factor to long term business success (Kuosmanen & Kousmanen, 2009).

Within the scientific community, climate change is increasingly seen as one of the most important long term issues facing humanity. It is also leading to business volatility, risk, and disruption (Hoffman & Forbes 2011). In a study done by MIT Sloan Management Review and Boston Consulting Group reveals that one-third of the companies studied believe that they are prepared for climate-change –related impacts (Kiron et. al, 2013). In a similar report on 'Climate Changes and Your Business' discloses that six major industry sectors (aviation, health care, tourism, transportation, oil & gas and financial services) are in particular danger from climate change risks and should pay more attention to the risks arising out of climate change (Anonymous, 2008; Clemo, 2008). According to PWC climate change/environmental issues are expected to impact companies in the next two to five years (Poirier, 2010). In order to deal with climate change related issues effective management decision-making, particularly in the appraisal of large-scale and long-term international projects, requires climate change risk to be incorporated into risk appraisals (Romilly, 2007).

Business organizations play an important role in addressing the challenges of climate change and global warming. General Electric's strategic focus on energy efficiency is not only beneficial to customers but also to the environment and the company (Romar, 2008). On the other hand Pinkse and Kolks (2011) reveal that climate change is characterised by institutional failures because there is neither an enforceable global agreement nor a market morality. It is an international business issue, as its institutional failures materialize differently in different countries and multi-national enterprises need to consider carefully their strategies to cope with non-market forces in multiple institutional setting. Business organisations face a number of obstacles in learning how to adapt to climate change impacts, especially in relation to the weakness and ambiguity of signals about climate change and the uncertainty about benefits flowing from adaptation measures. Organisations rarely adapt 'autonomously', since their adaptive behaviour is influenced by policy and market conditions, and draws on resources external to the organisation (Berkhout et.al, 2006). Climate-proofing operations are also becoming increasingly pertinent to businesses (Wittneben & Kiyar, 2009). Further, managing the risk of climate change and lowering the impact on the environment, leading companies are gaining investor confidence, reducing their exposure and increasing profits thereby changing the rules of doing business (John, 2007).

Sodano and Hingley (2013) argued that interventions to tackle climate change are political rather than economic and depend on power relationships among different actors, such as states and large corporations, involved in their implementation. Companies are making climate change action known and important throughout their business. The Carbon Disclosure Project disclosing companies not only integrate climate change into core business strategy but also have senior executives or board oversight of their climate change programme (McTague, 2012).

Scholars have also argued that sustainable development may offer a significantly more fruitful way to pursue climate policy goals than climate policy itself (Robinson et.al, 2006). Specific industries have adopted climate change as a model for rational behaviour. For the oil industry, it is a trade-off between a socially detrimental process (emitting GHGs, hence inducing a risk of climate change) and a self-interested consequence (profits) (LeMenestreal et.al, 2002). Climate change will have direct impact on infrastructure and investment companies. Famous brand of UK like B&Q, British Gas, Marks & Spencer, Royal & Sun Alliance, BskyB and Tesco have committed to offering simple, inexpensive low-carbon solutions to consumers (Howard and Bristow, 2007). The Irish companies viewed climate change as a significant opportunity for their business. The study further shared that if companies did not pursue a strategy of environmentally friendly products and projects the value that has been built up will rapidly dissipate (Connell, 2008). The Australian tourism sector is vulnerable to the impacts of climate change on natural areas and the destination choices of long-haul travellers concerned about carbon emissions. The industry is responding through the provision of information on climate change, sustainability and going green along with targeted carbon footprints or carbon offset programmes (Zeppel & Beaumont, 2012).

Against this backdrop, the present paper attempts to study the Tata group's carbon foot printing and abatement exercise undertaken in combating the menace of global warming.

MAIN FOCUS OF THE CHAPTER

The year 2007 will be known as the era of unprecedented human growth and prosperity, the last prosperous phase before the collapse of global financial institutions like Lehman brothers in 2008. In the same year, IPCC stated with 90 percent probability that most of the warming since the mid-20th century had been caused by human activity- primarily fossil fuel combustion, deforestation and concluded warming of the climate system in undeniable. The IPCC report further stated that since 1995 to 2006 i.e. eleven of the twelve years rank among the warmest years on record since 1850.

The issue of global warming also received attention in the 79th Academy Awards, where the documentary 'An inconvenient Truth' won the Academy Awards for Best Documentary and Best Original Song.

Company boardrooms across the world are discussing on the issues of climate change and trying to find answers to questions as to how a company must adapt its businesses to minimize the impact on the environment? Is there a solution to the emission problem? Can a business grow without turning the climate patterns into a jangle of nerves? Companies across the spectrum, from engineering major Siemens and financial players like Standard Chartered, Deutsche bank to chemical companies such as Yara and furniture makers Ikea are assessing the impact of their operations on climate change and developing effective carbon abatement and adaptation strategies.

Methodology

Qualitative research using the case study technique was adopted for the study. Social sciences has not succeeded in producing general, context-independent theory and has in the final instance nothing else to offer than concrete, context-dependent knowledge. And the case study is especially well suited to produce this knowledge (Flyvbjerg, 2010). *A priori* method was used to select the case of the Tata group for the study. Data was collected from various secondary documents like company reports, internal newsletters and other internal publications and was analysed using content analysis. Document quality was assessed using Scott's (1990, p.6) criteria.

SOLUTIONS AND RECOMMENDATIONS

The Tata group was founded in 1868 by Jamsetji Tata, is one of the oldest, largest and most respected business conglomerates in India comprising of over 90 operating companies in seven business sectors: communication and information technology, engineering, materials, services, energy, consumer products and chemicals. The group has operations in more than 80 countries and the total revenues for 2013-14 were $103.27 billion, with 67.2 percent of this coming from businesses outside India (Tata, 2015).

There are 32 publicly listed Tata Companies and they have a combined market capitalization of about $37 billion (as on February 2015) and a shareholder base of 3.9 million. The companies within the group operate independently and have their own board of directors and shareholders. The major Tata companies are Tata Steel, Tata Motors, Tata Consultancy Services, Tata Power, Tata Chemicals, Tata Global Beverages, Indian Hotels and Tata Communications (Tata 2015).

With the increasing international foot print of the Tata companies, the Tata brand is also gaining international recognition. In 2014, Brand Finance, a UK-based consultancy firm, valued the Tata brand at $21.1 billion and ranked it 34th among the top 500 most valuable global brands.

The conglomerate has always believed in returning wealth to the society they serve. The group's belief on sustainability has led to a corporate policy that emphasises environment preservation. The group companies have been working to minimise their carbon foot print by formulating strategies that balance growth with care for the environment. Within the group companies – Tata Steel, Tata Consultancy Services, Tata Motors, Tata Chemicals and Tata Power have leading the way in framing targets and devising ways to balance growth with care for the environment.

Dealing With climate Change: The Sustainable Way

According to the Stern Report (2008), responding to climate change is a business imperative as much as it is a social and an environmental one. With this perspective the climate change initiative was integrated with the Tata Business Excellence Model (TBEM). To focus the spotlight on the emerging phenomenon of climate change, it was tagged as the Tata area of emphasis (TAE). TAE was embedded

Figure 1. Tata Business Excellence Model
Source: Shetty 2010, pg. 50

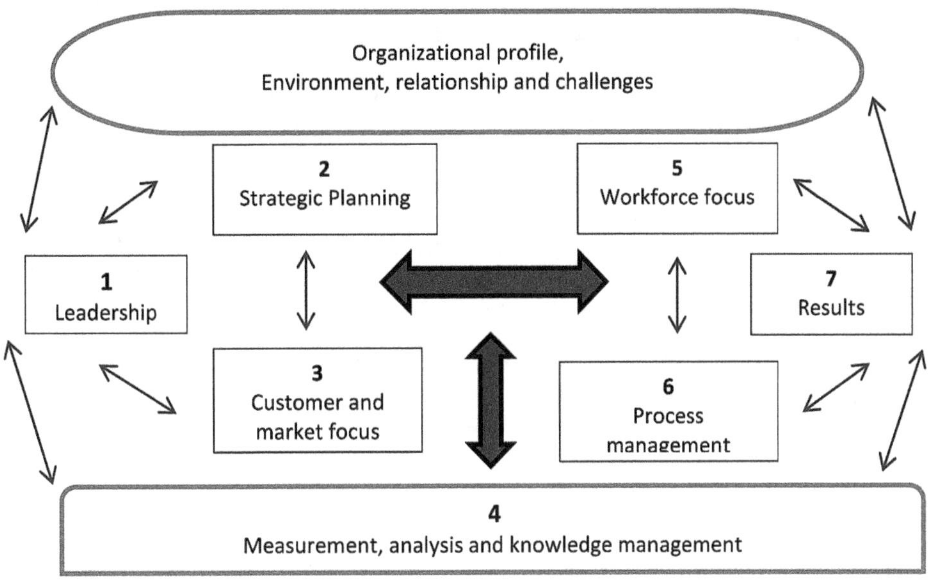

into categories of leadership and strategic management, both of which deal with the sustainability aspect of an organization (Figure 1).

Within the TBEM framework, sustainability is defined as the ability to address current business needs with agility and strategic intent. It empowers businesses to prepare successfully for the future business, market and operating environment. Climate change was measured and mapped as a TAE within this dimension of the model and was integrated with other issues that define sustainability. The objective of identifying climate change as a TAE was to (a) enable organizations and assessors to do a deep dive analysis and provide insights in alignment with the group's climate change policy; and (b) include a climate conscious behaviour among employees.

A specific roll-out process was designed to implement climate change as a TAE. The process kept balance between the host of external and internal economic, social and environmental factors that make up the sustainability dimensions while zooming in on climate change. The roll-out process began with mapping the climate change maturity levels (Figure 2) of the group companies.

The group companies were mapped to assess the maturity levels of the companies and were categorised as: Beginner, Reactive, Responsive, Proactive and Leader. This categorization helped in classifying companies as per their progress made along each of the five dimensions ie. results, capability and culture, support processes, planning and execution and climate change aspirations. Each dimension was broken down into a set of characteristics that assessors used to collect and collate data to score a company and assign it a maturity category.

Climate Change Roll Out Strategy

In order to deal with the issues of climate change, the Tata group began the journey by setting up a Centre of Excellence in Tata Quality Management Services, a Think Tank of the group.

Figure 2. Mapping Climate Change Maturity Levels
Source: Shetty, 2010, pg. 53.

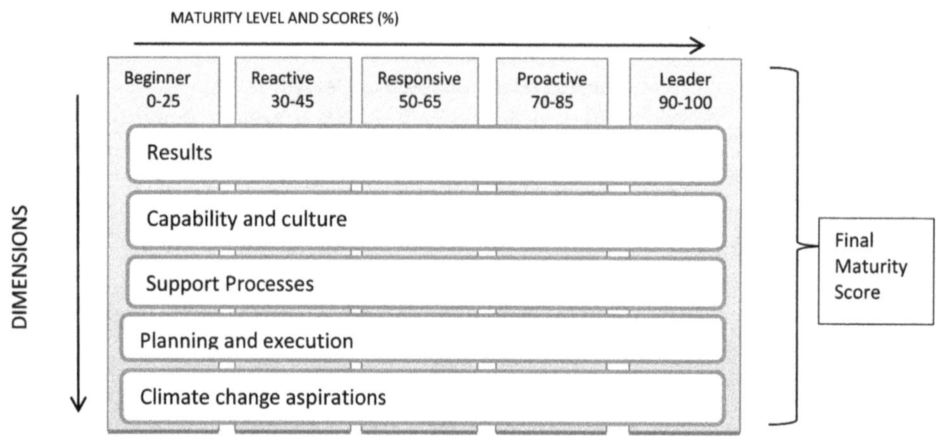

Figure 3. Key action Areas and Progress Status

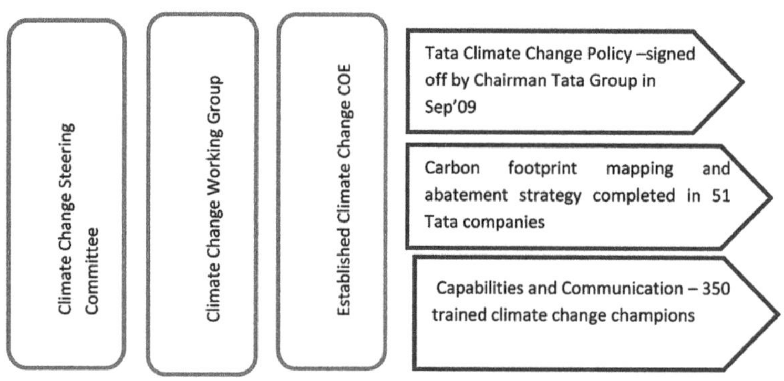

A Steering Committee (SC) of Senior Tata Executives was set up to guide the efforts across Tata Companies. SC decided the governance structure and the key action areas are given in Figure 3.

The SC constituted of a Working Group comprising the Chief Sustainability Officers of key Tata Companies and formed a Climate Change team under TQMS. Along with internal structuring, two international consulting firms were hired to assist the companies in: (a) measuring the carbon footprint of the companies; and (b) identifying opportunities to reduce carbon intensity and develop long term low carbon growth strategies.

In order to keep all the company on the same footing, a climate change policy was articulated which ensured not only a common framework for change but also warranted that the changes taking place were institutionalised and implemented in more companies. The policy enables the companies to play a proactive role in the development of emerging climate change regulations along with policy makers, government and other stakeholders like civil society, NGOs and customers. It also enabled the companies to base all key organizational decisions from a "low carbon lens", taking for example vendor selection which would be reflected in the performance management systems of the companies.

A robust roll out process was designed to implement climate change initiatives across companies (Figure 4).

Figure 4. Roll Out Strategy

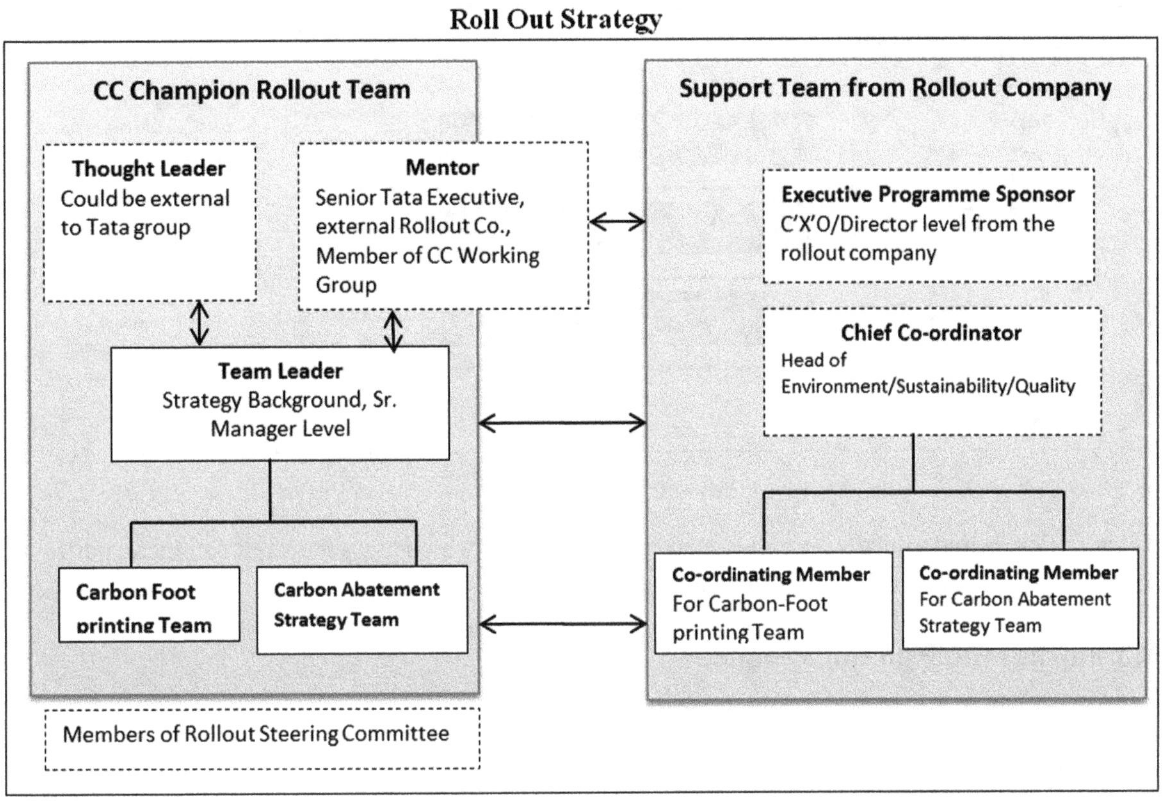

Source: Author

According to the data the rollout process began by building capacities of mangers in the group companies. The capacity building workshops were divided into a two part process. The first part focused on class room training and the second part lay emphasis on building skills by deploying the participants to do an actual carbon foot printing assignment. The roll out process trained 350 climate change (CC) champions. These CC champions undergo a three day intensive training programme at the end of which they have the skills to conduct carbon foot printing and abatement strategy exercise. The trained CC champions were subsequently assigned a Tata company to apply their newly acquired skills. The CC champions worked alongside with a mirror team from the host company thereby enhancing the pool of trained managers in the Tata group.

The third element of the roll out process was sustained communication aimed at all Tata employees (Figure 5).

The process of communicating the agenda of climate change began with an internal logo designing contest and out of 500 entries received from all over the world where the Tata group is present, the logo "grey to green" which came from Titan, Bangalore was chosen. Other channels of communication like a video, thematic posters, celebration of Tata

Environment Month, online competitions were organised. Apart from these, around the year activities like expert talks organized through the regional networking forum of Tata companies called the Tata

Figure 5. Communication with Employees

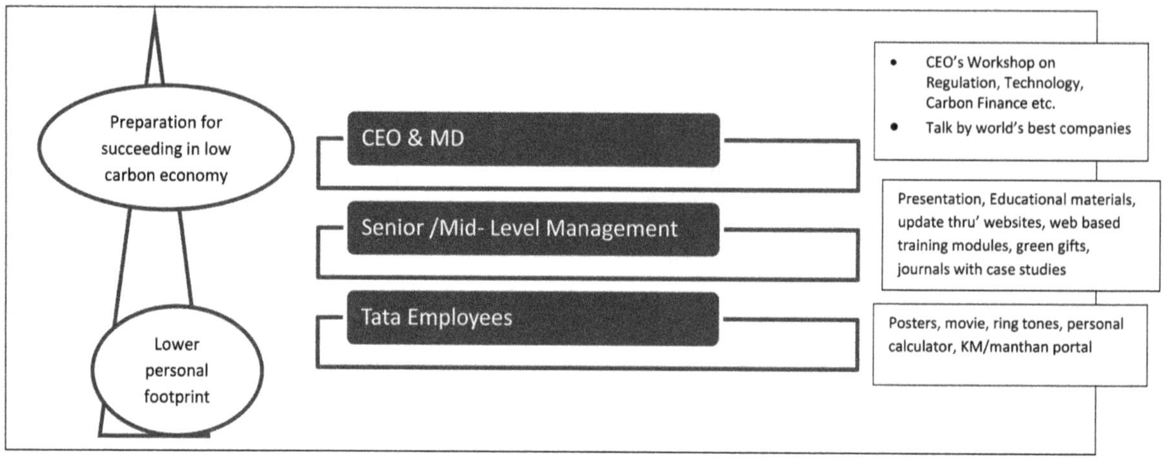

Network Forum and a detailed monthly electronic newsletter which gives the latest technical, policy and financial update on climate change and sustainability issues.

Monitoring the Roll out Process

A multi-level review and monitoring system was developed to measure the effect of the initiative (Figure 6).

A score card with lead and lag indicators was used to carry out the first 3 levels of review (Figure 6). The lead indicators include elements like investment in development of clean technology products. The lag indicators focus on their impact in reduction of emissions. The final level of review is through the annual assessment carried out under the Tata Business Excellence Model (TBEM). The TBEM is derived from the Malcolm Baldridge model tweaked to suit the unique character of the Tata group. It promotes a process view of the organization and measures its impact in terms of business results. Climate change/sustainability indicators have been woven into the existing model rather than creating a separate one. This enabled the integration of the process into the corporate strategy of the Tata companies rather than it being a standalone climate change/ sustainability strategy.

Figure 6. Review and Monitoring System

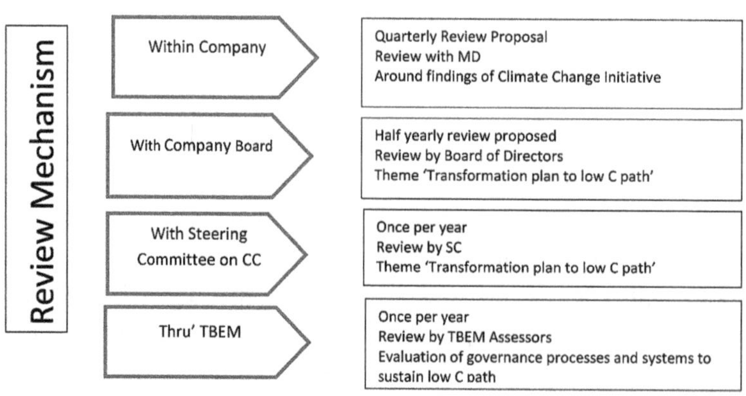

Table 1. Measuring the Footprint of TCL

Carbon Footprint of TCL	Years		
	2007-08	2008-09	2009-10
Direct emission (million mt CO_2e)	2.45	2.52	2.69
Indirect emission(million mt CO_2e)	0.13	0.15	0.07
Total emission (million mt CO_2e)	2.58	2.67	2.76
Product CO_2 intensity –specific emissions (million mt CO_2e per mt of product sale)	0.66	0.65	0.63

Source: Quest 2010, p.22

Dealing with Climate Change: Tata Companies

This section discusses the various initiatives taken by the Tata Group of companies namely Tata chemicals, Tata Steel Europe and Tata Motors in dealing with climate change.

Since it was chosen to join a core team of five companies that would lead the Tata group's battle against climate change, Tata Chemicals (TCL) has taken huge strides on the combat trail. It has adapted its businesses to become energy efficient and use less water, and is innovating with new chemical and fertilizer products and agri-services to take care of the environmental responsibilities. The footprint mapping process was initiated at all India operations and later extended to all geographies (Table 1).

The first step which TCL took in this regard was to quantify CO_2 equivalent emissions across manufacturing locations and measure the size and scope of its carbon footprint (Table 2). This was done by tracking emissions from manufacturing units including captive power plants, logistics operations and staff travel.

The data in Table 2 indicates that the total emissions have gone up as production has increased over the years, thus it is important to consider specific emission as given in Table 2.

It can be thus concluded that the TCl's specific carbon footprint across product categories is higher than that of the best industry performers. The data gathered enabled the company to understand the

Table 2. Specific Carbon Foot print Across TCL Operations

Location	Total Emissions (mt CO_2e per year) 2009-10	Mt CO_2e per mt products, 2009-10	Target mt CO_2e per mt products (2020)
TCL (India operations)	2.76	0.63	0.50
Rallis India	0.03	1.95	1.56
General Chemicals Industries Products, US	1.21	0.70	0.60
Brunner Mond, UK	0.84	0.65	0.42
Magadi Soda Company, Kenya	0.13	0.35	0.30
Total Global	4.97	0.62	0.50

Source: Quest, 2010, p.22

potential for energy efficiency, input usage and process efficiency for each activity and sub-activity that forms the backbone of the company's manufacturing units and helped TCL to identify business levers that would help with carbon abatement, which were built into a long-term strategic plan. This exercise helped the Babrala urea manufacturing unit to become a benchmark in energy efficiency and carbon footprint across the world (Choudhury, 2010).

Tata Steel Europe (TSE)

For TSE, the consequences of climate change are being assiduously analysed to help both, the company and the communities it serves, harness the opportunities and adopt a sustainable way of life. TSE used the life cycle assessment (LCA) tool to understand the impact of a process or a product on the environment (Figure 7). This is used to focus on carbon emissions across the complete value chain of a product and to measure and analyse the steel making process to bring down CO_2 emissions. LCA was used by TSE to gauge the impact of a product on the environment and develop applications that would help build green buildings, drive fuel-efficient vehicles and package products in an environmentally sensitive manner. This would not only enable TSE to build a sustainable future for the company but also for its customers.

LCA (Figure 7) allows the environmental performance of a product to be assessed over its full life cycle; from the extraction of raw materials through to material/component and product manufacture, the use phase of the final product and, finally the end of life considerations such as reuse, recycling and disposal. The assessment allows the potential environmental impact categories of a product over the life cycle to be calculated. The impact category most closely associated with climate change generally referred to as carbon footprint. Additionally LCA helped to assess other potential environmental impact categories such as 'water footprint' and' resource consumption' for TSE.

LCA is used by TSE to inform and convince customers about the efforts under taken by to minimise the impact of the company's products and services on the environment. The LSC studies show that most of the carbon emissions happen across the supply chain and are associated with the 'product use' phase. So driving of a car or the occupation of a building has greater impact on the climate than sourcing and production of materials that the products are made from. The recycling and reusing these products are

Figure 7. LCA in TSE
Source: Business ethics and sustainability in the steel industry

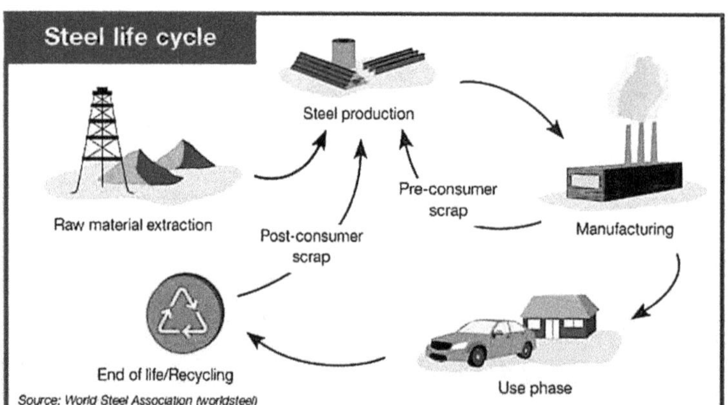

also found to have a significant benefit in terms of reducing emissions. The use of steel in low-carbon technologies, such as renewable energy generation, low-carbon vehicles and energy –efficient buildings provides significant opportunities for reducing carbon emissions. As an outcome of this LCA, TSE has produced new steel grades which reduce carbon emissions from energy generation in the wind power sector and in the distribution of electricity. The product can also be used to reduce electrical losses in transformers, which are essential for power generation and supply (Griffin & Brimacombe, 2010).

Efforts have also been made by TSE to reduce the environmental impacts of buildings. It has developed a brand of pre-finished steel called Colorcoat®, which does not corrode as fast as other products in the same category. The product is used in building cladding applications and was designed with the help of the Research, Development and Technology (RD&T) function, which undertook a comprehensive cradle-to- grave life cycle assessment of a range of cladding products. This process not only helped TSE to develop a new product but they were also used to spread awareness among consumers through Environmental Product Declarations.

TSE is using advanced high strength steels (AHSS) in new vehicle designs. Studies show that the use of one kilogram of AHSS saved eight kilograms of CO_2 over the life cycle of the vehicle. This has enabled TSE to evaluate the benefits of AHSS in new vehicle designs and tackle its quest for sustainable automotive structures.

TSE has also been active in recycling steel. Recycling metals after their use plays an important role in reducing GHG emissions, as each tonne recycled saves resources by avoiding primary production from raw materials. In its effort to reduce its overall carbon foot print, TSE along with other European steelmakers have collaborated in the ULCOs project. ULCOS stands for ultra-low carbon dioxide steelmaking. It is a consortium of 48 companies and organizations from 15 European countries that have launched a cooperative research and development initiative to enable a 50 percent reduction in CO_2 emissions compared to current best practice in primary steel making.

TSE is thus, involved with dealing with the issues of climate change on two fronts: at one level it is working with other European steelmakers to bring down the CO_2 emissions in the steel manufacturing process and on the other developing applications of steel that will help cut the emissions in user industries.

Tata Motors

Tata Motors (TML) fight against climate change began with mapping its carbon footprint and developing mitigation strategies to cut back its GHGs and devise a sustainable approach to growth. The company adopted a systematic plan to check its power usage and achieve power unity factor by (a) improvement in the quality of power distributed to the plant; (b) reduction in wastage of power, this is the amount of power lost to the system on account of inefficiencies. The company also initiated several environment management programmes to reduce energy and fuel consumption and to shift away from fossil fuels.

TML has put in place a system to institutionalise the practice of saving power and achieving unity power factor at the truck manufacturing plant at Pimpri and car manufacturing plant at Chikhali in Pune. As a first step the TML looked at the complexities of the electrical distribution system (Table 3).

Table 4. shows the results of a series of projects undertaken to improve the power factor to unity.

The real challenge for the company lay in sustaining the same on an on-going basis. To ensure this, TML decided to introduce a new design for all future substations and collaborated with external partners to make the Real Time Power Factor Correction methods more cost effective so that it can be universally applied.

Table 3. Complexities of Power Use in TML Manufacturing Plants in Pune

Power Systems	Factors
Multiple voltage levels	The Pune plant receives power from the state grind at 220kV. This is distributed internally at 22kV, 6.6kV and 415kV. Power factor has to be improved at each voltage level
Variable demand	The total electrical connected load increased from about 150 MW in 1999 to 194MW in 2009 while power demands varies from 14 MW to 50 MW over a 24-hour period. This variation poses over compensation problems during periods of low demand and could also lead to overvoltage problems
Substation spread	In 1999 there were 76 substations with transformers of varying capacity. The number has gone up to 100 substations today with the network spread over 700 acres. Each has different load patterns and voltage requirement and hence, needs to be accounted for separately.
Capacitor challenge	The distribution system had a mix of fixed capacitors, which had to be operated manually. Fixed capacitors also had to be switched off manually during power failure.

Source: Kulkarni, 2010, p. 43

Table 4. TML Unity Power Factor at Plant level

Stage	Event	Completed	Corresponding Improvement in Power Factor
1	Change from manually operated to automatic capacitors	March 2000	0.946 to 0.96
2	Changing fixed capacitors to Automatic Power Factor Correction panels	December 2000	0.96 to 0.98
3	Finalising fixed capacitor requirement and recalibrating the systems	November 2001	0.98 to 0.993
4	Reactive power compensation in weld shop using Real Time Power Factor Correction	March 2002	0.993 to Unity

Source: Kulkarni, 2010,p. 43

The unity power factor enabled TML to reduce its demand for power by 3MVA. The cumulative energy saved, on this account would amount to total GHG reduction of approximately 2.50 lakh tonnes of CO_2 equivalent.

Progress So Far

The carbon footprint of 51 Tata companies has been estimated and their carbon management strategy has been developed thereby addressing almost 99 percent of the group's emission (Table 5).

An early estimate of the group indicates that specific emissions can be brought down by 10-15 percent, if Tata companies adopt abatement measures instead of going down the business- as-usual path. In order to bring down emissions, the scope of carbon foot printing exercise is being expanded to include the complete value chain from the vendors to the disposal of the equipment also known as the LCA. It has also been realised among the group that water is increasingly becoming scares resource, a water foot printing sustainability impact and sustainability response exercise has commenced in the 'Big Four' manufacturing arm of the group namely, Tata Chemicals, Tata Motors, Tata Steel and Tata Power.

Table 5. Climate Change Initiatives of Tata Companies

Company Name	Decreasing Costs	Increasing Revenues
Tata Steel Ltd	Target of achieving 20% GHG emission reduction in the next decade. Portfolio of projects to recover waste heat from stove waste gas, coke dry quenching and blast furnace through toy gas heat recovery turbines, beneficiation of raw materials	PV coated metal cladding for power generating zero carbon emission buildings. Special steels for the wind power and automobile industry
Tata Power Ltd	Selection of energy efficient super critical stream generators at their Mundra power station	Development of leading edge technology projects in renewable power generation like floating concentrated solar power plants
Tata Motors Ltd	Use of regenerative test beds and benchmarking of utility systems such as plant air system	Development of hybrid and electric vehicles. Advanced fuel cells with on board hydrogen generation system.
Tata Chemicals Ltd	Target of reducing specific product emission by 20% over 10 years by energy efficiency, process integration, recovery of products from wastes, low carbon intensity co-products/derivatives.	Joint efforts by Tata power, Tata Steel and Tata Chemicals to manufacture algae from carbon emissions. Various nano-technology applications like customised fertilizers, coatings for glass used in buildings, etc.
Tata Communications	Reduce electricity/water usage, business travel	Telepresence solution to reduce/replace business travel.
Indian Hotels Ltd	Reduce electricity/water usage, business travel	Provide " Earth Rooms" to the environmentally sensitive guest(s)
TCS	Reduce electricity/water usage, business travel	Provide cutting edge IT solutions for smart grids, smart buildings, transportation, etc.

Source: Author

CONCLUSION

The consequences of a changing climate are clear to all and few doubt the need for sustained and concerted action to cope with the changes. Climate change is the hot topic of boardrooms across the world. Companies across the spectrum, from engineering to financial companies, to chemical companies and furniture makers are assessing the impacts of their operations on climate change and developing effective carbon abatement and adaptation strategies.

For the Tata companies, there are innumerable lessons to be learnt from the developments worldwide. Even as the companies' climate change plans are developed into multiple initiatives across Tata companies; the experiences, the successes and the failures that have come the way of other companies are providing key insights and helping to build a roadmap to take the agenda forward.

REFERENCES

Berkhout, F., Hertin, J., & Gann, D. H. (2006). Learning To Adapt: Organizational Adaptations to Climate Change Impact. *Climatic Change*, *78*(1), 135–156. doi:10.1007/s10584-006-9089-3

Berns, M., Townend, A., & Khayat, Z. et al.. (2009). Sustainability and Competitive Advantage. *MIT Sloan Management Review*, *51*(1), 19–26.

Choudhary, S. G. (2010). Planet first. *Quest. TQMS*, *5*, 20–26.

Clemo, K. (2008). Preparing for climate change: Insurance and small Business. *The Geneva Papers on Risk and Insurance. Issues and Practice*, *33*(1), 110–116. doi:10.1057/palgrave.gpp.2510160

Climate Change and Business: from Supply Chain Carbon Emissions to a Wider Risk Set2008*Business and the Environment*, 19(7), 1-3.

Elliott, J. (2012). *An Introduction to Sustainable Development*. USA: Routledge.

Faber, N., Jorna, R., & Van Engelen, J. (2005). The Sustainability of "Sustainability"- A Study into The Conceptual Foundations of the Notion of Sustainability. *Journal of Environmental Assessment Policy and Management*, *7*(1), 1–33. doi:10.1142/S1464333205001955

Flyvbjerg, B. (2010). Five misunderstandings about case-study research. In C. Seale, G. Gobo, J. F. Gubrium, & D. Silverman (Eds.), *Qualitative Research Proactive*. Los Angeles/London/New Delhi/Singapore/Washington D.C.: Sage.

Griffin, A., & Brimacombe, L. G. (2010). Steely resolve'. *Quest. TQMS*, *5*, 27–32.

How climate change affects India. (2011). *India climate Portal*. Retrieved from http://www.indiaclimateportal.org

Howard, S., & Bristow, S. (2007). Business and Climate Change: Rising Public Awareness Creates Significant Opportunity. *UN Chronicle*, *44*(2), 56.

Hult, G.T.M. (2011). Market –focused sustainability: Market orientation plus! *Journal of the Academy of Marketing Science*, *39*(1), 1–6. doi:10.1007/s11747-010-0223-4

John, B. (2007). The Business of Climate change. *CMA Management*, *81*(1), 48–49.

Khan, S. A., Kumar, S., Hussain, M. Z., & Kalra, N. (2009). Climate Change, Climate Variability and Indian Agriculture: Impacts, Vulnerability and Adaptation Strategies. In S. N. Singh (Ed.), *Climate Change and Crops. Environmental Science and Engineering*. Heidelberg: Springer-Verlag Berlin. doi:10.1007/978-3-540-88246-6_2

Kiron, D., Kruschwitz, N., Haanes, K., & Kehrbach, S. K. (2013). How serious is climate change to Business? *MIT Sloan Management Review*, *55*(1), 75–77.

Kulkarni, B. (2010). Powered by passion. *Quest. TQMS*, *5*, 41–43.

Kuosmanen, T., & N. Kuosmanen. (2009). How not to measure Sustainable Value (and how one might). *Ecological Economics*, 69(2), 235-243.

LeMenestrel, M., Le Menestrel, M., & de Bettignies, H.-C. (2002). Processes and Consequences in Business Ethical Dilemmas: The Oil Industry and Climate Change. *Journal of Business Ethics*, *41*(3), 251–266. doi:10.1023/A:1021237629938

Loucks, E. S., Martens, M. L., & Cho, C. H. (2010). Engaging Small- and Medium-sized Businesses in Sustainability. *Sustainability Accounting Management and Policy Journal, 1*(2), 178–200.

Mukherjee-Saha, J. (2011). Business Sustainability amidst global churning paradoxes and dilemmas. *GMJ, 5*(1/2), 19–24.

O'Connell, L. (2008). Driving change: KPMG, Studies Impact of Climate Change on Irish M & A Activity. *Accountancy Ireland, 40*(6), 36–38.

Pinkse, J., & Kolk, A. (2012). Multinational enterprises and climate change; Exploring institutional failures and embeddedness. *Journal of International Business Studies, 43*(3), 332–341. doi:10.1057/jibs.2011.56

Poirier, L. (2010). *PWC Survey: Emissions Reductions Top U.S. Business Concern in Climate change Debate. Global Refining & Fuels Reports*. Hart Energy Publishing.

Quest (n. d.). The New world order. *Quest*, 3, 35-39.

Robinson, J., Bradley, M., Busby, P., Connor, D., Murray, A., Sampson, B., & Soper, W. (2006). Climate Change and Sustainable Development: Realizing the opportunity. *Ambio, 35*(1), 2–7. doi:10.1579/0044-7447-35.1.2 PMID:16615692

Romar, E. J. (2009). Snapshots of the Future: Darfur, Katrina, and Maple Sugar (Climate Change, the Less Well –off and Business Ethics). *Journal of Business Ethics, 85*(Suppl. 1), 121–132. doi:10.1007/s10551-008-9932-8

Romilly, P. (2007). Business and climate change risk: A regional time series analysis. *Journal of International Business Studies, 38*(3), 474–480. doi:10.1057/palgrave.jibs.8400266

Scott, J. (1990). *A matter of Record – Documentary Sources in Social Research*. Cambridge: Polity.

Sisaye, S. (2011). Ecological systems approaches to sustainability and organizational development emerging trends in environmental and social accounting reporting systems. *Leadership and Organization Development Journal, 32*(4), 379–398. doi:10.1108/01437731111134652

Sodano, V., & Hingley, N. (2013). The food system, climate change and CSR from business to government case. *British Food Journal, 115*(1), 75–91. doi:10.1108/00070701311289885

Tata. (2015). Leadership with trust. Retrieved from www.tarta.co.in/aboutus/sub-index/Leadership–with–trust

Zeppel, H., & Beaumont, N. (2012). Climate change and tourism futures: Responses by Australian tourism agencies. *Tourism and Hospitality Research, 12*(2), 73–88. doi:10.1177/1467358412444807

Chapter 10
Sustainable Development and the Integration of Normative Standards

Eduardo Guilherme Satolo
UNESP - Univ Estadual Paulista, Brazil

Alexandre Tadeu Simon
UNIMEP - Universidade Metodista de Piracicaba, Brazil

Sergio Silva Braga Jr
UNESP – Univ Estadual Paulista, Brazil

ABSTRACT

The sustainability definition is still unclear, and requires more in-depth studies to expand their understandings and their effective implementations. One of the means by which to achieve organizational sustainability presented in literature is the use of the concept of Integrated Management System (IMS). Normative management standards such as ISO 9000, ISO 14000, OHSAS 18000, and SA 8000 stand out with large numbers of accreditation by the companies targeted for joint establishment of an IMS, and more recently, normative standards have been identified as mechanisms to ensure the sustainable development of organizations by the integration of economic development, social progress, and environmental quality. This chapter aims to show the evolution of concept of sustainable development; present a definition for the organizational environment; highlight the interaction term "organizational sustainability" with basic normative standards systems management; and guide organizations by describing how interactions between normative standards and the dimensions of sustainability can occur.

INTRODUCTION

In last two decades, sustainability concept has been expanding and gradually embraced by organizations and governments worldwide (Gibson, 2006; Mullerat, 2010).

DOI: 10.4018/978-1-5225-0350-7.ch010

Figure 1. Sustainability dimensions (adapted from Fresner & Engelhardt, 2004)

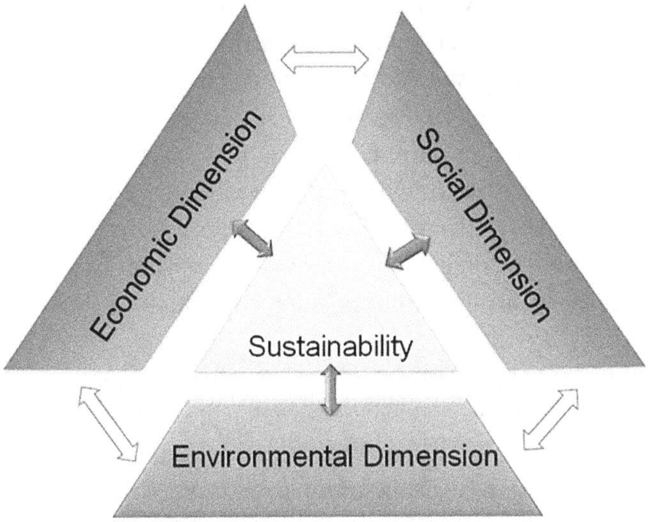

Sustainability is a relatively new, evolving concept and no consensus exists regarding its definition and a true meaning. A great deal of research has been committed to developing the theme, and researchers have created their own definitions and principles (Xu, Zao, Dawson, Hao, Zang & Tao, 2006; Siena, 2008; Singh, Murty, Gupta &Dikshit, 2009; Barron, 2010) what has caused difficulties in understanding and implementing the concept in the organizational environment.

The term sustainability has inflated, and it is becoming ambiguous with contradictory and obscure associations (Smeraldi, 2009). This conflation occurs due to different definitions and applications the term sustainability, sustainable development and organizational sustainability have received depending on the purpose of the study. Clearly, a precise definition of sustainability still lacks (XU et al., 2006).

Although the concept of sustainability does not present a common definition, John Elkigton's 1987's definition, is widely used. It describes sustainability as mechanisms to ensure current actions do not limit the economic, social, and environmental options for future generations (Elkington, 1998).

In the organizational environment, Barron (2010) explains that the key to sustainable development is to integrate economic development, social progress, and environmental quality. This integration is represented by the triple bottom line concept (Figure 1).

Economic dimension includes operating cost reduction through management of labor productivity, research and development expenses, and investments in training and other forms of human capital (Jamali, 2006). This dimension projects future prospects, as guided by the steady growth expectation; it also involves innovation concept as an essential element for its realization (Barron, 2010).

Environmental dimension primarily addresses the process, products and services impacts on the environment, biodiversity, and human health (Jamali, 2006). Performance improvement is associated to manufacturing and supply resources reductions in order to reduce pollution as much as possible (IISD, 2009; Dillard, Dujon & King, 2009).

Social dimension aims to guarantee workers' rights, promoting continuous improvement of workplace conditions through the establishment of adequate health conditions and safety. It involves effective engagement of stakeholders contributing to proactive participation of diverse actors, such as company, employees, associations, customers, government, and NGOs (Jamali 2006; IISD, 2009; Dillard et al. 2009).

These three dimensions are inter-related, and no single dimension should be prioritized over another. Current research does not allow one to state the three dimensions embrace all sustainability fields (Henriques, 2004). Nonetheless, Elkington (1998), emphasizes that companies managing these dimensions are more likely to succeed and have opportunities to gain new markets.

Sustainability applied to the business environment – called here organizational sustainability – is referenced in literature as corporate responsibility, and addresses mechanisms for displaying, measuring, and improving. Internationally, companies, governments, and non-governmental organizations have made many important improvements in various conduct code models, standards, and methods (Jamali, 2006; Siena, 2008).

Smeraldi (2009) presents data about sustainability's evolution between 2004 and 2009. The author used GetAbstract, considered the main service upon technical summaries for executives. Through this database, Smeraldi (2009) identified more than 5000 titles on sustainability. Between 2007 and 2009, the sustainability theme stood out as the most discussed topic, with over 450 new titles. 120 of them mentioned the organizational sustainability approach. This made sustainability the most popular theme in business literature, overcoming issues like trading, capital markets, marketing, recruitment, and project financing.

Expanding data collection conducted by Smeraldi (2009) for the period from 2010 to march 2013, the GetAbstract (2014) database published more than 266 new titles addressing the organizational sustainability theme keeping it in the leading position in business literature topics.

Linton, Klassen and Jayaraman (2007) identified, roughly 1300 scientific articles addressing sustainability, applied economics, business, and management, published in Scopus database from 2004 to 2005. Expanding the data collection from 2006 to 2013, Scopus database published 8337 articles, which shows an average annual growth of 23% between 2004 and 2013 (last 10 years). This recently collected data confirms and justifies the importance of organizational sustainability and researchers' growing interest.

It is very important to recognize that an organization should be prepared and well-structured to deal with sustainability issues. Organizations have basic needs that must be met in order to survive. These needs are met through interaction between the organization and the environment where it operates, which brings up the fact that an organization first needs to be structured internally before going to external environments. In other words, organizations should plan, implement, verify, and measure internal sustainability before deploying and measuring outward actions (Anderson, Bider, Johannensson & Perjons, 2005).

Lueneburger and Goleman (2010) recently confirmed these findings by indicating that achieving organizational sustainability relates primarily to operational reality; then comes public perception. Companies that develop an image unrelated to their actual achievements are at serious risk of harming their corporate reputation with impacts that go beyond the company's image.

Baron (2010) highlights that organizations, therefore, cannot understand sustainability as a mechanism only for external actions. It is important to get the necessary internal sustainability. Management systems standards have a strong synergy enabling organizations to achieve sustainable development and, therefore, it is important to measure the levels of sustainability in an organization.

Ferroli and Librelotto (2011) featuring sustainability by interdisciplinary, i.e. only through the integration of different disciplines and branches of knowledge it may be possible to obtain viable products that meet the demands of a constantly evolving market, without harming the environment and society.

One of the ways to make a sustainable system is performing its management in an integrated manner, ie, involving the environmental, economic and social dimensions. According Fresner and Engelhardt

(2004), the integrated management system allows us to understand the needs of employees and set goals, enabling improved training, communication, quality and service assurance, having thus the requirements for sustainability of the sector.

The updated normative standards versions of quality management, environmental management, health and safety, and social responsibility, show that there is great similarity and compatibility between them, which led to the trend of integrating them. This would allow creating a common regulatory standard to promote more sustainable management systems. The integration of management standards is discussed through the concept of Integrated Management System (IMS), which seeks to integrate commonalities of various management systems and respect the individuality of each standard. Normative management standards such as ISO 9000, ISO 14000, OHSAS 18000 and SA 8000 stand out in organizational with a large number of accreditation by the companies being targeted for a joint establishment of a IMS and more recently these normative standards have been identified in literature as a mechanism to ensure the sustainable development of the organization by performing the integration of economic development, social progress and environmental quality.

The concept of IMS can be understood as the integration of various management systems, which combine to create an overview of the process, a systemic approach and focus of employees in order to put all management practices in a single system (Pojasek, 2006). This allows the management system created to be more efficient in achieving the objectives of that when there are several individual overlapping systems (De Cicco, 2004).

Within this theme, this chapter aims to (i) show the evolution of the concept of sustainable development, (ii) present a definition for the organizational environment, (iii) highlight the interaction term organizational sustainability with basic normative standards systems management, and (iv) guide organizations describing how interactions between normative standards and the dimensions of sustainability can occur.

EVOLUTION OF THE CONCEPT OF SUSTAINABILITY

The timeline on the concept of sustainability has strands advocated by some authors. A run of the concept was medial fifteenth century (Marquardt, 2006), others approach the origin from the eighteenth century (Grober, 2007). A third group of authors reported the story only from the twentieth century, when the concept became relevant for today's society (Xu et al., 2006).

These various strands are also found when analyzing the origin of the term. Although the terms sustainability and sustainable have first appeared in the Oxford Dictionary of English only during the second half of the twentieth century, the equivalent term in French (Durabilité and durable), German (Nachhaltigkeit and nachhaltig) and Dutch (and duurzaamheid Duurzaam) have been used for centuries (Du Pisani, 2006).

The aspect most frequently cited in the literature is the origin of the term in German (Nachhaltigkeit), first used in the eighteenth century by German jurist Hans Carlowitz that inspired the theory of English author John Evelyn and the French statesman Jean Baptiste Colbert, began to plan the optimal use of forests, which serve as a feedstock for power generation of iron and silver industries. In this theory, the authors argued that the industrial production could not exceed the playback speed of the woods. Thus, Carlowitz became known for creating the term, but not in its concept (Marquardt, 2006; Grober, 2007; Djalali & Vollaard, 2008).

Only in the 1970s, the concept of sustainability is replaced presence and global importance, being supported by the context of a growing awareness of impending ecological crisis. This occurs in conjunction with the emergence of the term "sustainable development", credited to Barbara Ward, founder of the International Institute for Environment and Development (Du Pisani, 2006). From the emergence of terminology, several authors presented their definitions, such as Goldsmith et al, 1972; Meadows, 1972; Coomer, 1979 and Allen, 1980.

In 1987, the current World Commission on Environment and Development (Brundtland Commission called at the time) introduced the world to the concept of sustainable development as a mechanism for political and ethical guidelines for dealing with ecological and social crisis of the planet, through the report "Our Common Future", published by the United Nations (Djalali & Vollaard, 2008; Cerin & Scholtens, 2011).

The apex so that the concept of sustainability was widespread worldwide occurred during the execution of the UN Conference for Environment and Development, in 1992, which became the principle and concept of sustainability known through political and scientific discourses (Marquardt, 2006).

During the last two decades, the concept has evolved, particularly with global development forums, having reached consensus on the need to create procedures that contribute to societal change (Pope, Annandale & Morrison-Saunders, 2004).

ORGANIZATIONAL SUSTAINABILITY

Organizations are commonly referred to as economic entities whose main objectives relate to the production of goods and provision of services in the most efficient way possible.

In recent decades, business management, in addition to meeting the requirements and ensure customer satisfaction and customer loyalty, begin to consider the welfare of employees and the work environment and worry about the impact generated for society and the local community (Wilkinson & Dale, 1999).

These changes become noticeable with the advent of the paradigm of environmental development in the early 1980s, where organizations begin to move away from their narrow economic conception and make profound strategic adaptations in response to environmental pressures and social expectations (Robinson, 2000)

Inspired in part by a growing recognition of the need to transition from environmental management for more comprehensive management called sustainable, in the 1990s, a new change occurred (Jamali 2006). Pressures from other stakeholders in the business, so-called stakeholders - as shareholders, owners, partners, government agencies - influence companies demonstrating best practices in organizational management, compliance with laws and regulations and at the same time, reduce costs and impact on the environment (Bamber, Sharp & Castka, 2004).

This advancement and dissemination of concepts in organizations made them forced to serve beyond the traditional competitive advantages, other parameters to ensure satisfaction and customer loyalty (Karapetrovic, 2002).

These factors resulted in increased complexity of business management and the origin of new demands for transparency and corporate citizenship (Jamali, 2006).

This development took place to meet the expectations motivated from the 1960s, the development and introduction of management standards and standardization, which came to help companies meet these customer relations, supplier and society (Wang & Tsai, 2009; Santos, Mendes & Barbosa, 2011), resulting in the need for organizations to deploy and certify their environments in order to gain competitive advantages in the dispute and the markets in which they operate (Sohal & Zutshi, 2004).

Although these normative standards undergo certifications from renowned international bodies, no guarantee that the company remains on track towards sustainability, given the fact that there are no regulatory standards to assess what level of sustainability that an organization is.

Commonly, when it comes to management standards and certifications for organizational environments, perceived to implementation of the quality management system, environmental management, management of health and safety at work and, more recently, social responsibility standards those who have a synergy with the pillars of sustainability, i.e. with the environmental, economic and social dimensions.

In essence, the normative standards are documents that define the most effective and efficient way to make a product or perform a task. These normative standards are written by a rigorous and formal process, involving extensive consultation with all relevant agencies, and outlines the organizational "best practices" and regulatory obligations to fulfill. When treated in the context of sustainability, standards of management systems have an extremely useful role in providing a roadmap for the implementation of sustainable organizational development in a growing number of areas (Baron, 2010).

On this point, Vasques (2007) stresses that the absence of a normative standard for sustainable development hinders the beginning of the actions of organizations because they do not get guidelines developed.

Given this, companies wishing to start deploying a sustainable organizational policy end up finding difficulties, since no access to information, is little scientific content by writing about it, is the high cost that the information has, when obtained from consulting firms.

Recently, with the publication of the updated normative standards of quality management, environmental management, management of health and safety and social responsibility versions, there has been a high degree of compatibility and similarity between these (Jorgensen & Simonsen, 2002; Jorgensen, Remmen & Mellado, 2006; Bernardo, Casadeus, Karapetrovic & Heras, 2009), which spawned the ambition and tendency to integrate them, the fact that, according to Jorgensen (2008), which would allow to create a common normative standard to promote management systems more sustainable.

Such integration of management standards is discussed through the concept of IMS, which seeks to integrate commonalities of various management systems and respect the individuality of each standard (Celik, 2009). This reasoning can be used as the basis of a model of internal measurement of sustainability, given that regulatory standards, as well as the dimensions have a strong synergy between themselves and particular characteristics that need to be respected.

To achieve sustainability and the concept of the triple bottom line, Tsai and Chou (2009) advocate the need for organizations to implement and certify their systems through regulatory management standards such as ISO 9000, ISO 14000, OHSAS 18000 and SA 8000.

However, while the integration of the dimensions of sustainability cannot be deduced, manage aspects of the triple bottom line remains a challenge (GRI, 2010). Further integrate dimensions mean embracing the ambiguity in dealing with an elusive and diverse range of themes and values (Jamali, 2006).

For Jorgensen (2008), the mechanism for organizations to become sustainable is the need to analyze, holistically, production processes and products developed in order to improve them on an operational level, in all its parts and throughout the chain, you will see the quality, the environment and the health and safety of employees as to make them sustainable mechanism.

With an IMS, Pojasek (2011) argues that the management of sustainability can be achieved through communication between existing systems. The proposed Jorgensen (2008) and Pojasek (2011) meets the objective contained in this chapter, where achieving sustainability concerns, first, the internal reality of the organization, then come public perceptions (Lueneburger & Goleman, 2010), i.e. the internal structure through integrated management systems can be a powerful mechanism for achieving organizational sustainability.

The definitions of the concept of IMS, processes and outcomes, according to Bernard et al. (2009), vary not only depending on the situation of the organization, but also because of the particularity of the studies. Karapetrovic, Casadesus and Heras (2010) complement these few studies, when they occur, are based on theoretical or empirical cases, contributing to the need for debate on the subject.

The theoretical conceptualization of IMS in the literature, according to Zeng, Shi and Lou (2007), are still restricted to a few jobs. Can be cited as studies: Beckmerhagen, Berg, Karapetrovic and Willborn (2003), Karapetrovič (2003), De Cicco (2004), Pojasek (2006), Arifin, Aiyub, Awang, Jahi and Iteng (2009) and Bernard et al. (2009).

The latter, Bernard et al. (2009) combining the definitions of various authors, summarizes the concept of IMS as the process of connecting various management systems of the organization into a single system to common resources and continually improve stakeholder satisfaction.

This integration of the management system is more than simply adding elements of standards in an already structured system (Fresner & Engelhardt, 2004), and cannot be set to put different systems in a single manual (Wilkinson & Dale, 1999).

The IMS shape "system of systems", each system maintains its individual identity (Wilkinson & Dale, 1999), aiming to use the synergies between these rules during the creation and implementation of a management system (Mackau, 2003).

Although studies leading to the formulation of the theory of IMS are being made, there is no consensus among authors about the theoretical definition of the subject. What exists is a systemic view on the need for a consolidated management, but still at an early stage, where we seek the integration of systems (Karapetrovic, 2010; Simon, Karapetrovic & Casadesus, 2012; Satolo, 2012; Satolo, Calarge e Cauchick, 2013).

This lack of consensus on the definition of IMS can be seen in the various terms used to characterize it. In the international literature, the terminology of IMS is treated as Integrated Management System (Mackau, 2003; Holdsworth, 2003; Bamber, Sharp, &Castka, 2004; Zeng, Shi & Lou, 2007; Arifin, et al., 2009; Bernardo et al., 2009; Bernardo et al., 2010; Santos, et al., 2011; Hamidi, Omidvari & Meftahi, 2012), Integrated Management Systems (Wilkinson & Dale, 1999; Karapetrovic, 2002; Beckmerhagen et al., 2003; Fresner & Engelhardt, 2004; Jorgensen et al., 2006; Salomene, 2008; Karapetrovic, Casadesus & Heras, 2010; SIMON et al., 2012) and as a Unified Management System (Jorgensen & Simonsen, 2002).

This variety of classifications, sorted by Hacking and Guthrie (2008) as an "alphabet soup" of acronyms and terms, currently generates a confusing picture about the topic.

In this chapter, the term will be adopted Integrated Management System (IMS), it will be dealt with the integration of various systems into a single management system.

This integration of various management systems is treated in the literature linking up quality, environment, and health and safety at work (Wilkinson & Dale, 1999) and, more recently, social responsibility (Jorgensen et al., 2006).

Thus, it becomes necessary to understand the structure and evolution of the standards for IMS, which is the subject of the next section. It caveat, however, that this is a brief explanation of such rules, not being objective in this section or further explain any points for standards, but important for the purpose of this chapter points.

Basic normative standards and the Integrated Management System (IMS) to reach organizational Sustainability

As shown in the previous section and illustrated in Figure 2, in an IMS are integrated basically four normative standards: quality [ISO 9000 (ISO, 2008a)]; environment [ISO 14001 (ISO, 2008b)]; occupational health and safety [OHSAS 18001 (BSI, 2008)], and social responsibility [ISO 26000 (ISO, 2010)].

Other normative standards may select IMS organization. Are cited as examples specific standards for industries such as food [ISO 22000 (ISO, 2006)] aerospace [AS 9100 (SAE, 2004)], automotive[TS 16949 (ISO, 2002)], telecommunication [TL 9000 (ISO, 1998)], information systems [ISO 27001 (ISO, 2007a)], and supply chain management [ISO 28000 (ISO, 2007b)].

However, according to the author, these normative standards are rarely discussed in the context of IMS.

In order to clarify and facilitate the understanding of IMS, according to Wilkinson and Dale (1999), it is important to understand the history and structure relevant to IMS standards, because the need for clear definitions is always useful to establish good communication about a particular topic, especially when this is an emerging issue, as in this case.

In this chapter, the use of the concept of IMS features is restricted to the four management systems commonly employed by organizations (quality, environmental, occupational health and safety, and social responsibility) because they are central themes of the work and used as the basis for the proposal the measurement of organizational sustainability model.

Figure 2. Model of a IMS

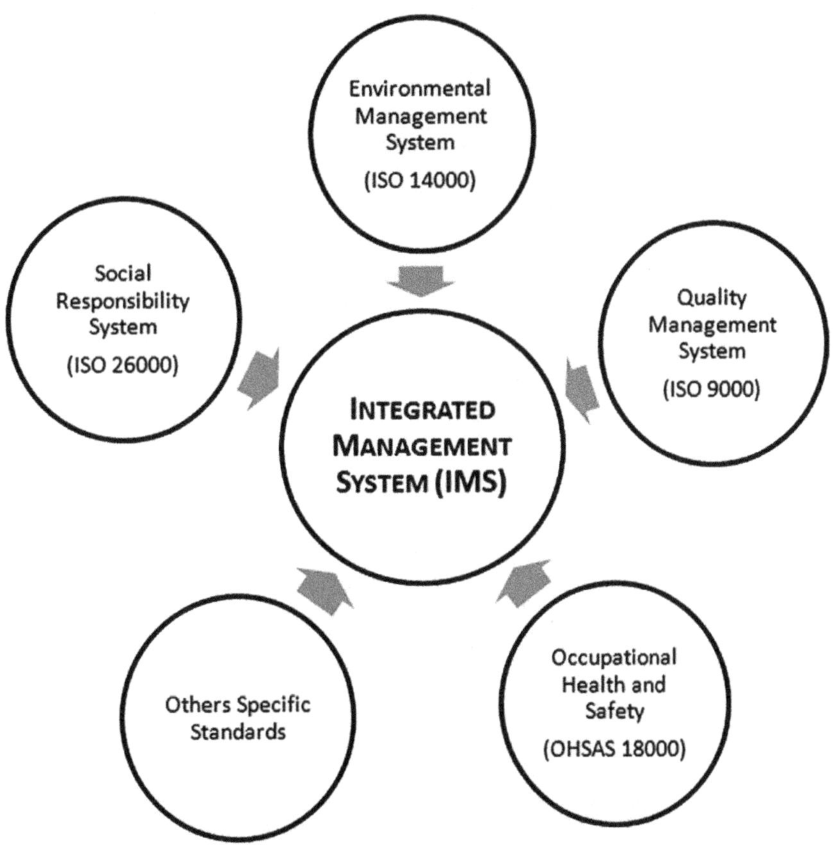

Quality Management System: ISO 9001

ISO 9001 describes a number of key elements to enable the design and implementation of the quality management system (Zeng, Shi & Lou, 2007).

It is considered one of the most successful and most deployed normative standards (Bayati & Taghavi, 2007), and the development of its first release occurred in 1987. Currently, the standard is presented in its fourth update, in force since December 2008, and which sought mainly increased compatibility with ISO 14000 families (Mollenkopf et al., 2010).

ISO 9001 consists of 18 requirements, namely: general requirements, documentation requirements; responsibility of management, planning, resource management, infrastructure, work environment, planning of product realization, customer-related processes, design and development; acquisition, product and service delivery, control of monitoring and measuring equipment, measurement, analysis and improvement, control of nonconforming product, analysis of data, and improving (ISO, 2008a).

The implementation of the requirements of the standard in the organization can occur in conjunction with the tool called PDCA cycle, and the elements involved shown in Figure 3.

Figure 3. Model of a quality management system based on process (ISO, 2008a)

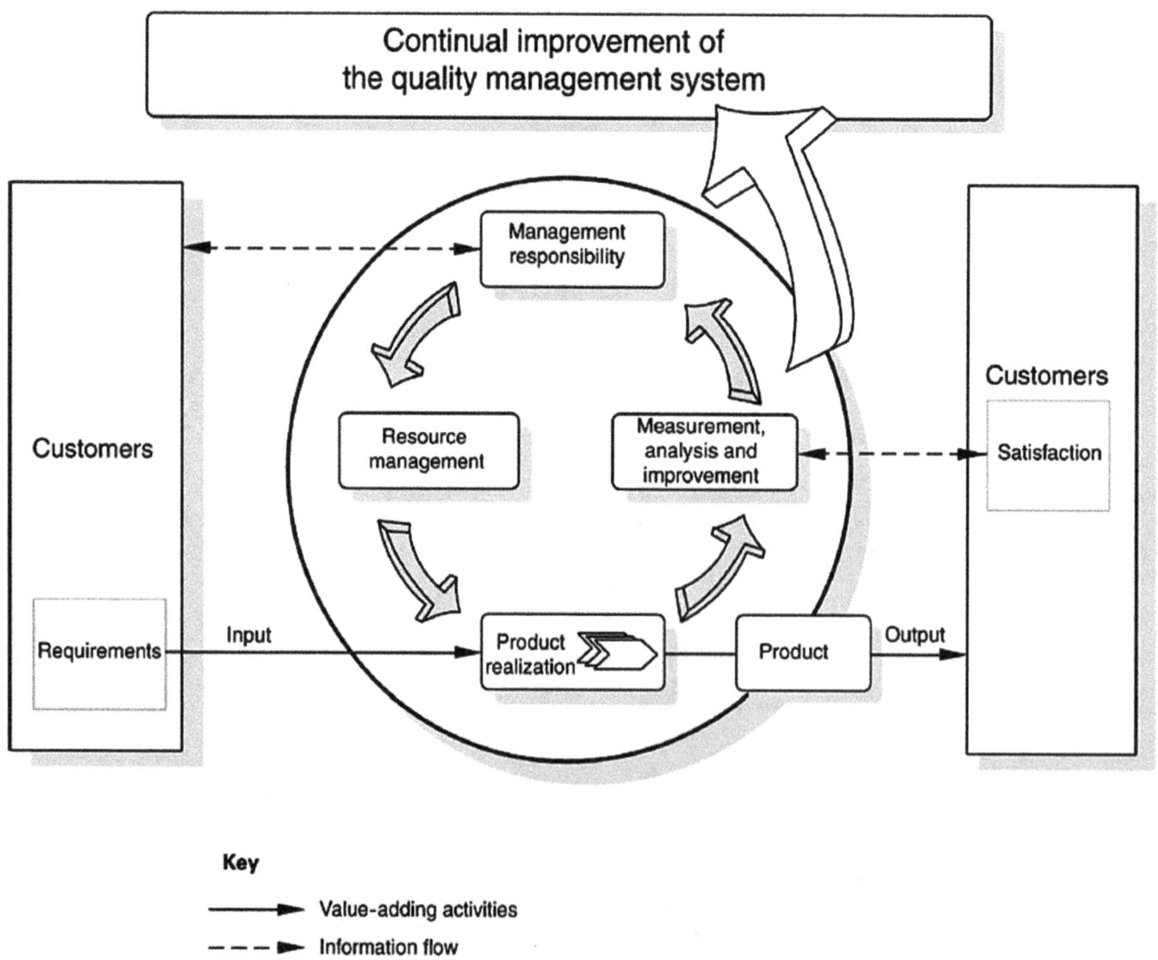

To Naveh and Marcus (2005), the implementation of ISO 9001 requires that the company has an internal standard procedure, replicable routines and processes for product development, manufacturing, delivery, service and support.

The benefits associated with implementing ISO 9000 reported in the literature are diverse, being featured: definition of consumer focus, increase access to international markets, increase the level of employee engagement, maintaining the consistency of the business processes, ensuring the reduction loss of time, material and effort, implementation of continuous improvement. Its implementation, however, only confirms that the company has an effective quality management system, but does not guarantee the quality of products manufactured by the company, this warranty is through an actively managed and implemented by the culture of quality (Jorgensen, 2008; Terziovski & Power, 2007).

The acceptance and success of this standard are significant, verifying that, worldwide, by December 2012, more than 1.1 million companies are certified in 184 countries (ISO, 2013).

Environmental Management System: ISO 14001

The objective of ISO 14001 is to create an environmental management system that allows the organization structuring mechanisms for continuous improvement in performance and reduction in resources consumed and, therefore, of environmental resources (Bernardo et al., 2009).

The ISO14000 series of standards, established in 1996, was reviewed in 2004 (Lagodimos, Chountalas & Chatzi, 2007) creating a higher level of similarity with the standards of ISO 9000 series (Karapetrovic & Casadesus, 2009).

The requirements covering the ISO 14001 comprises a total of 22, which are requirements of the environmental management system, general requirements, environmental policy, planning, environmental

Figure 4. Elements of an environmental management system (ISO, 2008b)

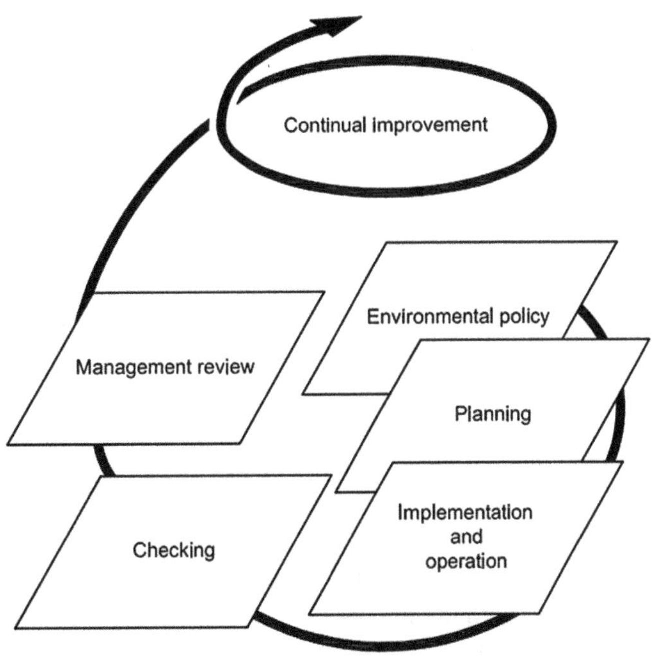

aspects, legal and other requirements, objectives and targets; management program (s) environmental, implementation and operation, structure and responsibility; training, awareness and competence, communication, environmental management system documentation, document control, operational control, emergency preparedness and response; checking and corrective action, monitoring and measurement, non- compliance and corrective and preventive actions; records; audit of the environmental management system, management review (ISO, 2008b).

The implementation of these requirements (shown in Figure 4), as well as ISO 9000, involves the PDCA cycle. According to Zutshi and Sohal (2004), during the deployment process, one should take into account five key elements: environmental policy, planning, implementation and operation, inspection and corrective action, and management review.

Oliveira and Serra (2010) list a number of associated with the deployment of an Environmental Management System, among these benefits: access to new markets, law-abiding management, regulatory incentives, risk reduction, access to more capital, the improvement of the production process, improvement of environmental performance improvement in the overall management of the company, improving the relationship with employees, improving public image, competitive advantage within specific segments, meeting the demands of customers, increased quality of life, achievement of clean operations (green), increasing the competitiveness of the product or service and public awareness.

There are approximately 285.000 companies in 167 countries worldwide ISO 14001 certified (ISO, 2013).

Occupational Health and Safety: OHSAS 18001

The Management System for Occupational Safety and Health is characterized by specifying procedures for organizations to manage their activities, products and services, in order to control and prevent circumstances that may result in accidents, injuries or occupational diseases.

This standard emerged in 1996 as BS 8800 - Guide to Occupational Health and Safety Management System Safety. In 1999, OHSAS 18001 (Occupational Health and Safety Assessment Series) was launched as the first specification of Health and Safety to submit character certificate (Zeng, Shi & Lou 2007), having passed in July 2007, for its first review.

Its creation was designed to allow a harmonious integration with the standards of quality management and environmental management. To do so, requires the preparation of a specific security plan to be updated on an annual basis, and is broad enough to include the entire organization.

This similarity can be seen by the 22 requirements that comprise the OHSAS 18001 standard, which include: general requirements planning; planning for hazard identification, risk assessment and risk control, legal and other requirements; occupational health and safety policy objectives; implementation and operation; resources, roles, responsibility, accountability and authority, competence training and awareness, communication, participation and consultation, documentation, document control, operational control, preparedness and emergency response, monitoring and performance measurement, evaluation of conformity; incident investigation, nonconformity, corrective and preventive actions; records and record management, internal audit, management review (BSI, 2008).

Representative elements of the implementation of OHSAS 18001 processes also use the concepts of PDCA cycle and are highlighted schematically in Figure 5.

Among the benefits associated with the implementation of OHSAS 18000 reported in the literature include: reducing the number of accidents through prevention and control of hazards in the workplace,

Figure 5. Elements of a management system for occupational health and safety (BSI, 2008)

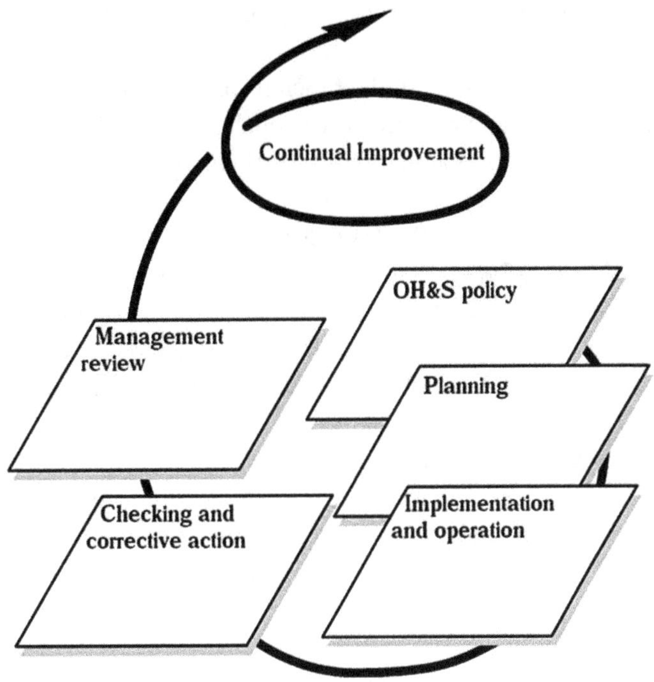

reducing the risk of serious accidents occur, reducing material losses caused by accidents and downtime, reducing severance costs and insurance, cost reduction with the withdrawal of workers, the potential reduction in downtime and associated costs, demonstrating to stakeholders of corporate commitment to health and safety (Zeng, Tam &Tam, 2008; Duijm, Fievez, Gerbec, Hauptmanns & Konstandinidou, 2008; Vinodkumar &Bhasi, 2011).

Jorgensen et al. (2006) describe the ISO voted twice times for the development of a standard in their patterns, both being denied. However, in 2013, began working for the creation of the new ISO standard, ISO 45001, on occupational health and safety management system requirements is being produced by the Project Committee, ISO PC 283, with the intention of publication in October 2016. The standard will be aligned with ISO 9001 (Quality Management) and ISO 14001 (Environmental Management), themselves which are undergoing revision and are due for publication in 2015.

According to the latest available survey, worldwide, approximately 55,000 companies are certified in 80 countries (BSI, 2011).

Social Responsibility: ISO 26000

ISO 26000 (ISO, 2010) standard was launched in 2010, resulting in a five-year trial involving 450 experts from 99 countries. His aim is to establish an international standard for organizational establishment of a Management System of Social Responsibility.

Its origin is based on SA8000, launched in October 1997, the Social Accountability 8000 (SA8000), which established the first global standard certification focused on corporate social responsibility. Created by CEPAA - Council on Economics Priorities Accreditation Agency, currently called the SAI

- Social Accountability International - U.S. non-governmental body - the standard focuses on securing the rights of workers involved in production processes, promoting standardization in all sectors business (Willians, 2010).

The ISO 26000 aims to guide companies to adopt socially responsible practices. During its preparation, rose the possible parallels with the standards already established Quality Management and Environmental Management, consisting of eight requirements, highlighted in Figure 6: organizational governance, human rights, labor policies, environment, fair practices operation, consumer issues, community involvement and development (Castka & Balzarova, 2008; ISO, 2010).

According Pojasek (2011), ISO 26000 is not a management system such as ISO 9000 and ISO 14000, but a roadmap for organizations that want to establish a system of Social Responsibility. According to the author, the generation of a system of organizational management, like ISO 9000, should be supported in the specification ISO 26000, which guides the structuring of a system of social responsibility management based on PDCA cycle.

As new rules, in order to guide organizations, is not yet scheduled for audits to obtain certifications (Castka & Balzarova, 2008; Pojasek, 2011). Similarly, point benefits are still limited to a few published studies, and further efforts need from researchers. Perera (2009) in one of these studies highlights benefits in terms of productivity gains, improved image and cost savings.

Figure 6. Elements of the social responsibility norm (ISO, 2010)

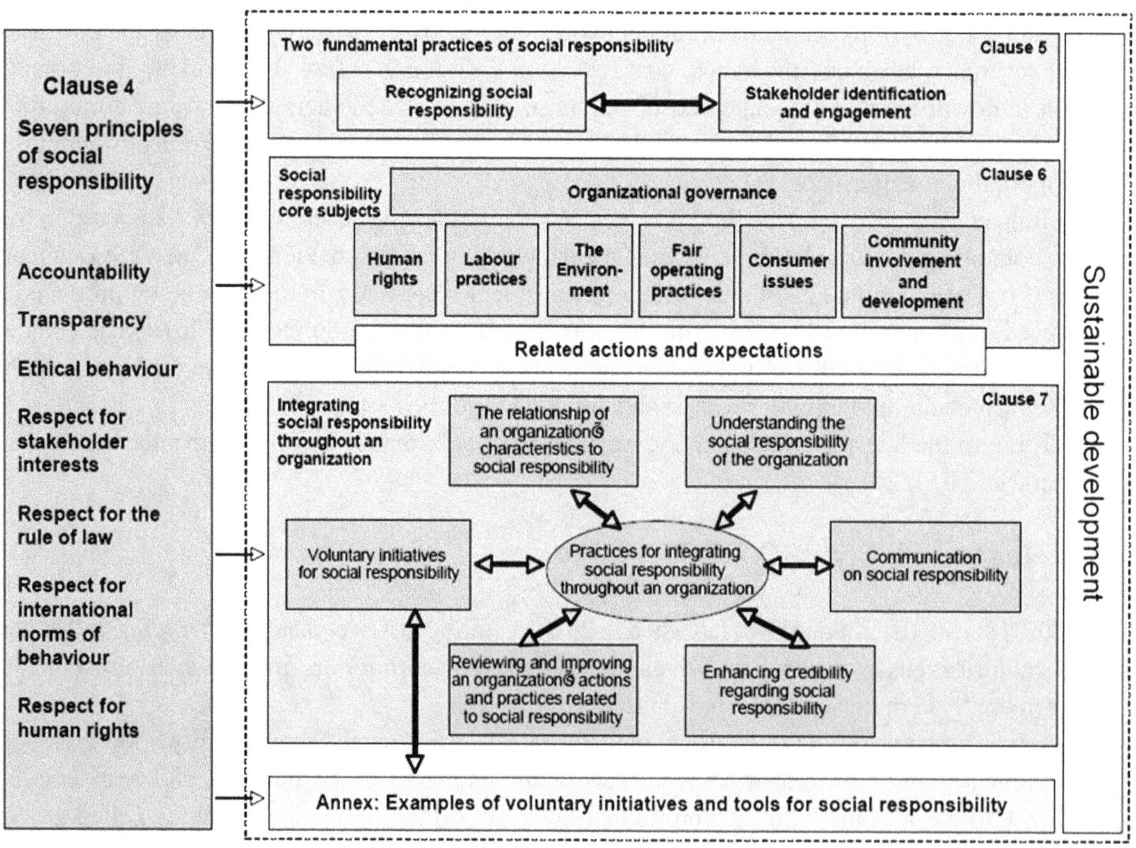

NORMATIVE STANDARDS FOR MANAGEMENT A INTEGRATED SYSTEM

Although this discussion on the creation of an integrated system governed by a single normative standard has become an obvious thread, currently, there is a worldwide accepted standard that covers an integration methodology (Bernardo, Casadesus, Karapetrovic & Heras, 2010).

Internationally, the ISO published in 2008 the book called The Integrated Use of Management System Standards, reporting and promotes a reference to sixteen experiences of implementing IMS (ISO, 2008c). This book, however, lacks a method for deployment and management of the system.

At the national level, different countries have developed their methods of integration, i.e., Australia and New Zealand - AS / NZS 4581: 1999 Denmark - DS 8001: 2005 Spain - UNE 66177: 2005 and the UK - PAS 99: 2012 (Bernardo, 2009; Bernardo et al., 2010.). These normative standards for deployment of IMS will be briefly discussed in the following subsections.

AS/NZS 4581: 1999

Was published AS / NZS 4581:1999 standard, developed by standardization bodies in New Zealand and Australia in 1999. According to Ferreira (2009), this normative standard addresses the integration of management systems guidelines for organizations with the purpose of business, government and community. The rules do not restrict its use only for quality management, environmental, safety and occupational health system and can be applied to all types of management systems.

According to the author, establishes the rules common to all systems components to be integrated. To this end, comprises nine components for a management system, which: responsibility and leadership management, identification and needs analysis, policy and objectives, planning and implementation of the system, resource allocation, communication and information systems, management processes and activities, monitoring and measurement, critical analysis of the direction and improvement plans (Arifin, 2009). Figure 7 shows the model of IMS standard AS / NSZ 4581:1999.

Figure 7. – IMS Model used in the standard AS / NZS 4581:1999 (Ferreira, 2009)

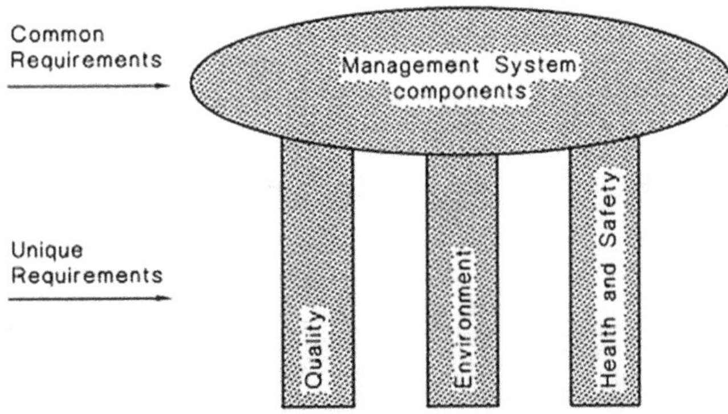

UNE 66177: 2005

In 2005, Spain issued its normative standard IMS, called UNE 66177:2005 - Systems gestión Guide to integración la Gestión de los systems. The normative standard defines IMS as the set formed by the combination of organizational structure, responsibilities, procedures, processes and resources that are established for the integrated management systems (Barral, 2007; Guillén & Carmenates, 2009).

The steps for implementation of a IMS -based normative standard UNE 66177 (2005) and the study of Guillén and Carmenates (2009) are divided into four:

1. Design: creating the objective and subjective conditions to begin to define the system and obtain the commitment of all levels.
2. Planning: determines the status of the organization for the implementation of the current regulatory standards for the development and integration of existing systems, set the integration method and program of activities to implement and effectively manage the changes.

Figure 8. Model proposed by UNE 66177 (AENOR, 2005) for the implementation of IMS

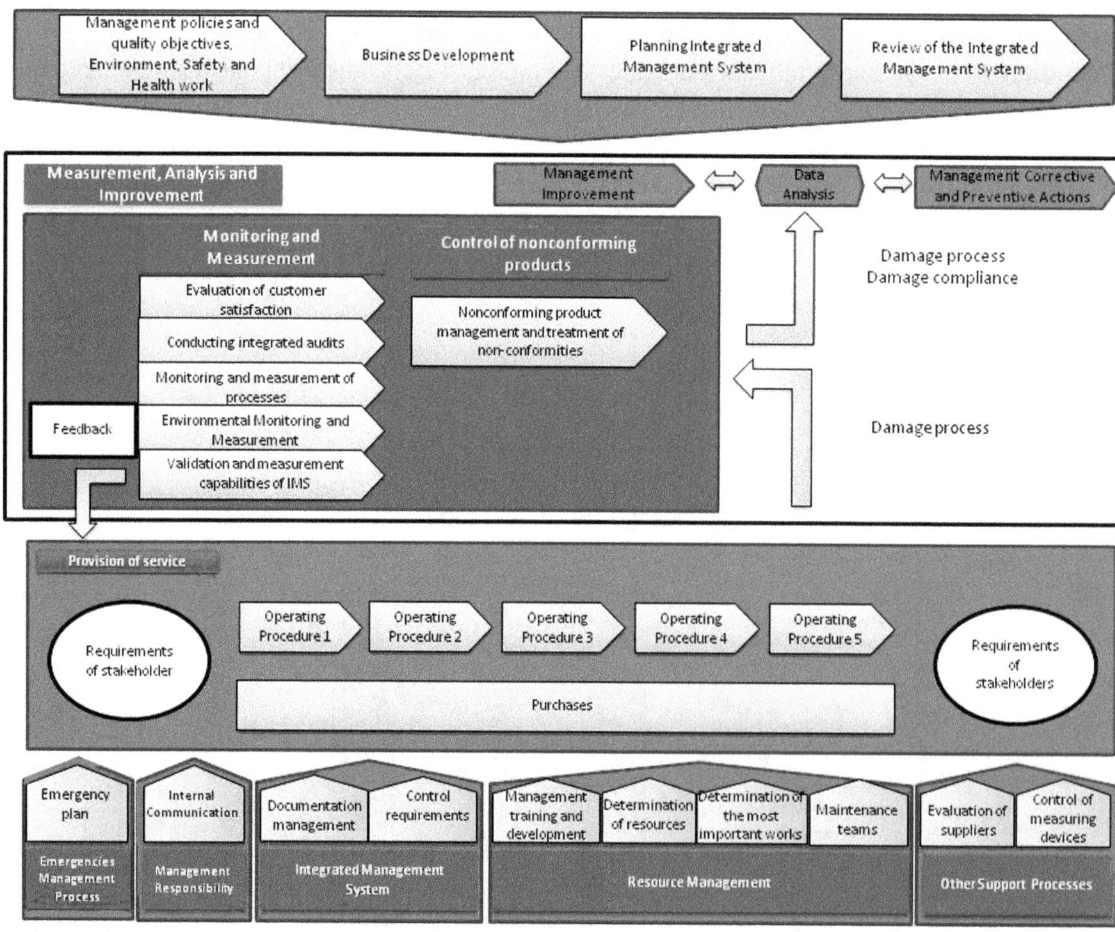

3. Monitoring and measurement: objective periodically evaluate and adjust the implementation plan and ensure the suitability, adequacy and effectiveness of the system since its implementation was performed.
4. Continuous Improvement: aims to identify opportunities for continuous improvement to help increase the performance of IMS, and the appropriate tools in each case.

Figure 8 is proposed by UNE 66177 (AENOR, 2005) for the implementation of a IMS model.

DS 8001: 2005

The normative standard Danish (Dansk Standard 8001, 2005) had started its production in 2003 from studies by a large working group under the motto'' an organization, a system "(Jorgensen, 2008). The standard describes the characteristics of a good management and common and individual elements that may involve a IMS (Rasmussen, 2007).

Figure 9 is proposed by DS 8001 (Dansk Standard, 2005) to implement a IMS model.

In the first level, the normative pattern possessed a strategic approach, with emphasis on corporate governance. In this step, we try to emphasize the importance of the management system, organizations

Figure 9. Model proposed by DS 8501 (2005) for the implementation of IMS

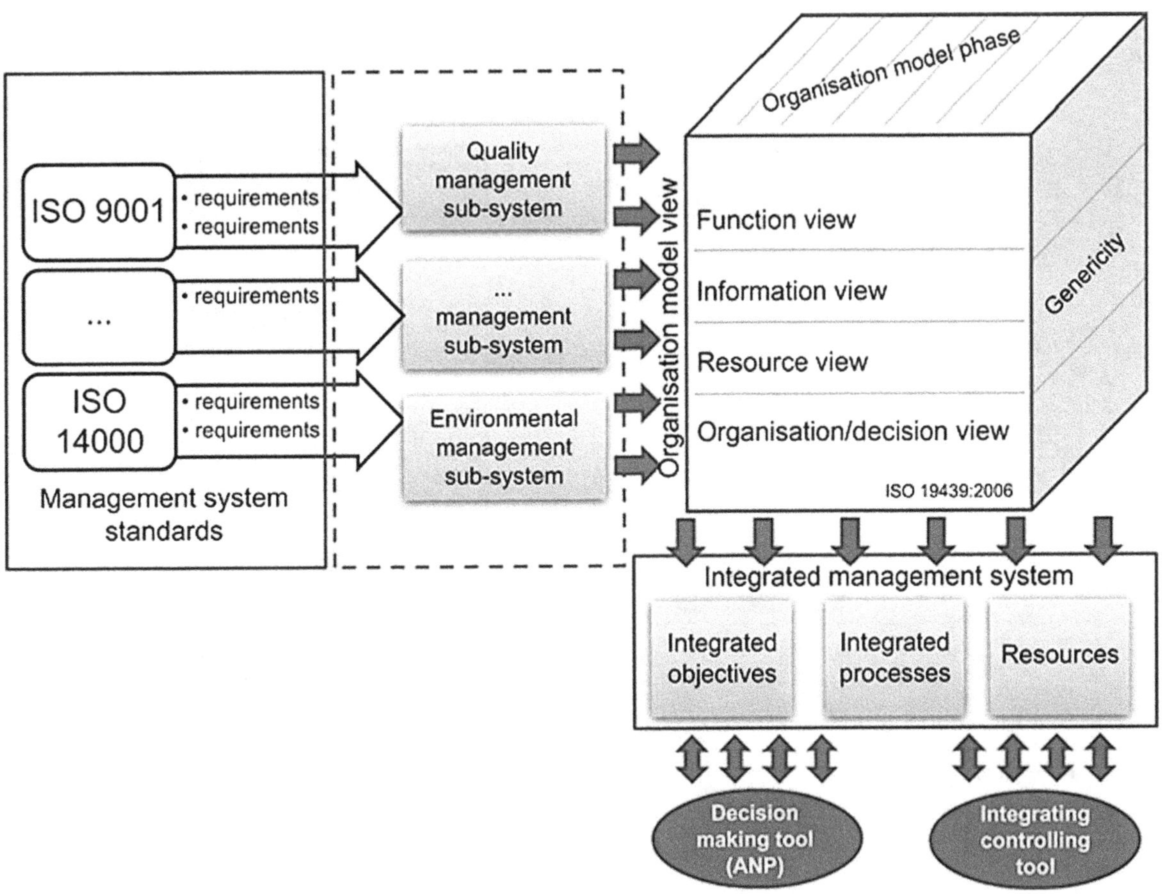

demonstrating the starting point for business strategies. The organization has developed a mission, vision and strategy, and create corresponding goals and policies, meeting the demands of stakeholders (Rasmussen, 2007; Jorgensen, 2008).

At the second level, the normative standard describes the common elements in an IMS and also provides standard tools for such integration.

Finally, the third level of the normative standard, discusses the different management areas and describes how an organization can choose to integrate your IMS, employing several security management systems such as quality, environment, energy, occupational health and safety, food, social responsibility, risk management, among others (Rasmussen, 2007; Jorgensen, 2008).

PAS 99:2012

In 2006, the United Kingdom through the British Standards (BSI), based on the ISO Guide 72:2001, developed his legal standard to meet the common requirements of an IMS, called PAS 99 (Publicly Available Specification), and which underwent its first revision in 2012. This objective normative standard to be used as a model for the implementation of common standards or specifications of requirements management system in an integrated manner (Vasconcelos, 2011).

To Vasconcelos (2011), the PAS 99:2006 provides a simple model for organizations to integrate into a single structure all the standards and specifications of management systems that adopt. However,

Figure 10. Requirements to delineate the structure of the IMS (BSI, 2006)

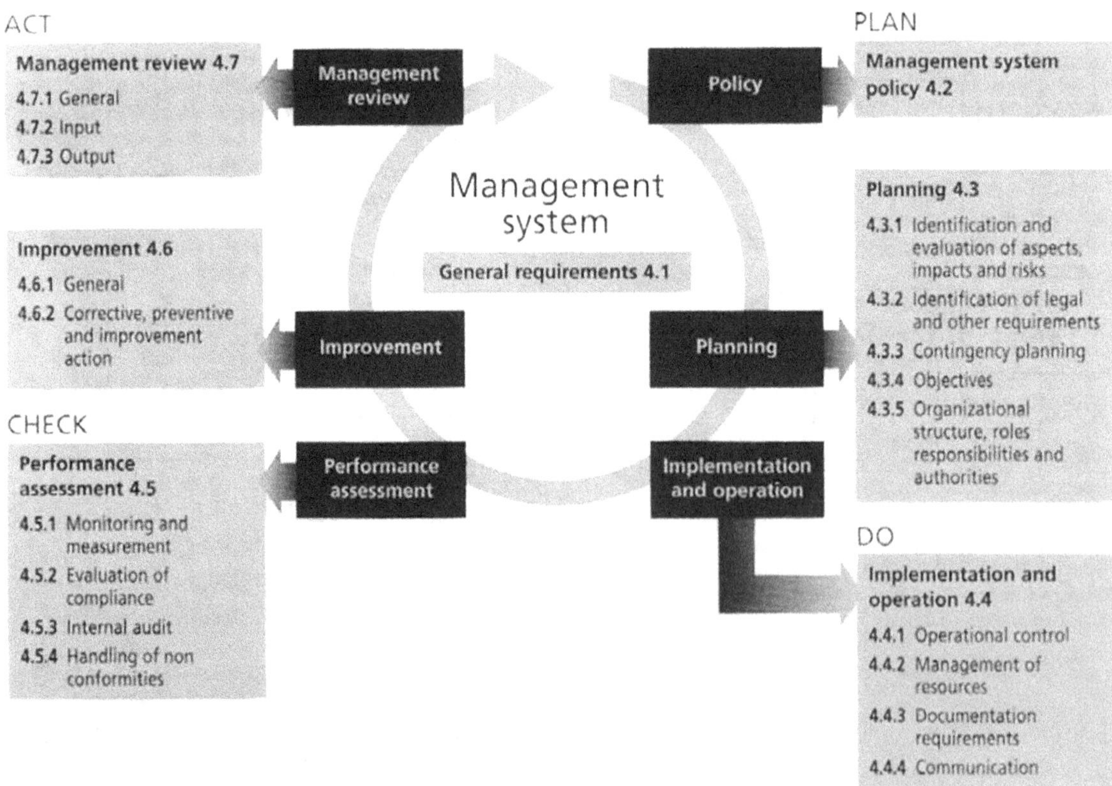

to properly deploy PAS 99:2006 does not guarantee the company than the other management systems comply with regulatory standards governing them, i.e., the joint implementation of the environmental management system and quality through PAS 99:2006 does not guarantee that it fully conforms to the required respectively by ISO 14001 and ISO 9001 standards. The structure created by the PAS 99:2006 is shown in Figure 10.

INTERRELATIONS BETWEEN THE REQUIREMENTS OF THE STANDARD PAS 99:2012 AND DIMENSIONS OF ORGANIZATIONAL SUSTAINABILITY

The design of an IMS is given for the use of synergies between the different normative standards and the elimination of redundancies within them.

In recent years, the revision of normative standards ISO 9001, ISO 14001, OHSAS 18001 and ISO 26000 increased the similarities between different patterns and made them more compatible. This improved the possibility of integrating the standards and supporting links between the necessary systems allowing target a sustainable management system (Karapetrovic, Casadesus & Heras, 2010; Bernardo et al, 2010). These standards form the integrated IMS and have high interaction with the dimensions of sustainability (triple bottom line).

The equivalence between the requirements of the normative standards ISO 9000, ISO14000 and ISO18000 is presented in Table 1, made from the proposed requirements for integration of management systems with the normative standard PAS 99.

Table 1. Equivalence between the elements of an IMS based on the requirements of the standard PAS

Requirements		Equivalence between the Normative Standards and PAS 99:2006	Requirement on			
Version 2006	Version 2012		ISO 9001	ISO 14001	OHSAS 18000	ISO 26000
4.1	4.3 4.4	General requirements	4.1.	4.1	4.1	5.1
4.2	5.2	Management system policy	5.3	4.2	4.2	5.2.1
4.3	6	Planning	5.4	4.3	4.3	3.3
4.3.1	4.1 6.1.1	Identification and evaluation of aspects, impacts and risks	4.2.2	4.4.4	4.5.1	3.5.2
4.3.2	4.2	Identification of legal and other requirements	7.2.1	4.3.1 4.3.2 4.4.6	4.3.2	5.2.1 5.2.2 5.2.3
4.3.3	6.1.2	Contingency planning	8.3	4.4.7	4.4.7	6.4.6.2. 6.5.3.2
4.3.4	6.2	Objectives	5.4.1	4.3.3	4.3.3	5.2.1
4.3.5	5.1 5.3	Organizational structure, roles, responsibilities and authorities	5.	4.4.1	4.4.1	6.2.1.1 6.2.1.2
4.4	8	Implementation and operation	7	4.4	4.4	-
				4.4.6	4.4.6	3.4.3

continued on following page

Table 1. Continued

Requirements		Equivalence between the Normative Standards and PAS 99:2006	Requirement on			
Version 2006	**Version 2012**		**ISO 9001**	**ISO 14001**	**OHSAS 18000**	**ISO 26000**
4.4.1	8.1	Operational control	4.2.3	4.4.5	4.4.5	5.2.1 5.2.2 5.2.3
			4.2.2	4.5.3	4.5.3	6.4
4.4.2	7.1	Management of resources	6	4.4.1	4.4.1	6.2
	7.2		6.2.2	4.4.2	4.4.2	6.4.7
4.4.3	7.5 7.5.1 7.5.2 7.5.3	Documentation requirements	4.2	4.4.4	4.4.4	7.3 7.4 7.5
4.4.4	4.2	Communication	7.2.3	4.4.3	4.4.3	6.2.2
	7.4		5.5.3	4.4.3	4.4.3	4.3
4.5	4.5	Performance assessment	8	4.5.1	4.5.1	6.2.3
4.5.1	9.1 9.1.1	Monitoring and measurement	8.2.3	4.5.1	4.5.1	6.5 6.4.6 6.4.7
4.5.2	9.1.2	Evaluation of compliance	8.2	4.5.1	4.5.2	6.4.6 6.4.7
4.5.3	9.2	Internal audit	8.2.2	4.5.3	4.5.4	-
4.5.4	10.1	Handling of nonconformities	8.3 8.5.2 8.5.3	4.5.2	4.5.2	6.7.6
4.6	10	Improvement	8	4.5	4.5	7.7
4.6.1	10.2	General	8.1	4.5.1	4.5.1	7.7.1
4.6.2	10.1	Corrective, preventive and improvement action	8.2.3	4.5.1	4.5.1	-
			8.2.4			
4.7	9.3	Management review	5.6	4.6	4.6	7.3
4.7.1	9.3	General	5.6.1.	4.6	4.6	7.3
4.7.2	9.3.1	Input	5.6.2.	4.6	4.6	7.3
4.7.3	9.3.2	Output	5.6.3	4.6	4.6	7.3

*Retired from Risk Tecnologia (2006)

Justifies the choice of the standard PAS 99 standard as the basis for establishing the inter-relationship with the sustainability, three important aspects: (i) this normative standard, although not universal, stands out for its further spread in relation to other models presented (ii) has as proposal for its creation, maintaining strong relationship to the normative ISO and OHSAS standards, and (iii) its implementation, as well as in other normative standards in its process uses the PDCA cycle, thus facilitating system management the high level of similarity.

It is observed from Table 1 the high level of synergy between the three normative standards. Verifies the existence of various sections / subsections that interrelate as well as items that are characteristic only a few patterns, highlighting the concept presented by several authors (Jamali, 2006; Celik, 2009; Bernardes et al., 2009; Pojasek, 2011) where the systems that make up an IMS feature common and particular points that need to be respected.

The establishment of the interrelationships between normative standards beyond demonstrates strong synergy, the possibility to carry out the implementation of the management system following the steps of the PDCA cycle, so this strongly defended by normative standards.

The PDCA cycle had its origin with Dr.W. Edwards Demings lectured in Japan in 1950. Deming, Father of total quality management has put forth the concept of Plan, Do, Check and Act cycle. The PDCA cycle was broadened to include strategies and methods to develop, test, and implement changes that would result in improvement. The PDCA-cycle is a never ending cycle designed to improve quality and efficiency. PDCA cycle approach to quality improvement helps on institute to focus on the goal and obtain the commitment of all the stakeholders to the accomplishment of the goal set (Singh, 2013).

Thus, the regulatory requirements presented in Table 2 can be grouped into six major areas, which are related to stage of the PDCA cycle. The six topics are: (i) policy management system; (ii) Planning, (iii) Implementation and operation; (iv) Performance assessment; (v) improvement and (vi) Management review. These relationships are shown in Figure 11 and described in Table 2.

Table 2. Relationship between Normative Standards and Requirements of IMS and PDCA Cycle steps

PDCA Cycle Steps	Axes		Requirements	
Plan	1	Management system policy	1.1	Management system policy
	2	Planning	2.1	Identification and evaluation of aspects, impacts and risks
			2.2	Identification of legal and other requirements
			2.3	Contingency planning
			2.4	Objectives
			2.5	Organizational structure, roles, responsibilities and authorities
Do	3	Implementation and operation	3.1	Operational control
			3.2	Management of resources
			3.3	Documentation requirements
			3.4	Communication
Check	4	Performance assessment	4.1	Monitoring and measurement
			4.2	Evaluation of compliance
			4.3	Internal audit
			4.4	Handling of nonconformities
Act	5	Improvement	5.1	General
			5.2	Corrective, preventive and improvement action
	6	Management review	6.1	General
			6.2	Input
			6.3	Output

Source: Elaborated by the authors

Figure 11. Relationship between Normative Standards of IMS and PDCA Cycle steps

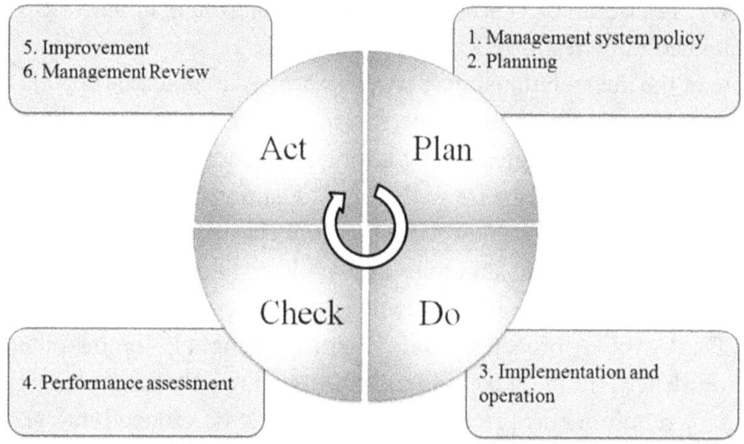

The implementation of IMS in this way now goes by meeting the requirements of PAS 99 standard, and that are directly related to the environmental, social and economic dimensions proposals for sustainable development, and these relationships below described.

PLAN

Axis 1: Management System Policy

Requirement 1.1: Management System Policy

It is understood that the policy of the management system of the organization should include in its mission / vision the social and environmental dimensions as well as the maintenance of its economic health, showing the internal and external stakeholders concern for the future of the organization and provide gains beyond the financial interests with ethical behavior and respect for their partners and the environment, envisioning long-term career plan to ensure more able and motivated human resources.

Axis 2: Planning

Requirement 2.1: Identification and Evaluation of Aspects, Impacts and Risks

The organization shall establish, implement and maintain procedures to identify areas of risk and impact of its activities, products and services within the defined scope of your management system. Should establish a set of procedures for social and environmental assessment to be followed and their interconnection with the economic evaluation of the project. These impacts and risks can be evaluated taking into account: the image of the organization, legal points, points financial, strategic points, operating points and outsourcing.

Requirement 2.2: Identification of Legal and Other Requirements

This requirement refers to the establishment, implementation and maintenance of procedures to determine the legal and other requirements related to its activities, products and services that are relevant to the scope of the management system requirements in order to take them into account in establishing, implementing and maintenance management system. It is important to understand the legal requirements in this (mainly environmental) regulation as a mechanism of competitive opportunity and not as a punishment mechanism. Identify which are the critical demands of society in relation to the business and likely predict which actions will be taken by relevant standards or policies and make investments for efficiency gains and provide significant competitive advantage. This will bring to the organization the opportunity to gain know-how coming from pioneering and reputation, which will reach it, the economic part of the organization.

Requirement 2.3: Contingency Planning

The organization shall establish, document and maintain procedures to identify and respond to any unplanned event, emergency or potential disaster. This procedure should aim to prevent or mitigate the consequences of any such occurrences and consider the continuity of business operations. Are among these contingencies from product recall actions until disaster management such as fire, flood or environmental disasters. An unplanned, emergency or disaster event occurs without the knowledge of such organization possesses and, in most cases, is directly related to its stakeholders in its various representations, i.e., an employee who may be an accident within the organization or client that might get hurt due to failure in a product. Whatever the contingency, the company brings the need for economic planning initially to invest in improvements to prevent these events and also to face the need for repairs to society and to the internal and external environment due to its occurrence.

Requirement 2.4: Objectives

Shall be established by the organization objectives that take into account significant aspects, legal obligations, other requirements and its commitment to continuous improvement by implementing their policy, which must be measurable, relevant and realistic. It is important that the objectives of the management system envisage and to target improvements for employees and the environment, through continuous improvement of existing conditions in the workplace and improve its performance with respect to the amount of resources that are used to purchase raw materials and products finished in order to reduce them to short, medium and long term.

Requirement 2.5: Organizational Structure, Roles, Responsibility and Authority

The organization must indicate specific representatives of management who, irrespective of other responsibilities, shall have a function, defined responsibility and authority for ensuring that the management system is established, implemented and maintained, and also report to top management performance management system for critical analysis, including recommendations for improvement. Must representatives have a knowledge of standards / specifications adopted in the organization's management system so that they can contribute significantly in the continuous improvement of social, environmental and economic dimensions. The organization also allow other employees and direct channels stakeholders within the structure to provide feedback and assist in the improvement process.

DO

Axis 3: Implementation and operation

Requirement 3.1: Operational Control

The management system must be structured to ensure that operations are performed in specific conditions, in order to meet the policies and objectives of the organization as well as the legal requirements and other requirements. It is understood that, for the social dimension, the organization must ensure that the operations in your production system offer employees safety, to prevent unsafe working conditions. From the point of view of the environment, you must ensure that the amount of resources used to manufacture and purchase of supplies appropriate, avoiding waste. At this point, you can also waste be included not only in the manufacturing environment, which directly affects the economic dimension, which can be inserted aspects of losses from a logistical, administrative, quality and energy. All the losses that occur in the manufacturing environment result in financial losses and significantly impact the outcome of the organization. Meet and measure these operating losses is of significant importance for the future of the organization.

Requirement 3.2: Resource Management

The organization shall ensure that all people who work for it and on its behalf are competent on the basis of appropriate education, training, skills and experience to the tasks assigned to them. To occur, the organization shall determine, provide and maintain the infrastructure and resources needed to achieve their goals. Thus, the organization of social and environmental point of view, must constantly train employees (internal or outsourced) so that they know their role for the organization and to society, performing with seriousness and responsibility, their activities. For such points occur, it is for the organization to plan and provide resources for training, understanding that these are investments for the organization and help the non-occurrence of accidents and incidents that may result in financial losses. The organization must also demonstrate the resources available to its employees and the correct way to use them, so as to maintain their integrity. It is also important for the management and maintenance of the resources that the organization presents employees with the opportunity for professional growth through democratic and transparent criteria for career advancement and to combat discrimination of any kind.

Requirement 3.3: Documentation Requirements

The documentation of the management system must have the description of the scope of the system, including standards / specifications adopted management systems; statements of policies and objectives of the organization; system manual; documented procedures and record that is required by the management system and its component standards and documents as determined by the organization to be necessary to ensure the planning, operation and effective control of its processes. The organization makes this documentation is known to all employees, as well as what is their role and responsibility in the design, implementation and improvement of this document.

Requirement 3.4: Communication

The organization shall establish, implement and maintain effective arrangements for internal communication between the various levels and functions of the organization, and for receiving, recording and responding to relevant communications from interested parties. It is worth at this point to present to internal and external community reports performance against the goals of the management system and its past, present and future to improve the system performance in social, environmental and economic dimensions shares.

CHECK

Axis 4: Performance Evaluation

Requirement 4.1: Monitoring and Measurement

The organization shall perform measurements and monitoring to determine the degree of compliance with the established goals. This should include the recording of information to monitor the performance of the relevant operational controls and to evaluate compliance with the organization's goals and the ability of the processes to achieve planned results. It is up to responsible management system perform the monitoring and measuring points defined in operational requirements and documentation control requirements, these actions should be proactive, to achieve the goals and plans, and reactive, responding to and reporting accidents and incidents. Such indicators should allow, where possible, the evaluation and measurement of economic gains and losses, we understand that this is part of the sustainability of the organization.

Requirement 4.2: Conformity Assessment

The organization shall conduct periodic evaluations of compliance with legal requirements, I understand, in this respect, that the organization must meet and review the scope of your system by removing, inserting or changing requirements from the update of the standards / specifications that make up your management system. This process should occur in partnership with the stakeholders of the organization by checking the economic, environmental and social impacts that the changes may bring, be they positive or negative.

Requirement 4.3: Internal Audit

The organization shall establish and maintain a program of regular audit of its management system as a mechanism of control and verification that all objectives outlined in its design are being met and understood by all, although within this process evaluation of social, environmental and economic organization.

It also, at the end of the audit, the presentation of results and their comparison with previous audits in order to ensure continuous and clear process to everyone in the organization.

Requirement 4.4: Treatment of Non-Conformities

When identified noncompliance must corrected, and the actions taken to mitigate their impact. It is important to verify the origin of non-compliance. If this non-compliance is the result of the operation on the part of the employee failures, it is important to conduct further training to prevent recurrence, denoting this point the social dimension. In the environmental point of view, nonconformities generate material losses, creating the need to reprocess the part or not being able to dispose of your place its reuse. It should first be aware of these things and, if you need to rework or disposal, possess mechanisms that do not harm the environment or minimize expenses for his recovery. Nonconformities cause economic losses and measure the losses of materials and parts, as well as the need to rework the same or proper disposal of your expenses, it is of extreme necessity, and also a point of assessing organizational improvement towards the latter.

ACT

Axis 5: Improvement

Requirement 5.1: General

The organization shall continually improve the effectiveness of the management system through the use of policy, objectives, audit results, analysis of data on the performance evaluation, corrective and preventive actions and management review and should define and allocate responsibility and authority the improvement of the GIS.

Requirement 5.2: Corrective, Preventive and Improvement Action

The organization must know and be prepared to take corrective, preventive and improvement actions, listing and defining how the organization will act on nonconformities or potential nonconformities (including feedback from stakeholders), as well as registering and reviewing the effectiveness and the results of actions performed by establishing appropriate mechanisms for evaluation and measurement of economic gains and losses.

Axis 6: Critical Analysis by the Direction

Requirement 6.1: General

It is up to senior management critically to analyze the organization's management system at planned intervals, to ensure its continuing suitability, adequacy and effectiveness, making assessing opportunities for improvement and the need for changes in the management system, including policy and objectives. It is understood, therefore, that the senior management, this requirement should act primarily as an evaluation mechanism (regulation) on the actions undertaken and as a controller requirement management system and should actively participate in the interaction with stakeholders (stakeholders), bringing points of improvements listed by them, whether social, environmental or economic order.

Requirement 6.2: Inputs

The inputs relate to the set of information available to senior management for monitoring management system. For these data are critically analyzed by senior management, should the information set of social, environmental and economic order to understand, respecting its characteristics, the results of audits, feedback from stakeholders, the status of preventive actions and provide corrective follow-up actions from previous management reviews, changes in circumstances, including developments in legal and other requirements related to aspects of the organization and risk requirements, recommendations for improvement, data and information on the organization's performance and the results of the evaluation of compliance with legal and other requirements.

Requirement 6.3: Outputs

Outputs relate to the whole of any decisions and actions related to the management system. Top management must present actions of social, environmental and economic comprising respecting its characteristics, the quest for improving the efficiency of the management system, the improvement related to stakeholder requirements, and the need for resources order to enable system improvement management and its processes.

FUTURE RESEARCH DIRECTIONS

Research involving the topic of sustainability and the subject IMS are growing in the scientific literature on various aspects such as the development of models of management, economic evaluation models, the degree of adhesion.

However, few studies envision the relationship between these two topics (IMS and sustainability), this being a wide field of research.

The organizational management in a sustainable manner, respecting the environment and society, and given the economic primordium of its existence, causes business processes become complex and require further research.

Another aspect is the development of normative standards, which undergo constant updates that has brought closer to the dimensions of sustainability, but is that organizations are following these changes? Does the concept of sustainability is really understood? Do organizations have focused on becoming sustainable internally?

We observe, therefore, that the theme has gaps to be exploited, may be mentioned: (i) lack of consensus on the definition / concept of sustainability; (ii) various approaches to applying sustainability in the organizational environment; (iii) the low level of literature and research in the area; (iv) absence of a universally accepted evaluation model; (v) review existing models have problems with selection, prioritization and weighting of dimensions (Siena, 2008); (vi) concern the valuation models of sustainability to the organization with the external aspects, omitting regarding the evaluation of the internal aspects (Lueneburger & Goleman, 2010).

One of the reasons for these gaps, according to Siena (2008), is the difficulty in materializing the concept of sustainability, as the conceptual bases on the issues involved are not consolidated, there is no consensus on what to measure, how to measure and mainly about how to weigh and combine data.

In this context, according to Henderson (2011), to conduct studies that contribute to the evolution of the theme, so it is necessary to identify hypotheses that can be tested and support the development of new research.

- What environmental, social and economic indicators are relevant to use?
- What are possible indicators to measure and report, and what obstacles exist to measure and report?
- How the indicators should be aggregated into a sustainability index based on a fairly simple model, but reliable, available resources in the real world?
- You can verify this index, to some extent, comparing the results of the index to the literature and own ambitions of the organization?
- What are the limitations and uncertainties of the conceptual structure of suggested content?
- What are the main opportunities and benefits of working with an aggregate index?

For this to occur in a relevant, comprehensive and consistent development model needs to have a methodological framework which involves answering questions, as pointed by Singh et al. (2009).

CONCLUSION

In this chapter, the concepts of organizational sustainability and IMS, in order to present their definitions and the main normative bases that characterize have been described.

Were also highlighted the initiatives of creating normative standards of validity for global management of an integrated organizational system. One key point is to demonstrate the synergy that exists between the IMS and the dimensions of sustainability.

It was found that the PAS 99:2006 model is presented as a normative standard most similar to ISO 9000, ISO 14000, ISO 26000 and OHSAS 18000 standards, having a methodology for implementing a IMS approaching these standards, employing the concept of PDCA for its management.

We notice, however, that evidence is necessary to establish a relationship between these patterns and normative dimensions of sustainability, demonstrating synergy.

This characterization of synergy between the IMS and sustainability enables organizations see through normative standards already known and reported in the literature, the possibility of improving their internal processes in order to achieve their actions to provide future survival without harming society, the maintaining the environment and economically alive.

REFERENCES

AS9100: Quality Management Systems: requirements for aviation, space and defense organizations – revision c. (2009). SAE International, London.

Allen, R. (1980). *How to save the world*. London: Kogan Page.

Andersson, B., Bider, I., Johannesson, P., & Perjons, E. (2005). Towards a formal definition of goal-oriented business process patterns. *Business Process Management Journal, 11*(6), 650–662. doi:10.1108/14637150510630846

Arifin, K., Aiyub, K., Awang, A., Jahi, J. m., & Iteng, R. (2009). Implementation of Integrated Management System in Malaysia: The level of organization's understanding and awareness. *European Journal of Scientific Research, 31*(2), 188–195.

Bamber, C. J., Sharp, J. M., & Castka, P. (2004). Third party assessment: The role of the maintenance function in an integrated management system. *Journal of Quality in Maintenance Engineering, 10*(1), 26–36. doi:10.1108/13552510410526848

Barral, M. (2007). La gestión integrada y la dinámica de sistemas: Criterios a aplicar en los laboratorios clínicos. *Acta Bioquímica Clínica Latinoamericana, 41*(3), 407–418.

Barron, V. (2010). How standards contribute to sustainable development. *BSI White Paper*. Retrieved from http://www.bsigroup.co.uk/en/Forms/Document-Downloads/How-standards-contribute-to-sustainable-development/

Bayati, A., & Taghavi, A. (2007). The impacts of acquiring ISO 9000 certification on the performance of SMEs in Tehran. *The TQM Magazine, 19*(2), 140–149. doi:10.1108/09544780710729980

Beckmerhagen, I. A., Berg, H. P., Karapetrovic, S. V., & Willborn, W. O. (2003). Integration of management systems: Focus on safety in the nuclear industry. *International Journal of Quality & Reliability Management, 20*(2), 209–228. doi:10.1108/02656710310456626

Bernardo, M., Casadesus, M., Karapetrovic, S., & Heras, I. (2009). How integrated are environmental, quality and other standardized management systems? An empirical study. *Journal of Cleaner Production, 17*(8), 742–750. doi:10.1016/j.jclepro.2008.11.003

Bernardo, M., Casadesus, M., Karapetrovic, S., & Heras, I. (2010). An empirical study on the integration of management system audits. *Journal of Cleaner Production, 18*(5), 486–495. doi:10.1016/j.jclepro.2009.12.001

BS OHSAS 18001: Occupational Health and Safety. (2008). British Standards Institution.

Castka, P., & Balzarova, M.A. (2008). ISO 26000 and supply chains - On the diffusion of the social responsibility standard. *International Journal of Production Economics, 111*(2), 274–286. doi:10.1016/j.ijpe.2006.10.017

Celik, M. (2009). Establishing an Integrated Process Management System (IPMS) in ship management companies. *Expert Systems with Applications, 36*(4), 8152–8171. doi:10.1016/j.eswa.2008.10.022

Cerin, P., & Scholtens, B. (2011). Linking responsible investments to societal influence: Motives, assessments and risks(editorial). *Sustainable Development*, *19*(5), 289–358.

Coomer, J. (1979). *Quest for a sustainable society*. Oxford: Pergamon.

De Cicco, F. (2004). *Sistemas Integrados de Gestão*: agregando valor aos sistemas ISO 9000 (in Portuguese). Retrieved from http://www.qsp.com.br/artigo.shtml

Dillard, J., Dujon, V., & King, M. C. (2009). *Understanding the social dimension of sustainability. Unitec Kington*. Routledge.

Djalali, A., & Vollaard, P. (2008). The complex history of sustainability: an index of trends, authors, projects and fiction. After Zero, 18(1), 160p.

Du Pisani, J. A. (2006). Sustainable development: Historical roots of the concept. *Environmental Sciences*, *3*(2), 83–96. doi:10.1080/15693430600688831

Duijm, N. J., Fiévez, C., Gerbec, M., Hauptmanns, U., & Konstandinidou, M. (2008). Management of health, safety and environment in process industry. *Safety Science*, *46*(6), 908–920. doi:10.1016/j.ssci.2007.11.003

Elkington, J. (1998). *Cannibals with Forks: The Triple Bottom Line of 21st Century Business*. Canada: New Society Publishers.

Ferreira, S.S. (2009) Modelo de sistema integrado de gestão em eficiência energética e segurança e saúde no trabalho. Brasília: PPGENE, DM.

Ferroli, P.C.M. & Librelotto, L.I. (2011). Integração da sustentabilidade em ferramenta projetual: feapsus. *Produção online*, 11(2), 447-475.

Fresner, J., & Engelhardt, G. (2004). Experiences with integrated management systems for two small companies in Austria. *Journal of Cleaner Production*, *12*(6), 623–631. doi:10.1016/j.jclepro.2003.09.013

Getabstract. (2014). *The World's Largest Library of Business Book Summaries*. Retrieved from http://www.getabstract.com/

Gibson, R. B. (2006). Sustainability assessment: Basic components of a practical approach. *Impact Assessment and Project Appraisal*, *24*(3), 170–182. doi:10.3152/147154606781765147

Global Reporting Initiative. (2014). Retrieved from http://www.globalreporting.org

Goldsmith, E., Allen, R., Allaby, M., Davoll, J., & Lawrence, S. (1972). *A blueprint for survival*. Harmondsworth: Penguin.

Grober, U. (2007). *Deep roots: a conceptual history of "sustainable development" (Nachhaltigkeit)*. Berlin: WZB.

Guillén, O. P., & Carmenates, Y. P. (2009). Sistema de Gestión Integrado: Calidad, Medio Ambiente y Seguridad y Salud en el Trabajo como herramienta del Sistema de Dirección y Gestión en la Empresa de Investigaciones y Proyectos Hidráulicos de Ciego de Ávila. *Enlace. Inter-American Commission of Women*, *87*(5), 1–20.

Hacking, T., & Gurthrie, P. (2008, February). A framework for clarifying the meaning of Triple Bottom-Line, Integrated, and Sustainability Assessment. *Environmental Impact Assessment Review*, *28*(2-3), 73–89. doi:10.1016/j.eiar.2007.03.002

Hamidi, N., Omidvari, M., & Meftahi, M. (2012). The effect of integrated management system on safety and productivity indices: Case study, Iranian cement industries. *Safety Science*, *50*(5), 1180–1189. doi:10.1016/j.ssci.2012.01.004

Henderson, S. (2011). The development of competitive advantage through sustainable event management. *Worldwide Hospitality and Tourism Themes*, *3*(3), 245–257. doi:10.1108/17554211111142202

Henriques, A. (2004). CSR, Sustainability and Triple Bottom Line. In A. Henriques & J. Richardson (Eds.), *The Triple Bottom Line: does it all add up?* (pp. 17–25). London: Earthcan.

Holdsworth, R. (2003). Development and implementation of an integrated management system: Applications approach to design. *Journal of Hazardous Materials Practical*, *104*(1), 193–205. doi:10.1016/j.jhazmat.2003.08.001 PMID:14602409

Sustainability report. (2009). IISD. Retrieved from http://www.iisd.org/about/sdreporting/2002/eco_dim.asp

ISO. (1998). *TS 9000 – Quality management and quality assurance standards-part 3: guidelines for the application of ISO 9001 to the development, supply and maintenance of software*. Switzerland: International Organization for Standardization.

ISO. (2002). *TS 16949:2002 – Quality Management System – Particular requirements for the application of ISO9001:2000 for automotive production and relevant service part organizations*. Switzerland: International Organization for Standardization.

ISO. (2006). *ISO 22000: Food safety management systems - Requirements for any organization in the food chain*. Switzerland: International Organization for Standardization.

ISO. (2007a). *ISO 27001: Information technology - Security techniques-Information Security management systems-Requirements*. Switzerland: International Organization for Standardization.

ISO. (2007b). *ISO 28000: Specifications for security risk management systems for the supply chain*. Switzerland: International Organization for Standardization.

ISO. (2008). *ISO 9001: Quality management systems — Requirements*. Switzerland: International Organization for Standardization.

ISO. (2008b). *ISO 14001: Environmental management systems—Requirements with guidance for use*. Switzerland: International Organization for Standardization.

ISO. (2008c). *The Integrated Use of Management System Standards*. Switzerland: International Organization for Standardization.

ISO. (2009). *ISO 26000: Guidance on social responsibility*. Switzerland: International Organization for Standardization.

ISO. (2013). *The ISO survey of management system standard certifications – 2012*: Executive summary. Retrieved from http://www.iso.org/iso/iso_survey_executive-summary.pdf

Jamali, D. (2006). Insights into triple bottom line integration from a learning organization perspective. *Business Process Management Journal*, *12*(6), 809–821. doi:10.1108/14637150610710945

Jørgensen, T. H. (2008). Towards more sustainable management systems: Through life cycle management and integration. *Journal of Cleaner Production*, *16*(10), 1071–1080. doi:10.1016/j.jclepro.2007.06.006

Jørgensen, T. H., Remmen, A., & Mellado, M. D. (2006). Integrated management systems: Three different levels of integration. *Journal of Cleaner Production*, *14*(8), 713–722. doi:10.1016/j.jclepro.2005.04.005

Jorgensen, T. H., & Simonsen, G. (2002). Prospects of a unified management system. *Corporate Social Responsibility and Environmental Management*, *9*(2), 91–98. doi:10.1002/csr.11

Karapetrovic, S. (2002). Strategies for the integration of management systems and standards. *The TQM Magazine*, *14*(1), 61–68. doi:10.1108/09544780210414254

Karapetrovic, S. (2003). Musings on integrated management systems. *Measuring Business Excellence*, *7*(1), 4–13. doi:10.1108/13683040310466681

Karapetrović, S., Casadesus, M., & Heras, I. (2010). Analysis of integration within the standards-based Integrated Management Systems.*Proceedings of International Quality Conference* (pp. 1-12).

Lagodimos, A. G., Chountalas, P. T., & Chatzi, K. (2007). The state of ISO 14001 certification in Greece. *Journal of Cleaner Production*, *15*(18), 1743–1754. doi:10.1016/j.jclepro.2006.06.015

Linton, J. D., Klassen, R., & Jayaraman, V. (2007). Sustainable supply chains: An introduction. *Journal of Operations Management*, *25*(6), 1075–1082. doi:10.1016/j.jom.2007.01.012

Lueneburger, C., & Goleman, D. (2010). The change leadership sustainability demands *MIT. Sloan Management Review*, *51*(4), 49–55.

Mackau, D. (2003). SME Integrated management system: A proposed experiences model. *The TQM Magazine*, *15*(1), 43–51. doi:10.1108/09544780310454448

Marquardt, B. (2006). Historia de la sostenibilidad: Un concepto medioambiental en la historia de Europa central (1000-2006). *Historia Critica*, *32*(2), 172–197.

Meadows, D. H. (1972). *The limits to growth: a report of the Club of Rome's project on the predicament of mankind*. New York: Universe Books.

Mollenkopf, D., Stolze, H., Tate, W. L., & Ueltschy, M. (2010). Green, lean, and global supply chains. *International Journal of Physical Distribution & Logistics Management*, *40*(1), 14–41.

Mullerat, R. (2010). *International Corporate Social Responsibility: the role of corporations in the economic order of the 21st century*. Netherlands: Kluwer Law International.

Naveh, E., & Marcus, A. (2005). Achieving competitive advantage through implementing a replicable management standard: Installing and using ISO 9000. *Journal of Operations Management*, *24*(1), 1–26. doi:10.1016/j.jom.2005.01.004

Oliveira, O. J., & Serra, J. R. (2010). Benefícios e dificuldades da gestão ambiental com base na ISO 14001 em empresas industriais de São Paulo. *Production*, *20*(3), 429–438. doi:10.1590/S0103-65132010005000013

Pojasek, R. (2006). Is your integrated management system really integrated? *Environmental Quality Management*, *16*(2), 89–97. doi:10.1002/tqem.20124

Pojasek, R. (2011). ISO 26000 guidance on Social Responsibility. *Environmental Quality Management*, *20*(3), 85–93. doi:10.1002/tqem.20292

Pope, J., Annandale, D., & Morrison-Saunders, A. (2004). Conceptualizing sustainability assessment. *Environmental Impact Assessment Review*, *24*(6), 595–616. doi:10.1016/j.eiar.2004.03.001

Rasmussen, J. M. (2007). *Integrated Management Systems* – an analysis of best practice in Danish companies (Thesis). Aalborg University.

Robinson, S. (2000). Key survival issues: Practical steps toward corporate environmental sustainability. *Corporate Environmental Strategy*, *7*(1), 92–105. doi:10.1016/S1066-7938(00)80118-9

Salomone, R. (2008). Integrated management systems: Experiences in Italian organizations. *Journal of Cleaner Production*, *16*(16), 1786–1806. doi:10.1016/j.jclepro.2007.12.003

Satolo, E. G. (2012). *Measurement model for assessment of organizational sustainability* [Thesis in Portuguese]. Santa Bárbara d'Oeste: Unimep.

Satolo, E.G., Calarge, F.A. & Miguel, P.A.C. (2013). Experience with an integrated management system in a sugar and ethanol manufacturing unit: possibilities and limitations. *Management of Environmental Quality*: *an international journal*, 24(6), 710-725.

Siena, O. (2008). Método para avaliar desenvolvimento sustentável: Técnicas para escolha e ponderação de aspectos e dimensões. *Production*, *18*(2), 359–374. doi:10.1590/S0103-65132008000200012

Simon, A., Karapetrovic, S., & Casadesus, M. (2012). Evolution of Integrated Management Systems in Spanish firms. *Journal of Cleaner Production*, *23*(1), 8–19. doi:10.1016/j.jclepro.2011.10.025

Singh, R. K., Murty, H. R., Gupta, S. K., & Dikshit, A. K. (2009). An overview of sustainability assessment methodologies. *Ecological Indicators*, *15*(1), 189–212. doi:10.1016/j.ecolind.2008.05.011

Singh, V.K. (2009). PDCA Cycle: a quality approach. *Utthan – The Journal of Management Sciences*, 1(1), 89-96.

Smeraldi, R. (2009). *The new manual of sustainable business (in Portuguese)*. São Paulo: Publifolha.

Dansk Standard. (2005). *Management systems*: guidelines for integrated management system. DS 8001.

Tecnologia, R. (2007). Especificação de requisitos comuns de sistemas de gestão com estrutura para a integração. São Paulo: Risk Tecnologia Editora.*Sistemas Integrados de Gestão PAS, 99.*

Terziovski, M., & Power, D. (2007). Increasing ISO 9000 certification benefits: A continuous improvement approach. *International Journal of Quality & Reliability Management*, *24*(2), 141–163. doi:10.1108/02656710710722266

Tsai, W.-H., & Chou, W.-C. (2009). Selecting management systems for sustainable development in SMEs: A novel hybrid model based on DEMATEL, ANP, and ZOGP. *Expert Systems with Applications*, *36*(2), 1444–1458. doi:10.1016/j.eswa.2007.11.058

UNE 66177. *Management systems: guide for the integration of management systems.* (2005). AENOR, Madrid.

Vasconcelos, D. S. C. (2011). Integrated Management System and the PAS 99.

Vasques, e. (2007). How thing in sustainability (in portuguese). Revista B2B.

Vinodkumar, M. N., & Bhasi, M. (2011). A study on the impact of management system certification on safety management. *Safety Science*, *49*(3), 498–507. doi:10.1016/j.ssci.2010.11.009

Wang, C.-H., & Tsai, D.-R. (2009). Integrated installing ISO 9000 and ISO 27000 management systems on an organization.*Proceedings of International Carnahan Conference On Security Technology*, (265–267). doi:10.1109/CCST.2009.5335527

Wilkinson, G., & Dale, B. G. (1999). Integrated management systems: An examination of the concept and theory. *The TQM Magazine*, *11*(2), 95–104. doi:10.1108/09544789910257280

Williams, P. (2010). SA 8000 - The first decade: Implementation, influence and impact. *Development in Practice*, *20*(2), 298–300. doi:10.1080/09614520903564280

Wolfe, A. (2002). Questionnaire design. In R. Birn (Ed.), *The international hand book of marketing research techniques* (pp. 87–102). London: Market Research Society.

Xu, F. L., Zhao, S. S., Dawson, R. W., Hao, J. Y., Zhang, Y., & Tao, S. (2006). A triangle model for evaluating the sustainability status and trends of economic development. *Ecological Modelling*, *195*(3-4), 327–337. doi:10.1016/j.ecolmodel.2005.11.023

Zeng, S. X., Shi, J. J., & Lou, G. X. (2007). A synergetic model for implementing an integrated management system: An empirical study in China. *Journal of Cleaner Production*, *15*(18), 1760–1767. doi:10.1016/j.jclepro.2006.03.007

Zeng, S. X., Tam, V. W. Y., & Tam, C. M. (2008). Towards occupational health and safety systems in the construction industry of China. *Safety Science*, *46*(8), 1155–1168. doi:10.1016/j.ssci.2007.08.005

Zutshi, A., & Sohal, A. S. (2004). Adoption and maintenance of environmental management systems: Critical success factors. *Management of Environmental Quality: An International Journal*, *15*(4), 399–419. doi:10.1108/14777830410540144

ADDITIONAL READING

Asi, F. M., Searcy, C., Zutshi, A., & Fisscher, O. A. M. (2013). An integrated management systems approach to corporate social responsibility. *Journal of Cleaner Production*, *56*(10), 7–17.

Asif, M., Bruijn, E. J., & Fisscher, O. A. M. (2010). Meta-management of integration of management systems. *The TQM Journal*, *22*(6), 570–582. doi:10.1108/17542731011085285

Atkinson,g., Dietz, S., & Neumayer, E. (2007). *Handbook of sustainable development*. Northampton: Elgar Original Reference Series.

Berardi, U. (2011). Sustainability assessment in the construction sector: Rating systems and rated buildings. *Sustainable Development*, *19*(5), 71–76.

Bossel, H. (1999). *Indicators for sustainable development: theory, method, applications: a reporter to the Balaton Group*. Canada: International Institute for Sustainable Development.

BSI -British Standards Institution. (2014). Retrieved from http://www.bsigroup.com/

Ciers, D. C. (2011). *Integrated Management Systems implementation tool: a guide for organizational management systems conformant to ISO quality, environmental and occupational health & safety standards*. Bloomington: Xlibris Corporation.

Douglas, A., & Glen, D. (2000). Integrated management systems in small and medium enterprises. *Total Quality Management*, *11*(4-6), S686–S690. doi:10.1080/09544120050008075

Geraldo, F. (2008). Sustainability: What does it mean for the operations manager? *Journal of Operations and Supply Chain Management*, *1*(2), 1–16.

Guo, H. & SUO, Z. (2011). Structural model of the capability of enterprises sustainable development.*Proceedings of International Conference on Management Science and Industrial Engineering* (pp. 962-965).

ISO - International Organization for Standardization. (2014). Retrieved from http://www.iso.org/iso/home.html

Jerneck, A., Olsson, L., Ness, B., Anderberg, S., Baier, M., Clark, E., & Persson, J. et al. (2011). Structuring sustainability science. *Sustainability Science*, *6*(1), 69–82. doi:10.1007/s11625-010-0117-x

Kates, R. W., Clark, W. C., Corell, R., Hall, M. J., Jaeger, C. C., Lowe, I., & Dickson, N. M. et al. (2001). Sustainability science. *Science*, *292*(5517), 641–642. doi:10.1126/science.1059386 PMID:11330321

Labodova, A. (2004). Implementing integrated management systems using a risk analysis based approach. *Journal of Cleaner Production*, *12*(6), 571–580. doi:10.1016/j.jclepro.2003.08.008

Lee, T. H. Shiba, s. & Wood, R.C. (1999). Integrated Management Systems: a practical approach to transforming organizations. New Jersey: John Wiley & Sons.

Little, A. D. (2008). Integrity + Innovation = Sustainable performance: the sustainability value formula. Retrieved from http://www.adl.com/iplusi

López-Fresno, P. (2010). Implementation of an integrated management system in an airline: A case study. *The TQM Journal*, *22*(6), 629–647. doi:10.1108/17542731011085311

Lundin, U. (2003). Indicators for Measuring the Sustainability of Urban Water Systems—a Life Cycle Approach [PhD Thesis, Department of Environmental Systems Analysis]. Chalmers University of Technology, Göteborg, Sweden.

Nee, A. Y. H. (2009). Implementation of Integrated Management System: environmental and safety performance and global sustainability.*Proceedings of International Conference on Energy and Environment* (pp. 232-241).

Nowicki, P., Kafel, P., & Sikora, T. (2013). Selected requirements of integrated management systems based on pas 99 specification. *International Journal for Quality Research*, *7*(1), 97–106.

Pardy, W. & Andrews, t. (2009). Integrated Management Systems: leading strategies and solutions. Maryland: Government Institutes.

Phillis, Y. A., & Andriantiatsaholiniaina, L. A. (2001). Sustainability: An ill-defined concept and its assessment using fuzzy logic. *Ecological Economics*, *37*(3), 435–456. doi:10.1016/S0921-8009(00)00290-1

Pohl, E. (2006). Towards corporate sustainable development – ITT Flygt sustainability index [Licentiate Thesis]. Malarden University.

Valdés, M. C. M., & Bretaña, G. R. (2009). La implantación de sistemas integrados de gestión: Un reto a la empresa cubana. *Revista La Habana*, *1*(1), 33–42.

William Lewis, W. (1998). *PDCA/Test*. Florida: CRC Press; doi:10.1201/9781420048131

KEY TERMS AND DEFINITIONS

Integrated Management System: Represents the unification of various management systems into a single system, taking advantage of the many synergies between them, without ignoring the peculiarities of each.

Normative Standards: A document that contains a technical specification or other precise criteria designed to be used consistently as a rule, guideline, or definition.

Organizational Sustainability: Related to the concept of sustainability, advocating the care the social, environmental and economic dimensions, is mainly focused on the internal processes of the organization, understanding that it is necessary to first have internal sustainability for later view external actions.

PDCA Cycle: Created in the 1950s by Deming, is a systematic process for continuous improvement, accomplished through four steps: Plan, Do, Check and Act.

Requirements: Documented definition of a property or behaviors that a particular product or service must meet.

Sustainability: Concept of extensive discussion. His concept seeks to promote longevity through the attainment of economic profit, respecting the environment and society.

Sustainability Dimensions: Known as social, economic and environmental dimensions, the tripod represent the ideals of sustainability (triple bottom line).

Chapter 11
Sustainable Non Traditional Manufacturing Processes:
A Case with Electrochemical Machining (ECM)

Abhishek Tiwari
Birla Institute of Technology, India

Kaushik Kumar
Birla Institute of Technology, India

Supriyo Roy
Birla Institute of Technology, India

ABSTRACT

Sustainable operations strategy, an important field of research for engineering optimization, is posing an increasing challenge to operations managers. Sustainable operations can become a key to competitive advantage. Consequently, there is a need for research for the complex interfaces between the dimensions of sustainable operations strategy. The companies need to accommodate environmental demands, emphasized by media and government, along with social aspect in day to day operations for formulating strategies. Non-traditional manufacturing processes involves mechanical, thermal, electrical or chemical energy or combinations of these energies but do not use a sharp cutting tools as required for traditional manufacturing processes. In this chapter a non-traditional machining process namely Electro chemical Machining (ECM) has been considered. Material Removal Rate (MRR) and Surface Roughness (SR) has been optimized using Taguchi, Grey Relation, Artificial Neural Network, Fuzzy Logic and Genetic Algorithm to achieve the best.

DOI: 10.4018/978-1-5225-0350-7.ch011

1. INTRODUCTION

Sustainability is a recent increasingly global issue to all business led organizations; management strategies and practices towards sustainability have been established to cater to a growing interest to business environment. Impacts on sustainability are strongly influenced by shop floor decisions in the supply chain; it is vital that the functions of operations management embrace the requirements of sustainability management. This has implications for decision making in associated with all aspects of operations management in the supply chain. Scope of any given 'operation' is, thus, expanded considerably with the nature of operations management perspective. Generally accepted definition of sustainability is considered in the line of development as *'...development that meets the needs of the present without compromising the ability of future generations to meet their own needs'* (Brundtland Commission, 1987).

In economic perspective, every human action or societal undertaking is purely depending on basic objective of profit maximization or loss minimization. This requires the trade-off between increase in net profit and decrease in available energy (Fan & Tengyan, 2012). As per their view, notion of sustainability arises from the need to balance between increase in economic benefit and decrease in energy. Sustainability engineering can be regarded as an engineering discipline to craft frameworks or establish paradigms for attaining sustainability in any human or social perspective; an emerging discipline of engineering. Obviously, sustainability assessment constitutes the initial phase of sustainability engineering (Fan & Tengyan, 2012).

Sustainability in long term requires that artifacts, materials, systems and processes be designed to maximize reuse and utility (Tsahalis et al., 2012). Sustainability optimization task is basically an attempt to integrate mutually as well as conflicting goals like minimize negative impact and maximize quality, minimize cost and maximize profit, etc. Apart from this, holistic view of sustainability must cover entire life of a product from material procurement, design, production, distribution to its end-of-life. Introduction of sustainability issues, therefore, complicates the problem of process optimization more by adding more objectives to meet.

As per the latest report of 'FoFdation: Foundation for the Sustainable Factory of the Future', manufacturing sustainability optimization is by its definition a Multi-Objective Optimization problem. It integrates sustainability methods, goals, targets, indexes, indicators and product properties that need to be optimized, by adjusting a large number of product / process / tool parameters, thus leading to a large number of alternative solutions. In order to optimize the performance of manufacturing process, a more complex model is required. This enhanced process and sustainability model should be able to calculate the production performance on sustainability, as well as, production performance regarding time, cost, quality or any other indicator of interest (Tsahalis et al., 2012). A general sustainability optimization for the entire life cycle of the product is not the same as the sustainability optimization of a machine tool stage, although they are related to each other.

Sustainability in optimization involves many metrics / indicators and constitutes a structure of multiple-objective optimization problem. Even if overall indices are used; objectives are contradicting. The basic methodology is described in the following steps (Tsahalis, Moussas, & Tsahalis, 2012):

- Index and Indicator goals: will provide objective functions for the optimization problem. The optimization tool will try to reduce the overall target-indicator distance based on the adopted sustainability methodology.

Figure 1. ECM process

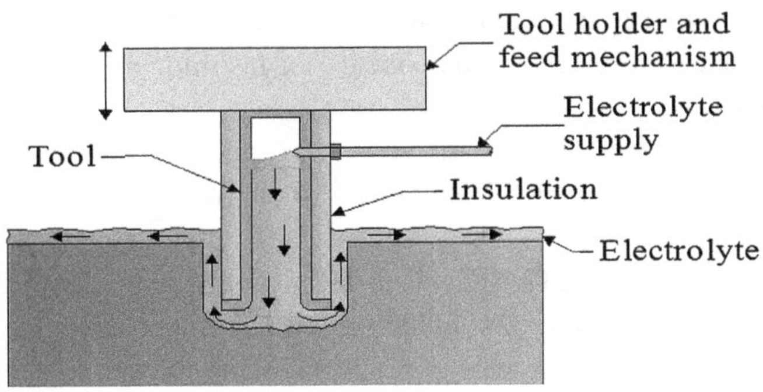

- **Optimization Search Space:** is composed by all potential changes of the configuration proposal elements. Acceptable combinations of the elements of search space will populate optimization problem space.
- **Optimization Problem Space:** is composed of all potential product, process and tool configuration alternatives. Sustainability Methods and Indicator functions are applied to the Search Space elements in order to assign the corresponding sustainability index.
- **Evolutionary Optimization Tool:** iteratively '*search the best*' elements as per the objective functions and finally come out with the 'best' one.

Non-traditional manufacturing processes is considered as a group of processes which remove excess material by different techniques involving combination of energies in terms of mechanical, electrical or chemical processes etc. (Jagadeesa, 2013). Like traditional manufacturing processes, no sharp cutting tools are used here. Extremely hard and brittle materials are difficult to machining by traditional machining processes. Nontraditional machining processes, popularly known as advanced manufacturing processes, are employed where traditional machining processes are either unfeasible or economical due to: extremely hard fragile materials difficult to set for traditional machining; work piece is too flexible or slender; and finally, complex shape of part. Different types of non-traditional machining processes have been developed to achieve required machining conditions (Jagadeesa, 2013). Proper fitting of these processes may come out with many advantages over traditional machining processes.

2. SUSTAINABILITY IN ELECTROCHEMICAL MACHINING (ECM)

ECM is basically a metal-removal process based on the '*principle of reverse electroplating*'. Here, particles travel from anodic material towards cathode material. A current of electrolyte fluid carries away the depleted material before it has a chance to reach the machining tool. The cavity produced is the female mating image of the tool shape.

A schematic representation of ECM process is shown in Figure 1. ECM tool is positioned very close to work piece and a low voltage, high amperage DC current is passed between work piece and electrode.

The usage of the aforesaid technique by the industrial community falls under *three* categories:

1. The products under consideration are for mass production and hence the Material Removal Rate is to be maximised to reduce the production time. In this the Surface Roughness does not play an important role. Usually the machine components for fast moving consumer goods (FMCG) fall under this category.
2. The products in this are usually tailor made and has specific application in which SR plays the vital role and MRR is unimportant. The components used in aviation e.g. space shuttle, aircraft etc. comes under this.
3. The machine component used in FMCG where SR is equally important comes in this category. In this both MRR and SR are considered simultaneously. The engine components in a car, for example piston, cylinder etc., are some examples of products in this section.

For any decision maker, out of the above cases, the '*best sets of values of the operating parameters*' are always desired. In this paper, an attempt is made to provide the best possible solution to the each of the above problem.

3. LITERATURE REVIEW

Sustainability is today's growing topic. Large Industry houses with their shop floor performance have also shown keen responsibility for preserving planet and people. Present study is motivated by the basic rule of three P's in sustainability - people, planet and profit. Sustainability continues moving from the periphery to the heart of business and aligned to core business and value drivers. As organizations' focus on sustainability, we believe the lack of maturity of sustainability performance management tools and solutions is a contributing reason to companies' failure to integrate sustainability performance management processes.

Sustainability and business performance must be tightly integrated if sustainability performance management is to be effective. When organizations carry out sustainability performance management activities as a separate functional area not aligned to larger business objectives, it is virtually impossible to identify and quantify the true impact of performance management initiatives on various parts of the business. Sustainability is a complex multi-stakeholder problem that must be addressed at different organizational levels. Improving sustainability usually implies hindering other areas of the system. The trade-offs of each alternative must be carefully handled and also evaluated to reconcile by the stakeholders before making any agreement.

Long term trends in managing sustainability include ongoing globalization and increasing intensity of competition, growing demands of environmental protection and resource scarcity. Therefore, there is a need for reliable, flexible and cost-efficient business systems capable of supporting customer differentiation. With a differentiation strategy, a company seeks to develop products and services that are perceived as unique in its industry that create a value advantage for its end users. This emphasizes the importance of focusing on one or more attributes that customers perceive as important, which usually leads to higher cost levels. In cost leadership, a company aims to be in lowest cost producer in its industry. The sources of cost advantage vary between industries. Usually they include economies of scale, proprietary technology, and preferential access to raw materials.

Sustainability is a strategic issue; require top management commitment, responsibility of the same being assigned to management at board level. This is integrated in the corporate level and policy should be adopted in the form of guiding principles and visions. Based on a top down approach, this is rooted in the supply chain strategy. Sustainability goals are therefore defined in supply chain KPI systems.

The growing concern towards protecting the environment has now transferred into green policies (Sarkis et al., 2011, Gunasekaran & Gallear, 2012). Today, green manufacturing has become common jargon used by almost one-third of manufacturing specialists and researchers. This is, perhaps, inevitable. There is a dire need for companies, and especially those engaged in manufacturing activities, to take a proactive, rather than are active, approach in this regard. According to a report released by UNEP (2011), global manufacturing industry consumes 35 percent of the total electricity consumed worldwide and is responsible for 20 percent of the world's CO_2 emissions, which is detrimental to all life on earth. The effect of this argument is to draw attention towards green manufacturing.

In a globalized era where firms are focusing on optimizing product quality and product cost, the average distance between markets and manufacturing units has increased (Dubey et al., 2015c). For the last two decades, we have seen increasing trends among western countries to shift their manufacturing units to Asian countries due to low cost of production per unit. In recent years, there has been increased competition among these countries to attract foreign direct investment as well as conforming to global standards in a stringent environmental norms (Dubey et al, 2015c).

In recent years, the sustainability theme has attracted increasing attention from academia and researchers. Manufacturing is one of the human activities that have caused major damage to the environment (UNEP, 2011). Manufacturing activities consume nearly 35% of total global production of electricity and are responsible for nearly 20% of total global carbon emissions (Graedel et al., 2011). According to Industrial Development Report, 2013, the manufacturing sector contributes one in six jobs globally.

Since the financial crisis of 2008, there has been increased debate on maintaining sustained growth. In this context, the manufacturing sector has a potential role to play. Sustainable manufacturing has attracted increasing attention from scholars (Azzone & Noci, 1998; Garbie, 2013, 2014; Gunasekaran & Spalanzani, 2012; Kannegiesser & Gunther, 2014). While there is a rich body of literature on sustainable manufacturing, research on theory building in sustainable manufacturing is scant. The current literature on sustainable manufacturing has attempted to use traditional approaches as well as alternative methods like case studies to build theory. However, techniques like grounded theory (with notable exception Binder & Edwards, 2010), action research (with notable exception Coughlan & Coghlan, 2002) and ethnographic studies were not completely exploited in the field of operations and supply chain management literature. Garetti et al. (2012) defined sustainable manufacturing as 'a set of technical and organisational solutions contributing to the development and implementation of innovative methods and practices and technologies for mitigating the excess of environmental load and for enabling and environmentally benign life cycle of products.'

International Trade Administration (2013) has defined sustainable manufacturing as *'creation of goods that uses series of processes which minimizes the negative environmental impacts, conserve energy and n*atural resources, are safe for employees, communities and consumers and are economically sound'. These two definitions provide an insight into the 4P's of sustainable manufacturing (performance, processes, practices and people) that are important for reducing negative environmental impacts, securing better working conditions, and achieving sustainable growth in terms of revenue generation, profit margin and employment generation.

Zubir and Habidin (2012) developed a sustainable manufacturing practices model and tested the impact on performance using structural equation modelling. Ghadimi et al. (2012) attempted to develop a weighted fuzzy approach for product sustainability assessment using a case study approach. Garbie attempted, in two studies (2013, 2014), to provide a theoretical model for sustainable manufacturing. Garbie (2013) developed a model for sustainable manufacturing enterprises. The model includes such dimensions as international issues, contemporary issues, innovative products, reconfigurable manufacturing systems, manufacturing strategies, and performance measurements. Mani et al. (2014) attempted to build a performance measurement system for sustainable manufacturing enterprises. However, the model failed to address social and economic dimensions, in the absence of which the model may not be regarded as a sustainable manufacturing model. Mittal and Sangwan (2014a, 2014b) presented a model for sustainable manufacturing enterprises. In the study, the authors classified drivers and barriers into three categories: internal drivers, policy drivers, and economic drivers, and further tested these using structural equation modeling. The preceding discussion on recent contributions to sustainable manufacturing reflects increasing attention from researchers. However, there are still many methodological gaps which call for diversity in research methodology (Dubey et al., 2015b).

Sustainable manufacturing has been the subject of debate in the last few years, amongst both researchers and industrialists (Dubey et al., 2015a). The world-class manufacturing (WCM) is a defined as, *'the recognition of an organization as a benchmark by its industry sector and, for some aspects, by other industry sectors'*. Brown et al. (2007) aligned manufacturing strategy to world-class performance, which further extended the literature (Hayes & Wheelwright, 1984; Schonberger, 1986; Roth, Giffi, & Seal 1992; Steudel & Desruelle 1992; Flynn, Schroeder, & Flynn 1999; Schroeder & Flynn, 2002). Manufacturing strategies have attracted much attention from academia (Garbie & Shikdar 2011; Garbie 2013, 2014a) and practitioners. However, the discipline requires more attention due to the rapidly changing scenario. Economic criteria have been replaced rapidly by environment, social and economic criteria, which are popularly, termed the '*Triple Bottom Line*' approach. In the near future, the demand for research focusing on sustainable manufacturing practices, which is the missing link in extant literature, will attract the serious attention of academia and policy-makers (Dubey, Gunasekaran, & Chakrabarty, 2015). In the past, economic criteria and customer satisfaction were the sole criteria for benchmarking manufacturing firms. However, in recent years, sustainability has been embraced by almost every manufacturing firm in the world (Sharma & Kodali 2008).

Sustainable manufacturing is defined as creation of manufactured goods through the use of a series of processes that minimise the negative environmental impacts, conserve energy and natural resources, are safe for employees, communities and consumers, and are economically sound (International Trade Administration 2013; Mani et al. 2014). Similarly, Garetti, Mummolo, and Taisch (2012) defined sustainable manufacturing as a set of technical and organisational solutions contributing to the development and implementation of innovative methods, practices and technologies, in the manufacturing field, for mitigating the excess of environmental load and for enabling an environmentally benign life cycle of products. The integration of environmental requirements throughout the entire lifetime of a product needs a new way of thinking and the application of new decision tools (Kaebernick & Sun 2003; Hong, Roh, & Rawski, 2012). Thus, sustainable manufacturing involves green product design, green procurement, green technology and green production (Noci 1997; Azzone & Noci 1998; Vachon 2007). Manufacturing practices have evolved in the last two decades from traditional manufacturing, which revolved around cost, quality, delivery and flexibility (Sánchez & Pérez 2001), to sustainable manufacturing, which aims to achieve a balance between environmental, social and economic benefits (Carter & Rogers 2008).

Recent study on sustainable manufacturing has attracted burgeoning interest. The majority of studies have attempted to explore the drivers of sustainable manufacturing or sustainable manufacturing practices. However, most of the literature has failed to delineate the causes, decisions and effects. Based on an analysis of these papers, sustainability in manufacturing begins with design for environment, life cycle assessment, the selection of raw materials and appropriate manufacturing strategies which impact the environment and social and economic performance. Garbie (2013, 2014a) presents an attempt to develop a comprehensive model for sustainable manufacturing to address the global economic crisis, while Garetti and Taisch (2012) make an attempt to outline the state-of-the-art sustainable manufacturing.

Work related to ECM process optimization starts from last decades. In recent years, Das et al. (2014) developed optimization of Surface Roughness and MRR in Electro-chemical Machining of EN 31 by using Grey-Taguchi Approach. Results show that optimal combination for maximization of MRR and minimization of surface roughness is at electrolyte concentration 10%, voltage 10 V, feed rate 0.25 mm / min and inter-electrode gap 0.2mm. Vikas et al. (2014) carried out the Effect and Optimization of Machine Process Parameters on MRR for EN19 & EN41 materials using Taguchi. Result showed that for EN41 material and EN19 material, Discharge current had a greater impact over MRR as compare to other machining parameters. Neto et al. (2006) conducted an experiment to predict the Intervening variables in electrochemical machining. Material removal rate, roughness and over-cut were studied. Results show that feed rate was main parameter affecting material removal rate.

The electrochemical machining shows best results with nitride sodium over surface roughness and over-cut. Bhattacharyya et al. (1999) highlight the Investigation for controlled electrochemical machining through Response Surface Methodology. This study also emphasized on various experiment data that confirm the validity and creativeness of developed mathematical models for dominated process parameters on material removal rate and overcut phenomena. In order to achieve maximized material removal rate optimal combination of dominated parameters determined. Fang et al. (2014) presented the effects of pulsating electrolyte flow in electrochemical machining. Experiments were conducted to verify the possibility of the proposed process and study the effects of pulsating flow on material removal rate. Using optimal pulsating parameters, the material removal rate and surface profile were significantly improved. Liu et al. (2014) presented as a new approach on Abrasive enhanced electrochemical slurry jet micro-machining: Comparative experiments and synergistic effects. Ebeid et al. (2014) address his work towards higher accuracy for ECM hybridized with low-frequency vibrations using the response surface methodology by developing mathematical models for analysis of influence of various process parameters on overcut and concavity. Klocke et al. (2013) conduct an Experimental research on the electrochemical machining on titanium and nickel-based alloys for aero engine components. Researchers have done a lot of work on ECM, works refer to De Silva et al. (1998), Siebertz et al. (2010), Ebeid et al. (2004) etc.

4. EXPERIMENTATION FOR ELECTRO CHEMICAL MACHINING PROCESSES

4.1 Machine Used

Experiments were conducted on METATECH electrochemical machining setup supplied by Metatech Industry. Figure 2 shows experimental setup used for ECM machining. The ECM setup comprises three main functional units: machining chamber, control panel and electrolyte circulation system. The work-piece (anode) is fixed inside the machining chamber in fixture and the tool (cathode) is attached to the main screw which is controlled by a servo feed system. To avoid short-circuit; a current sensing circuit is interfaced between tool and stepper motor controller circuit. If the current exceed the acceptable limit, a signal is sent to the stepper motor controller circuit which immediately reverses the downward motion

Figure 2. Electrochemical Machine chamber, Control Panel

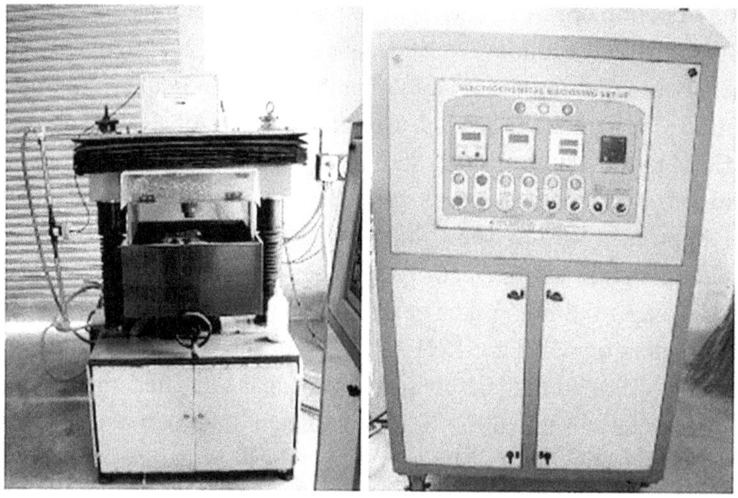

of the tool or shut off the machining. Electrolyte is pneumatically pumped through an electrolyte (KCl) tank to machining chamber.

4.2 Material Used

Cylindrical cross section of dimension EN 19 tool steel block of dimension Φ 25mm and 15 mm height with good ductility and shock resisting property was used as work piece. It is renowned for its wear resistance properties and where high strength property is required, shown in Figure 3. Commonly used for axels, drive shaft, crankshaft, connecting rod, high tensile bolt, stud, propeller shaft joint, rifle barrels, and induction hardening pin. Tool used for ECM process is made of copper with round cross-section of dimension Φ 17 mm shown in Figure 2. Electrolyte fed axially in machining zone between tool and work piece through centrally hole drilled in tool. Potassium chloride (KCl) is select as an electrolyte, since it has no passivation effect on the surface of work piece.

Figure 3. Copper Tool, Machined EN 19 material

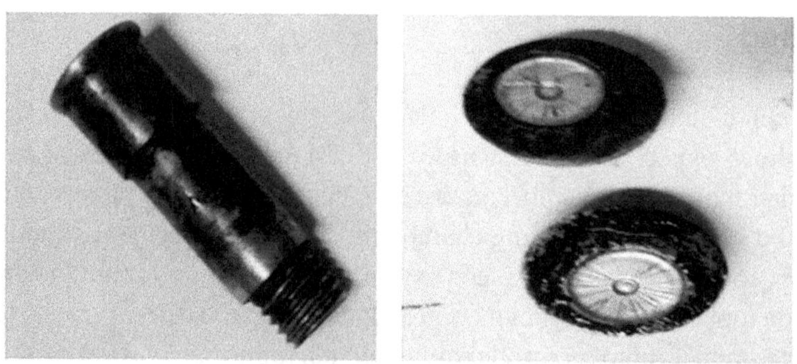

Table 1. Input Parameters with Levels

Process Parameter (unit)	Symbol	Level 1	Level 2	Level 3
Electrolyte conc. (%)	A	10	15	20
Voltage (V)	B	8	10	12
Feed rate (mm/min)	C	0.1	0.21	0.32
Inter-electrode gap, mm	D	0.2	0.25	0.3

4.3 Input Considered with their Levels

The independent input parameters viz. Electrolyte concentration (%), Voltage (V), Feed rate (mm/min) and Inter-electrode gap (mm) were considered. Selection of the parameters was done based on the literature survey which suggested that the above are the most important factors for MRR and SR. The table 1 indicates the input parameters along with its levels.

4.4 Output Considered

4.4.1 Material Removal Rate (MRR)

Material removal rate in electrochemical machining can be defined as:

MRR = C .I. h (cm3/min)

C: specific (material) removal rate (e.g., 0.2052 cm3/amp-min for nickel);

I: current (amp);

h: current efficiency (90-100%).

The rates at which metal can electrochemically remove are in proportion to the current passed through electrolyte and elapsed time for that operation. Many factors other than current influence the rate of machining like electrolyte type, rate of electrolyte flow, etc.

4.4.2 Surface Roughness (SR)

Surface roughness is usually defined as a deviation of a surface from an ideal level and is defined according to international standard (ISO 4287:1997). Arithmetic mean surface roughness Ra is defined as deviation of a surface from a theoretical center line (Rmean) over a measuring length Lm (Figure 4).

4.5 Design of Experiment

In the present research work experiment is planned according to Taguchi's design of experiment method. Taguchi (1990) suggested that, total Degree of freedom (DOF) of selected orthogonal array must be greater

Figure 4. Arithmetic mean surface roughness Ra

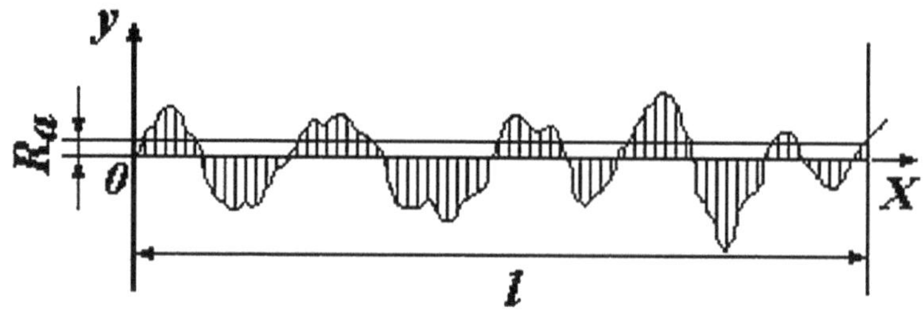

Table 2. Orthogonal Array

Exp. No.	A	B	C	D	Exp. No.	A	B	C	D
1	1	1	1	1	15	2	2	3	2
2	1	1	2	2	16	2	3	1	1
3	1	1	3	3	17	2	3	2	2
4	1	2	1	2	18	2	3	3	3
5	1	2	2	3	19	3	1	1	3
6	1	2	3	1	20	3	1	2	1
7	1	3	1	3	21	3	1	3	2
8	1	3	2	1	22	3	2	1	1
9	1	3	3	2	23	3	2	2	2
10	2	1	1	2	24	3	2	3	3
11	2	1	2	3	25	3	3	1	2
12	2	1	3	1	26	3	3	2	3
13	2	2	1	3	27	3	3	3	1
14	2	2	2	1					

than or equal to the total DOF of standard OA. Since total DOF of four factor and their interactions is 20 (4X2+3X4). And standard OA suggested for 4 factor with 3 level are L9 (DOF 8) and L27 (DOF 26), hence the L27 (DOF 26) OA has been selected for experimental work and tabulated as shown in table 2.

5. OPTIMIZATION OF SUSTAINABILITY FOR NON-TRADITIONAL MANUFACTURING PROCESSES OF MATERIAL REMOVAL RATE BY

5.1 Using Taguchi Method

This research is dedicated to find out optimum combination of machining process parameters within a given range for better outputs in ECM, with the application of Taguchi approach. L27 orthogonal array design was used and data obtained from the experiments were analyzed to assess the effect of various controllable machining parameter variations on material removal rate. ANOVA is also carried out to identify the level of significance of each factor and their interactions. Emphasis is on the development of comprehensive mathematical models for material removal rate describing the interactive and higher-order influences of the various machining process parameters. Finally, three dimensional surface plots for MRR and machining parameter generated using statistical software Minitab 16. A regression equation is also generated to speculate the values of MRR outside the considered level.

In the late 1940s *Dr. Genichi Taguchi* accomplished many convincing research with optimization techniques as a researcher in *Electronic Control Laboratory in Japan*. Taguchi method, which is also popular as the Robust Design eminently, enhances engineering productivity. The Robust Design method assists assure customer satisfaction, by premeditative taking in to account the noise factors and the cost of failure in the field. It aims at enhancing the basic action of the product or process, thus expediting flexible design and concurrent engineering. Undoubtedly, it is the most dynamic methodology usable enhanced quality and to lower product cost, simultaneously also diminish development interval.

In the past many authors have used this approach for optimization of ECM process. Vikas et al. carried out the Effect and Optimization of Machine Process Parameters on MRR for EN19 & EN41 materials using Taguchi. Result showed that for EN41 material and EN19 material, Discharge current had a greater impact over MRR as compare to other machining parameters. Neto et al. (2006) conducted an experiment to predict the Intervening variables in electrochemical machining. The material removal rate, roughness and over-cut were studied. The results show that feed rate was the main parameter affecting the material removal rate. The electrochemical machining shows best results with nitride sodium over surface roughness and over-cut. Xiaolong Fang et al. (2013) presented the Effects of pulsating electrolyte flow in electrochemical machining. Experiments were conducted to verify the possibility of the proposed process and study the effects of pulsating flow on material removal rate (MRR). Using optimal pulsating parameters, the material removal rate and surface profile were significantly improved. Liu et al. (2014) presented as a new approach on Abrasive enhanced electrochemical slurry jet micro-machining: Comparative experiments and synergistic effects. Klocke et al. (2013) conducted an Experimental research on the electrochemical machining on titanium- and nickel-based alloys for aero engine components. Experimental results of feed rate as a function of current density for an ECM sinking operation with a tool electrode and external flushing are analysed to the theoretical anodic dissolution behaviour considering Faraday's law.

5.1.1 Taguchi's Approach to Parameter Design

Designers can utilize standard and systematic approach for executing experimentation to obtain optimum settings of design parameters for quality and cost very efficiently. The method gives priority to move quality back to the design stage, pursuing to design a/process or product. Orthogonal arrays are utilized in course of Taguchi method, to analyse a large number of variables with a fewer number of experiments. The conclusions made from small-scale experiments are genuine over the complete experimental domain consists of the control factors and their corresponding level settings. Taguchi method utilizes a statistical measure of performance called Signal-to-Noise (S/N) ratio, in order to examine the results. The S/N ratio takes into account both the mean and the variability of the response data. After performing the statistical analysis of S/N ratio, an ANOVA requires to be applied for estimating the relative importance of various factors and for computing error variance. The predicted optimum setting need not correspond to one of the rows of the matrix experiment in Taguchi method for parameter design. Therefore, an experimental confirmation is run utilizing the predicted optimum levels for the process parameters being investigated.

The intention is to confirm that the optimum conditions suggested by the matrix experiments do indeed contribute the projected improvement. If the projected improvements and observed match, the suggested optimum conditions will be adopted.

5.1.2 Taguchi's S/N ratio for Performance Evaluation

There is a loss function which describes the deviation from the target and further transformed into S/N ratio. The transformed S/N ratio is also defined as quality evaluation index. The least variation and the optimal design are obtained by analyzing S/N ratio. The higher the S/N ratio, the more stable the achievable quality. There are three Signal-to-Noise ratios of common interest for optimization of Static Problems;

(i) Nominal-the-Best (NB): In this approach, the closer to the target value, the better and the deviation is quadratic. The formula (1) for these characteristics is:

$$\frac{S}{N}\,Ratio = -10\log\left(\frac{y}{S_y^2}\right) \tag{1}$$

(ii) Lower-is-Better (LB): The Lower-is-Better approach held when a company desires smaller values. The formula (2) for these characteristics is:

$$\frac{S}{N}\,Ratio = -10\log\left(\frac{1}{n}\sum_{i=1}^{n} y^2\right) \tag{2}$$

(iii) Higher-is-Better (HB): Higher-is-Better (HB) is required when a manufacturer desires higher values of a characteristic. The formula (3) for these characteristics is:

$$\frac{S}{N}\,Ratio = -10\log\left(\frac{1}{n}\sum_{i=1}^{n}\frac{1}{y^2}\right) \tag{3}$$

Here,

y = Average of observed values;

s_y^2 = Variance of y;

N = Number of observations

However, Taguchi method is considered only for single objective optimization problems. It cannot be utilized for getting the single optimal setting of process parameters considering more than one performance parameter.

5.2 Using Fuzzy Logic

This part is concerned with the design of fuzzy logic based model to simulate the material removal rate. Main objective to develop fuzzy model is its practical application, such that a design engineer can use it in process planning time to time. The efficiency of the model is measured by comparing the predicted with the practical results.

Fuzzy inference is the method in which the mapping from a given input to an output is formulated utilizing fuzzy logic. Fuzzy logic can potentially acquire human decision making, common sense reasoning, and other perspectives of human apprehension. The fuzzy-rule based methodology is a core reasoning process where experts" experience and subject knowledge can possibly be implemented and translated into the machine language. The following elements involve the process of fuzzy inference: Membership Functions, If-THEN Rules and Logical Operations. Generally two kinds of fuzzy inference systems can be utilized: *Mamdani* type and *Sugeno* type. Depending on the way in which outputs are determined, these two types of inference systems are varied. The most commonly utilized fuzzy methodology is *Mamdani's* fuzzy inference method. This was developed by *Ebrahim Mamdani* in 1975 as an effort to control a boiler and stem engine combination by incorporating a set of linguistic control rules accessed from human operators' experiences. *Mamdani* fuzzy model is developed on the basis of the combinations of IF-THEN rules taking account both consequent predicts and fuzzy antecedent.

A fuzzy system generally consists of four components i.e. fuzzifier, an inference engine, a knowledge base and a defuzzifier. The membership functions are first utilized by fuzzifier to convert the crisp inputs into fuzzy sets. Then the fuzzy values are generated by the action of inference engine on fuzzy rules performing fuzzy reasoning. After that the defuzzifier defuzzifies these fuzzy values into crisp outputs. Number of research work has been focused on optimization of electrochemical machining using fuzzy logic. Jagdev et al. presented fuzzy model to simulate the material removal rate in ultrasonic drilling of porcelain ceramic. Prabhu et al. predicted the surface roughness for without CNT based cutting fluid and with CNT. Further, a fuzzy logic system is used to investigate the relationship between the control factors and response.

Present work comprise of four input and one output (Figure 5). And the *Mamdani* inference method based on fuzzy rules is utilized in this present study for fuzzy inference reasoning. The fuzzy rule base consists of a group of if–then control rules with the two inputs, $x1$ and $x2$, and one output y, that is:

Rule: if $x1$ is $A1$ and $x2$ is $B1$ then y is $C1$ else

Ai, Bi and Ci are fuzzy subsets defined by the corresponding membership functions. Nine fuzzy rules are directly derived based on effect of input variable on the removal rate. In the paper, three fuzzy subsets are assigned in the four inputs (Figs. 6-9). Five fuzzy subsets are assigned in the output (Figure 10).

Then fuzzy set for each output variable requires defuzzification, after the aggregation process. Fuzzy values can be converted into one single crisp output value utilizing the defuzzification method. One of the well-known techniques *the centre of gravity method* is utilized for defuzzifying fuzzy output functions is used in this study.

5.3 Using Artificial Neural Network

ANN is a powerful tool for study and analyse high degree of nonlinear system. One of the major advantages of using neural network approach is that it can be modelled easily for the given set of input output data base. Thus a trained ANN model accurately justifies the complex behaviour of dynamic

Figure 5. Structure of the four-input-one-output fuzzy logic unit

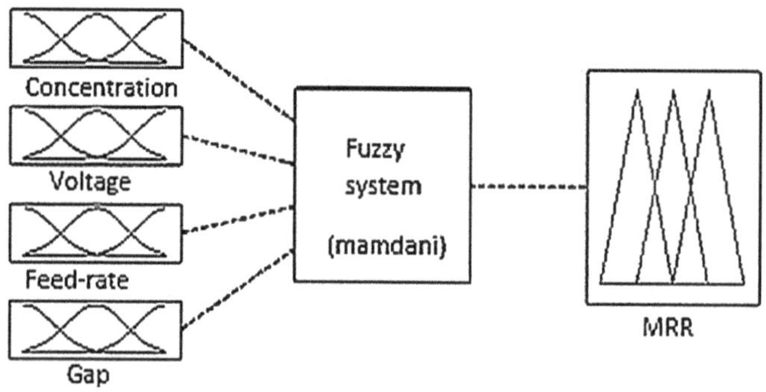

Figure 6. Membership functions for Concentration

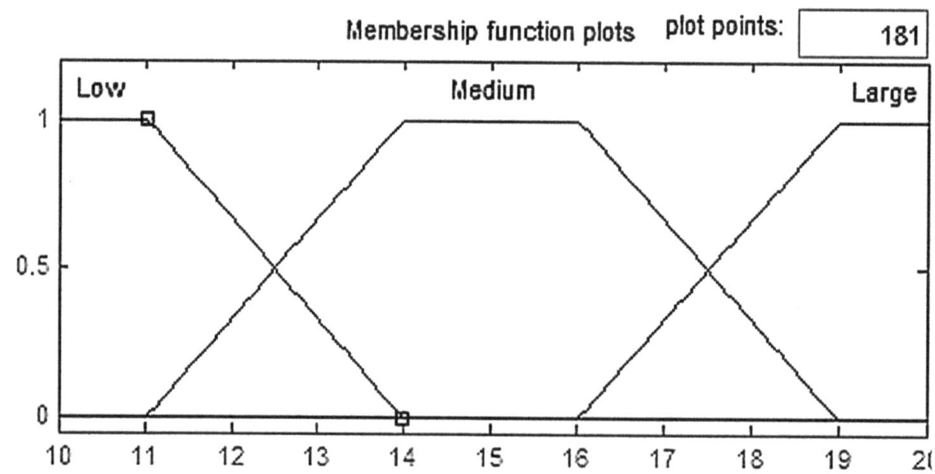

process. This technique is appreciated in ECM processes which show the high degree of complexity. Then it is difficult to generalise the process accurately with its physical mechanism. Many researchers have developed the mathematical models for response data set using statistical tools. Zhiyong Li et al. (2009), predict the accuracy ECM process with the help of BPNN for Aero-engine blade considering the effect of applied voltage, concentration, feed rate and inter-electrode gap (IEG). Panda et al. (2012) reported an intelligent approach for the modelling of DS–ECSM process using FEM and ANN in integrated manner. Kuo Tsai et al. (2001) revealed the capability ANN approach to determine the surface finish of the product on varying electrode polarity during EDM machining. G. Q. Shang, et al. (2008) developed a neural model with BPP to understand the variation of anode accuracy in ECM with shaped electrode. G. Ugrasen et al. (2014) outlined the development of model and its application to estimation of machining performances using Multiple Regression Analysis (MRA), Group Method Data Handling Technique (GMDH) and Artificial Neural Network (ANN). P.G. Benardos et al. (2003) reviewed the surface roughness predicted in machining ANN employed several approaches and practices. Nayak et al. (2012) performed an experimental work on ECM for the responses (i.e. average surface roughness and MRR) according Taguchi design of experiment (DOE) and the result were analysed by artificial neural network.

Figure 9. Membership functions for Gap

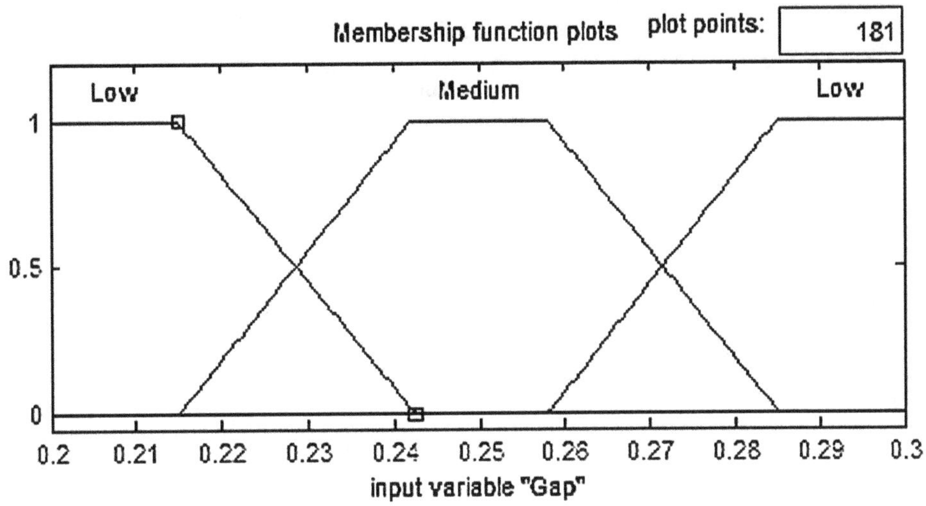

The present study deals with the development of artificial neural network to determine the suitable machining process parameters in order to obtain optimum surface roughness characteristics in ECM process. Artificial neural network with Error Back Propagation Training Algorithms (EBPTA) is used for the training and testing purpose of the experimental data. The initially a data set is required to train the ANN model. Then developed experimental database was used to train and simulated the ANN model to determine the predicted value. The EBPTA is based on universal delta-rule that involves dynamic weight updates system so as to reduce the mean squared error (MSE).

In this part, ANN model will be used for prediction of surface roughness. The experiment study would illustrate the error for the roughness average (Ra) and roughness maximum (Rt) properties. So with reference to the mid-level quality characteristics supported by Artificial Neural Network the degree of non-linearity that exists in nature between the output and input variable is identified.

Figure 10. Membership functions for Metal Removal Rate

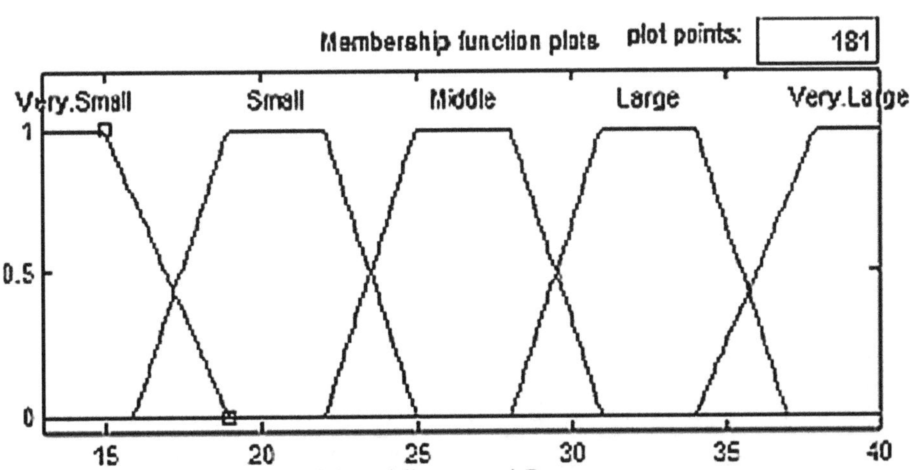

Figure 7. Membership functions for Voltage

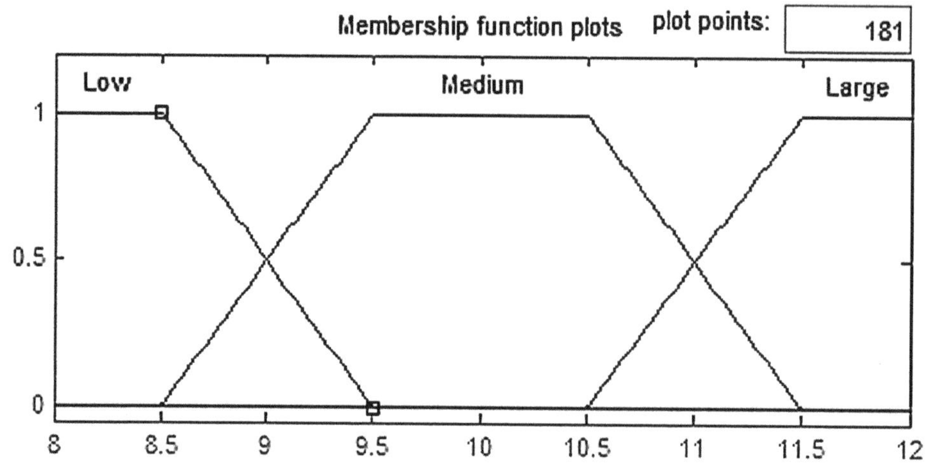

The multi-layer ANN network shown in Figure 11 comprises of input, output and hidden neurons layer. Multi-layer ANN architecture consists of 4-30-2 network model is used for simulation. Four neurons in the input layer corresponding to 4 inputs variable (i.e. concentration, voltage, feed rate, inter-electrode gap), two neurons in the output layer (corresponding to 2 outputs, Ra and Rt). One hidden layer with 30 neurons was found suitable employed in the present study. For training and testing 'nntool' is used which is available in optimization tool box in 'MATLAB' software (Math Works). When maximum number of epoch and minimum MSE is achieved, training of ANN model is stopped. The mean square error (MSE) is computed using following equation (4).

$$MSE = \frac{1}{NP}\sum_{P=1}^{NP}\sum_{k=1}^{k}\left(d_{kp} - o_{kp}\right)^{2} \quad \mathrm{MSE} = \frac{1}{\mathrm{NP}}\sum_{\mathrm{p}=1}^{\mathrm{NP}}\sum_{\mathrm{k}=1}^{\mathrm{k}}\left(\mathrm{d}_{\mathrm{kp}} - \mathrm{o}_{\mathrm{kp}}\right)^{2} \tag{4}$$

Figure 8. Membership functions for Feed-rate

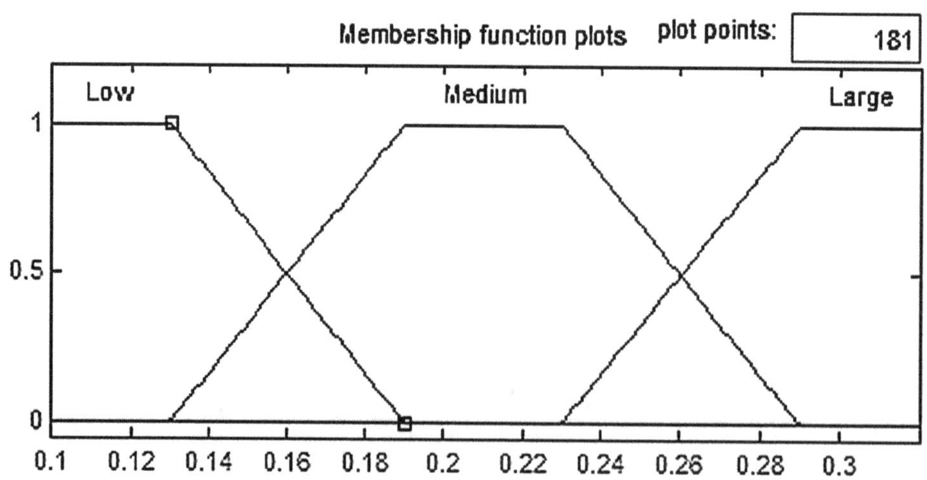

Figure 11. Multi-layer feed forward artificial neural network

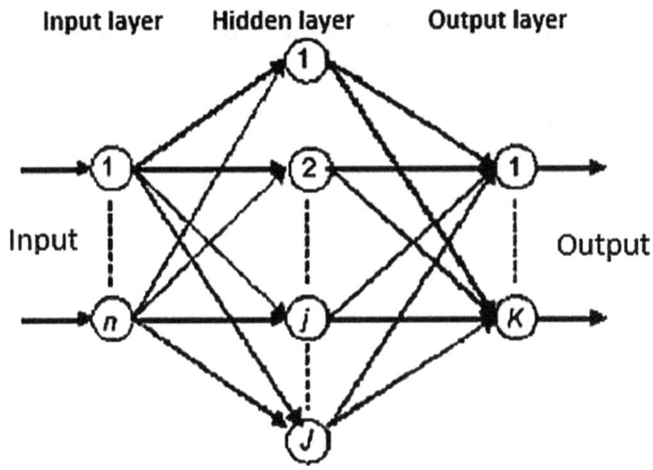

Where

NP =number of training patterns,

dk,p = desired output for the pth pattern and Okp is predicted output for the pth pattern.

5.3.1 Artificial Neural Network Training and Testing

The training of ANN has been carried out for 27 input–output patterns with the help of 'nntool' available in *MATLAB* software (2012a). The factors are used for training purpose are:

- Learning rate = 0.05;
- Momentum factor = 0.85,
- Maximum number of epochs = 1000,
- Tolerance for MSE= 0.0001.

The ANN training and simulation was carried out with the help of "*traingdx*" function. The variation of MSE with number of epoch is shown in Figure 12.

From the Figure 12 it can be depict that best validation performance is 0.064573 at epoch 71. Regression correlation of the Ra and Rt are presented in Figure 13. The correlation coefficient shows that up to what extent network trained for target values. For RC value is 1, best correlation between experimental and predicted value. In the present study, the RC value is 0.99041 (Figure 13) for both the given targets (Ra and Rt), indicating a very good correlation.

Figure 12. The variation of mean squared error (MSE)

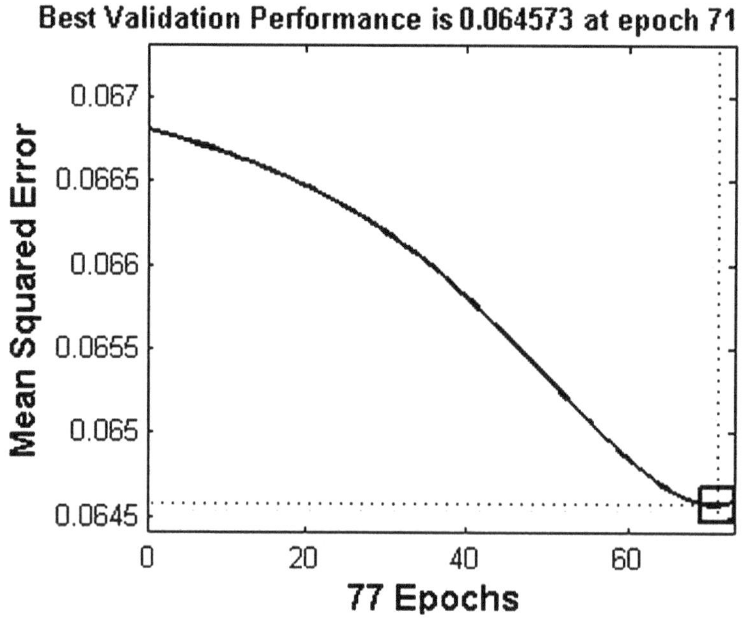

5.4 Using Grey Relation Approach

Surface roughness characteristics and metal removal rate were taken as the responses in this part. Optimum level of input variables was identified for multi-response parameter using grey relational grade. Also analysis of variance technique is to be implemented to understand the significant contributions of each input process variables and finally a confirmation test is to be conducted to validate the work.

The grey system theory was initiated by Deng in 1982 and subsequently modified in 1989 (Deng, 1989). This method has been convinced as an advantageous methodology for approaching with incomplete, poor and uncertain information. To resolve the complex interrelationships among the multiple performance characteristics efficiently, the grey relational analysis can be implemented which is based on grey system theory. It also gives an effective solution to the discrete data problem. With the employment of grey relational analysis, the relation between machining parameters and performance can be procured.

Therefore, number of research work have been focused on electrochemical machining to enhance the machining process since; the development of experimental ECM process by W. Gussef in 1929. Recently Das et al. (2014) did an experiment on ECM for optimization of multi-quality characteristics using grey relation approach. MRR and SR are taken as the response characteristics. Jabbaripour et al. (2012) conducted an EDM to investigate the MRR, tool wear ratio and aspects of surface veracity for Ti–6Al–4V using grade approach. Lijo paul et al. have conducted an experimental work on silicon wafer for valuations of process parameter of ECDM using grade relation approach. Manimaran et al. (2013) also have made an investigation on grinding AISI 316 steel by the Taguchi and grey relational analysis. The performance was measured for MRR, SR and grinding force (Ft). Schuster et al. (2000) have made an attempt to use a combination of electrolyte solution of $CuSO_4$ and $HClO_4$ during micro electrochemical (ECM) machining. Asokan et al. (2008) have also find out the optimal combination for responses (MRR and SR) using grey relational approach with control variable in ECM of hardened steel.

Figure 13. Correlation of the training patterns for Ra and Rt with Number of epochs

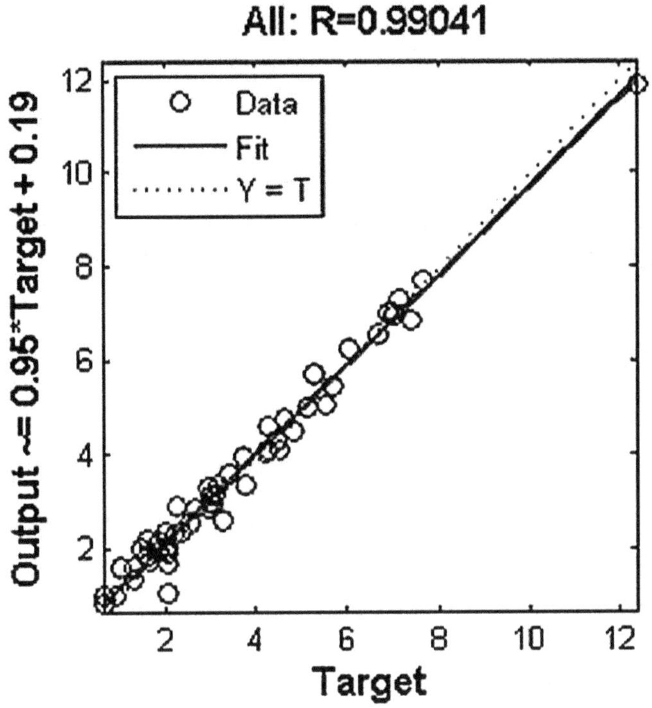

The principal steps of grey relational approach (GRA) are firstly the performances of all alternatives are converted into a comparability sequence. This step is known as grey relational generation. A reference sequence (ideal target sequence) is developed according to these sequences. Then, calculation of the grey relational coefficient between the reference sequence and all the comparability sequence is done. Finally, the calculation of grey relational grade between every comparability sequences and the reference sequence is performed based on these grey relational coefficients. The alternative will be chosen as best choice, if the comparability sequence derived from that alternative has the highest grey relational grade between itself and the reference sequence.

The effect of some attributes may be over sighted, when the units or dimensions in which performance is measured are different for different attributes. This may also occur that some performance attributes have a very large range. In addition, if the directions and goals of these attributes are different, it will develop unreliable outcomes in the analysis. Therefore, it is essential for translation of all performance values for every alternative into a comparability sequence, in a method equivalent to normalization. This processing is termed as grey relational generating in GRA. In present research work Higher-the-best conditions are applied for stabilization of MRR data-set (equation 5), and for SR parameters, smaller-the-better criterion (equation 6) is used.

$$X_i\left(k\right) = \frac{y_i\left(k\right) - min y_i\left(k\right)}{max y_i\left(k\right) - min y_i\left(k\right)} \tag{5}$$

$$X_i(k) = \frac{\max y_i(k) - y_i(k)}{max\ y_i(k) - min\ y_i(k)} \tag{6}$$

5.4.1 Grey Relation Coefficient

The calculation of the grey relational coefficient from the stabilized data-set to represent the correlation between the desired and actual experimental data is performed using equation 7.

$$\varepsilon_i(k) = \frac{\Delta_{min} + \psi\Delta_{max}}{\Delta_{oi}(k) + \psi\Delta_{max}} \tag{7}$$

Where $\Delta_{oi}(k)$ is difference of the absolute value between $x_o(k)$ and $x_i(k)$. Δ_{min}, Δ_{max} are respectively the minimum and maximum values of the absolute differences $\Delta_{oi}(k)$. Ψ is a distinguishing coefficient, $0 \leq \psi \leq 1$, the purpose of which is to weaken the effect of Δ_{max} when it gets too big. In the present research work for MRR Ψ is taken 0.5 and 0.1 for each surface characteristic.

5.4.2 Grey Relational Grade

The grey relational grade is treated as the overall response of the process instead of the multiple responses like MRR and surface roughness. The grey relational coefficients are calculated for the experimental data using $\Psi = 0.5$. The grey relational grade γ_i is obtained (equation 8) by averaging the grey relational coefficient as follows:

$$\gamma_i = \frac{1}{n}\sum_{k=1}^{n}\varepsilon_i(k) \tag{8}$$

Where n is the number of process responses. Higher value of grey relational grade implies stronger relational degree between the ideal sequence *x0(k)* and the given sequence *xi(k)*. Thus multi-response optimization problem is converted into single response optimization problem.

In this section Optimization of ECM Process for machining of EN-19 with Multi quality Characteristics via Grey–Taguchi Approach. Optimum level of input variables have been identify for multi-response parameter using grey relational grade. Also analysis of variance (ANOVA) technique has been implemented to understand the significant contributions of each input process variables. Finally confirmation test is conducted to validate the work.

5.5 Using Non-Dominated Sorting Genetic Algorithm

In the present section multiple regression equation was developed for SR and MRR, and the same was utilized for simultaneous optimization using genetic algorithm with hybrid function. A set of optimal combination of parameter will be proposed for quality and cutting efficiency from the results of genetic algorithms.

Genetic algorithms belong to the larger class of evolutionary algorithms (EA), which generate solutions to optimization problems using techniques inspired by natural evolution, such as inheritance, mutation, selection, and crossover. Decision making situation occurs in all fields like science, technology and management, etc. where GA is applied with an objective to maximize or minimize a task. In order to solve the problems related to inventory, transportation, queuing, scheduling etc., many optimization procedures have been developed over the past six decades. Most of the traditional optimization procedures end its search in the 'local optima' rather than finding the 'global optima'.

Keeping such constraints constantly in view, several researchers have tried to optimize the machining performance by adopting different optimization techniques. Acharya et al. (2013) have optimized the electro-chemical machining of hardened steel using NSGA- II, for metal removal rate and surface roughness. In his work current, voltage, electrolytic flow rate and inter-electrode gap is taken as input parameters. Jain and Jain (2007) have optimized the three most important process parameter i.e. feed rate, electrolyte flow velocity, and applied voltage to minimize geometrically inaccuracy subjected to temperature, choking, and passivity constraints using real-coded genetic algorithms. Senthilkumar et al. (2012) optimize the Electro-chemical machining process for Metal removal rate and Surface roughness and found the optimal parameter by NSGA-II. Electrolytic concentration, electrolytic flow rate, applied voltage and tool feed rate is take as the process parameter. An experimental investigation and multi-optimization of wire EDM for surface roughness, metal removal rate and white layer is conducted by Rao et al. (2014) using genetic algorithm.

GA mimics the principal of natural genetics and natural selection to constitute search and optimization procedure. The non-dominated sorting genetic algorithm has been criticized for its high computational complexity, lack of elitism and its choice of the optimal parameter value for sharing parameter σ. The NSGA-II is a modified version, which has a better sorting algorithm, incorporates elitism and does not require the choosing of a sharing parameter a priori. There are two key concepts in NSGA-II: a fast non-dominated sorting of the population and a crowding distance. This algorithm uses the elite-preserving operator, which favours the elites of a population by giving them an opportunity to be directly carried over to the next generation.

Algorithms for NSGA-II are shown below:

1. A fitness function is chosen to represent the problems, a selection operator, a crossover operator, and a mutation operator. Select population size, n, crossover probability, p_c, and mutation probability, p_m. initialize the random population of strings of size L. Assume a maximum allowable generation number t_{max}. Set t=0.
2. Evaluate each string in the population.
3. If $t > t_{max}$ or other termination criteria is satisfied, then *Terminate*.
4. Pre-form reproduction on the population.
5. Perform, crossover on random pairs of strings.
6. Perform mutation on every string.
7. Evaluate strings in the new population. Set $t = t + 1$ and go to step 3.

In the present work a set of optimal combination of parameter are proposed for quality and cutting efficiency from the results of genetic algorithms.

6. RESULT AND DISCUSSION - OPTIMIZATION OF SUSTAINED MATERIAL REMOVAL RATE AND SURFACE ROUGHNESS

6.1 Using Taguchi Approach

The optimal combination of process parameters was obtained by using the analysis of signal-to-noise (S/N) ratio in order to achieve maximized material removal rate by using Minitab 16 software (2003). Signal-to-noise ratio was tabulated in Table 3 for different input combination. Response was analyzed using Analysis of Variance and Main effect plot. Optimum values for responses of different combination were evaluated by considering the highest value of S/N ratio. This was done as the basic need was to maximize material removal rate.

After generation of signal-to-noise ratio, response table for mean S/N ratio generated. With the help of response table the rank of different parameter were predicted as depicted in table 4, which was used to understand the level of importance of each machining factor (i.e. concentration, voltage, feed rate and gap).

Table 3. Design of experiments and results

Exp. No.	A	B	C	D	MRR	SNRA1	Exp. No.	A	B	C	D	MRR	SNRA1
1	1	1	1	1	21.6561	26.7116	15	2	2	3	2	26.9427	28.6088
2	1	1	2	2	17.2399	24.7306	16	2	3	1	1	24.1401	27.6547
3	1	1	3	3	19.7452	25.9092	17	2	3	2	2	25.3715	28.0869
4	1	2	1	2	14.1189	22.9960	18	2	3	3	3	22.8238	27.1677
5	1	2	2	3	14.8370	23.4269	19	3	1	1	3	23.0998	27.2721
6	1	2	3	1	16.1359	24.1558	20	3	1	2	1	21.8259	26.7794
7	1	3	1	3	19.7707	25.9204	21	3	1	3	2	23.5669	27.4468
8	1	3	2	1	24.2038	27.6776	22	3	2	1	1s	22.1806	26.9194
9	1	3	3	2	23.0828	27.2657	23	3	2	2	2	25.2529	28.0462
10	2	1	1	2	13.2484	22.4432	24	3	2	3	3	32.3931	30.2090
11	2	1	2	3	16.8153	24.5140	25	3	3	1	2	22.2930	26.9633
12	2	1	3	1	16.4968	24.3479	26	3	3	2	3	24.7134	27.8586
13	2	2	1	3	16.1783	24.1786	27	3	3	3	1	37.7070	31.5284
14	2	2	2	1	25.4777	28.1232							

Table 4. Response Table for Signal to Noise Ratios (Larger is better)

Level	A	B	C	D
1	25.42	25.57	25.67	27.10
2	26.13	26.30	26.58	26.29
3	28.11	27.79	27.40	26.27
Delta	2.69	2.22	1.73	0.83
Rank	1	2	3	4

Figure 14. Main Effect plot for MRR

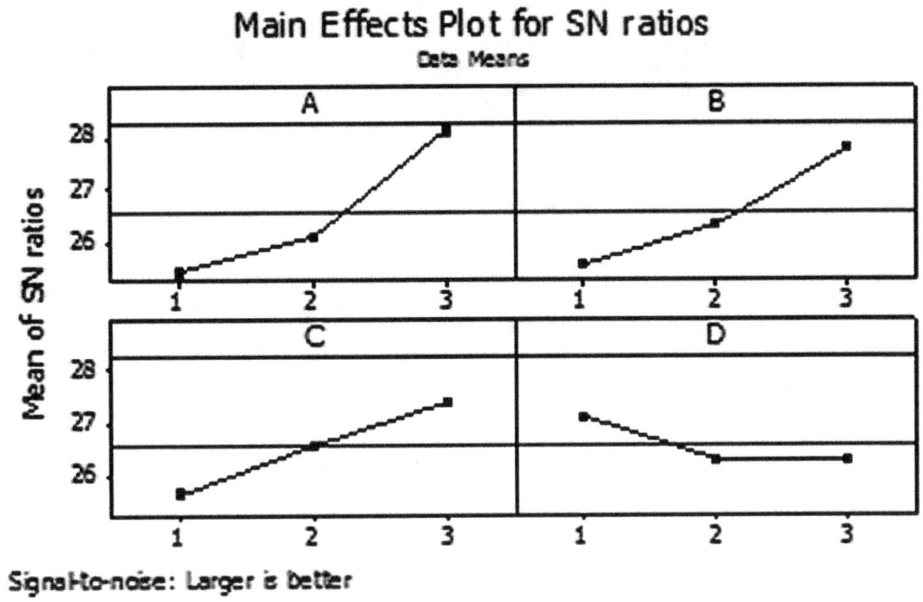

6.1.1 Main Effect Plot

Main effect graph was plotted using response table for S/N ratio; it uses the mean average of the factor considering each level response. Here main effect plot (Figure 14) shows that as the concentration, voltage and feed rate increases, MRR also increases. The higher inclination of the line in the main effect plot indicates that the parameter is highly significant during machining and hence from it can also be identified which one of the factor have higher significant. From the main effect plot, optimal combination was ascertained as A3B3C3D1 for maximum material removal rate during ECM machining of EN19 tool steel.

6.1.2 Analysis of Variance (ANOVA)

Analysis of Variance is a statistical tool and the table (table 5) for the same was generated using Minitab 16 software. The ANOVA table indicates the importance of input process parameter toward MRR. The entire result was conformed at 95% confidence level and it is found that concentration followed by voltage and feed rate have significant impact over MRR. Only Interaction of Voltage and electrolyte concentration shows impact over MRR. Inter-electrode gap have least impact over MRR.

ANOVA results are based on the F-ratio, it is also known as variance ratio, and it is the ratio of variance between the experiments to the variance within the sample. Here standard F-value taken at 95% confidence level ($F_{0.05, 2, 6}$=5.14), thus it is clear from the above table that electrolyte concentration has most significance influence over MRR followed by voltage and feed rate. Interaction of concentration and voltage (A*B) also has significant influence over material removal rate (14.55%).

Table 5. Analysis of Variance for MRR, using Adjusted SS for Tests

Source	DF	Seq SS	Adj SS	Adj MS	F	P	contribution %
A	2	230.624	23.624	115.312	13.0.7*	0.007	29.42
B	2	143.333	143.33	71.666	8.12*	0.020	18.28
C	2	99.287	99.287	49.643	5.62*	0.042	12.66
D	2	26.989	26.989	13.494	1.53	0.291	3.44
A*B	4	114.039	114.039	28.510	3.23	0.097	14.55
A*C	4	73.882	73.882	18.470	2.09	0.200	9.42
B*C	4	42.576	42.576	10.644	1.21	0.398	5.43
Error	6	52.956	52.956	8.826			6.75
Total	26	783.685					100
S = 2.97085 R-Sq = 93.24% R-Sq(adj) = 70.72%							

6.1.4 Development of Regression Model

Regression analysis is statistical tool for analysing the relation between variable. It is one of the commonly used techniques to predict the behaviour of response with respect to process variable. With the help of Minitab 16 regression equation 9 was generated for MRR and the various variables. It is expressed as:

$$MRR = 19.4804 - 2.57463\,A + 3.7971\,B - 7.06879\,C + 0.743546\,D + 0.392631\,A*B + 2.42502\,A*C + 0.198613\,A*D + 1.25078\,B*C - 2.14163\,B*D + 1.03104\,C*D \qquad (9)$$

6.1.5 Confirmation Test

Confirmation test was conducted to validate the result or to see whether the improvement in result is obtained or not. In confirmation table S/N ratio for experiment and predicted value were presented. Predicted S/N ratio is determine with the help of minitab 16 and experimental was obtained using the following relation (equation 10):

$$\hat{\gamma} = \gamma_m + \sum_{i=1}^{0}\left(\hat{\gamma}_i - \gamma_m\right) \qquad (10)$$

Where γ is the optimal level of process parameter, γ_m is the total mean S/N ratio, γ_i is the mean S/N ratio and O is the number of main design parameter. The improvement in S/N ratio is found to be 3.833 db as depicted in table 6.

Table 6. Confirmation table

Parameter	Mean parameter for EN 19	Optimal parameter for EN 19	
	Mean	Predicted	Experimental
	A2B2C2D2	A3B3C3D1	
S/N RATIO	27.6941	30.9556	31.52843962
MRR	27.851		37.707
Improvement in S/N ratio = (31.528) - (27.6941)= 3.833 db			

6.1.6 Surface Plot for MRR with Effect of Process Parameter

Three-dimensional Surface plot shows the estimated response as function of controllable process parameter. Figure 15 shows the variation of MRR with two of the four variables, it is clear from the graph as the concentration and voltage increases material removal rate increase. Since at higher electrolytic concentration, more electrochemical reaction take place between the copper tool and work piece due to increasing amount of negative ion to perform electrochemical reactions with the metallic ions. Voltage also play a significant role, as the voltage increase it cause a greater amount of machining current available in machining gap, thereby causing the increase in MRR.

In the Figure 16, variation of MRR with electrolytic concentration and feed rate shown. The result clearly indicates that as the feed rate and concentration increases metal remove rate increases. The main reason for the effect is, as the electrolytic concentration increase, it increases the concentration of electrolytic negative ion within the machining gap, which increases the material removal rate and as the feed rate increases the inter electrode gap decreases which lead to increase in the current density within the machining gap with the consequent anodic dissolution and hence increases the material removal rate.

Figure 15. Surface plot of between concentration and voltage

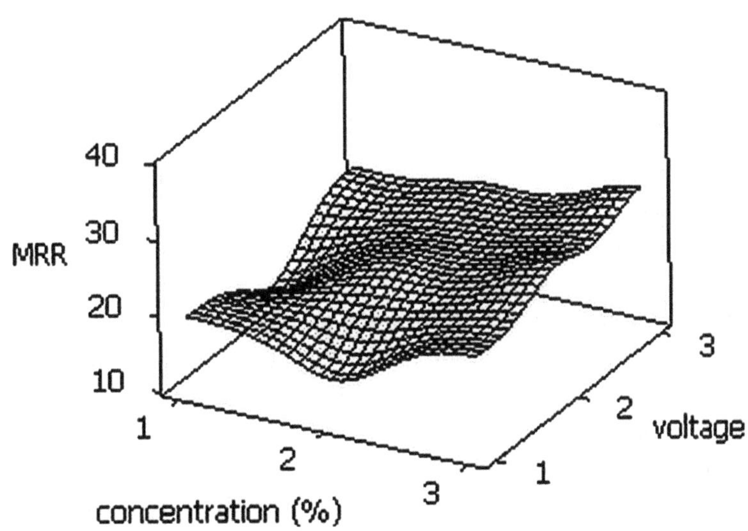

Figure 16. Surface plot of between concentration and feed rate

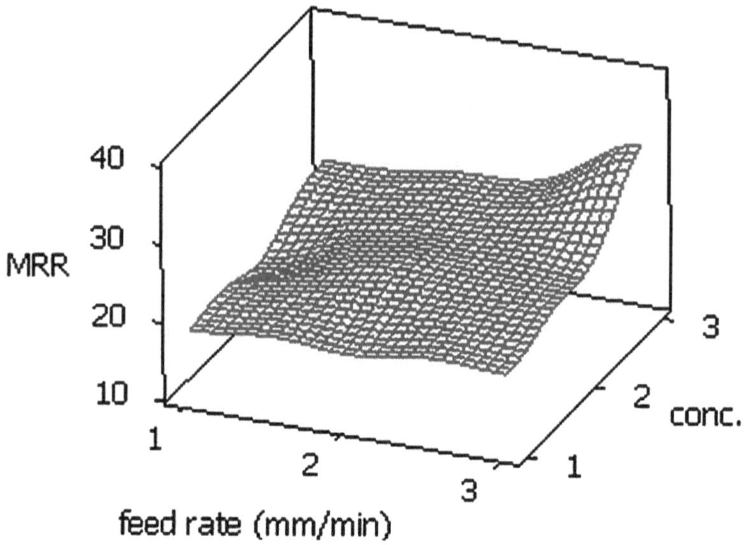

6.2 Using Fuzzy Logic

6.2.1 The Evaluation of Optimal Setting

Scatter plots for the experimental and predicted values with fuzzy modelling is depicted in Figure 17. The efficiency of the developed fuzzy models was evaluated by R2 values. In fuzzy models, R2 values for MRR were found as 96.5%, which meant the high correlation that existed between the experimental and predicted values. The predicted values achieved by fuzzy logic modelling were compared with experimental values and results were plotted as shown in Figure 18.

Optimal combination of process variable estimated to enhance the performance. Optimal parameter setting is measured on the basis of mean parameter. In this regard residual values for response has been plotted (Figure 19), it shows at test run 24 (concentration 20%, voltage 10V, feed-rate 0.32mm/min, and inter-electrode gap 0.3) has lowest residual value respectively 0.5069.

6.2.2 Confirmation Test

In order to ascertain the extent of the improvement as indicated by the optimum parameters analysis, a validation test was required to be undertaken. The predicted MRR was tallied with the experimental ones respectively (Table 7). The mid-level combination of process parameters is assumed as the initial condition. It can be easily seen that the residual error for A3B2C3D3 is less as compared to the A2B2C2D2 From the table, it give the improvement of MRR at the optimal condition is about 2.5149 dB (Table 8).

The 3D plot for MRR from the fuzzy logic model was constructed as shown in Figures 20, 21, 22 and 23. As per Figure 20, MRR is optimum at the higher concentration (20%) and significantly got enhanced by increasing the voltage. From Figure 21, it is clearly seen that as the feed rate increase MRR, but increase in removal rate quit less initially. When feed rate close to mid-level MRR increases sharply. In Figure 22 it can be observed that as the gap decreases MRR increase. Since in small inter-electrode

Figure 17. Scatter plot for experimental and

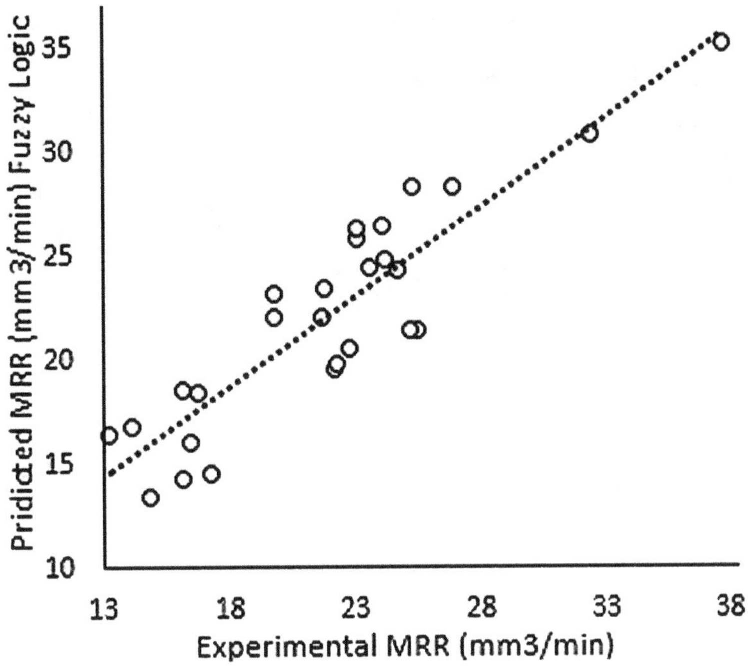

Table 7. Experiment and predicted MRR using fuzzy model result for response parameter

Exp. No.	A	B	C	D	MRR	S/N (MRR)	Predicted value	Absolute error	Exp. No.	A	B	C	D	MRR	S/N (MRR)	Predicted value	Absolute error
1	1	1	1	1	21.6561	26.7116	22.8	1.1439	15	2	2	3	2	26.9427	28.6088	28.4	1.4573
2	1	1	2	2	17.2399	24.7306	14.6	2.6399	16	2	3	1	1	24.1401	27.6547	26.5	2.3599
3	1	1	3	3	19.7452	25.9092	22.1	2.3548	17	2	3	2	2	25.3715	28.0869	28.4	3.0285
4	1	2	1	2	14.1189	22.9960	16.8	2.6811	18	2	3	3	3	22.8238	27.1677	20.6	2.2238
5	1	2	2	3	14.837	23.4269	13.5	1.337	19	3	1	1	3	23.0998	27.2721	26.4	3.3002
6	1	2	3	1	16.1359	24.1558	18.6	2.4641	20	3	1	2	1	21.8259	26.7794	23.5	1.6741
7	1	3	1	3	19.7707	25.9204	23.2	3.4293	21	3	1	3	2	23.5669	27.4468	24.5	0.9331
8	1	3	2	1	24.2038	27.6776	24.8	0.5962	22	3	2	1	1	22.1806	26.9194	19.6	2.5806
9	1	3	3	2	23.0828	27.2657	25.8	2.7172	23	3	2	2	2	25.2529	28.0462	21.5	3.7529
10	2	1	1	2	13.2484	22.4432	16.5	3.2516	**24**	**3**	**2**	**3**	**3**	**32.3931**	**30.2090**	**32.9**	**0.5069**
11	2	1	2	3	16.8153	24.5140	18.5	1.6847	25	3	3	1	2	22.293	26.9633	19.8	2.493
12	2	1	3	1	16.4968	24.3479	17.1	0.6032	26	3	3	2	3	24.7134	27.8586	25.4	0.6866
13	2	2	1	3	16.1783	24.1786	14.3	1.8783	27	3	3	3	1	37.707	31.5284	35.2	2.507
14	2	2	2	1	25.4777	28.1232	21.4	4.0777									

gap current density increase which increase the anodic dissolution rate. Similarly in Figure 23, shows the voltage and feed rate plot, as the voltage increases current density increases between the gap which increase the MRR, higher feed-rate also help the MRR.

Figure 18. Variation of Experimental and predicted MRR predicted MRR

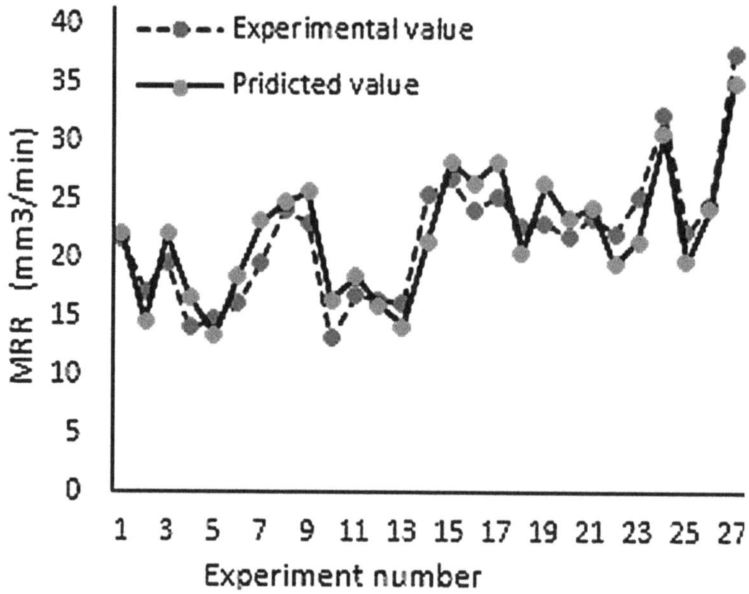

Figure 19. Variation of residual with training pattern

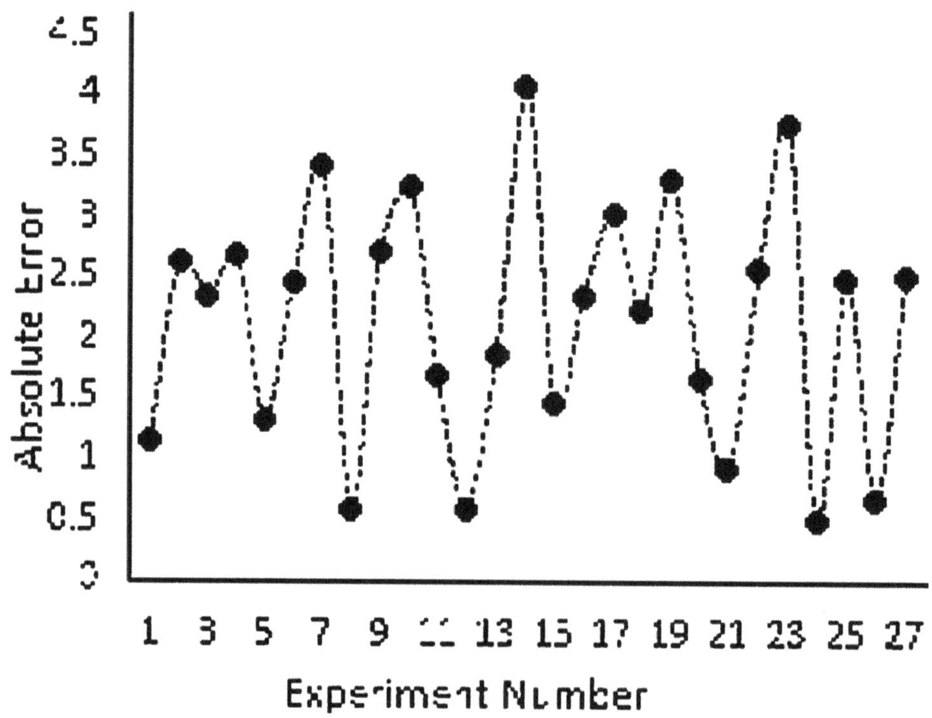

Figure 20. 3D surface plot for effect of voltage and concentration

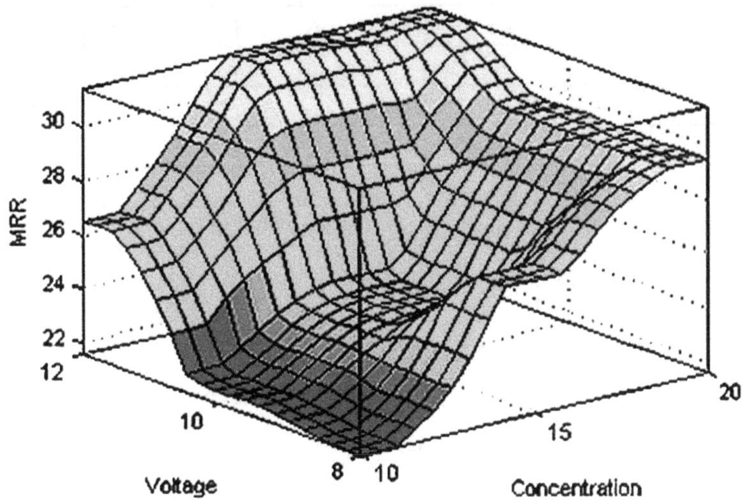

Figure 21. 3D surface plot for effect of feed-rate and concentration

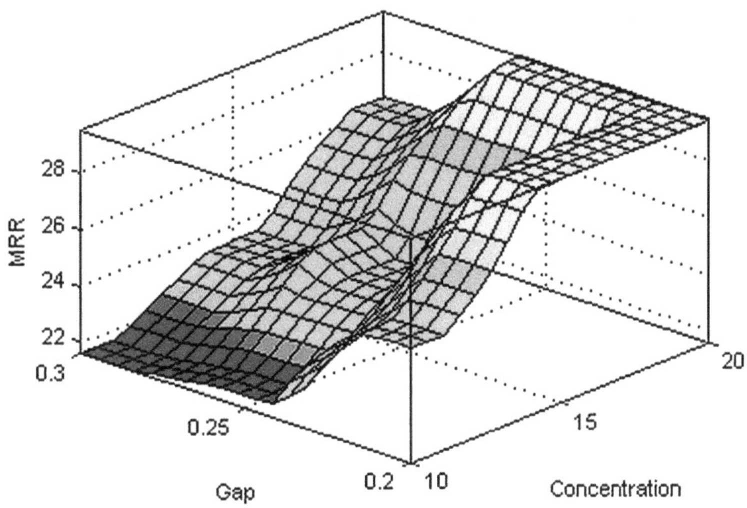

6.3 Using Artificial Neural Network

Trained ANN was simulated by presenting same input patterns, which were employed for the training and testing purpose. For each set of input, the predicted value of *R*a and *R*t are compared with the respective measured values and the residual error is computed according to the equation 11 and residual error is listed in table 9.

$$\%Absolute\,error = \left| y_{i,expt} - y_{i,pred} \right| \tag{11}$$

Table 8. Confirmation runs of optimal parameters for MRR

Parameter	Mean parameter for EN 19	Optimal parameter for EN 19	
	Mean	Predicted	Experimental
	A2B2C2D2	A3B2C3D3	
S/N RATIO	27.6941		30.2090
MRR	27.851	32.9	32.3931
Improvement in S/N ratio = 2.5149 dB			

Where *yi*: expt is the measured value, and *yi* pred is the ANN predicted value of the response for *i*th trial.

6.3.1 The Evaluation of Optimal Setting

After having finished the ANN train, ANN was tested and the experimental results are graphically compared with the results obtained from training network (Figure 24).

It was found that the predicted and experimental values were very fairly close to each other. Optimal combination of process variable estimated to enhance the performance. Performance enhancement of electrochemical machining for optimal parameter setting is measured on the basis of mean parameter. In this regard residual values for both response have been plotted. Figure 20 depict shows at test run 15 (concentration 15%, voltage 10V, feed-rate 0.32mm/min, and inter-electrode gap 0.25) both Ra, and Rt have lowest residual value respectively 0.0104 and 0.0046.

Figure 22. 3D surface plot for effect of concentration and gap

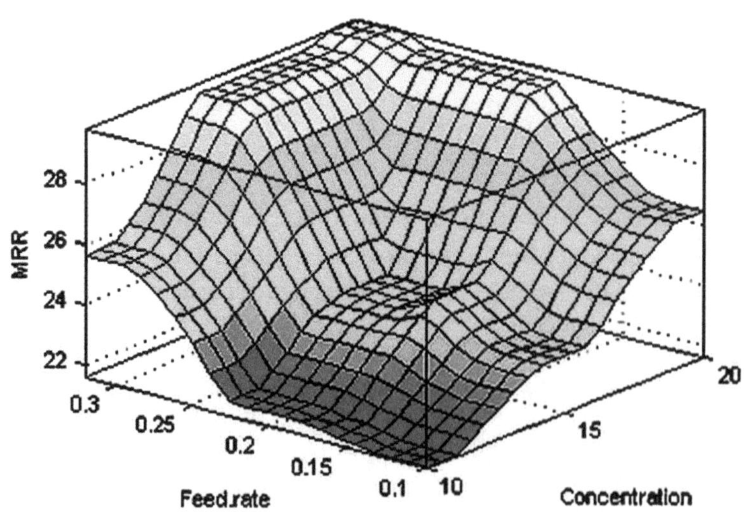

Figure 23. 3D surface plot for effect of voltage and feed-rate on MRR

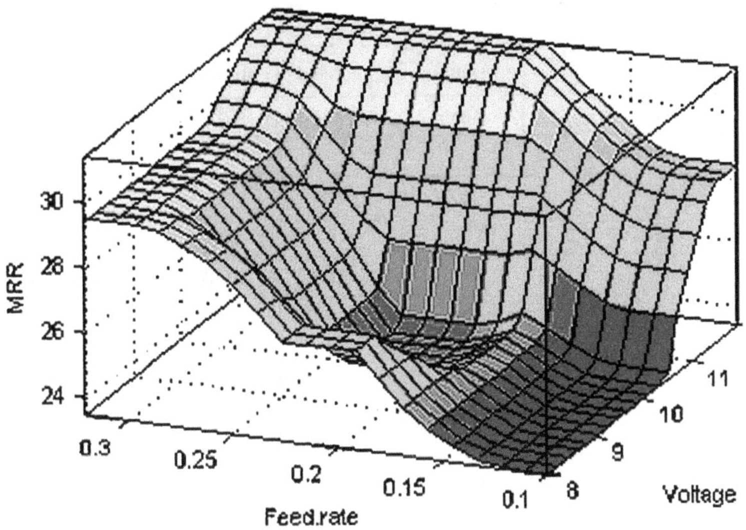

6.3.2 Conformation Test

In order to ascertain the extent of the improvement as indicated by the optimum parameters analysis, a validation test was required to be undertaken. The predicted average roughness and the maximum roughness values were tallied with the experimental ones respectively (Table 10). The mid-level combination of process parameters is assumed as the initial condition. It can be easily seen that the residual error for A2B2C3D2 is less as compared to the A2B2C2D2 From the table, it give the improvement of roughness average (Ra) and roughness maximum (Rt) at the optimal condition is about 0.077 and 0.591.

6.4 Using Grey Relation Analysis

S/N ratio for grade is calculated using higher-the-best (HB) condition, as grey-grade is to be maximized. Response Table for S/N of grade are showed in table 11. Table 12 gives the normalised value and table 13 the responses.

It is clear from the response table feed rate have highest rank. Main effect plot for S/N ratio has been shown in Figure 25. Optimal input variable arrangement for maximum MRR and minimum SR for ECM process is found as A1B1C3D1.

6.4.1 Analysis of Variance (ANOVA)

ANOVA test help to determine the level of contribution of each parameter. This result based on the F-ratio, here standard F-value taken at 90% confidence level (F0.1, 2, 6=3.46). If the standard F-value is more than the calculated F-value than that machining factor play significant at the desired confidence level. It is clear from the ANOVA table (table 14) that feed-rate have highest contribution of 36.80%, followed by inter-electrode gap having 13.93% contribution.

Table 9. Residual results of R_a and R_t by the BPNN

Run no.	Conc. A	Voltage B	Feed-rate C	Gap D	Roughness Average (Ra)			Roughness Maximum (Rt)		
					Actual	Predicted	Residual	Actual	Predicted	Residual
1	1	1	1	1	2.666	2.6128	0.0532	6.058	5.98192	0.07608
2	1	1	2	2	2	1.9555	0.0445	4.488	4.43292	0.05508
3	1	1	3	3	1.693	1.5635	0.1295	3.637	3.52904	0.10796
4	1	2	1	2	3.089	3.0673	0.0217	7.044	6.99776	0.04624
5	1	2	2	3	2.07	2.0027	0.0673	4.629	4.59732	0.03168
6	1	2	3	1	1.085	1.0183	0.0667	1.986	1.94393	0.04207
7	1	3	1	3	1.888	1.7957	0.0923	4.288	4.1661	0.1219
8	1	3	2	1	1.403	1.3845	0.0185	2.262	2.22656	0.03544
9	1	3	3	2	2.062	1.9504	0.1116	4.851	4.77466	0.07634
10	2	1	1	2	1.616	1.5976	0.0184	3.707	3.66734	0.03966
11	2	1	2	3	1.653	1.6145	0.0385	3.773	3.70894	0.06406
12	2	1	3	1	1.919	1.8438	0.0752	2.069	2.03612	0.03288
13	2	2	1	3	2.35	2.2958	0.0542	5.282	5.21066	0.07134
14	2	2	2	1	1.255	1.2304	0.0246	3.407	3.35788	0.04912
15	2	2	3	2	1.33	1.3196	0.0104	2.977	2.9724	0.0046
16	2	3	1	1	2.957	2.9116	0.0454	6.684	6.64134	0.04266
17	2	3	2	2	3.022	2.9751	0.0469	6.893	6.85634	0.03666
18	2	3	3	3	2.225	2.2042	0.0208	5.136	5.0896	0.0464
19	3	1	1	3	5.514	5.41	0.104	12.376	12.24756	0.12844
20	3	1	2	1	3.176	3.135	0.041	7.127	7.0688	0.0582
21	3	1	3	2	2.02	1.9685	0.0515	4.52	4.48042	0.03958
22	3	2	1	1	3.422	3.3756	0.0464	7.636	7.57948	0.05652
23	3	2	2	2	2.062	1.9926	0.0694	4.217	4.1436	0.0734
24	3	2	3	3	1.882	1.8405	0.0415	4.261	4.20892	0.05208
25	3	3	1	2	3.125	3.0735	0.0515	7.011	6.9452	0.0658
26	3	3	2	3	2.553	2.4771	0.0759	5.661	5.56744	0.09356
27	3	3	3	1	1.32	1.2316	0.0884	3.01	2.94358	0.06642

6.4.2 Confirmation Test

A confirmation test is performed to validate the result and to see the improvement in result. Also a prediction of the grey relational grade γ is calculated using equation 12

$$\widehat{\gamma} = \gamma_m + \sum_{i=1}^{0}\left(\widehat{\gamma_i} - \gamma_m\right) \tag{12}$$

Figure 24. Variation of residual with training pattern

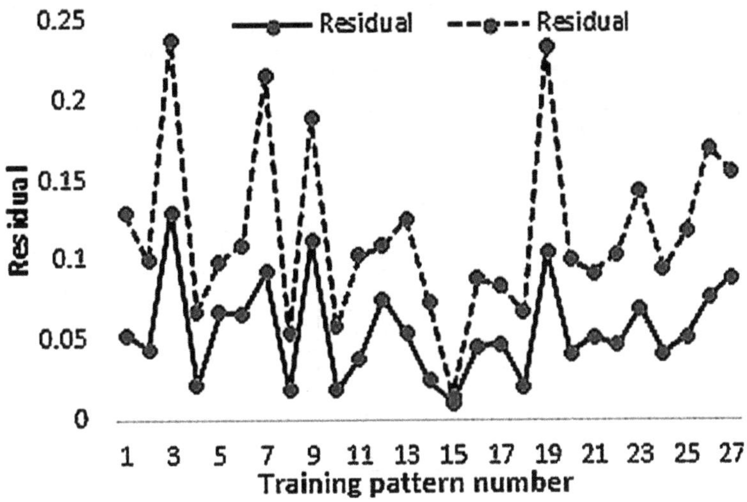

Table 10. Confirmation runs of optimal parameters for Ra, and Rt

Optimal Parameter				**Experimental/ Predicted Ra**	**Experimental Predicted Rt**
A (%)	B (V)	C (mm/min)	D (mm)	(μm)	(μm)
15	10	0.32	0.25	1.33/1.3196	2.977/2.9724
Mean parameter					
A	B	C	D		
15	10	0.21	0.25	1.253	2.386
Difference b/w mean and optimal combination				0.077	0.591

Where γ_m is the total mean grey relational grade, $\breve{\gamma}_i$ is the mean grey relational grade at the optimal level. The assessment of the expected and investigated grade is shown in Table 15.

From the table 14, it is found that the improved grey-relational-grade at the optimal condition is about 48.13% with respect to mean and multi-quality characteristics are greatly improved through this study.

6.5 Using Genetic Algorithm

6.5.1 Mathematical Model for MRR and SR

After taking the measurement, with the help of MINITAB.16 regression equation is generated for each response. It is use to represent the relation between machining parameter and output variables, namely metal removal rate and surface roughness. The statistical regression models are given in following equations (13 & 14:

Table 11. Experiment result for response parameter

Exp. No.	A	B	C	D	MRR	Ra (mm)	Rq (μm)	Rsk (mm)	Rku (μm)	Rsm (mm)
1	1	1	1	1	21.6561	2.666	3.392	0.208	4.049	0.182
2	1	1	2	2	17.2399	2.000	2.488	0.096	3.491	0.118
3	1	1	3	3	19.7452	1.693	0.944	0.098	1.753	0.059
4	1	2	1	2	14.1189	3.089	3.955	0.655	5.582	0.181
5	1	2	2	3	14.837	2.070	2.559	0.716	5.702	0.158
6	1	2	3	1	16.1359	1.285	0.841	0.065	1.016	0.046
7	1	3	1	3	19.7707	1.888	2.400	0.753	4.118	0.130
8	1	3	2	1	24.2038	1.403	1.259	0.092	2.579	0.058
9	1	3	3	2	23.0828	2.062	2.789	0.291	3.288	0.119
10	2	1	1	2	13.2484	1.616	2.091	0.075	2.851	0.091
11	2	1	2	3	16.8153	1.653	2.120	0.555	2.895	0.089
12	2	1	3	1	16.4968	1.919	1.150	0.267	1.234	0.034
13	2	2	1	3	16.1783	2.350	2.932	0.367	3.446	0.121
14	2	2	2	1	25.4777	1.255	4.152	0.557	5.275	0.236
15	2	2	3	2	26.9427	1.330	1.647	0.228	2.645	0.121
16	2	3	1	1	24.1401	2.957	3.727	1.090	6.657	0.141
17	2	3	2	2	25.3715	3.022	3.871	0.583	5.187	0.125
18	2	3	3	3	22.8238	2.225	2.911	0.562	3.892	0.121
19	3	1	1	3	23.0998	5.514	6.862	1.428	8.981	0.335
20	3	1	2	1	21.8259	3.176	3.951	0.217	2.509	0.142
21	3	1	3	2	23.5669	2.020	2.500	0.224	1.910	0.060
22	3	2	1	1	22.1806	3.422	4.214	0.110	3.112	0.150
23	3	2	2	2	25.2529	2.062	2.155	0.328	3.458	0.120
24	3	2	3	3	32.3931	1.882	2.379	0.306	2.575	0.100
25	3	3	1	2	22.293	3.125	3.886	0.101	3.992	0.159
26	3	3	2	3	24.7134	2.553	3.108	0.342	2.439	0.128
27	3	3	3	1	37.707	1.320	1.690	0.315	2.901	0.077

$$\textit{Material Removal Rate} = 19.480 - 2.5746A + 3.797B - 7.068C + 0.7435D + 0.392631A*B + 2.42502A*C + 0.198613\,A*D + 1.25078\,B*C - 2.14163\,B*D + 1.03104\,C*D \quad (13)$$

$$\textit{Surface Roughness} = 1.87777 + 1.4305\,A + 0.0094\,B - 0.41265C - 0.4631\,D - 0.2576A*B - 0.35624A*C + 0.096333\,A*D + 0.126\,B*C + 0.079644\,B*D + 0.12848\,C*D \quad (14)$$

Table 12. Normalized valve of experiment data and grey relation coefficient

Exp no.	Normalized data						Grey relation coefficient						Grade
	MRR	Ra	Rq	Rsk	Rk	Rsm	MRR	Ra	Rq	Rsk	Rku	Rsm	
1	0.3437	0.6687	0.576	0.8950	0.6192	0.508	0.4324	0.2318	0.190	0.4880	0.2079	0.169	0.2867
2	0.1631	0.8250	0.726	0.9772	0.6892	0.720	0.3740	0.3637	0.267	0.8147	0.2434	0.263	0.3879
3	0.2656	0.8971	0.982	0.9757	0.9074	0.916	0.4050	0.4929	0.853	0.8050	0.5194	0.546	0.6037
4	0.0355	0.5693	0.482	0.5671	0.4267	0.511	0.3414	0.1884	0.162	0.1876	0.1485	0.169	0.1996
5	0.0649	0.8086	0.714	0.5223	0.4116	0.588	0.3484	0.3432	0.259	0.1731	0.1452	0.195	0.2441
6	0.1180	0.9929	1	1	1	0.960	0.3618	0.9341	1	1	1	0.714	0.8351
7	0.2666	0.8513	0.741	0.4952	0.6105	0.681	0.4054	0.4022	0.278	0.1653	0.2043	0.238	0.2824
8	0.4479	0.9652	0.930	0.9801	0.8037	0.920	0.4752	0.7421	0.590	0.8346	0.3375	0.556	0.5893
9	0.4020	0.8105	0.676	0.8341	0.714	0.717	0.4554	0.3454	0.236	0.3762	0.2595	0.261	0.3223
10	0	0.9152	0.792	0.9926	0.7696	0.810	0.3333	0.5412	0.325	0.9316	0.3026	0.345	0.4632
11	0.1458	0.9065	0.787	0.6404	0.7640	0.817	0.3692	0.5169	0.320	0.2176	0.2977	0.353	0.3458
12	0.1328	0.8440	0.948	0.8517	0.9726	1	0.3657	0.3907	0.660	0.4028	0.7851	1	0.6008
13	0.1197	0.7428	0.652	0.7784	0.6949	0.710	0.3622	0.2800	0.225	0.3109	0.2468	0.257	0.2801
14	0.5	1	0.450	0.6390	0.4652	0.328	0.5	1	0.153	0.2169	0.1575	0.129	0.3596
15	0.5598	0.9823	0.866	0.8804	0.7954	0.710	0.5318	0.8502	0.427	0.4553	0.3283	0.257	0.4750
16	0.4453	0.6003	0.520	0.2478	0.2917	0.644	0.4740	0.2001	0.172	0.1173	0.1237	0.219	0.2179
17	0.4956	0.5851	0.496	0.6199	0.4763	0.697	0.4978	0.1942	0.165	0.2083	0.1603	0.248	0.2458
18	0.3914	0.7722	0.656	0.6353	0.6389	0.710	0.4510	0.3051	0.225	0.2152	0.2168	0.257	0.2784
19	0.4027	0	0	0	0	0	0.4556	0.0909	0.090	0.0909	0.0909	0.090	0.1517
20	0.3506	0.5489	0.483	0.8884	0.8125	0.641	0.4350	0.1814	0.162	0.4727	0.3478	0.217	0.3028
21	0.4218	0.8203	0.724	0.8833	0.8877	0.913	0.4637	0.3576	0.266	0.4615	0.4711	0.536	0.4261
22	0.3651	0.4911	0.439	0.9669	0.7368	0.614	0.4406	0.1642	0.151	0.7517	0.2753	0.206	0.3315
23	0.4908	0.8105	0.781	0.8070	0.6934	0.714	0.4954	0.3454	0.314	0.3413	0.2459	0.259	0.3336
24	0.7827	0.8527	0.744	0.8231	0.8042	0.780	0.6970	0.4045	0.285	0.3612	0.3381	0.312	0.3992
25	0.3697	0.5609	0.494	0.9735	0.6263	0.584	0.4423	0.1855	0.165	0.7910	0.2111	0.194	0.3315
26	0.4687	0.6952	0.623	0.7967	0.8213	0.687	0.4848	0.2470	0.209	0.3297	0.3588	0.242	0.3121
27	1	0.9847	0.858	0.8169	0.7633	0.857	1	0.8675	0.414	0.3528	0.297	0.411	0.5573

6.5.2 Problem Formulation

In the present work, objectives are to minimizing the average SR and to maximizing the MRR. Genetic algorithm leads the minimizing behaviour, Therefore some modification is needed for MRR model. Objective function for MRR is suitably minimized, also required constrain are formulated as shown below:

Objective 1 = 1/MRR

Objective 2 = SR

Figure 25. Main effect plot for S/N ratio

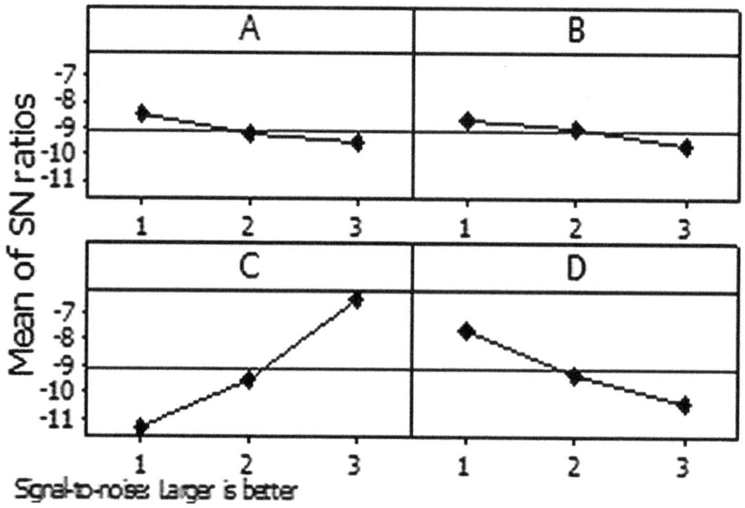

Table 13. Response table for signal to noise ratios

Level	A	B	C	D
1	**-8.520**	**-8.662**	-11.374	**-7.622**
2	-9.252	-9.035	-9.488	-9.326
3	-9.561	-9.636	**-6.472**	-10.385
Delta	1.041	0.974	4.902	2.763
Rank	3	4	1	2

Table 14. Analysis of variance for grade

Source	DF	Seq SS	Adj SS	Adj.MS	F	Contribution %
A	2	0.02280	0.02280	0.01140	0.75	**3.75**
B	2	0.01117	0.01117	0.00558	0.37	**1.9**
C	2	0.22383	0.22383	0.11192	7.36*	**36.80**
D	2	0.08467	0.08467	0.04234	2.79	**13.93**
A*B	4	0.08228	0.08228	0.02057	1.35	**13.54**
A*C	4	0.03471	0.03471	0.00868	0.57	**5.7**
B*C	4	0.05695	0.05695	0.01424	0.94	**9.3**
Error	6	0.09118	0.09118	0.01520		**15.0**
Total	26	0.60761				**100**
S = 0.123278 R-Sq = 84.99% R-Sq(adj) = 34.97%						

Table 15. Validation table

Parameter	Mean parameter for EN 19 A2B2C2D2	Optimal parameter for EN 19 A1B1C3D1	
	Mean	Predicted	Experimental
S/N ratio of grade	-10.6733		-4.84601
Grade	0.261022	0.6970331	0.5724
Improvement in S/N ratio = 5.827 db			

Subjected to:

$10 \leq A \leq 20$

$08 \leq B \leq 12$

$0.1 \leq C \leq 0.32$

$0.2 \leq C \leq 0.30$

With the help of fitness function modelled, non-dominated parameter setting obtained within the upper and lower bound for MRR and SR, as shown in table 16. Variation of metal removal rate and surface roughness is plotted as shown in Figure 26. This Figure shows the Pareto front, for the optimal response at different optimal parameter setting. Since optimal setting for the process is selected according to the product requirement and its service condition, therefore manufacturing engineer have to select the optimal setting for the product from the available solutions in Pareto front.

From the table 15, for experiment run 14 average surface roughness have minimum value of 1.255 µm and MRR correspond to the 25.477 mm³/min, where the electrolyte concentration, applied voltage, tool feed rate and inter-electrode gap are 15%, 19 V and 0.21 mm/min and 0.3 mm respectively.

From the optimality table 4, optimized roughness value very close to the experimental value is selected. For experiment run 10, For trail No. 10, the SR and MRR are 1.251 µm and 27.149 mm³/min respectively, and the pertinent machining variables are electrolyte concentration, applied voltage, tool feed rate and inter-electrode gap are 12%, 8 V and 0.31 mm/min and 0.29 mm respectively. Therefore solution set obtained from the NSGA-II are in close agreement with the experiment result and for the input parameters as well. The confirmation test is conducted, for the selected optimum conditions for the cases of MRR and *R*a as presented in Table 17. The performance for the optimum machining condition and the actual machining condition is measured. From the verification Table (table 17), it can be observed that the calculated percentage error is within acceptable limit. The percentage error between the experimental and optimum values for MRR and SR are 7.4% and 5.2%, respectively. Clearly, this confirms excellent reproducibility of the experimental conclusions.

Table 16. Optimal combination of parameter in ECM for EN 19 tool steel

Exp. No.	Conc. (%)	Voltage (V)	Feed rate (mm /min)	Gap (mm)	MRR	Ra	Exp. No.	Conc. (%)	Voltage (V)	Feed rate (mm /min)	Gap (mm)	MRR	Ra
1	19	8	0.32	0.297	33.210	1.236	15	15	8	0.31	0.297	32.352	1.238
2	17	8	0.32	0.296	31.333	1.241	16	10	12	0.10	0.269	17.315	1.419
3	14	11	0.30	0.297	20.094	1.274	17	14	10	0.31	0.297	23.287	1.261
4	14	10	0.31	0.296	24.213	1.259	18	12	9	0.31	0.296	28.117	1.249
5	16	9	0.31	0.297	26.629	1.253	19	13	11	0.29	0.295	18.903	1.287
6	14	11	0.31	0.297	21.757	1.266	20	10	11	0.30	0.296	19.342	1.278
7	18	8	0.32	0.298	33.232	1.235	21	10	10	0.31	0.296	24.551	1.259
8	10	12	0.11	0.236	17.428	1.402	22	12	9	0.31	0.296	28.881	1.247
9	10	12	0.18	0.298	17.955	1.330	23	11	12	0.15	0.295	17.583	1.365
10	12	8	0.31	0.297	27.149	1.251	24	10	11	0.31	0.298	20.674	1.270
11	12	11	0.31	0.298	22.500	1.263	25	12	12	0.25	0.293	18.451	1.302
12	14	8	0.32	0.296	30.760	1.242							
13	10	10	0.31	0.297	22.583	1.263							
14	12	9	0.31	0.296	27.786	1.251							

Figure 26. Optimal chart obtained through NSGA-II

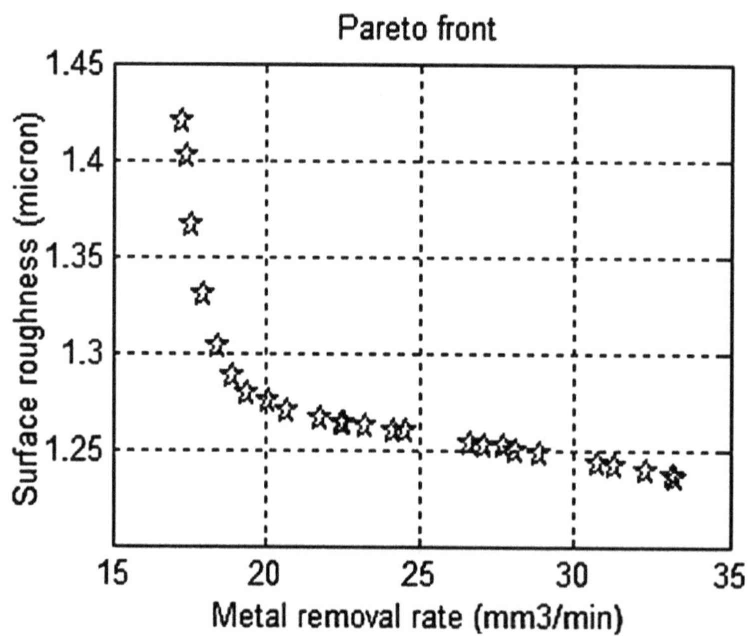

Table 17. Validation test results for electrochemical machining of EN 19.

Concentration %	Voltage (V)	Feed rate (mm/min)	Gap (mm)	MRR (mm3/min)			Surface roughness (μm)		
				Predicted	Experimental	Error %	Predicted	Experi-mental	Error %
12	8	0.31	0.297	27.149	25. 269	7.4	1.251	1.32	5.2

7. CONCLUSION

Growing business with sustainability is one of the biggest challenges of operations managers in the twenty-first century. Large commercial and industry houses have also shown responsibility for keeping environmental stability while maintaining profit. This study primarily focused to *'aims to sustain the three P's - people, planet, and profit'* with manufacturing. Sustainability continues moving from the periphery to the heart of business and aligned to core business and value drivers. As organizations' focus on sustainability, we believe general lack of maturity of sustainability performance management tools where solutions are a contributing reason to companies' failure to integrate sustainability performance management processes. With this new information, consumers and governments are increasingly aware and demanding about the environmental performance of companies. But, one must realize that operational performance correlates with long-term economic metrics, but so much with short-term metrics. It is crucial to secure top management support to allow this integration to evolve. Only then, it will mature and support the firm's sustainability strategy, delivering superior triple bottom line performance. To operationalize the organization's sustainable strategy, operations manager must integrate quality, environmental and safety management and strive to eliminate all non-conformances with the adoption of continuous process improvement. Finally this optimization may be *'sustained'* to various industrial applications for fast product realization without iterating the process parameters. The paper concludes that the various process parameters are dependent of the required output and so for the particular requirement the various optimization techniques employed in Electro Chemical Machining can be employed and the system would be tailor made as per the ultimate objective.

Future direction of this research may be further developed by using more advanced and recent heuristics like Simulated Annealing, Ant Colony Optimization or Particle swarm Optimisation. Again hybridization of two or more techniques like SAGA (Simulated Annealing within Genetic Algoritm) or SAPSO (Simulated Annealing within Particle Swarm Optimisation), etc. may lead to even better solutions- further study is envisaged.

REFERENCES

Acharya, B. G., Mohanty, C. P., & Mahapatra, S. S. (2013). Multi-objective Optimization Machining of Hardened Steel Using NSGA- II. *Procedia Engineering*, *51*, 554–560. doi:10.1016/j.proeng.2013.01.078

Asokan, P., Kumar, R. R., Jeyapaul, R., & Santhi, M. (2008). Development of Multi-objective Optimization Models for Electrochemical Machining Process. *International Journal of Advanced Manufacturing Technology*, *39*(1-2), 55–63. doi:10.1007/s00170-007-1204-8

Azzone, G., & Noci, G. (1998). Identifying effective PMSs for the deployment of 'green' manufacturing strategies. *International Journal of Operations & Production Management, 18*(4), 308–335. doi:10.1108/01443579810199711

Benardos, P. G., & Vosniakos, G. C. (2003). Predicting surface roughness in machining: A review. *International Journal of Machine Tools & Manufacture, 43*(8), 833–844. doi:10.1016/S0890-6955(03)00059-2

Bhattacharyya, B., & Sorkhel, S. K. (1999). Investigation for controlled electrochemical machining through response surface methodology-based approach. *Journal of Materials Processing Technology, 86*(1-3), 200–207. doi:10.1016/S0924-0136(98)00311-2

Binder, M., & Edwards, J. S. (2010). Using grounded theory method for theory building in operations management research: A study on inter-firm relationship governance. *International Journal of Operations & Production Management, 30*(3), 232–259. doi:10.1108/01443571011024610

Brown, S., Squire, B., & Blackmon, K. (2007). The contribution of manufacturing strategy involvement and alignment to world-class manufacturing performance. *International Journal of Operations & Production Management, 27*(3), 282–302. doi:10.1108/01443570710725554

Carter, C. R., & Rogers, D. S. (2008). A framework of sustainable supply chain management: Moving toward new theory. *International Journal of Physical Distribution & Logistics Management, 38*(5), 360–387. doi:10.1108/09600030810882816

Coughlan, P., & Coghlan, D. (2002). Action research for operations management. *International Journal of Operations & Production Management, 22*(2), 220–240. doi:10.1108/01443570210417515

Das, M. K., Kumar, K., Barman, T. K., & Sahoo, P. (2014). Optimization of Surface Roughness and MRR in Electrochemical Machining of EN31 Tool Steel using Grey-Taguchi Approach. *Procedia Materials Science, 6*, 729–740. doi:10.1016/j.mspro.2014.07.089

Deb, K. (2002). *Multi-Objective Optimization using Evolutionary Algorithms*. Singapore: John Wiley & Sons.

Deb, K., Pratap, A., Agarwal, S., & Meyarivan, T. (2002). A fast and elitist multi-objective genetic algorithm: NSGA-II. *IEEE Transactions on Evolutionary Computation, 6*(2), 182–197. doi:10.1109/4235.996017

Deng, J. (1989). Introduction to Grey System. *Journal of Grey System, 1*, 1–24.

Dubey, R., Gunasekaran, A., & Chakrabarty, A. (2015a). World-class sustainable manufacturing: Framework and a performance measurement system. *International Journal of Production Research, 53*(17), 5207–5223. doi:10.1080/00207543.2015.1012603

Dubey, R., Gunasekaran, A., Singh, S., and Singh, T., (2015b). Building theory of sustainable manufacturing using total interpretive structural modeling. *International Journal of Systems Science Operations & Logistics*. DOI:.10.1080/23302674.2015.1025890

Dubey, R., Gunasekaran, A., & Stephen, J. C. (2015c). The design of a responsive sustainable supply chain network under uncertainty. *International Journal of Advanced Manufacturing Technology, 80*(1-4), 427–445. doi:10.1007/s00170-015-6967-8

Ebeid, S. J., Hewidy, M. S., El-Taweel, T. A., & Youssef, A. H. (2004). Towards higher accuracy for ECM hybridized with low-frequency vibrations using the response surface methodology. *Journal of Materials Processing Technology, 149*(1-3), 432–438. doi:10.1016/j.jmatprotec.2003.10.046

Fan, L. T., & Tengyan, Z. (2012). Estimation of energy dissipation and cost: the foundation for sustainability assessment in process design. In D.C.Y. Foo, M.M. El-Halwagi, & R.R. Tan (Eds.), Recent Advances in Sustainable Process Design and Optimization. Singapore: World Scientific Publishing Co. Pte Ltd.

Fang, X., Qu, N., Zhang, Y., Xu, Z., & Zhu, D. (2013). Effects of pulsating electrolyte flow in electrochemical machining. *Journal of Materials Processing Technology, 214*(1), 36–43. doi:10.1016/j.jmatprotec.2013.07.012

Flynn, B. B., Schroeder, R. G., & Flynn, E. J. (1999). World class manufacturing: An investigation of Hayes and Wheelwright's foundation. *Journal of Operations Management, 17*(3), 249–269. doi:10.1016/S0272-6963(98)00050-3

FoFdation: Foundation for the Sustainable Factory of the Future. (2012).

Garbie, I. H. (2013). DFSME: Design for sustainable manufacturing enterprises (an economic viewpoint). *International Journal of Production Research, 51*(2), 479–503. doi:10.1080/00207543.2011.652746

Garbie, I. H. (2014). An analytical technique to model and assess sustainable development index in manufacturing enterprises. *International Journal of Production Research, 52*(16), 4876–4915. doi:10.1080/00207543.2014.893066

Garbie, I. H., & Shikdar, A. (2011). Analysis and estimation of complexity level in industrial firms. *International Journal of Industrial and Systems Engineering, 8*(2), 175–197. doi:10.1504/IJISE.2011.041368

Garetti, M., Mummolo, G., & Taisch, M. (2012). (Ed.). Special issue on-sustainable manufacturing. Production Planning & Control, 23(2-3), 79-82

Garetti, M., & Taisch, M. (2012). Sustainable manufacturing: Trends and research challenges. *Production Planning and Control, 23*(2-3), 83–104. doi:10.1080/09537287.2011.591619

Ghadimi, P., Azadnia, A. H., Mohd Yusof, N., & Mat Saman, M. Z. (2012). A weighted fuzzy approach for product sustainability assessment: A case study in automotive industry. *Journal of Cleaner Production, 33*, 10–21. doi:10.1016/j.jclepro.2012.05.010

Graedel, T. E., Allwood, J., Birat, J. P., Reck, B. K., Sibley, S. F., Sonnemann, G., & Hagelüken, C. (2011). *Recycling Rates of Metals-A Status Report. A Report of the Working Group on the Global Metal Flows to the International Resource Panel.* United Nations Environment Programme.

Gunasekaran, A., & Gallear, D. (2012). Special issue on sustainable development of manufacturing and services. *International Journal of Production Economics, 140*(1), 1–6. doi:10.1016/j.ijpe.2012.07.005

Gunasekaran, A., & Spalanzani, A. (2012). Sustainability of manufacturing and services: Investigations for research and applications. *International Journal of Production Economics, 140*(1), 35–47. doi:10.1016/j.ijpe.2011.05.011

Hayes, R. H., & Wheelwright, S. C. (1984). *Restoring Our Competitive Edge: Competing Through Manufacturing*. New York: John Wiley.

Hong, P., Roh, J. J., & Rawski, G. (2012). Benchmarking sustainability practices: Evidence from manufacturing firms. *Benchmarking: An International Journal*, *19*(4/5), 634–648. doi:10.1108/14635771211258052

Sustaining Employment Growth: The Role of Manufacturing and Structural Change (Industrial Development Report). (2013). United Nation Industrial Development Organization (UNIDO).

International Trade Administration. (2013). Retrieved from http://trade.gov/competitiveness/sustainablemanufacturing/how_doc_defines_SM.asp

Jabbaripour, B., Sadeghi, M. H., Faridvand, S., & Shabgard, M. R. (2012). Investigating the effects of EDM parameters on surface integrity-MRR and TWR in machining of Ti-6Al-4V. *Mining Science and Technology*, *16*, 419–444.

Jagadeesa, T. (2013). *Non Traditional Manufacturing processes-an introduction. Lecture notes, Sept., 4th*. India: NIT Calicut.

Jain, N. K., & Jain, V. K. (2007). Optimization of electro-chemical machining process parameter using genetic algorithms. *Mining Science and Technology*, *11*, 235–258.

Kaebernick, H., Kara, S., & Sun, M. (2003). Sustainable product development and manufacturing by considering environmental requirements. *Robotics and Computer-integrated Manufacturing*, *19*(6), 461–468. doi:10.1016/S0736-5845(03)00056-5

Kannegiesser, M., & Günther, H. O. (2014). Sustainable development of global supply chains-part 1: Sustainability optimization framework. *Flexible Services and Manufacturing Journal*, *26*(1-2), 24–47. doi:10.1007/s10696-013-9176-5

Klocke, F., Zeis, M., Klink, A., & Veselovac, D. (2013). Experimental research on the electrochemical machining of modern titanium and nickel-based alloys for aero engine components. *Procedia CIRP*, *6*, 368–372. doi:10.1016/j.procir.2013.03.040

Liu, Z., Nouraei, H., Papini, M., & Spelt, J. K. (2014). Abrasive enhanced electrochemical slurry jet micro-machining: Comparative experiments and synergistic effects. *Journal of Materials Processing Technology*, *214*(9), 1886–1894. doi:10.1016/j.jmatprotec.2014.04.003

Mani, M., Madan, J., Lee, J. H., Lyons, K. W., & Gupta, S. K. (2014). Sustainability characterization for manufacturing processes. *International Journal of Production Research*, *52*(20), 5895–5912. doi:10.1080/00207543.2014.886788

Manimaran, G., & Kumar, M. P. (2013). Multi-response Optimization of Grinding AISI 316 Stainless Steel Using Grey Relational Analysis. *Materials and Manufacturing Processes*, *28*(4), 418–423. doi:10.1080/10426914.2012.709347

Minitab User Manual Release 14. (2003). State College, PA, USA: MINITAB Inc.

Mittal, V. K., & Sangwan, K. S. (2014a). Development of a structural model of environmentally conscious manufacturing drivers. *Journal of Manufacturing Technology Management*, *25*(8), 1105–1134.

Mittal, V. K., & Sangwan, K. S. (2014b). Development of a model of barriers to environmentally conscious manufacturing implementation. *International Journal of Production Research*, *52*(2), 584–594. doi:10.1080/00207543.2013.838649

Nayak, K. C., & Tripathy, R. K. (2012). Relevance vector machine based prediction of MRR and SR for electrochemical machining process. *International Journal of Mechanical Engineering and Technology*, *3*(3), 394–403.

Neto, J., Silva, E., & Silva, M. (2006). Intervening variables in electrochemical machining. *Journal of Materials Processing Technology*, *179*(1-3), 92–96. doi:10.1016/j.jmatprotec.2006.03.105

Noci, G. (1997). Designing 'green' vendor rating systems for the assessment of a supplier's environmental performance. *European Journal of Purchasing & Supply Management*, *3*(2), 103–114. doi:10.1016/S0969-7012(96)00021-4

Our Common Future (Brundtland Report). (1987). World Commission on Environment and Development, Oxford University Press, Oxford.

Panda, M. C., & Yadava, V. (2012). Intelligent Modeling and Multi-objective Optimization of Die Sinking Electrochemical Spark Machining Process. *Materials and Manufacturing Processes*, *27*(1), 10–25. doi:10.1080/10426914.2010.544812

Paul, L., & Hiremath, S. S. (2014). Evaluation of process of ECDM using grey relation analysis. *Procedia Materials Science*, *5*, 2273–2282. doi:10.1016/j.mspro.2014.07.446

Prabhu, S., Uma, M., & Vinayagam, B. K. (2015). Surface roughness prediction using Taguchi-fuzzy logic-neural network analysis for CNT nano fluids based grinding process. *Neural Computation & Application*, *26*(1), 41–55. doi:10.1007/s00521-014-1696-8

Rao, P. S., Ramji, K., & Satyanarayana, B. (2014). An Experimental Investigation and Multi-Optimization of Wire EDM Parameter for Surface Roughness- Metal Removal Rate and White Layer in Machining of Aluminium Alloy. *Procedia Material Science*, *5*, 2197–2206. doi:10.1016/j.mspro.2014.07.426

Roth, A. V., Giffi, C. A., & Seal, G. M. (1992). *Operating strategies for the 1990s: elements comprising world-class manufacturing. In Manufacturing Strategy. Process and Content* (pp. 133–165). London: Chapman & Hall.

Sánchez, A. M., & Pérez, M. P. (2001). Lean indicators and manufacturing strategies. *International Journal of Operations & Production Management*, *21*(11), 1433–1452. doi:10.1108/01443570110407436

Sarkis, J., Zhu, Q., & Lai, K. H. (2011). An organizational theoretic review of green supply chain management literature. *International Journal of Production Economics*, *130*(1), 1–15. doi:10.1016/j.ijpe.2010.11.010

Schonberger, R. J. (1986). *World Class Manufacturing: The Lessons of Simplicity Applied.* New York: Free Press.

Schroeder, R. G., & Flynn, B. B. (Eds.). (2002). *High performance manufacturing: Global perspectives.* NY: Wiley.

Schuster, R., Kirchiner, V., Allongue, P., & Etrl, G. (2000). Electrochemical micro machining. *Inder Science*, *289*(5476), 98–101. doi:10.1126/science.289.5476.98 PMID:10884233

Senthil Kumar, C., Ganesan, G., & Karthikeyan, R. (2012). Optimization of ECM Process Parameters using NSGA-II. *Journal of Minerals and Materials Characterization and Engg*, *11*(10), 931–937. doi:10.4236/jmmce.2012.1110091

Shang, G. Q., & Sun, C. H. (2008). Application of BP neural Network for Predicting Anode Accuracy in ECM. *Proceedings of theInternational Symposium on Information Science and Engineering* (pp. 20-25). doi:10.1109/ISISE.2008.55

Sharma, M., & Kodali, R. (2008). TQM implementation elements for manufacturing excellence. *The TQM Journal*, *20*(6), 599–621. doi:10.1108/17542730810909365

Siebertz, L., Van Bebber, D., & Hochkirchen, T. (2010). *Design of Experiments (DoE)*. London: Springer Heidelberg Dordrecht.

Silva, A. K. M., & McGeough, J. (1998). Process monitoring of electrochemical micromachining. *Journal of Materials Processing Technology*, *76*(1-3), 165–169. doi:10.1016/S0924-0136(97)00334-8

Singh, J., & Gill, S. S. (2009). Fuzzy Modelling and Simulation of Ultrasonic Drilling of Porcelain Ceramic with Hollow Stainless Steel Tools. *Materials and Manufacturing Processes*, *24*(4), 468–475. doi:10.1080/10426910802714407

Srinivas, N., & Deb, K. (1995). Multi-objective optimization using non-dominated sorting in genetic algorithms. *Evolutionary Computation*, *3*, 221–248.

Steudel, H. J., & Desruelle, P. (1992). *Manufacturing in the nineties, how to become a mean, lean, world-class competitor*. New York: Von Nostrand Reinhold.

Taguchi, G. (1990). *Introduction to Quality Engineering*. Tokyo: Asian Productivity Organization.

Tsahalis, J., Moussas, V. C., & Tsahalis, H. T. (2012, July 4-7). Towards a holistic optimization of production sustainability: a multi-level approach. *Proceedings of the5th International Conference from Scientific Computing to Computational Engineering (5th IC-SCCE)*, Athens.

Tsai, K. M., & Wang, P. J. (2001). Predictions on Surface Finish in Electrical Discharge Machining based upon Neural Network Models. *International Journal of Machine Tools & Manufacture*, *41*(10), 1385–1403. doi:10.1016/S0890-6955(01)00028-1

Ugrasen, G., Ravindra, H. V., Naveen Prakash, G. V., & Keshavamurthy, R. (2014). Estimation of machining performances using MRA, GMDH and Artificial Neural Network in Wire EDM of EN-31. *Procedia. Materials Science*, *6*, 1788–1797.

Vachon, S. (2007). Green supply chain practices and the selection of environmental technologies. *International Journal of Production Research*, *45*(18-19), 4357–4379. doi:10.1080/00207540701440303

Vikas, S., Shashikant, , Roy, A. K., & Kumar, K. (2014). Effect and Optimization of Machine Process Parameters on MRR for EN19 & EN41 materials using Taguchi. *Procedia Technology*, *14*, 204–210. doi:10.1016/j.protcy.2014.08.027

Zhiyong, L., & Hua, J. (2009). Machining accuracy prediction of Aero-engine Blade in electrochemical Machining based on BP Neural network. *Proceedings of theInternational Workshop on Information Security and Application (IWISA)* (Vols. 21-22, pp. 244-247).

Zubir, A. F. M., & Habidin, N. F. (2012). The development of sustainable manufacturing practices and sustainable performance in Malaysian automotive industry. *Journal of Economics and Sustainable Development*, *3*(7), 130–138.

Chapter 12
Sustainable Supply Chain:
An Institutional Perspective

Susheela Girisaballa
Symbiosis International University, India

Sonali Bhattacharya
Symbiosis International University, India

ABSTRACT

Sustainable supply chain management (SSCM) is an evolving discipline, for which many theories are being proposed. It is dependent on other disciplines, of which organizational theories are of tremendous importance. One such theory that has great relevance to SSCM is Institutional Theory. The purpose of this study is to develop a model that relates the enablers of SSCM to Institutional Theory. This is achieved in three steps: First, SSCM enablers were identified through a literature review. These internal enablers go through certain external challenges when supply chain players practice Sustainability. Second, an attempt is made to explain these challenges from the perspective of Institutional Theory. The role of Institutional Isomorphism in regulating and creating a direction for achieving the sustainability goal is identified. Finally, a theoretical model is developed to describe the link between supply chain players, institutional isomorphism, and the support and participation of the top management in creating an enabling environment for legitimization of sustainable supply chain.

INTRODUCTION

The concepts of sustainable supply chain management (SSCM) and green supply chain management (GSCM) are in the nascent stage and are likely to have an impact for centuries to come. Because supply chains involve both inter-organizational and intra-organizational dynamics among partners, some of the organizational theories have a role in establishing Sustainability in supply chains. The integration (Wu, 2014; Yu et al., 2014) and coordination of external (Darnall et al., 2008) and internal factors in implementing GSCM practices from the viewpoint of Environment (Azevedo et al., 2011) and Economics have been discussed earlier (walker et al., 2008; Green et al., 2012; Shi et al., 2012; Zhu et al., 2012). However, with the growing importance of these concepts, there is a need for application-based interdis-

DOI: 10.4018/978-1-5225-0350-7.ch012

ciplinary research (Linton et al., 2007; Sanders & Wagner, 2011; Sarkis et al., 2011). One such theory that has relevance to SSCM in the current scenario is the institutional theory (DiMaggio & Powell, 1983; Zhang & Dhaliwal, 2009).

The legitimization of institutional pressures in sustainability practices leads to a more predictable and more acceptable behaviour among supply chain partners (Connelly et al., 2011; Glover et al., 2014). By adopting sustainability, and simultaneously catering to Institutional pressures, organizational efficiency can be improved both operationally and strategically. Operational efficiency can be improved by concentrating on the successful philosophies of supply chain management (SCM) such as Total Quality Management (TQM) and Ethics. Strategic efficiency can be improved by overall adoption of sustainability, thus gaining a competitive edge and advantage in the market (Kang et al., 2012).

This paper attempts to review the existing literature and develop a theoretical model that forms a link between the external Institutional pressures and internal enablers of a sustainable supply chain with focus on practice-oriented utility (Corley & Gioia, 2011). The literature review has been organized sequentially in three steps: To start with, SSCM is defined and its enablers are identified. The Institutional theory and its relevance to sustainability are discussed next. Finally, a theoretical model is developed to link SSCM enablers with Institutional Isomorphism, which stabilizes sustainability across the chain. Hypotheses are then framed.

BACKGROUND

The available literature on applications of Organizational Theories to SSCM suggests that this area is still in its nascent stage, with several theories being put forward to explain SSCM (Sarkis et al., 2011). Although the relevance of institutional theory to SSCM has been mentioned briefly (Seuring et al., 2008, Sarkis et al., 2011, Glover et al., 2014). Earlier research on GSCM seeks to provide solutions to the environmental dimension of sustainability, and the research carried out on corporate social responsibility (CSR) provides answers to the social dimension of a more holistic concept of sustainability. Therefore, literature is also reviewed from other related areas such as GSCM (Zhu et al., 2013), CSR (Huq et al., 2013), and relevance of Institutional theories in other disciplines. An attempt is made to understand and relate the enablers of SSCM to external Institutional pressures.

Literature Review

A systematic literature review (Tranfield et al., 2003) is carried out on SSCM for the period 1984-2014 with more than 80% of articles published during 2008-2014. Institutional theory was extensively reviewed from the viewpoint of its application to SSCM. A structured approach is used in this concept-centric review (Webster & Watson, 2002). The concepts considered in this review are 'Sustainability Supply Chain Management' and 'Institutional Theory.' The literature review is used to develop a model to indicate the relationship between enablers of a sustainable supply chain and institutional theory based on content analysis (Seuring & Gold, 2012). Therefore the chapter is derived through deductive method based on the literature review.

Google Scholar, Emerald Insight, JSTOR, and Scopus were used as search engines. In Emerald Insight, articles with the keywords Sustainable Supply Chain Management, were considered, and in Google Scholar, the most cited and most relevant articles with the keywords Sustainability, Supply Chain

Management, Sustainable Supply Chain Management, Institutional Theory, Institutional Pressures, and Institutional Isomorphism were considered.

A preliminary search was made for SSCM, and it yielded 6, 58,000 results initially. The scope of this review is limited to articles belonging to supply chains and logistic Journals. To ensure a high quality, only seminal articles were chosen for the review. The following important journals were widely referred to:

- International Journal of Physical Distribution & Logistics Management
- Journal of supply chain management
- Journal of cleaner production
- Supply Chain Management: An International Journal
- IIM--B Management Review
- International Journal of Production Economics
- International Journal of Operations and Production Management
- International Journal of Production Economics
- Journal of Business Logistics
- MIT Sloan Review
- Harvard Business Review
- Journal of Academy of Marketing Science
- Academy of Management Review
- Decision Support Systems
- Journal of Purchasing and Supply Management
- Journal of Operations Management
- European Journal of Purchasing & Supply Management
- Ecological Economics
- European Journal of Operational Research
- American Sociological Review
- International Journal of Logistics Systems and Management
- International Journal of Productivity and Performance Management
- Journal of Business Logistics

Although research on sustainable and green supply chain management has been performed since the 1990s, more holistic SSCM models were constructed only after 2007 (Seuring, 2013), and this has tremendously improved since 2011 (Brandenburg et al., 2014). Hence, the literature search was narrowed down to a manageable time frame by restricting 80-90 Percent of the review to the period 2008-2014.

1. SUSTAINABILITY

Sustainable development has originated as a policy in the Brundtland report of 1987 and has become the "*Buzzword*" in every field. Efforts were made to understand the difference between development and growth in terms of sustainability. One school of thought believes that development emphasizes quality, and growth emphasizes quantity (Brundtland, 1987). This therefore means treating human economy as a finite and significant part of an ecosystem and because economy is finite, it can only be developed but cannot have growth (Daly, 1990). Another school of thought believes that development and growth are synonymous (Robinson, 2004).

Sustainability has been defined as "development that meets the needs of the present -without compromising the ability of the future generations to meet their own needs" (World Commission on Environment and Development, 1987). Since then, the term 'sustainability' has evolved from a vague human versus nature approach to a more vivid three-dimensional triple bottom line approach (Elkington, 1997; Carter & Rogers, 2008).

Although it is definitely understood that sustainability is by far the preferable solution compared to non-sustainable practices, the question arises as to how sustainable practices can be carried out on a larger scale. There has always been a lack of clarity in understanding the word sustainability. A substantial growth in economy would mean immense utilization of scarce environmental and social resources. Conservation of natural resources and social responsibilities would mean lower economic growth, the 'economic degrowth' (Schneider et al., 2010) that questions the fundamental basis of business and management. The major challenge that organizations face is to find an optimal practice or a trade-off practice that suits a robust economic growth without affecting the environment and that which is socially responsible (Giminez et al., 2012). This can be achieved by increasing reuse and recycling of products (Kang et al., 2012) and by increasing job opportunities for the poor and the jobless by consistently educating and training them on the concepts of sustainable practices (Dos Santos et al., 2013).

2. SUSTAINABILITY IN A SUPPLY CHAIN CONTEXT

There is debate regarding the status of supply chains: whether it is a function, structure, discipline, process, philosophy, or whether it is an approach. SCM is a discipline with a cross disciplinary approach and not a function (Croom et al., 2000; Ellram & Cooper, 2014). A sustainable supply chain is an integrated way of fulfilling environmental considerations, social development, and economic growth of stakeholders and customers, and it includes information (Pereira, 2009), material, and capital needs (Seuring, 2013). Along with the integration at the macro level, certain practices in each of these dimensions individually have become critical for fulfilling the objective of sustainability. The three dimensions of SSCM are:

a. The Environmental Dimension

The environment is one of the most widely researched dimensions of SSCM. The environmental dimension of a supply chain entails a holistic association of the organization with its stakeholders, especially its suppliers in enhancing the environmentally safe and quality product (Ashby et al., 2012), by restricting emissions within the cap and conserving natural resources and energy (Gimenez & Tachizawa, 2012). The scope of the environment in a supply chain concerns product safety, environmental risk management, occupational health and safety, pollution prevention, and resource conservation to waste management (Srivastava, 2007; Azevedo et al., 2011; Gunasekharan & Spalanzani, 2012).

b. The Social Dimension

The social dimension of sustainability includes making employees imbibe values and ethos; providing them with information, skill, and knowledge (Sarkis et al., 2010 in Ashby et al., 2012); networking within players; and including labour orientation that improves the quality of life (Gimenez & Tachizawa, 2012). For a long time, social responsibility has been treated as a non-profit responsibility outside the arena of business functions (Walley & Whitehead, 1994). The implementation of SSCM practices by one partner alone in a supply chain and not by all tiers has resulted in the failure of the sustainability concept. This has led to the emergence of CSR, wherein the focal firm sets the guidelines for its suppliers and they work together to produce a sustainable product. Therefore, the need for assessment of suppliers and collaboration with them arose (Gimenez & Tachizawa, 2012). The concept of collaboration and its profiling (Giminez et al., 2012; Blome et al., 2014) has become essential especially after the evolution of a supply chain from the inter-firm perspective to inter-supply chain perspective (Kang et al., 2012). This can especially have strategic implications for firms (Gold et al., 2010).

c. The Economic Dimension

The economic dimension of sustainability aims to increase economic growth through operational efficiency (Gimenez & Tachizawa, 2012), return on investment (Ageron et al., 2012), earnings per share, and firm's revenue (Dos Santos et al., 2013). However, the question that arises is "How can one decide whether a product is environmentally non-threatening, socially responsible, or economically viable and how does one quantify Sustainability?"

The establishment and practice of sustainability methods in a supply chain are often associated with high operational costs; there was thus the notion that the economic dimension is often a hurdle encountered in SSCM (Collins et al., 2007; Wu & Pagell, 2011). It is therefore beneficial for organizations to accept and move on with the notion that sustainability practices are not always meant for profit making (Carter and Rogers, 20-08). However, it was proven that, with time, the long-term impact of such practices is economically beneficial for organizations and shareholders (Bose & Pal, 2011; Ageron et al., 2012; Dam & Petkova, 2014). This in fact has been a cost-effective and successful strategy for organizations. The best option would be to develop a practice that is profit oriented for the organization and economical for the buyer (Chouinard, 2011).

Although the term sustainability has been in existence for quite some time, its relevance and practice have been clearly identified in SCM only recently (Carter & Rogers, 2008). SSCM emphasizes four business areas: Strategy, Organizational Culture, Risk Management, and Transparency (Carter & Easton, 2011).

Several quantitative and qualitative models have been developed for reverse logistics and closed loop supply chains. However, only very few have been developed in the associated areas of sustainable supply chain and green supply chain. A substantial number of quantitative studies (Croom et al., 2000; Zhu & Sarkis, 2004; Dekker et al., 2011) have been performed on the environmental and economic aspects of SCM as compared to the social aspect (Brandenburg et al., 2014). Hence, there is a research gap in the areas of partial and full sustainability practices (Lozano, 2008). Partial sustainability lacks extensive study and the one that is least researched is the social dimension of sustainability. Full sustainability integrates environmental, social, and economic criteria. The combination of theoretical and empirical analysis is yet to be explored for sustainable supply chains. Creation of an integrated model using all the three criteria of sustainable supply chains still remains a challenge (Seuring, 2013).

Gold et al. (2010) insist on focusing on inter-partner supply chain to achieve sustainability. The government and industries have a major role in guiding players with definite policies rather than leaving it to them to develop their own strategies, based on market variables alone (Hassini et al., 2012). The major challenge that organizations face is to find an optimal practice or a trade-off practice that suits a robust economic growth without affecting the environment and is socially responsible (Cetinkaya et al., 2011). This can be achieved in two ways:

3. PRACTICES OF SUSTAINABILITY

a. Optimizing the Existing Scenario

This includes every unit of economic activity (economic growth) that can be enhanced by decreasing the energy or raw material use (the environmental dimension) and increasing the stakeholder's welfare (social dimension) (Robinson, 2004). These dimensions therefore overlay each other (Ashby et al., 2012), and thus result in the optimization of resources and reduction of costs, which is a major concern for organizations adopting sustainability practices.

b. Re-Visualizing the Future Scenario

Development of sustainability techniques of environmental and social standards ensures the growth of economy throughout the supply chain (Hassini et al., 2012). This can be achieved as follows:

- Set the standards and indicators (Veleva & Ellenbecker, 2001) for sustainability.
- Ensure a certification process to follow these standards.
- Detect the focal area of supply chain activity.
- Identify driving factors of all the stakeholders of the supply chain.
- Build a strategy that has long-term sustainability.
- Convert the strategy into an industry strategy and a global strategy.

4. ENABLERS OF SUSTAINABLE SUPPLY CHAIN MANAGEMENT

There is strong evidence in the literature for identification of enablers of SSCM. The literature includes factors affecting SSCM (Carter & Rogers, 2008; Seuring & Müller, 2008; Walker et al., 2008; Zhang & Dhaliwal, 2009; Pagell & Wu, 2009; Zhu & Liu, 2010; Gupta & Palsule-Desai, 2011; Acosta et al., 2013; Huq et al., 2013; Tachizawa et al., 2014; Glover et al., 2014; Varsei et al., 2014; Grosvold et al., 2014), in a specific industry (Dos Santos et al., 2013) and value chain analysis (Fearne et al., 2012). There is little evidence of these factors being associated with institutional theory. Some of the enablers have been discussed in the context of policies and approaches and how organizations have moved from adhering to rigid government rules to more flexible accountability and market-related pressures (Gupta & Palsule-Desai, 2011).

The difference between drivers and enablers has been very clearly specified by Gimenez & Tachizawa (2012). Drivers are factors that help an organization before implementation of SCM, and enablers are factors that aid in implementation of SCM. The authors have also mentioned that the literature has mostly not differentiated these factors on the basis of drivers and enablers. Drivers and enablers have been considered without any demarcation. Therefore, even articles related to sustainable supply chains include both drivers and enablers of sustainable supply chains. Similarly, all enablers do not have the same impact on the sustainability of all industries. The influence of these enablers varies sector-wise. Manufacturing sectors require more of environmental enablers than the other two dimensions (Dubey et al, 2014); service sectors need more of extensive social enablers than environmental enablers, and financial sectors need more of economic enablers.

5. INSTITUTIONAL THEORY

The homogeneity of organizations across a business environment is achieved despite there being an initial diversity in aspects such as output, structure, and culture. Organizations tend to exhibit similarities despite having varied interests. Three types of isomorphic pressures emerge in the process: coercive pressures, normative pressures, and mimetic pressures (DiMaggio & Powell, 1983). Institutional theory helps in identification and resolution of issues caused by external pressures (Varsei et al., 2014).

The term 'pressures' in "institutional pressures," coined by DiMaggio & Powell (1983), has been used synonymously as legitimization (Zhang & Dhaliwal, 2009). For the purpose of uniformity and to retain the original terms, the term 'pressure' will be used throughout as in institutional pressures, environmental pressures, social pressures, and economic pressures'.

The institutional theory has been viewed from the perspective of 'New Institutionalism.' It includes not only the organization but also the supply chain players and individuals. Institutional pressures make organizations adopt strategies or approaches required for SCM (Lavassani & Movahedi, 2010).

6. INSTITUTIONAL THEORY AND SSCM

Institutional Isomorphism has successfully guided supply chain players in the adoption of technology in supply chains (Zhang & Dhaliwal, 2009), Electronic Supply Chain Management adoption (Ke et al., 2009), enterprise resource planning (Liang et al., 2007), supply chain integration and collaboration (Reefke & Trocchi, 2013), and supply chain security practices (Williams et al., 2009). A theory of institutional isomorphism has been developed for green supply chains (Zhu et al., 2013; Dubey et al., 2014; Jain and Sharma, 2014), and social sustainability (Huq et al., 2013; Huq et al., 2014), from the supplier's point of view (Acosta et al., 2013) or by restricting to an industry or sector (Glover et al., 2014). However, there is a need to study its relevance to sustainable supply chains by taking into account the 'Triple Bottom Line' (Carter & Rogers, 2008) to include all the three dimensions of sustainability.

The importance of institutional theory in legitimizing the ecological sustainability in an organization through development of concepts and assimilating them within the organization was studied by Jennings Zandbergen (1995). These authors identified a biosphere model that comprised ecological and social systems that contribute eventually to a strong economic system. They however did not discuss the role of supply chain players.

Seuring and Muller (2008) used a Delphi technique to identify certain stressors faced by organizations from the government, customers, legal organizations and Non-Governmental Organizations (NGOs) in implementation of sustainability in supply chains.

Zhu et al. (2013) identified the effect of institutional pressures from the environmental dimension that could lead to operational advantages, with more emphasis on green supply chains. These authors related all the three types of institutional pressures to the environment and indirectly to the economic dimension.

Huq et al. (2013) conducted a study on institutional theory isomorphism with respect to the social responsibility of SSCM of an apparel industry in developing countries. These authors identified certain barriers and enablers for SSCM with respect to institutional isomorphism. However, the study was restricted to the social dimension only.

Glover et al. (2014) discussed institutional pressures with respect to sustainability of a dairy supply chain. These authors in particular studied the importance of environmental sustainability with respect to energy efficiency and integrated internal factors of sustainability with external sources especially in terms of suppliers and customers.

Jain and Sharma (2014) identified 14 drivers to initiate the GSCM, which was further grouped into three major categories – Government Regulations, Customer Requirements, and Competition. In this context, these authors mentioned the institutional theory. However, drivers and pressures were discussed interchangeably.

Although there is evidence of research done on SSCM and institutional pressures, there is still a requirement to bring in similarity among the supply chain players with respect to sustainability practice.

MAIN FOCUS OF THE CHAPTER

To achieve this similarity among all the players of a particular supply chain, supply chain players, enablers of sustainability, and institutional pressures have to be brought to a common platform. The model (Figure 1) is an effort toward bringing all players together to achieve an isomorphism.

7. RESEARCH MODEL

The model shown in Figure 1 has three main components: First, the enablers of SSCM that are internal factors, through which sustainability practices can be implemented within an organization. Second, three institutional pressures act as external factors and they determine, challenge, or allow internal enablers across the industry. Third, the supply chain diffusion network is a medium through which both enablers and institutional pressures pass to reach the goal of sustainability.

Figure 1. Model Showing Legitimization of Sustainable Supply Chain

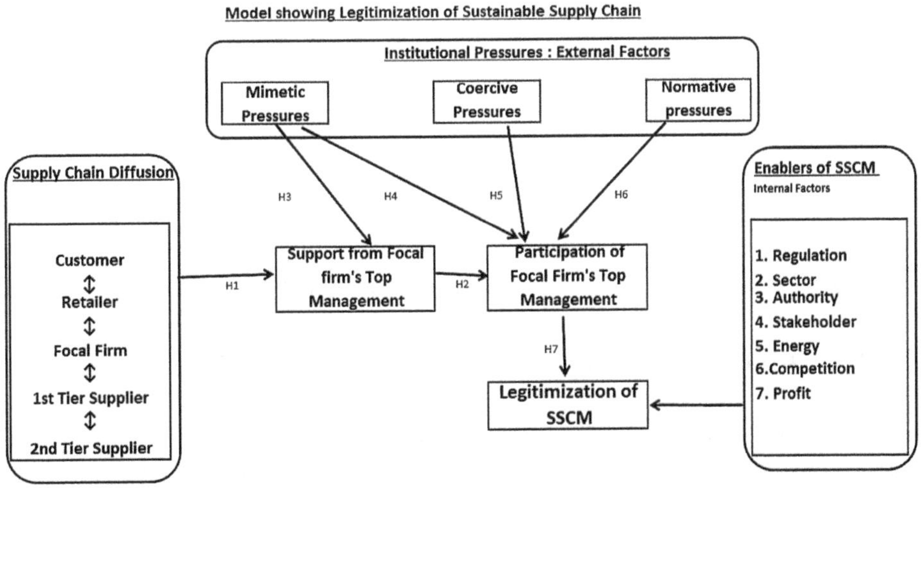

Extensive research has been performed in identifying SSCM enablers (Carter & Rogers, 2008; Seuring & Müller, 2008; Pagell & Wu, 2009; Gupta & Palsule-Desai, 2011; Fearne et al., 2012; Acosta et al., 2013; Huq et al., 2013; Dos Santos et al., 2013; Tachizawa et al., 2014). A review has been conducted to specifically recognize the enablers identified earlier. Although these enablers act as internal factors for the adoption of sustainability in the supply chain by the focal firm, there are certain external pressures that could aid in further legitimization of sustainability. These external pressures have been discussed as institutional pressures (DiMaggio & Powell, 1983).

"Legitimization" (Zucker, 1987; Zhang & Dhaliwal, 2009) is the term used for acceptance of pressures over a long period of time. These pressures arise due to institutional isomorphism among firms and are beneficial because they increase the chances of successful implementation or assimilation (Liang et al., 2007) of SSCM.

a. Supply Chain Diffusion

Firms cannot compete with one another in a supply chain. Rather, they have to coordinate, integrate (Azevedo et al., 2011), and work together to achieve the goals of profit maximization or cost minimization (Croom et al., 2000). Until a product or service is delivered, firms have to work in unison to meet customer requirements and sustainability requirements (Ageron et al., 2012; Caniels et al., 2013). Therefore, all players are inevitably influenced by highly volatile external stressors and internal factors. Hence, the model suggested in Figure.1 considers all the players of the supply chain. However, the focal firm keeps varying among players. Mostly, it is the retailer who is in direct contact with the customer, and acts as a focal firm.

Sustainability has to be adopted across the partners of a supply chain for it to be successfully legitimized. The Environmental, Social, and Economic criteria of sustainability can be achieved through certain enablers. Although enablers assist in reducing the internal pressures that diffuse across supply chain partners, organizations tend to generate situations of uncertainty (Liang et al., 2007) that can lead to either reduction or failure of adoption of sustainability. In such circumstances, external institutional pressures aid in providing a solution through isomorphism. The link between enablers and isomorphic pressures increases the probability of legitimization and paves the way for sustainability among players.

The importance of an environmentally safe, socially viable, and economical product or service is well understood by the end users and by the stakeholders of a supply chain (Hassini et al., 2012). It is definitely a challenge for the stakeholders to retain the customer or focal firm's requirements of sustainable services. This involves a tremendous effort for identification and mitigation of risk that needs extreme dependency among supply chain players and support from the top management (Linton et al., 2007; Closs et al., 2011) of the focal firm (Spekman et al., 1998).

H_1: The greater the isomorphism of the supply chain partners, the stronger the support from the top management of the focal firm toward the legitimization of sustainability in Supply Chain Management.

b. Focal Firm's Top Management Support And Participation

Sustainability in certain cases is less visible and less tangible in terms of profits. The reduced pollution emissions and conservation of resources lead to cost reduction (Varsei et al., 2014), which may not be very apparent as a profit to the stakeholders. Thus, this may not always be of great significance to the 'Top Management' (Kiron et al., 2012; Grosvold et al., 2014). Legitimization of sustainability, however, requires acceptance and decision making by the 'Top Management' (Pagell & Wu, 2009) to achieve the firm's long-term goals.

The faith and dependence among supply chain partners and the leadership commitment (Eccles et al., 2012) of strategic managers in turn lead to a strong embedding of the sustainability concept and its legitimization across the supply chain (Paulraj, 2011). For a real transformation to take place, the top management support (Hoejmose et al., 2012) in assimilating sustainability depends on the motivation, preparedness (Law & Gunasekharan, 2012), and operating principles (Wu & Pagell, 2011) of the focal firm and supply chain players both upstream and downstream.

The implementation of sustainability in a firm and in supply chains is a strategic decision and needs the approval of the top management (Chen & Paulraj, 2003). The management's support is required for collaborating and integrating sustainability practices within the firm and the chain. At times, the top management is supportive of the implementation, but the firm may not have enough expertise on handling sustainability issues within and among supply chain players. This leads to stronger top management support but weaker top management participation. Therefore, both support and participation of the top management are crucial, and the greater the support, the greater would be the effort made toward strategic implementation and legitimization of sustainability. Thus, participation depends on the top management support and its corporate governance (Richart et al., 2005).

H_2: Stronger support of the top management of the focal firm leads to stronger top management participation in the legitimization of sustainability in supply chain management.

c. Institutional Pressures: External Factors

Institutional theory has a lot of relevance to supply chains. It helps in "embedding" (Jennings & Zandbergen 1995) sustainability practices among supply chain partners. Although enablers aid in implementation of sustainability in supply chains, the influence of external pressures from various stakeholders cannot be ignored. Zhang & Dhaliwal (2009) explained the influence of institutional pressures on supply chain enablers. Organizations, over a period of time, accept these pressures, understand the need, and legitimize them to cater to such needs. Therefore, external institutional pressures from the Government, front and back supply chain partners, competitors, and customers always exist. These play an important role and pave the way for achieving the goal of sustainability. The three Institutional Pressures that affect the organizations are:

i. Mimetic Pressures

Mimetic pressures are the outcome of similar responses to unforeseen situations or stressors. Mimetic pressures are mostly caused by imitative actions that emulate an already successful competitor (Sarkis et al., 2011). Although mimetic pressures compel organizations to adopt various sustainability practices, certain inter-supply chain traits, such as knowledge, skills, social networks, and assets, are difficult to be copied. Such traits definitely lead to a competitive advantage, and thereby retain the firm's originality and provide an economic edge over its competitors despite the costs of legitimization of sustainability

(Gold et al., 2010). Therefore, the top management should definitely support sustainability practices that have emerged due to mimetic pressures, which in turn leads to their participation.

H_3: Higher levels of mimetic pressures lead to stronger support from the top management of the focal firm toward the legitimization of sustainability in supply chain management.

H_4: Higher levels of mimetic pressures give rise to more participation from the top management of the focal firm in the legitimization of sustainability in supply chain management.

ii. Coercive Pressures

Coercive pressures exist because of inter-organizational pressures that arise due to policies and legal issues of powerful organizations and the power exerted by people in power. Laws and certain regulations are factors that drive environmental actions (Sarkis et al., 2011). Organizations may face various forms of coercive pressures, such as trade barriers exerted by the government, regulations, and the industry itself. Such pressures force firms to accept and implement sustainability practices in areas such as energy and CO_2 emissions for environment and labour policies and regulations for the social dimension. Hence, the top management's desire to participate in sustainability practices increases with coercive pressures.

H_5: Higher levels of coercive pressures give rise to more participation from top management of the focal firm in the legitimization of Sustainability in supply chain management.

iii. Normative Pressures

Normative pressures result from appropriate daily business activities that follow similar norms among various firms. The increasing expectation levels of customers in developed -countries toward the availability of sustainable products can create normative pressures (Sarkis et al., 2011). Stakeholders' expectations are another type of normative pressures faced by a firm.

H_6: Higher levels of normative pressures give rise to more participation of the top management of the focal firm in the legitimization of sustainability in supply chain management.

Although there can be enough trust and support for sustainability from the top management, firms that are strategically strong in terms of knowledge, skill, and expertise tend to participate more, practice, and make sustainability their normal production requirement (Hoejmose et al., 2012).

H_7: More participation from top management of the focal firm leads to a stronger legitimization of sustainability in supply chain management.

d. Enablers of Sustainable Supply Chain

There are two major advantages of enablers of sustainability. Primarily, enablers act as factors through which a supply chain can adopt and practice sustainability. The adoption of sustainability by supply chain players includes a more complex human resource variable. The comprehension, adoption, and assimilation of higher sustainability levels at every stage and with every player are extremely difficult to achieve. In such circumstances, enablers can be used as performance indicators (Bai & Sarkis, 2014; Stephen & Roger, 2014) for measurement of the association between the focal firm and supply chain partners (Grosvold et al., 2014) to obtain a holistic sustainable product or service.

Certain essential SSCM enablers have been extracted from the available literature (Zhang & Dhaliwal, 2009) based on their relevance to the three dimensions of sustainability (environmental, social, and economic) and the kind of institutional pressures (coercive, normative, and mimetic) they exert.

i. Regulation

The regulatory factor exerts coercive pressure on the organization to ensure adherence to legislation with respect to the environment dimension of sustainability. There is no difference between public sector and private sector company compliance (Walker et al., 2008). However, there is a shift in the kind of pressure exerted by the government with respect to regulations and policies, from forcing the organizations to adhere to minimal regulatory standards to adopting a more flexible target and customer-based approach (Caniels et al., 2013; Gupta & Palsule-Desai, 2011). As a part of the environmental protection program, some organizations have set upper limits for greenhouse gas emission or pollution levels for factories. Although an organization has sufficient flexibility in terms of the number of units of pollutants emitted, it faces pressure in the resultant cost increase (Varsei et al., 2014). Further, apart from setting upper limits, certain certifications are made mandatory. These concern the environmental dimension of manufacturing firms, especially design, component usage, waste management, packaging, etc. (Gunasekharan & Spalanzani, 2012). Certifications with respect to the social dimension of sustainability insist on workers' welfare and general conditions of labour employed (Pagell & Wu, 2009).

ii. Sector

This has a holistic effect on sustainability and includes the firm, government, distributors, workers, trade unions, retailers, customers, media, and NGOs. Firms with good brand names are always under pressure to comply with -sectorial pressure due to public scrutiny that they face. Compliance with standards, however, in the long run reduces the overall costs incurred by both public and private sector firms (Walker et al., 2008). Sector has environmental, social, and economic impacts on supply chain firms. The players of the sector particularly exert coercive pressure by fixing standards and enforcing their compliance. Coercive pressures are generally high in sectors with high environmental pollution, and guidelines become more stringent and vice versa in lesser polluted sectors (Tachizawa et al., 2014). Various firms provide services wherein they advise manufacturing firms to reduce energy consumption, pollution levels or increase sustainable productivity. This is achieved through outsourcing services for implementing sustainability (Pagell & Wu, 2009).

iii. Authority

This is an internal driver that has the potential to affect both economic and noneconomic stakeholder practices in a supply chain. Authority is a great enabler especially when there is a huge resource base and when the firm in focus has a strong hold on the market. The highest authority lies with a retailer in the supply chain (Glover et al., 2014). The retailer firm in direct contact with the customer establishes the requirements of sustainability with other stakeholders (Huq et al., 2013). It places both coercive and normative pressure on its stakeholders. The influence, source, and distribution of authority of the retailer ensure compliance with sustainability norms (Tachizawa et al., 2014).

iv. Stakeholder

Visibility, transparency, and accountability can give rise to environmental and social pressures when abiding by certain guidelines (Huq et al., 2014) for stakeholders of a supply chain. This can have an impact on regular supply chain practices (Wu & Pagell, 2011; Tachizawa et al., 2014). The number of customers also has an impact on suppliers. Organizations with a large customer base are expected to produce environmentally safe products, and this exerts normative pressure on supplier firms (Walker et al., 2008). The onus is on the focal firm to convince the manufacturer to consider and adhere to sustainability practices and guidelines (Gupta & Palsule-Desai, 2011). If the supply chain demands a high level of integration and planning among its stakeholders, normative pressures tend to increase due to enhanced coordination concerns. Such pressures tend to reduce with the challenges among stakeholders (Varsei et al., 2014).

v. Energy Consumption

Measures such as reducing transportation fuel consumption, fuel costs, packaging material used and improving the working conditions of the employees contribute to cost cutting strategies of an organization (Carter & Rogers, 2008). Therefore, energy sources especially those related to carbon emissions should be replaced with renewable energy sources (Glover et al., 2014). The cost issues that arise due to enhanced environmental regulations can be overcome using models, tools, and techniques that integrate few or several operations in the manufacturing sector (Gunasekharan & Spalanzani, 2012). However, managers are yet to acquire knowledge and expertise to develop such successful models (Gupta & Palsule-Desai, 2011). There is a belief that even age old supply chain operations, such as continuous improvement and TQM, which have always been good profit-driven models, would not be able to withstand sustainability requirements. However, eventually, organizations are bound to come up with improved techniques based on these philosophies (Linton et al., 2007; Reefke & Trocchi, 2013; Dubey et al., 2014) on the lines of sustainability, and thereby add to the economic perspective of SSCM (Pagell & Wu, 2009).

vi. Competition

This causes a mimetic pressure and results from the scarcity of all types of resources, namely, man, materials, machines, and money. Organizations also compete for clients' orders (Huq et al., 2013). Organizations, while obtaining access to environmental information from their suppliers and following social responsibility, gain a competitive advantage over other firms (Walker et al., 2008). 'Exemplars' (Pagell & Wu, 2009) are leading organizations of a sector that are way ahead of their industry average, socially, environmentally, and economically. Exemplars put mimetic pressures on other organizations and lead the sustainability perspective. The presence or the absence of knowledge decides the degree of mimetic pressures that a firm experiences. Firms with high technical and knowledge expertise tend to be less dependent on either the parent organization or other partners of the supply chain. In contrast, firms with lower expertise levels tend to imitate its members and this sometimes causes a synergistic effect due to coordination among partners (Tachizawa et al., 2014). Firms also tend to compete with one another based on the quality of the product available with the supplier. This further puts mimetic pressure on suppliers to acquire good workmanship for greater quality and obtaining more orders (Huq et al., 2013). The quality Vis a Vis the cost of a product is a deciding factor for social responsibility. Competitors tend to succumb to social pressures relating to quality variations (Tachizawa et al., 2014).

vii. Profit

Adoption of sustainability practices, such as reduction of energy consumption and conducting more audits, has been instrumental in organizations getting a good return on investment, shared value (Porter & Kramer, 2011; Bose & Pal, 2011), savings, and in turn revenue (Dam & Petkova, 2014) and profits (Glover et al., 2014). This explains the economic advantage of sustainability resulting from mimetic pressures. Also, the establishment of joint ventures enables organizations to imitate eco-friendly strategies of the head office (Zhu & Liu, 2010).

LIMITATIONS AND FUTURE RESEARCH DIRECTIONS

This paper by no means tries to overlook the attempts made by previous authors toward the construction of the theory. Rather, it is an attempt to add another step to the existing literature in working toward conceptualization of sustainability theories. An effort is made to develop a model for Isomorphic Sustainability of Supply Chains. Certain limitations encountered during the process are highlighted, and the scope for future study is identified. The limitation of this paper is that it is restricted to a literature survey and construction of a model. The literature survey is mostly restricted to the period 2008-2014. The model however has to be tested and validated for practical application.

The scope for future research can be extended in two ways: The first is by building the conceptual theory using exploratory factor analysis. This may be followed by validating the model using confirmatory factor analysis, structural equation modelling, or partial least square regression technique. The second is by applying interpretive structural modelling and Mic Mac to enablers to establish strong constructs with a high driving power and discarding weak constructs with a low driving power and low dependence power. These strong constructs can be applied to various supply chain partners across sectors to empiricall-y evaluate the model using either structural equation modelling or partial least square regression techniques.

CONCLUSION

Sustainability has always been the keyword in this new era of manufacturing and services. Although sustainability is a relatively new theory, institutional theory has proven itself in stabilizing various practices of business with its ability to create isomorphism among stakeholders. The idea is to move from a non-sustainable practice or less sustainable practice to a more sustainable supply chain oriented practice to start with and eventually making this a partially or completely sustainable supply chain based practice that focuses wholly on these dimensions of sustainability. SSCM practices are very crucial. It must be emphasized that these enablers make sustainability practice achievable, when subject to institutional pressures, and persuade organizations to adopt Isomorphic Sustainability practices from both the economic and noneconomic perspectives. All supply chain players should together take advantage of these enablers and withstand institutional pressures in their march toward 'Isomorphic Sustainability.' This can happen only when all partners of a supply chain embed sustainability dimensions with similar specifications to meet their customer needs.

REFERENCES

Acosta, P., Acquier, A., & Delbard, O. (2013). Adoption of sustainable supply chain management programs from a supplier perspective. Proceedings of theXXII Conférence Internationale de Management Stratégique.

Ageron, B., Gunasekaran, A., & Spalanzani, A. (2012). Sustainable supply management: An empirical study. *International Journal of Production Economics*, *140*(1), 168–182. doi:10.1016/j.ijpe.2011.04.007

Ashby, A., Leat, M., & Hudson-Smith, M. (2012). Making connections: A review of supply chain management and sustainability literature. *Supply Chain Management: An International Journal*, *17*(5), 497–516. doi:10.1108/13598541211258573

Azevedo, S. G., Carvalho, H., & Cruz Machado, V. (2011). The influence of green practices on supply chain performance: A case study approach. *Transportation Research Part E, Logistics and Transportation Review*, *47*(6), 850–871. doi:10.1016/j.tre.2011.05.017

Bai, C. A., & Sarkis, J. (2014). Determining and Applying Sustainable Supplier Key Performance Indicators. *Supply Chain Management: An International Journal*, *19*(3), 5–5.

Blome, C., Paulraj, A., & Schuetz, K. (2014). Supply chain collaboration and sustainability: A profile deviation analysis. *International Journal of Operations & Production Management*, *34*(5), 639–663. doi:10.1108/IJOPM-11-2012-0515

Bose, I., & Pal, R. (2012). Do green supply chain management initiatives impact stock prices of firms? *Decision Support Systems*, *52*(3), 624–634. doi:10.1016/j.dss.2011.10.020

Burton, I. (1987). Report on Reports: Our Common Future: The World Commission on Environment a-nd Development. *Environment: Science and Policy for Sustainable Development*, *29*(5), 25–29. doi: 10.1080/00139157.1987.9928891

Caniëls, M. C., Gehrsitz, M. H., & Semeijn, J. (2013). Participation of suppliers in greening supply chains: An empirical analysis of German automotive suppliers. *Journal of Purchasing and Supply Management*, *19*(3), 134–143. doi:10.1016/j.pursup.2013.02.005

Carter, C. R., & Easton, P. L. (2011). Sustainable supply chain management: Evolution and future directions. *International Journal of Physical Distribution & Logistics Management*, *41*(1), 46–62. doi:10.1108/09600031111101420

Carter, C. R., & Rogers, D. S. (2008). A framework of sustainable supply chain management: Moving toward new theory. *International Journal of Physical Distribution & Logistics Management*, *38*(5), 360–387. doi:10.1108/09600030810882816

Cetinkaya, B., Cuthbertson, R., Ewer, G., Klaas-Wissing, T., Piotrowicz, W., & Tyssen, C. (2011). *Sustainable supply chain management: practical ideas for moving towards best practice*. Springer.

Chen, I. J., & Paulraj, A. (2004). Towards a theory of supply chain management: The constructs and measurements. *Journal of Operations Management*, *22*(2), 119–150. doi:10.1016/j.jom.2003.12.007

Chouinard, Y., Ellison, J., & Ridgeway, R. (2011).The sustainable economy. *Harvard Business review, 89*(10), 5-2-62.

Closs, D. J., Speier, C., & Meacham, N. (2011). Sustainability to support end-to-end value chains: The role of supply chain management. *Journal of the Academy of Marketing Science*, *39*(1), 101–116. doi:10.1007/s11747-010-0207-4

Connelly, B. L., Ketchen, D. J. Jr, & Slater, S. F. (2011). Toward a "theoretical toolbox" for sustainability research in marketing. *Journal of the Academy of Marketing Science*, *39*(1), 86–100. doi:10.1007/s11747-010-0199-0

Corley, K. G., & Gioia, D. A. (2011). Building theory about theory building: What constitutes a theoretical contribution? *Academy of Management Review*, *36*(1), 12–32. doi:10.5465/AMR.2011.55662499

Croom, S., Romano, P., & Giannakis, M. (2000). Supply chain management: an analytical framework for critical literature review. *European journal of purchasing & supply management, 6*(1), 67-83.

Daly, H. E. (1990). Toward some operational principles of sustainable development. *Ecological Economics*, *2*(1), 1–6. doi:10.1016/0921-8009(90)90010-R

Dam, L., & Petkova, B. N. (2014). The impact of environmental supply chain sustainability programs on shareholder wealth. *International Journal of Operations & Production Management*, *34*(5), 586–609. doi:10.1108/IJOPM-10-2012-0482

Darnall, N., Jolley, G. J., & Handfield, R. (2008). Environmental management systems and green supply chain management: Complements for sustainability? *Business Strategy and the Environment*, *17*(1), 30–45. doi:10.1002/bse.557

Dekker, R., Bloemhof, J., & Mallidis, I. (2012). Operations Research for green logistics–An overview of aspects, issues, contributions and challenges. *European Journal of Operational Research*, *219*(3), 671–679. doi:10.1016/j.ejor.2011.11.010

DiMaggio, P. J., & Powell, W. W. (1983). The iron cage revisited: Institutional isomorphism and collective rationality in organizational fields. *American Sociological Review*, *48*(2), 147–160. doi:10.2307/2095101

Dos Santos, M. A., Svensson, G., & Padin, C. (2013). Indicators of sustainable business practices: Woolworths in South Africa. *Supply Chain Management: An International Journal*, *18*(1), 104–108. doi:10.1108/13598541311293212

Dubey, R., Bag, S., & Ali, S. S. (2014). Green supply chain practices and its impact on organisational performance: An insight from Indian rubber industry. *International Journal of Logistics Systems and Management*, *19*(1), 20–42. doi:10.1504/IJLSM.2014.064029

Dubey, R., Gunasekaran, A., & Ali, S. S. (2014). Exploring the Relationship between Leadership, Operational Practices, Institutional Pressures and Environmental Performance: A Framework for Green Supply Chain. *International Journal of Production Economics.*

Eccles, R. G., Perkins, K. M., & Serafeim, G. (2012). How to become a sustainable company. *MIT Sloan Management Review*, *53*(4), 43–50.

Elkington, J. (1997). Cannibals with forks. *The triple bottom line of 21st century.*

Ellram, L. M., & Cooper, M. C. (2014). Supply Chain Management: It's All About the Journey, Not the Destination. *Journal of Supply Chain Management*, *50*(1), 8–20. doi:10.1111/jscm.12043

Fearne, A., Martinez, M. G., & Dent, B. (2012). Dimensions of sustainable value chains: Implications for value chain analysis. *Supply Chain Management: An International Journal*, *17*(6), 575–581. doi:10.1108/13598541211269193

Gimenez, C., Sierra, V., & Rodon, J. (2012). Sustainable operations: Their impact on the triple bottom line. *International Journal of Production Economics*, *140*(1), 149–159. doi:10.1016/j.ijpe.2012.01.035

Gimenez, C., & Tachizawa, E. M. (2012). Extending sustainability to suppliers: A systematic literature review. *Supply Chain Management: An International Journal*, *17*(5), 531–543. doi:10.1108/13598541211258591

Glover, J. L., Champion, D., Daniels, K. J., & Dainty, A. J. D. (2014). An Institutional Theory perspective on sustainable practices across the dairy supply chain. *International Journal of Production Economics*, *152*, 102–111. doi:10.1016/j.ijpe.2013.12.027

Gold, S., Seuring, S., & Beske, P. (2010). Sustainable supply chain management and inter-organizational resources: A literature review. *Corporate Social Responsibility and Environmental Management*, *17*(4), 230–245.

Green, K. W. Jr, Zelbst, P. J., Meacham, J., & Bhadauria, V. S. (2012). Green supply chain management practices: Impact on performance. *Supply Chain Management: An International Journal*, *17*(3), 290–305. doi:10.1108/13598541211227126

Grosvold, J., Hoejmose, S., & Roehrich, J. (2014). Squaring the Circle: Management, Measurement and Performance of Sustainability in Supply Chains. *Supply Chain Management: An International Journal*, *19*(3), 6–6.

Gunasekaran, A., & Spalanzani, A. (2012). Sustainability of manufacturing and services: Investigations for research and applications. *International Journal of Production Economics*, *140*(1), 35–47. doi:10.1016/j.ijpe.2011.05.011

Gupta, S., & Palsule-Desai, O. D. (2011). Sustainable supply chain management: Review and research opportunities. *IIMB Management Review*, *23*(4), 234–245. doi:10.1016/j.iimb.2011.09.002

Hassini, E., Surti, C., & Searcy, C. (2012). A literature review and a case study of sustainable supply chains with a focus on metrics. *International Journal of Production Economics*, *140*(1), 69–82. doi:10.1016/j.ijpe.2012.01.042

Hoejmose, S., Brammer, S., & Millington, A. (2012). "Green" supply chain management: The role of trust and top management in B2B and B2C markets. *Industrial Marketing Management*, *41*(4), 609–620. doi:10.1016/j.indmarman.2012.04.008

Huq, F., Stevenson, M., Zorzini, M., & Hendry, L. (2013). Social sustainability in the Apparel Supply Chain: an Institutional Theory perspective. *Proceedings of the 20th European Operations Management Association (EurOMA)*.

Huq, F. A., Stevenson, M., & Zorzini, M. (2014). Social sustainability in developing country suppliers: An exploratory study in the ready made garments industry of Bangladesh. *International Journal of Operations & Production Management*, *34*(5), 610–638. doi:10.1108/IJOPM-10-2012-0467

Jennings, P. D., & Zandbergen, P. A. (1995). Ecologically sustainable organizations: An institutional approach. *Academy of Management Review*, *20*(4), 1015–1052.

Kang, S. H., Kang, B., Shin, K., Kim, D., & Han, J. (2012). A theoretical framework for strategy development to introduce sustainable supply chain management. *Procedia: Social and Behavioral Sciences*, *40*, 631–635. doi:10.1016/j.sbspro.2012.03.241

Ke, W., Liu, H., Wei, K. K., Gu, J., & Chen, H. (2009). How do mediated and non-mediated power affect electronic supply chain management system adoption? The mediating effects of trust and institutional pressures. *Decision Support Systems*, *46*(4), 839–851. doi:10.1016/j.dss.2008.11.008

Kiron, D., Kruschwitz, N., Haanaes, K., & von Streng Velken, I. (2012). Sustainability nears a tipping point. *MIT Sloan Management Review*, *53*(2), 69–74.

Lavassani, K. M., & Movahedi, B. (2010). Critical Analysis of the Supply Chain Management Theories: Toward the Stakeholder Theory. *Proceedings of the POMS 21st Annual Conference,* Vancouver.

Law, K. M., & Gunasekaran, A. (2012). Sustainability development in high-tech manufacturing firms in Hong Kong: Motivators and readiness. *International Journal of Production Economics*, *137*(1), 116–125. doi:10.1016/j.ijpe.2012.01.022

Liang, H., Saraf, N., Hu, Q., & Xue, Y. (2007). Assimilation of enterprise systems: The effect of institutional pressures and the mediating role of top management. *Management Information Systems Quarterly*, 2007, 59–87.

Linton, J. D., Klassen, R., & Jayaraman, V. (2007). Sustainable supply chains: An introduction. *Journal of Operations Management*, *25*(6), 1075–1082. doi:10.1016/j.jom.2007.01.012

Lozano, R. (2008). Envisioning sustainability three-dimensionally. *Journal of Cleaner Production*, *16*(17), 1838–1846. doi:10.1016/j.jclepro.2008.02.008

Pagell, M., & Wu, Z. (2009).Building a more complete theory of sustainable supply chain management using case studies of 10 exemplars. *Journal of supply chain management, 45*(2), 37-56.

Paulraj, A. (2011). Understanding the relationships between internal resources and capabilities, sustainable supply management and organizational sustainability. *Journal of Supply Chain Management*, *47*(1), 19–37. doi:10.1111/j.1745-493X.2010.03212.x

Pereira, J. V. (2009). The new supply chain's frontier: Information management. *International Journal of Information Management*, *29*(5), 372–379. doi:10.1016/j.ijinfomgt.2009.02.001

Porter, M. E., & Kramer, M. R. (2011). Creating shared value. *Harvard Business Review*, *89*(1/2), 62–77.

Reefke, H., & Trocchi, M. (2013). Balanced scorecard for sustainable supply chains: design and development guidelines. *International Journal of Productivity and Performance Management, 62*(8), 805-826.

Ricart, J. E., Rodríguez, M. Á., & Sánchez, P. (2005). Sustainability in the boardroom: An empirical examination of Dow Jones Sustainability World Index leaders. *Corporate Governance*, *5*(3), 24–41. doi:10.1108/14720700510604670

Robinson, J. (2004). Squaring the circle? Some thoughts on the idea of sustainable development. *Ecological Economics*, *48*(4), 369–384. doi:10.1016/j.ecolecon.2003.10.017

Sanders, N. R., & Wagner, S. M. (2011). Multidisciplinary and multimethod research for addressing contemporary supply chain challenges. *Journal of Business Logistics, 32*(4), 317-323.

Sarkis, J., Zhu, Q., & Lai, K. H. (2011). An organizational theoretic review of green supply chain management literature. *International Journal of Production Economics*, *130*(1), 1–15. doi:10.1016/j.ijpe.2010.11.010

Schneider, F., Kall-is, G., & Martinez-Alier, J. (2010). Crisis or opportunity? Economic degrowth for social equity and ecological sustainability. Introduction to this special issue. *Journal of Cleaner Production*, *18*(6), 511–518. doi:10.1016/j.jclepro.2010.01.014

Seuring, S. (2013). A review of modelling approaches for sustainable supply chain management. *Decision Support Systems*, *54*(4), 1513–1520. doi:10.1016/j.dss.2012.05.053

Seuring, S., & Gold, S. (2012). Conducting content-analysis based literature reviews in supply chain management. *Supply Chain Management: An International Journal*, *17*(5), 544–555. doi:10.1108/13598541211258609

Seuring, S., & Müller, M. (2008). Core issues in sustainable supply chain management–a Delphi study. *Business Strategy and the Environment*, *17*(8), 455–466. doi:10.1002/bse.607

Sharma, S., & Jain, V. K. (2014). Drivers Affecting the Green Supply Chain Adaptation: A Review. *The IUP Journal of Operations Management*, *9*(1), 54–63.

Shi, V. G., Koh, S. L., Baldwin, J., & Cucchiella, F. (2012). Natural resource based green supply chain management. *Supply Chain Management: An International Journal*, *17*(1), 54–67. doi:10.1108/13598541211212203

Spekman, R. E., Kamauff, J. W. Jr, & Myhr, N. (1998). An empirical investigation into supply chain management: A perspective on partnerships. *Supply Chain Management: An International Journal*, *3*(2), 53–67. doi:10.1108/13598549810215379

Srivastava, S. K. (2007). Green supply-chain management: A state-of-the-art literature review. *International Journal of Management Reviews*, *9*(1), 53–80. doi:10.1111/j.1468-2370.2007.00202.x

Stefan, S., & Roger, B. (2014). Measuring and managing sustainability performance of supply chains. *Supply Chain Management International Journal (Toronto, Ont.), 19*(3).

Tachizawa, E. M., & Wong, C. Y. (2014). Towards a theory of multi-tier sustainable supply chains: A systematic literature review. *Supply Chain Management: An International Journal, 19*(5/6).

Tranfield, D., Denyer, D., & Smart, P. (2003). Towards a methodology for developing evidence-informed management knowledge by means of systematic review. *British Journal of Management, 14*(3), 207–222. doi:10.1111/1467-8551.00375

Varsei, M., Soosay, C. A., Fahimnia, B., & Sarkis, J. (2014). Framing sustainability performance of supply chains with multidimensional indicators. *Supply Chain Management: An International Journal, 19*(3), 3–3.

Veleva, V., & Ellenbecker, M. (2001). Indicators of sustainable production: Framework and methodology. *Journal of Cleaner Production, 9*(6), 519–549. doi:10.1016/S0959-6526(01)00010-5

Walker, H., Di Sisto, L., & McBain, D. (2008). Drivers and barriers to environmental supply chain management practices: Lessons from the public and private sectors. *Journal of Purchasing and Supply Management, 14*(1), 69–85. doi:10.1016/j.pursup.2008.01.007

Webster, J., & Watson, R. T. (2002). Analyzing the past to prepare for the future: Writing a literature review. *Management Information Systems Quarterly, 26*(2), 3.

Williams, Z., Lueg, J. E., Taylor, R. D., & Cook, R. L. (2009). Why all the changes?: An institutional theory approach to exploring the drivers of supply chain security (SCS). *International Journal of Physical Distribution & Logistics Management, 39*(7), 595–618. doi:10.1108/09600030910996279

Wu, G. C. (2013). The influence of green supply chain integration and environmental uncertainty on green innovation in Taiwan's IT industry. *Supply Chain Management: An International Journal, 18*(5), 539–552. doi:10.1108/SCM-06-2012-0201

Wu, Z., & Pagell, M. (2011). Balancing priorities: Decision-making in sustainable supply chain management. *Journal of Operations Management, 29*(6), 577–590. doi:10.1016/j.jom.2010.10.001

Yu, W., Chavez, R., Feng, M., & Wiengarten, F. (2014). Integrated green supply chain management and operational performance. *Supply Chain Management: An International Journal, 19*(5/6).

Zhang, C., & Dhaliwal, J. (2009). An investigation of resource-based and institutional theoretic factors in technology adoption for operations and supply chain management. *International Journal of Production Economics, 120*(1), 252–269. doi:10.1016/j.ijpe.2008.07.023

Zhu, Q., & Sarkis, J. (2004). Relationships between operational practices and performance among early adopters of green supply chain management practices in Chinese manufacturing enterprises. *Journal of Operations Management, 22*(3), 265–289. doi:10.1016/j.jom.2004.01.005

Zhu, Q., Sarkis, J., & Lai, K. H. (2012). Examining the effects of green supply chain management practices and their mediations on performance improvements. *International Journal of Production Research, 50*(5), 1377–1394. doi:10.1080/00207543.2011.571937

Zhu, Q., Sarkis, J., & Lai, K. H. (2013). Institutional-based antecedents and performance outcomes of internal and external green supply chain management practices. *Journal of Purchasing and Supply Management*, *19*(2), 106–117. doi:10.1016/j.pursup.2012.12.001

Zucker, L. G. (1987). Institutional theories of organization. *Annual Review of Sociology*, *13*(1), 443–464. doi:10.1146/annurev.so.13.080187.002303

KEY TERMS AND DEFINITIONS

De-Growth: A strategy which believes in partial or complete stoppage in use of non-renewable resources that cause damage to environment.

Enabler: The factor that possess the ability to make something happen. Enablers aid in achievement of a goal.

Focal Firm: The Firm that is primarily associated with defining and elaborating the specifications and features of a product or service. Focal Firm defines the requirements of a product or service to the other supply chain partners.

Isomorphism: The ability of organizations to display similarities in terms of a function or methodology.

Legitimization: Legitimization is a general, socially accepted code of conduct that is widely accepted among a group of common interests.

Stakeholder: An Individual who is directly or indirectly involved in a business. An organization has an impact on a stakeholder in various economic and non-economic perspectives.

APPENDIX

Appendix includes three tables constructed as a part of literature review. The tables -are for reader's reference and have not been called out in the chapter body (Tables 1-3).

Table 1. Literature Review based on Author's work on Sustainability of Supply Chain

Reference	Enabler	Environmental Dimension	Social Dimension	Economic Dimension
Coercive Pressures				
Rivera, (2004)	Fines and Trade Barriers	x		
	Laws and Regulations	x		
Tachizawa et al, (2014)	Industry	x	x	x
Walker et al., (2008)	Regulatory	x		
	Organization	x	x	x
Gupta & Palsule-Desai, (2011)	Regulation and Government policies	x		
	Cap and Trade	x		
Dos Santos et al, (2013)	Water Indicator	x		
	Preferential Procurement		x	
	Corporate social Investment contributor		x	
	Return on Equity			x
	Adjusted Headline EPS			x
	Share Price Appreciation			x
	Food Market Share			x
	Clothing, footwear and accessories			x
Glover et al (2014)	Power of retailer (supermarkets)	x		
Tachizawa et al (2014)	power	x		
Seuring & Müller, (2008)	Government Regulation	x		
	NGOs	x		
	Legal	x		
Pagell & Wu, (2009)	Certification		x	
	NGOs	x	x	
	Service	x		
Huq et al., (2013)	Compliance		x	x
	Better enforcement of law		x	
	buyers		x	
	media		x	
	end customers		x	
	NGOs		x	
	Trade Unions		x	
	workers		x	
	Government		x	

continued on following page

Table 1. Continued

Reference	Enabler	Environmental Dimension	Social Dimension	Economic Dimension
Acosta et al., (2013)	compliance with local and national legislation	x		
	Labour and Human Rights		x	
	Ethics and CSR Management		x	
(Varsei et al.,2014)	GHG Emissions	x		
	Hazardous and Toxic substances	x		
	Human Rights		x	
(Grosvold et al., 2014)	Legistlation and Compliance	x		
Normative Pressures				
Sarkis et al (2011)	Environmental expectation from the stakeholders and the customers	x	x	
Ball and Craig, (2010)	Ethical values		x	
	Ecological Thinking		x	
Tachizawa et al (2014)	Stakeholder Pressure	x	x	
	power	x		
	Distance		x	
Walker et al (2008)	Customers	x		
	Society	x	x	
Carter and Rogers (2008)	operational efficiency			x
Gupta &Palsule-Desai, (2011)	Extended Producer Responsibility	x	x	
	Integrative Models and Decision Support Tools	x		x
Dos Santos et al, (2013)	Good Business Journey	x		
	Energy Indicator	x		
	Carbon Foot Print Indicator	x		
	Food Packaging Indicator	x		
	Customer Tracking Indicator		x	
	No. of Permanent Employees		x	
	Training and Development		x	
	Health and Safety		x	
Glover et al., (2014)	Low Energy consumption	x		x
	Sustainable capital expense			x
	Carbon footprint audits	x		
	Installing renewable energy	x		
	value system		x	
Seuring & Müller, (2008)	lack of customer demand for sustainable products			x
Pagell& Wu, (2009)	Operational philosophies such as TQM and Continuous improvement			x
	Collaboration	x	x	x
	Community members		x	
	Integration	x		
	Training	x	x	x

continued on following page

Table 1. Continued

Reference	Enabler	Environmental Dimension	Social Dimension	Economic Dimension
Huq et al., (2013)	code of conduct		x	
	Socio-economic conditions		x	x
	culture		x	
	supplier development		x	
	education and training of suppliers		x	x
Acosta et al., (2013)	Energy and water resource use	x		
	waste management and recycling practices	x		
	community CSR programs		x	
	Health and food safety	x	x	
	Procurement	x	x	
(Varsei et al.,2014)	Water Usage	x		
	Waste Generation	x		
	service level			x
	Society		x	
	Product responsibility		x	
(Grosvold et al., 2014)	Customers and Stakeholders Network		x	
	Company Visibility		x	
Mimetic Pressures				
Zhu and Liu, 2010	Joint Ventures			x
Tachizawa et al (2014)	Material Criticality	x		
	Dependency		x	
	Knowledge Resource	x		
Walker et al (2008)	Competition	x		
Dos Santos et al, (2013)	Revenue			x
Glover et al (2014)	Dedication		x	
	Return on Investment			x
Pagell& Wu, (2009)	Competitors	x	x	
	Exemplars	x	x	x
	Innovation	x	x	x
Huq et al., (2013)	Motivation		x	x
	Expertise	x	x	x
	competition		x	
Varsei et al.,(2014)	Supply chain cost			x
Grosvold et al., (2014)	External Industry Environment	x		x
	Internal commitment and top management support	x	x	x

Table 2. Literature Review based on Author's work – Three Dimensions of Sustainability

		Review on Enablers of Sustainability					
Reference	**A**	**B**	**C**	**X**	**Y**	**Z**	**α**
	Environmental	**Social**	**Economic**	**Environmental and Social**	**Environmental and Economic**	**Social and Economic**	**Sustainability**
Tachizawa et al (2014)	Material Criticality	Dependency		Stakeholder pressure			Industry
	Power	Distance					
	Knowledge Resource						
Dos Santos et al, 2013	Water Indicator	Preferential Procurement	Return on Equity				
	Good Business Journey	Corporate social Investment contributor	Adjusted Headline EPS				
	Energy Indicator	Customer Tracking Indicator	Share Price Appreciation				
	Carbon Foot Print Indicator	No. of Permanent Employees	Food Market Share				
	Food Packaging Indicator	Training and Development	Clothing, footwear and accessories				
		Health and Safety	Revenue				
Gupta &Palsule-Desai, (2011)	Regulation and Government policies			Extended Producer Responsibility	Integrative Models and Decision Support Tools		
	Cap and Trade						
Sarkis et al (2011)				Environmental expectation from the stakeholders and the customers			
Ball and Craig, 2010		Ethical Values of the customer					
		Ecological thinking of the customer					
Zhu and Liu, 2010			Joint Ventures				
Walker et al (2008)	Regulation			Society			Organization
	Competition						
	Society						
Carter and Rogers (2008)			operational efficiency				
Rivera, 2004	Fines and Trade Barriers						
	Laws and Regulations						
Glover et al (2014)	Carbon footprint audits	value system	Sustainable capital expense		Low Energy consumption		
	Installing renewable energy	Dedication	Return on Investment				

continued on following page

Table 2. Continued

		Review on Enablers of Sustainability					
Reference	**A**	**B**	**C**	**X**	**Y**	**Z**	**α**
	Environmental	**Social**	**Economic**	**Environmental and Social**	**Environmental and Economic**	**Social and Economic**	**Sustainability**
Pagell& Wu, (2009)	Service	Certification	Operational philosophies such as TQM and Continuous improvement	NGOs		Competitors	Collaboration
	Integration	Community members					Exemplars
							Innovation
							Training
Seuring& Müller, (2008)	Government Regulation	lack of customer demand for sustainable products					
	NGOs						
	Legal						
Huq et al., (2013)		Better enforcement of law	education and training of suppliers			Compliance	Expertise
		buyers				Socio-economic conditions	
		media				education and training of suppliers	
		end customers				Motivation	
		NGOs					
		Trade Unions					
		workers					
		Government					
		culture					
		supplier development					
		code of conduct					
		competition					
Acosta et al., (2013)	compliance with local and national legislation	Labour and Human Rights		Health and food safety			
	Energy and water resource use	Ethics and CSR Management		Procurement			
	waste management and recycling practices						
		community CSR programs					

continued on following page

Table 2. Continued

		Review on Enablers of Sustainability					
Reference	**A**	**B**	**C**	**X**	**Y**	**Z**	**α**
	Environmental	**Social**	**Economic**	**Environmental and Social**	**Environmental and Economic**	**Social and Economic**	**Sustainability**
(Varsei et al.,2014)	GHG Emissions	Human Rights	service level				
	Hazardous and Toxic substances	Society	Supply chain cost				
	Water Usage	Product responsibility					
	Waste Generation						
(Grosvold et al., 2014)	Legistlation and Compliance	Customers and Stakeholders Network			External Industry Environment		Internal commitment and top management support
		Company Visibility					

Table 3. Review based on Literature work: Institutional Pressures Vs Enablers of SSCM

		Literature Review: Institutional Pressures Vs Enablers of SSCM							
Author	Coercive Pressures			Normative Pressures			Mimetic Pressures		
	Environmental	Social	Economic	Environmental	Social	Economic	Environmental	Social	Economic
Tachizawa et al (2014)	Industry	Industry	Industry	Stakeholder pressure	Stakeholder pressure		Material criticality	Dependency	
				Power	Distance		Knowledge Resource		
Dos Santos et al, 2013	Water Indicator	Preferential Procurement	Return on Equity	Good Business Journey	Customer Tracking Indicator				Revenue
		Corporate social Investment contributor	Adjusted Headline EPS	Energy Indicator	No. of Permanent Employees				
			Share Price Appreciation	Carbon Foot Print Indicator	Training and Development				
			Food Market Share	Food Packaging Indicator	Health and Safety				
			Clothing, footwear and accessories						
Gupta &Palsule-Desai, (2011)	Regulation and Government policies			Extended Producer Responsibility	Extended Producer Responsibility				
	Cap and Trade			Integrative Models and Decision Support Tools		Integrative Models and Decision Support Tools			
Sarkis et al (2011)				Environmental expectation from the stakeholders and the customers	Environmental expectation from the stakeholders and the customers				

continued on following page

Table 3. Continued

		Literature Review: Institutional Pressures Vs Enablers of SSCM							
Author	Coercive Pressures			Normative Pressures			Mimetic Pressures		
	Environmental	Social	Economic	Environmental	Social	Economic	Environmental	Social	Economic
Ball and Craig, 2010					Ethical Values of the customer				
					Ecological thinking of the customer				
Zhu and Liu, 2010									Joint Ventures
Walker et al (2008)	Organization	Organization	Organization	Customers	Society		Competition		
	Regulation			Society					
Carter and Rogers (2008)									operational efficiency
Rivera, 2004	Fines and Trade Barriers								
	Laws and Regulations								
Glover et al (2014)				Low Energy consumption	value system	Low Energy consumption		Dedication	Return on Investment
				Carbon footprint audits		Sustainable capital expense			
				Installing renewable energy					
Pagell& Wu, (2009)	Service	Certification		Collaboration	Collaboration	Collaboration	Competitors	Competitors	
	NGOs	NGOs		Integration	Community members	Operational philosophies such as TQM and Continuous improvement	Exemplars	Exemplars	Exemplars
				Training	Training	Training	Innovation	Innovation	Innovation

continued on following page

Table 3. Continued

	Literature Review: Institutional Pressures Vs Enablers of SSCM								
Author	**Coercive Pressures**			**Normative Pressures**			**Mimetic Pressures**		
	Environmental	**Social**	**Economic**	**Environmental**	**Social**	**Economic**	**Environmental**	**Social**	**Economic**
Seuring& Müller, (2008)	Government Regulation					lack of customer demand for sustainable products			
	NGOs								
	Legal								
Huq et al., (2013)		Compliance	Compliance		Socio-economic conditions	Socio-economic conditions	Expertise	Expertise	Expertise
		Better enforcement of law			education and training of suppliers	education and training of suppliers		Motivation	Motivation
		buyers			culture			competition	
		media			supplier development				
		end customers			code of conduct				
		NGOs							
		Trade Unions							
		workers							
		Government							
Acosta et al., (2013)	compliance with local and national legislation	Labour and Human Rights		Energy and water resource use	community CSR program				
		Ethics and CSR Management		waste management and recycling practices					
				Health and food safety	Health and food safety				
				Procurement	Procurement				

continued on following page

Table 3. Continued

		Literature Review: Institutional Pressures Vs Enablers of SSCM							
Author	**Coercive Pressures**			**Normative Pressures**			**Mimetic Pressures**		
	Environmental	**Social**	**Economic**	**Environmental**	**Social**	**Economic**	**Environmental**	**Social**	**Economic**
(Varsei et al.,2014)	GHG Emissions	Human Rights		Water Usage	Society	service level			Supply chain cost
	Hazardous and Toxic substances			Waste Generation	Product responsibility				
(Grosvold et al., 2014)	Legistlation and Compliance				Customers and Stakeholders Network		External Industry Environment		External Industry Environment
					Company Visibility		Internal commitment and top management support	Internal commitment and top management support	Internal commitment and top management support

Chapter 13
Sustainable Supply Chain Management:
Literature Review

Shirish C. Jeble
SIU, India

ABSTRACT

In the recent years, interest in Sustainable Supply Chain Management has grown as the effects of global warming can be seen through various scientific researches. In a global economy as supply chains are spreading across continents – economic, social and environmental sustainability – commonly referred as triple bottom line (TBL) assumes significance. Focal firms in supply chain are held accountable for their economic performance by shareholders, social and environmental performance by society. Even the actions of members of their extended supply chain come under scrutiny. Growing global supply chains facilitate economic growth, improve standard of living of people and provide business opportunities to distant locations across the world. However, they have unfavorable environmental impact. In order to preserve natural resources for future generations, sustainability needs to be embedded across the supply chains. This literature review - covers different aspects of greening the supply chain. Supply chain is viewed as a system consisting of focal firm, chains of suppliers, distributors, logistics partners, retailers etc. Role of each member of supply chain in sustainability goals is discussed in detail. Extant literature has covered individual parts of supply chain, whereas this paper integrates this knowhow to facilitate learning all interrelated concepts in SSCM. Some of the supply chain companies have already taken proactive steps towards SSCM and have certified in ISO14000. Through extensive literature review, important factors which contribute for better environmental performance for an organization are studied and a model for designing and implementing SSCM strategy is presented.

DOI: 10.4018/978-1-5225-0350-7.ch013

INTRODUCTION

Sustainable Supply Chain Management has gained significance in the recent years due to growing concern about environment. Extreme weather, rising temperature, scarcity of natural resources – all these call for a different strategy towards environment. (Winston, 2014). With globalization, growing economies and consumerism supply chains have grown quite big. In order to preserve natural resources for future generations, sustainability needs to be embedded across the supply chains.

As shown in the Figure 1, in addition to profit maximization, social and environmental sustainability are being considered necessary in what we refer as Triple Bottom Line (TBL). (Elkington, 1994). Moving one step further Caradona (2014), suggests economy as a subset of society and society as a subset of environment. Thinking beyond quarterly profits, business cannot sustain on the planet that fails. (Winston, 2014). Every member of supply chain needs to show concern towards TBL. Incremental investments into sustainability are no longer sufficient. As climate is changing, different long term approach spanning decades and generations will be needed. (Winston, 2014).

Wu and Dunn (1994) suggest a systems approach to evaluate environmental impact of entire supply chain. This approach sees supply chain as a system and recommends integrated strategy for greening the entire supply chain.

This literature review covers different aspects of greening the supply chain. Supply chain is viewed as a system consisting of focal firm, chains of suppliers, distributors, logistics partners, retailers etc. Role of each member of supply chain in sustainability goals is discussed in detail. Extant literature has covered individual parts of supply chain, whereas this paper integrates this knowhow to facilitate learning all interrelated concepts in SSCM.

Research Methodology

Most of the extant literature on Sustainable Supply chain has evolved over last two decades. Extant literature from leading journals around the world was referenced for this study. After review of over 150 articles on Sustainable Supply Chain Management from top journals in the field of Sustainable Supply Chain Management and Operations Management, 100 best and relevant articles were shortlisted in consultation with experts researching in this field. In addition 2 recent books and United Nations landmark report on sustainability "Our common future" were referenced for gaining historical perspective about sustainability. Papers from top journals, corporate websites, videos, website of supply chain professionals were referenced as and when additional information was required.

Figure 1. New paradigm: Corporate, social, and environmental responsibility

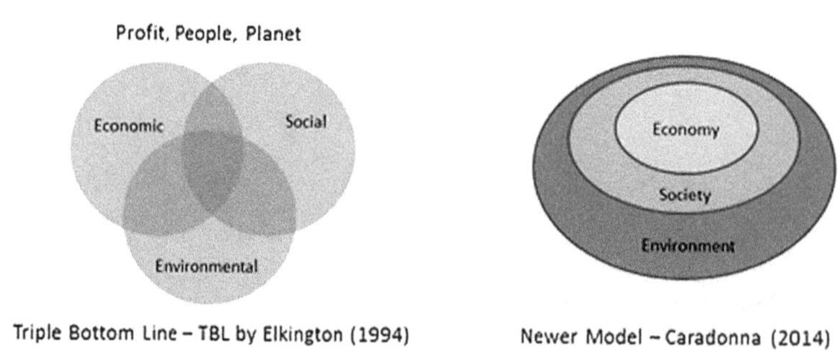

Supply Chain Management

Supply Chain consists of network of entities which are involved directly or indirectly in converting raw material to intermediate components or semi-finished goods, converting semi-finished goods to finished goods, transport of materials, inventory management, information management and distribution. Optimization at the supply chain level provides long term economic benefits for each member of the supply chain as opposed to maximizing benefits of the most dominant member of supply chain. In the long run supply chains which are competing and not the individual firms. Focal firms play the role of integrating and developing members of their supply chain in order to compete and sustain in long term in global business (Basu, 2011). For long term economic success, focal firms have to oversee and manage the Supply Chains. (Nelson et al. 2012)

Supply chains models have evolved significantly over last 100 years. (Refer to Figure 2) Henry Ford's supply chain in early 19th century was vertically integrated to efficiently manage mass production of single product in high demand period. Ford introduced standardization of parts and moving assembly line for automobile production. (Shah, 2009) (Ford, 1988)

Post World War II, Toyota's supply chain evolved to deal with slow growth period and fragmented demand for variety of products. Toyota approached supply chains with objective of waste reduction at every stage to reduce cost and maintain profitability. They pioneered inventory reduction techniques such as JIT, close integration of supplier operations with theirs and viewed suppliers as an extension of their own operations. (Ono, 1988) (Womack et.al., 1990)

In retailing Walmart focused on sourcing high volumes to reduce cost per unit, which ultimately leads to higher sales at stores. Walmart introduced cross docking as a strategy for managing inventory efficiently for time sensitive and high selling goods.

With advancement in technology leveraging ERP system and internet, companies such as Dell introduced make to order direct shipment model. This model empowers customers to choose different options in building a product as per their choice and shipping them in few days. This strategy allows them to maintain inventory of components instead of finished goods. In technology this provides competitive advantage as rate of obsolescence of such products is high.

Amazon.com supply chain was designed to ship millions of different retail products to individual customers directly from their fulfilment centers with the help of UPS and FedEx.

Sustainable Supply Chain Management

Social and Environmental Concerns in Supply Chains

Primary motives of each of the above SCM strategies were driven by economic and marketing objectives. There was no consideration towards social or environmental objectives. As per current trend manufactur-

Figure 2. Evolution of supply chains

ing and service work from developed countries (US, Canada, EU etc.) is getting outsourced to developing countries such as China, India, Malaysia, Indonesia etc. Though this strategy provides cost advantage, its carbon footprint is significant – intercontinental logistics (sea or air) and increased manufacturing / logistics activity in these countries. In a shared planet, activity in one region can have its consequence in other regions. During last few decades, with growing awareness about social and environmental issues, sustainability is gaining significance in supply chain management.

Fossil fuels such as coal and oil have finite supplies coming from earth. With increasing energy requirements and consumption of fossil fuels, most of the world's coal and oil reserves are depleting at a faster rates. It is important that members of supply chains plan to move towards renewable sources at the earliest. United Nations Brundtland Commission published report "Our Common future" in 1989 seeking "*Development meeting the needs of the current generation without compromising the ability of future generations to meet their needs*". UN efforts have given much needed impetus at political level for developing the field of sustainability. Further awareness within industry and society was built by popular video "*The Inconvenient Truth*" by US Vice President, Al Gore. (Al Gore, 2006). With political awareness, developed nations introduced regulations that induce companies to follow environmental standards for their supply chains. In the recent past, Kyoto protocol is putting together a framework to reduce GHG emissions.

Sustainability in SCM needs to focus on following areas and address them over the entire life cycle of the product:

1. Material consumption – (Mining activity)
2. Energy – power generation and consumption
3. Pollution of Air and Water
4. Solid wastes going to landfills

SSCM FRAMEWORK

Extant SSCM literature has dealt with individual elements such as sustainability, SSCM drivers, role of upstream suppliers, logistics, operational practices and framework for performance management. This review paper consolidates this information and gives a bird's eye view of the knowledge base as depicted in the Figure 3.

SSCM Drivers

Literature suggests 5 different drivers which persuade organizations to move towards SSCM. (Holt et.al, 2008), (Hsu et.al 2012), (Dubey et.al. 2014), (Glover et.al. 2014). One or more of these forces influence large, medium or small companies to move towards sustainable practices across supply chain. For example, large companies concerned for their brand image, social responsibility are either driven internally or by coercive pressure such as regulatory compliance requirements and agitations by environmental NGOs. Medium to small companies are driven by their large customer requirements or by the Mimetic (competitive) pressure to develop Green products and processes.

As the government, society and corporates are getting awareness about environmental impact of supply chains, momentum is building towards sustainable practices.

Figure 3. SSCM framework

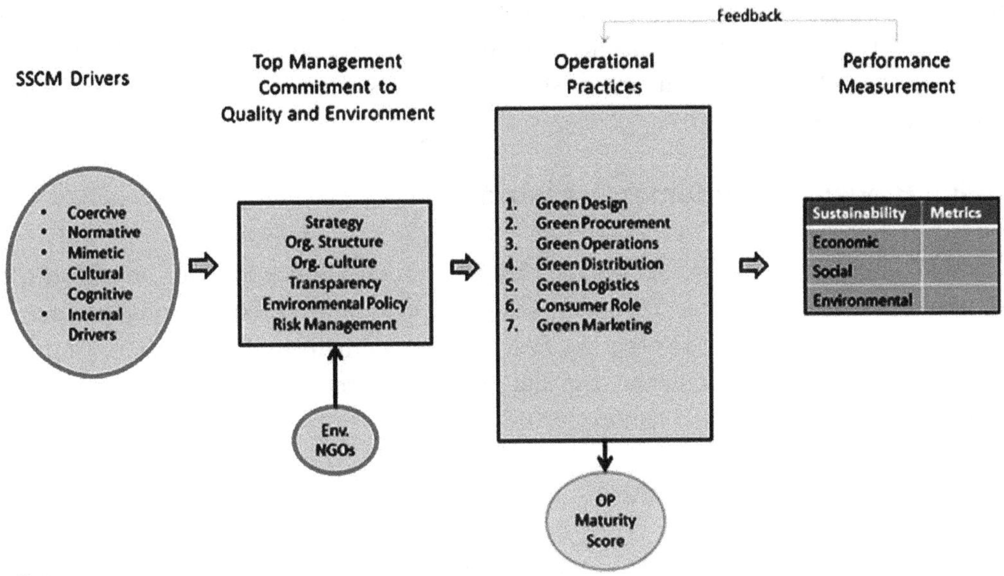

Coercive Pressure (Regulatory Pressure)

Most of developed countries have well defined regulatory framework for monitoring environmental compliance by companies. These regulatory requirements act as ultimate incentive for companies to go Green. (Hsu et.al. 2012). Corporations strive to meet the regulatory requirements due to risk of penalty as well as impact on brand image due to negative publicity. These requirements turn them to adopt environmental practices such as Green Purchasing, Green Design, Reverse Logistic etc. (Hsu et.al. 2012). US have setup Environmental Protection Agency (EPA) back in 60s to monitor, control and regulate emissions and wastages into environment. In Germany Topfer's law introduced several regulations that ban use of environmental harmful substances, law on the life-cycle economy and the packaging recycling requirement. Chinese rapid growth over last 30 years has raised millions out of poverty, but has raised concerns about long term sustainability as the country has to deal with rapid depletion of natural resources, pollution and intensive energy use. To address increasing wealth disparity across society China has adopted "inclusive growth and implementing strict environmental regulations to address the challenge of sustainability. (China, KPMG, 2011) Chinese government maintains nationwide census of pollution sources (Plambeck et. al. 2012). In developing countries especially in Asian region (developing countries such as India, Malaysia) GSCM practices and regulations are in an embryonic stage. There is lack of proactive attitude to reduce source of wastage or pollution at source. (Dubey et. al. 2014).

There are numerous cases wherein large branded companies are held responsible for actions of their suppliers in developing countries. Environmental NGOs often highlight violations through global media. Such pressures induce companies to address the issues. McDonalds when challenged by a media report, responded swiftly by working with multiple stakeholders (Farmers, intermediaries, government, key supplier and NGO) to find a resolution to everyone's satisfaction. Their initiative has led to stoppage of deforestation, long term sustainability of supply chain and a better brand image for McDonalds. (Goldberg, 2007)

Nike, when faced with negative publicity in the U.S. media, moved to make many strategic long term changes. Nike approached sustainability beyond compliance requirements by redesigning product, process and working with suppliers to set highest environmental standards. It worked with an NGO and suppliers to substitute harm chemicals and waste reduction, resulting in gain on both environmental and cost fronts. (York & Larson, 2006).

Normative Pressure: Customer Pressures

Environment conscious consumers demand eco-friendly products, they look for brands with good environmental track record. Corporate customers in developed countries require compliance with higher level of environmental standards as a necessity to do business with them. Walmart as a customer requires its suppliers to self-report environmental data and set future targets annually to understand and control environmental impact. (Denend & Plambeck, 2010)

Mimetic Pressure: Competitors

Companies try to mimic or copy competitor's best practices. Companies invest in R&D to innovate their products and services. When one company introduces green products and is well received in the market, competitors are motivated to follow with similar strategy.

Cultural Cognitive Isomorphism: Socio Cultural Responsibilities

Firms may perceive voluntary obligation to society in which they operate to meet social expectations. Corporates also set their vision towards sustainable practices as a part of balanced scorecard. Therefore, Companies perceive this as their moral and social obligation to make their products and supply chains sustainable. The pressure of socio cultural responsibility has motivated firms in Malaysia to adopt green operational practices. (Hsu et.al.2012)

Internal Drivers within Organization

In addition to above drivers, Holt and Ghobadian (2008) suggest internal drivers originating within the company. Top management interests in eco-friendly products, influential persons in a company such as CEO, nature friendly culture of the organization, general employee initiatives also drive greening programs in some companies. Starbucks led a unique social change wherein they worked closely with environmental NGO to set C.A.F.E practices across its supply chain, creating a long term environment friendly and sustainable source of quality coffee beans. Instead of using its dominant position in the supply chain for negotiating better prices, it persuaded customers to pay slightly higher cost for quality coffee. Through these practices it has engaged farmers towards sustainable farming practices, minimizing soil erosion. (Lee, 2007)

Each of above forces individually and collectively influences the greening of one or more members of supply chain. As a net effect of these pressures, firms are moving towards sustainable practices along supply chain. Focal firms play significant role in greening the supply chain.

How these Drivers Affect Supply Chain Members

There is good scope to study how these drivers affect adoption of green practices across supply chain members. Manufacturing companies in China are adopting green practices due to pressure from their western customers, which ask for environmental data and transparency. (Denend, 2010) (York et.al 2006) Manufacturing companies based in UK feel most of the pressure for greening their supply chain due to regulatory pressure and internal drivers such as organizational champions or influential CEO. The least pressure is felt due to societal drivers and individual customers. (Holt et.al. 2008) In Malaysia, regulatory measures, competitive pressures, customer pressure, socio-cultural responsibility are the prime drivers that affect adoption of green operational practices in the supply chain. These drivers lead to highest improvement in Green Procurement practices, Design for Environment and lowest in Reverse Logistics practices. (Hsu, et. al. 2012)

Adoption of SSCM

In spite of different pressures, firms have different approach towards adoption of greening their supply chain. According to study conducted by Haanaes et.al. (2011) there are 2 distinct groups - Embracers and Cautious adopters. Embracers have a long term view. They already consider sustainability as a core. They view investments in greening their supply chain as a source of competitive advantage, opportunity to improve brand image, getting access to newer markets and improving margins. Embracers are found among large and global companies (automobiles, commodities, consumer products, industrial goods, machinery, retail etc.). They tend to be resource intensive industries as they have higher levels of environmental footprint. It turns out that embracers are the highest performers in the business.

Cautious adopters look for greening the supply chain primarily for economic and risk mitigation reasons (reducing costs due to energy efficiency, reduced material usage and reduced risk of penalty). These are smaller size companies.

Azzone et.al. (1998) provide a more details taxonomy of strategic patterns of environmental behavior and recommend corresponding performance measurement framework for each strategy (Table 1).

Top Management Commitment to SSCM

For success of any strategic initiatives such as implementing Sustainability, top management commitment is important. (Digalwar, 2013), (Dubey et.al. 2014) established that leadership of an organization

Table 1. Taxonomy of Strategic Patterns

Strategy	Nature	Description
Evangelist	Internally Driven	Strategy following an ethical objective and radical approach towards environment
Proactive	Internally Driven	Systemic efforts to affect the whole value chain
Responsive	Externally Driven	Internally driven changes, low regulatory pressure
Reactive	Externally Driven	More to comply with environmental regulators
Nonresponsive	Externally Driven	Smaller firms with limited financial resources, trying to delay green investments as much as possible

Source: Azzone et.al (1998)

has tremendous influence on success of managing supplier relationship (SRM) and implementation of Total Quality Management (TQM). SRM includes maintaining relationship, integrating supplier operations with firm's operations using technology best practices and high level coordination with suppliers.

TQM is a philosophy which consists of culture, attitude and organization of a company that strives to provide customers with products and services that satisfy their needs.

Top management commitment alone is not sufficient for greening supply chain. Sustainability needs to be driven from top by embedding it into business strategy, developing organization culture, providing transparency to stakeholders about current state and plans and managing risks proactively (Carter & Rogers, 2008). Last but not the least organization structure should reflect strategic importance of SSCM (Toffel, 2014). Along with economic parameters, social and environmental sustainability should be part of top management review and goal setting process.

Top management has to integrate sustainability into business strategy so that it gets significance in almost all important functions such as marketing, product and process design, operations, human resources, quality etc. Top management has to educate and build an "Organization Culture" that cares for nature, promotes sustainability with each business partner. Once the culture is ingrained to the rank and file of the organization, it becomes part of working vocabulary in business.

Sustainability function should be embedded within top management, so that all important projects being evaluated on the basis of their sustainability score along with social and economic criteria. HP assigns senior executive CSO - Chief Sustainability Officer to monitor operations and important projects. (Toffel, 2014). Environmental Sustainability is part of job responsibility for a senior management member in Walmart. (Denend, 2010)

Maintaining transparency by engaging stakeholders, listening to their ideas, sharing information and securing their buy-ins for environmental projects is in the best interest of the firms. This helps to build a trust and brand image for the company. Companies such as Walmart, Starbucks and Nike have successfully worked with environmental NGOs to implement Green initiatives across their supply chains.

Corporates need to recognize and manage risks in global supply chain originating from human rights violations, environmental practices, unsustainable farming practices or weather related disruptions. Identifying, monitoring-resolving risks saves lot of future time and money spent in damage control. Focal firms brand image get affected by any violations across their supply chain, hence they need to proactively monitor and manage these risks.

SSCM OPERATIONAL PRACTICES

Success of Green Strategy depends on successful implementation of various practices within firm's manufacturing or service operations or across members of supply chain. There is a great opportunity to build eco-friendly products from design to end of life stage.

Green Engineering

Maximum benefits come from embedding Sustainability into designing new products or processes. Green Design is an approach where the environmental impact of a product and process is minimized at the design stage. One of the key inputs for designing products is consideration for eco-friendliness. While defining key parameters for next generation products, product design team must specify target energy consumption, emissions - far superior than existing products beyond regulatory needs. Designing products with alternate technologies with lowest possible environmental impact is a way forward.

Approaches such as continuous improvement by optimizing existing solution may provide incremental benefits for existing investment; however maximum benefits come when the problem statement is redefined. (McDonough et.al.2003) For example, instead of "improving fuel efficiency by 30%", problem may be restated as "how to use 100% renewable energy source". This approach is similar to breakthrough thinking to improve quality by several points in one go which is explained by quality guru Dr. Juran in his Quality Handbook. (Juran et.al. 1999)

McDonough et.al. (2003) suggest a Cradle to Cradle (C2C) Design instead of usual Cradle to Grave (C2G) for products lifecycle. C2C framework consists of 3 tenets and 12 principles described below. The 3 tenets are derived from the way natural ecosystem works. Natural systems which have evolved over millions of years are self-sustaining.

In nature, 1) waste of one entity is a food for another; 2) solar energy is the prime source of energy which is captured by plants and 3) nature maintains diversity. These 3 tenets can be mimicked for designing supply chain systems. For example, Principle 1 recommends avoiding use of non-hazardous materials. Walmart asks its suppliers to declare all the ingredients going into their products and monitors them against a negative list of hazardous materials (Denend, 2010). Energy sources should be renewable for planning new products (principle 12). For designing new automobiles, fuel cells or battery operated cars or any other cleaner energy sources should be the only options considered as an environmentally responsible auto company. For powering new or existing manufacturing plants, similar approach of using solar, wind, biomass or any other renewable resources should replace depleting energy sources such as coal or diesel.

Product designers should plan the product design for reuse or remanufacturing most of the components or material. Idea is to minimize mining activity for the same material. (Principles 9 and 11)

In quality management, 14 principles defined by of Dr. Deming continue to provide long term guidance to create value for customers. Similarly for environment management 3 tenets and 12 green design principles provide guidelines for testing new designs or proposed changes to existing designs and processes (Table 2).

There are many practitioners of above principles in the industry, governments and societies. Battery powered cars by Mahindra Reva, Tesla Motors, fuel cell powered cars by Toyota, Otis Elevators innovation to generate electricity while travelling down, solar powered street lights, solar water heaters - these are leading examples of above green design principles in practice. Nike has embedded Sustainability function in its design, manufacturing and marketing functions. Toyota considers its hydrogen powered cars with zero emissions to steer a revolution in personal mobility. Recently Indian Railways have launched CNG powered train on Rewari-Rohtak section.

Above principles can be applied while designing buildings or remodeling existing buildings, designing industrial parks or manufacturing facilities. Design should incorporate maximum availability of natural sunlight for most part of the day, use of renewable sources for running the operations, designing spaces for minimizing movement and transport of material.

Hoek (1999) recommends designing modular products which can be maintained or upgraded. As new features or modules are available, those can be plugged in to existing product. A particular module can be maintained or replaced as per need rather than replacing the entire product. This leads to huge saving of material and energy in manufacturing new product such as car, refrigerator etc. and extends useful life of a product purchased by consumer.

Table 2. Design principles

Principle 1: Inherent Rather Than Circumstantial: Designers need to strive to ensure that all materials and energy inputs and outputs are as inherently nonhazardous as possible.
Principle 2: Prevention Instead of Treatment - It is better to prevent waste than to treat or clean up waste after it is formed.
Principle 3: Design for Separation - Separation and purification operations should be designed to minimize energy consumption and materials use.
Principle 4: Maximize Efficiency - Products, processes, and systems should be designed to maximize mass, energy, space, and time efficiency.
Principle 5: Output-Pulled Versus Input-Pushed - Products, processes, and systems should be "output pulled" rather than "input pushed" through the use of energy and materials.
Principle 6: Conserve Complexity - Embedded entropy and complexity must be viewed as an investment when making design choices on recycle, reuse, or beneficial disposition.
Principle 7: Durability Rather Than Immortality - Targeted durability, not immortality, should be a design goal.
Principle 8: Meet Need, Minimize Excess - Design for unnecessary capacity or capability (e.g., "one size fits all") solutions should be considered a design flaw.
Principle 9: Minimize Material Diversity - Material diversity in multicomponent products should be minimized to promote disassembly and value retention.
Principle 10: Integrate Material and Energy Flows - Design of products, processes, and systems must include integration and interconnectivity with available energy and materials flows.
Principle 11: Design for Commercial "Afterlife" - Products, processes, and systems should be designed for performance in a commercial "afterlife."
Principle 12: Renewable Rather Than Depleting - Material and energy inputs should be renewable rather than depleting.

Source: Anastas et. al. (2003)

Some Additional Eco-Friendly Design Principles

- **Design to Consume Lesser Resource:** During design phase plan product and component design to consume lesser material, less energy during manufacturing phase, consume lesser energy during consumer usage phase. Value engineering technique can be adopted with green products in mind. Unilever is producing laundry products that use less water in rinsing, generating considerable water savings for people whom it sells its products. (Haanaes et.al. 2011) Maruti Suzuki introduced cars running on CNG in India which has significantly reduced emissions as compared to diesel or petrol powered cars.
- **Packaging Design**: Packaging needs to be designed in such a way to minimize the use of material or recycle packaging material.
- **Remanufacturing**: Design of assembly consisting of discrete components can provide for recycling or reusing components for remanufacturing purpose. Fiat, BMW and Volkswagen have formed an alliance to take back end of life cars for recycling purpose. (Azzone et.al. 1998).
- **Process Design**: Process design should take into account cradle to grave aspects of the products. Hazardous chemicals need to be neutralized properly and disposed of safely. For example, end of life batteries are sent back to manufacturers to reuse some of the materials such as lead, plastic and any acid is neutralized.
- **Designing Eco Industrial Parks:** Natural ecosystems (refer to Figure 4 and Figure 5) can provide valuable lessons for designing new eco industrial parks or restructuring existing ones. (Cote et.al. 1998) These eco systems work in a cycle wherein the resources are recycled and reused as shown in the Figure 4. Eco industrial park consists of a mix of manufacturing companies wherein byproduct of one firm becomes inputs to another firm, leading to minimum wastage and economic benefits to each of the firm. Besides, these companies share as many resources as possible. There are many success stories of such eco industrial parks established all over the world.

Figure 4. Circular natural eco-system

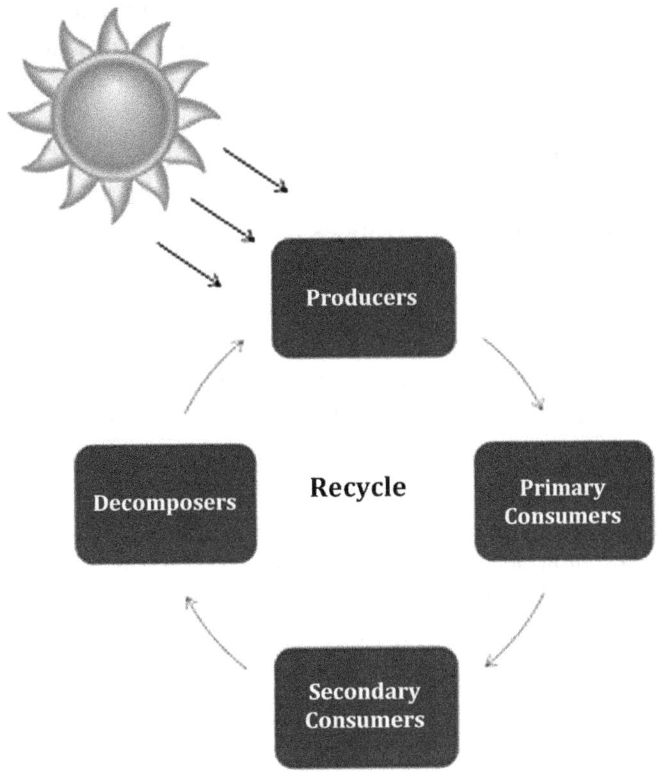

Figure 5. Linear industrial eco-system

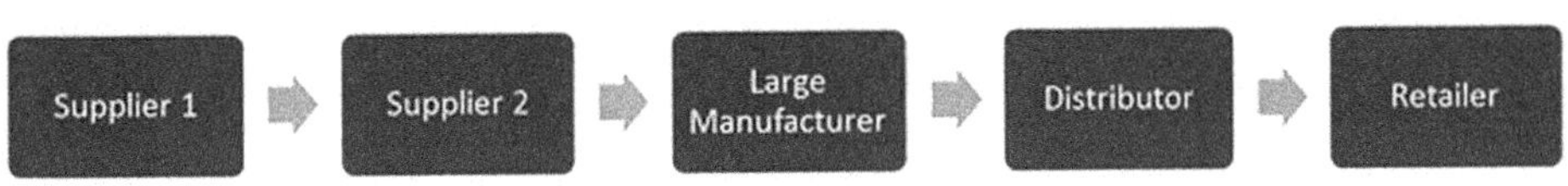

Further these parks can be enhanced with rooftop gardens (McDonough, 2003) for extending natural ecosystem, rain water harvesting for conserving water, flow of natural air for reducing dependence on air conditioning.

Kalundborg Eco-industrial Park in Denmark is considered one of the first such industrial symbiotic networks, wherein companies collaborate to share energy and resources. In this park - Power Company, Oil Company, Cement manufacturing company, District Heating Company, Farmers and Fisheries all demonstrate symbiotic relationship.

Green Procurement

Procurement deals with sourcing the material from suppliers. Many industries as well as government departments source various industrial and consumer durable goods from suppliers. Procurement function has responsibility to ensure supply side is aligned with TBL objectives, gradually reduce carbon footprint and hazardous material from their products. (Rao, 2008) They need to design strategy to strike a balance among TBL objectives. In addition to reducing cost, improving quality and delivery reliability, they need to monitor suppliers for their social and environmental performance.

Large manufacturing companies outsource their components and subassemblies to local and overseas suppliers to take advantage of lower overheads of SME companies. Many consumer durable goods and food supplies are sourced in large quantities from overseas suppliers based in low cost countries such as China, Malaysia, Indonesia, India, Brazil etc. Manufacturing practices of suppliers and farmers in relation to social equity and environmental violations in these countries often come under scrutiny from local governments and NGOs. Branded focal firms are held responsible for actions of members of their inbound supply chain.

Small and medium size companies are less advanced in environmental awareness and compliances (Haanaes, 2011). Therefore it becomes important to get the suppliers on board in reducing environmental impact. Procurement departments have responsibility to develop green sources. Green Procurement takes environmental aspects into consideration along with cost, quality, reliability, service, timeliness while making sourcing decisions with ultimate objective of reducing environmental impact of sourcing and improving resource efficiency. (Dubey et. al. 2013) Some of the leading companies such as Walmart, Apple and Nike have ongoing programs for environmental and social sustainability at their suppliers. After embracing sustainability principles in its own operations, Walmart has created supplier sustainability index, rating supplier performance on sustainability. Apple has successfully persuaded their suppliers to adopt sustainable business and labor practices.

Roadmap for Green Procurement

Figure 6 illustrates a roadmap for implementing sustainability in the Procurement stage of supply chain.

Figure 6. Green procurement roadmap

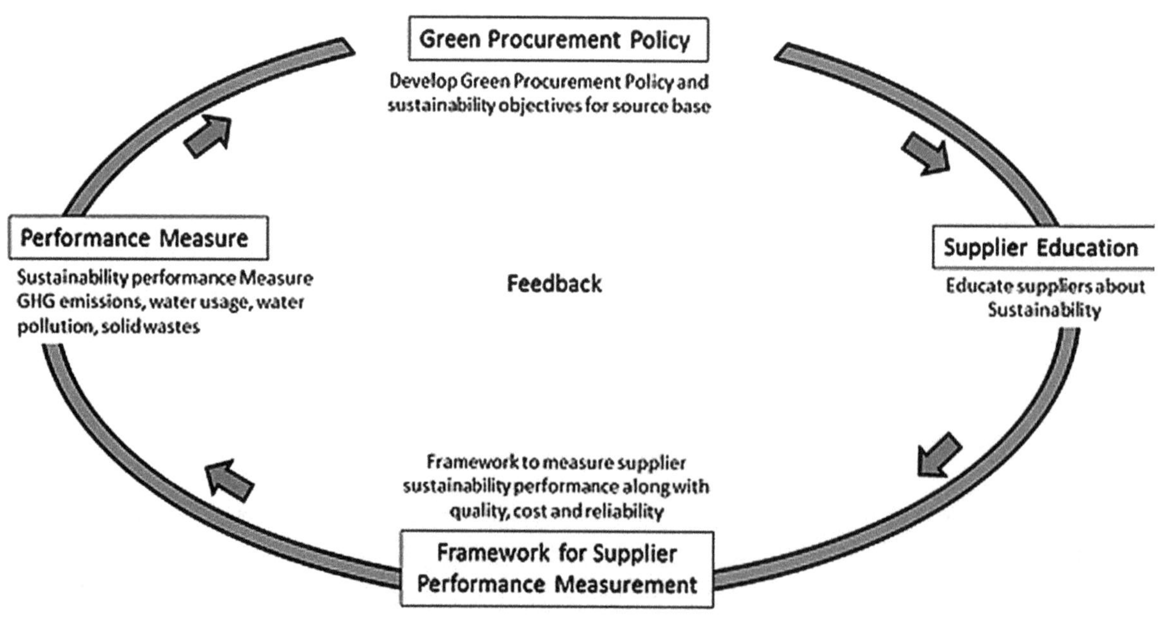

1. **Green Procurement Policy:** Organizations need to develop Green Procurement Policy and sustainability objectives for local and overseas suppliers. Policy should state expectations from suppliers with regard to social equity and environmental compliances. Policy should require suppliers to report current state data about various environmental parameters based on facts and encourage them to set their own targets towards greener practices and materials.
2. **Supplier Education about Sustainability Strategies:** There is a great scope of potential greening activities along the supply sources. Procurement department can interface and educate suppliers about best practices in the field of green engineering, energy sources, recycling etc.
   Conferences can be arranged to share best practices among suppliers. Nike has implemented intensive environmental engineering program at its supplier base in China, Vietnam and Indonesia. This program established performance baseline, sets improvement targets, focuses on improving efficiency of material use and reducing hazardous wastes. Nike engineers provide their expertise to supplier base during 80% of their time rather than just auditing them. (Plambeck et.al. 2012). This level of involvement with suppliers give significant cost saving and better environmental track record for supply chain.
   Some companies arrange visit of suppliers from developing countries to their plants to showcase their green initiatives. It was established from a study conducted by Dubey et.al. (2014) that supplier collaboration and Technology and Best Practices were the key factors in improving environmental performance of a firm. Some MNCs facilitate formation of supplier groups to share best green practices among themselves. (Plambeck, 2012)
3. **Framework for Supplier Performance Measurement:** Traditionally supplier performance is measured on parameters such cost, quality (defects), timeliness and reliability of supply. Metrics for measuring sustainability (GHG Emissions, Water consumption, Quantity of solid wastes going to landfill) also need to be incorporated in this framework.
4. **Performance Measure - Supplier Reporting and Audit:** Proactive companies such as Walmart ask self-reporting of key environmental data from suppliers such as GHG Emissions, Water consumption, quantity of solid waste, ingredients of products. Plambeck, 2012) recommend different strategies to get supplier information from developing countries. Using database of active NGOs in supplier country, increasingly improving government data for negative listing of companies, and working with those suppliers to enhance their compliances.

Additional Strategies for Green Procurement

Figure 6 provides a structure to facilitate improving sustainability on sourcing side of supply chain. In addition, procurement department needs to provide additional value by getting to know the supply chain deeper, single sourcing and learning best practices from other organizations.

1. **Knowing your Supply Chain:** (Choi et.al, 2011) advise not to lose control over your supply chain. They recommend not relying entirely on tier 1 supplier for managing lower tier suppliers. This has been the practice with some of the companies. Having a deeper understanding of supply chain including tier 2 or tier 3 suppliers can be helpful. Lower tier suppliers spot shift in technology or demand much earlier than their larger counterparts. Also, any strategic decisions (such as replacement of intermediate supplier or change in logistics strategy, mitigation of risks related to

environmental or social issues) are less disruptive if we know the lower tier suppliers and logistics network.

2. **Single Sourcing and Supplier Relocation:** Lean companies take supplier relationship to the next level. Suppliers with superior performance on quality, cost and delivery reliability are rewarded with 100% of the business by the organization following Lean philosophy. With single sourcing, and relocation of supplier closer to focal firms facility overall logistics costs are reduced with much better economy of scale and improved quality. (Womack and Jones, 2008). This strategy helps on both the economic and environmental fronts due to reduction of cost and environmental footprint. Focal firms can find it much easier to green the supply chain with single source collocated right next to them.

Green Operations

Transforming internal operations for a firm form an important component in greening the supply chain. This includes production as well as service operations. Firms can setup examples of green practices that can become model for rest of the members of supply chain. UK based manufacturing firms give first priority to greening their internal operations and then on logistics. It is much easier and economical to green internal operations first. Subsequently companies propagate green practices within rest of the members of supply chain such as suppliers. (Holt et.al. 2008) Green operational practices consist of several opportunities for manufacturing or service organizations:

1. **Employee Training:** First awareness and concern needs to be built in the top management team and employees towards greening the operations. Many companies have implemented TQM, Lean, Six Sigma or ISO 9000 initiatives to develop processes and quality mindset among employees. Similarly the need of the hour is developing environmental concern among employees, training and empowering them towards implementing green practices.
2. **Waste Reduction:** Firms can utilize lean philosophy to eliminate 7 types of wastages identified by Toyota. By implementing programs such as "Reduce-Reuse-Recycle" firms can reduce wastage of any resource such as energy, raw materials or water. Replacing paper towels with reusable cloth towels, recycling (metal scrap, water, plastic, used papers in offices), treating water coming out of plant and reusing it for suitable purposes such as watering plants, using energy efficient lamps etc. - these are some of the techniques towards reducing wastes.
3. **Energy Consumption:** Reducing use of energy during production process by redesigning the processes. Replace fossil fuels with renewable sources of energy by setting up a target of 100% renewable energy within few years. Companies such as Walmart have installed rooftop solar cells to power its stores and plans to provide 100% energy required for their operations by year. Apple has announced investment of $850 million 280 MW solar power plant recently. This solar plant will eventually produce enough electricity to power entire operations of Apple in California. Adani and Government of India are planning biggest solar power plant in the world generating 10,000 MW of electricity.
4. **Pollution Reduction:** As a socially responsible company, any air or water pollution which can occur needs to be reduced and gradually eliminated by use of latest technologies or proper maintenance.
5. **Reduce Resource Consumption:** Raw materials originating from mines, water, chemicals can be reduced through "Reuse", "Remanufacturing" and "Lifecycle Analysis" programs.

6. **Reduce Hazardous Materials:** In recent times, some of the countries have passed laws to regulate disposal of hazardous material. Many products such as cars, inverters and UPS make use of batteries which have sulfuric acid, lead and plastic. Sulfuric acid is quite corrosive and led is harmful in all its forms to environment and humans. Battery companies have implemented battery recovery programs to recover all 3 materials from batteries. Led and plastic is reused in manufacturing, sulfuric acid is neutralize. (Rao, 2003) Walmart controls hazardous substances in products sold or materials used for cleaning through a negative list and self-reporting process for suppliers. (Denend, 2010)
7. **Plantation:** Trees provide natural cooling, oxygen and moisture in the air. Manufacturing and service companies can plant trees around factories as well as on rooftop to develop the natural ecosystem around factories. Information technology services companies such as Infosys, Wipro are known for their well-developed campuses which are surrounded by greenery. Leading wind energy producing company - Suzlon has designed their head office in Pune, India on some of these principles. It is LEED certified green building.
8. **ISO 14001 – ISO:** 14001 is standard developed for implementing green operations (Refer Figure 7). It starts with deciding environmental goals and objectives within the company. It deals with emissions to air, water and land pollution, use of natural resources, energy sources, energy emitted (heat, sound, radiation), wastes and physical structures etc. Since ISO programs are driven from top management and have a system of audit for certification, it provides a good framework for implementing green operations practices.

Figure 7. ISO 14001

Green Distribution

There is room for reducing water consumption, energy usage and unnecessary movement in the distribution system through various opportunities as listed below. In fact this can lead to reduced use of energy, wastage and improvement in economic performance. Warehouses can be designed or redesigned for maximum use of natural light. Layouts should be reviewed with work study / motion study for optimizing movement of material and resources. Use of proper inventory management system and technology such as barcoding, ERP systems, PDAs helps in efficient management of inventory across distribution system. Most of the green operations practices can be adopted for warehouse locations in distribution system (recycling, water harvesting, use of solar energy etc.).

Green Logistics

Logistics industry is an important element of supply chain which enables movement of goods and services from source to customer. This industry is a major contributor to environmental degradation through its modes, infrastructure and traffic. (Banister et. al. 1993; Whitelegg, 1993). It can be classified as Inbound, Outbound and Reverse Logistics. The common methods for physical movement of goods are Railways, Seaways, Roadways, Airways and pipelines for liquids.

Railways and sea routes are suitable for bulk transport when duration of delivery is not a constraint. These are the relatively greener options of transport though in reality most of the transport takes place using road carriers, with much higher GHG emissions.

For time bound shipments airways and for shipments to distant locations road transport are most suitable. Road transport provides most efficient transport due to network of cross country highway system that has been developed in most of the countries. However trucking industry is a highest source of GHG emissions. Lack of proper maintenance, overloading, environmental risks caused by carrying hazardous material – these are some of concern areas for environmental impact from this industry.

The approach of logistics sector so far has been reactive to meet the legislative requirements of GHG emissions by pollution under control certification of vehicles. This approach to control pollution is an end of pipeline solution. (Hoek, 1999) Need of the hour is a strategic approach to eliminate pollution at the source by designing vehicles that do not pollute or alternate transportation methods.

Logistics for JIT and Cross Docking: In demand driven supply chains, to minimize inventory carrying costs, firms implement JIT and cross docking. These lead to multiple trips of trucks to move inventory from source to destination based on actual consumption. There is increase in transport cost but savings in inventory carrying costs of supply chain. However, there is overall higher level of GHG emissions and road traffic due to higher level of transport activity using trucks.

Traditionally, distribution network consists of Manufacturer, Distributor, Retailer and the transporter that connects them to move bulk of items. Customer transports goods form retailer shops to his home. With advancement of e-commerce and internet the distribution model is changing. E-Commerce companies such as Amazon.com, flipkart.com, Dell Computers take online orders from customers and ship them from their fulfilment center or manufacturing plant using logistics services from Fedex, UPS etc. This movement of goods in small quantities increases overall packaging, transport and ultimately GHG emissions. (Wu and Dunn (1994)

Green Paradox: Rodrigue et al. (2001) illustrate the green paradox of transport system. The goals of logistics system designers are to reduce cost, improve speed of delivery and reliability, ease of storage and use of e-Commerce solutions for better customer service. However these objectives are not aligned with sustainable supply chain objectives. Cost reduction of transport is often trucking solutions which increase pollution.

To improve delivery speed, more trips, smaller trucks or air transport are used which have higher GHG emissions. Rail and sea routes with lower overall environmental impact are considered least reliable among all transport options. Similarly e-commerce strategy of sending shipments from fulfilment centers to individual customers using services of UPS or Fedex provides convenience but have overall higher emissions compared with traditional distribution system.

Omni-channel retailing, shipments are done from the stores of brick and mortar companies such as Gap, Walmart, Macy's, Home Depot and Best Buys - they are trying to bridge the gap between themselves and online e-commerce retailers such as Amazon.com

Reverse Logistics: In closed loop supply chains, reverse logistics provides an important link to move products in reverse direction of supply chain from customer to manufacturer. These could be product returns from customer, end of life products for remanufacturing, refurbishing or safe disposal of products. Reverse logistics facilitates overall reduction of material consumption resulting in lesser need of mining activities.

Moving towards Green Logistics: To summarize different strategies towards green logistics:

1. Advancement in technology for electric and hydrogen powered cars need to reach transport sector. Truck manufacturing companies need to invest in research for design next generation trucks with alternate green fuels as well as improve efficiency of trucks. They can develop engines running on alternate cleaner fuels such as CNG and LPG. Chesapeake Energy, second largest producer of natural gas in USA, has in last few years converted its fleet of vehicles to CNG reducing environmental impact and significant reduction in fuel cost.
2. Vehicle Manufacturers need to compete on the basis of fuel efficiency as key order winning characteristics.
3. Truck operators need to be educated for regular maintenance of engines, driving under ideal conditions. (Dubey et.al. 2014)
4. Implementation of environmental compliance by vehicles will help reduce air pollution. (Dubey et.al. 2014).
5. Supply chains should consider use of reverse logistics to return products back to focal firms for "Remanufacturing", "Reusing", "Recycling" or "Disposing off" hazardous material in an eco-friendly process.
6. Leveraging information technology for logistics planning wherein transporters can collaborate, share trucks and improve utilization of trucks. By sharing information, instead of returning empty trucks after delivery they can be utilized for shipments on return routes.

Consumer Role

Some of the products have their environmental impact during product usage by consumers. Here consumers need to be educated about operating conditions for maximum efficiency. Educational advertisements through communication mediums TV, Radio will help build awareness with citizens.

Green Marketing

Marketing is one of the 3 line functions in the organization along with operations and finance. (Stevenson, 2009) The role of marketing is to choose target markets, get, keep and grow customers by creating, delivering and communicating superior value for them. (Kotler, 2000) Marketing team of an organization gathers information about demand, consumer trends and recommends marketing strategies for the company. It is a customer facing function, they understand consumer requirements, trends and can influence demand with strategies such as advertising, pricing, direct marketing etc.

Polonsky (1994) defines Green or Environmental Marketing as: "Green or Environmental Marketing consists of all activities designed to generate and facilitate any exchanges intended to satisfy human needs or wants, such that the satisfaction of these needs and wants occurs, with minimal detrimental impact on the natural environment."

Green Marketing can be a way of differentiating and positioning. Firms can exploit environmental concerns of consumers to differentiate and to position their products in order to seize new green markets. The real challenge for companies is to incorporate environmental mission into their business strategy rather than just promote the green products (Chen et.al. 2012). Companies which can develop green products and market them effectively can gain market share of this segment.

Green Marketing also includes recommendation of changes to product and service design based on changes to government regulations or customer feedback in order to improve the environmental performance of company's products.

As effects of global warming are visible and there is lot of research information on display in different media, consumer awareness about need for green products is growing. However not all consumers may be willing to sacrifice their lifestyle or pay additional price for ecofriendly products.

Pricing Strategy based on Consumer Degree of Greenness

Consumer demographics, income levels and purchasing power are major factors that influence their price sensitivity and willingness to buy green products. Ginsberg and Bloom (2004) categorize consumers into 5 groups based on their intention to buy green products. (Refer Figure 8) Their analysis shows that only 15% customers are genuinely concerned about environment and willing to pay more for green products. Rest of the 85% either do not understand the environmental issues or consider green products as too expensive and hence are not willing to purchase them.

Price is a major barrier preventing consumers from purchasing green products. (Weisstein et.al. 2014). Consumers with low degree of greenness are more likely to respond to price discount when there is high level of benefit in the promotion of green product. (Palazon et. al. 2009) Price sensitive consumers look for savings, quality and desired features when they compare green products with other options available in the market. Only when the green products are more economical due to lower fuel consumption they are inclined to buy them. India based leading automobile company - Maruti Suzuki has a fleet of car models offering CNG option for petrol cars which is popular with car buyers.

Consumers with a high degree of greenness are fully prepared to pay more for green products due to their comparatively high commitment to saving and protecting the environment. (Laroche et. al 2001) Tesla and Toyota have launched their electric and hydrogen powered cars in the premium luxury segments which are relatively less sensitive to pricing.

Green marketing strategy can be designed as per "Green" market segment indicated in the above graph.

Figure 8. Ginsberg and Bloom's consumer categories (2004)

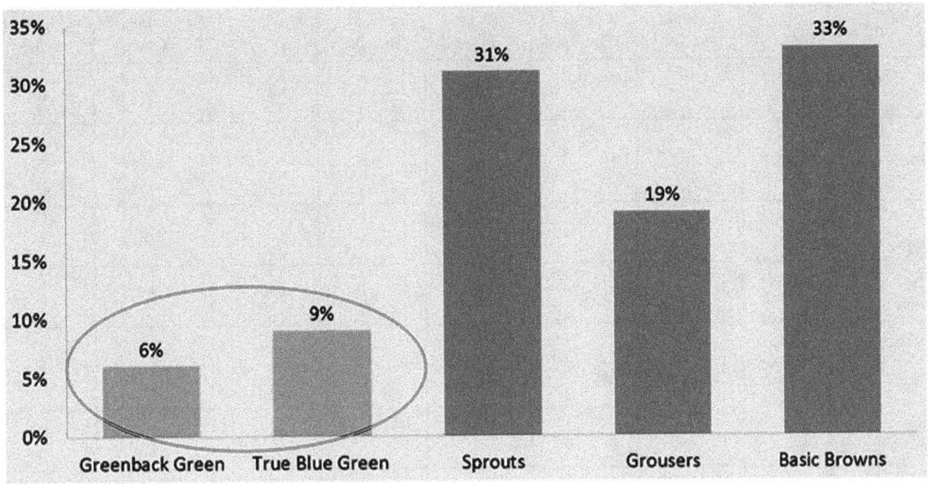

Operational Practices Maturity Score: (OPM Score)

Using above constructs (Green Design, Green Procurement etc.) environmental maturity score for an organization can be measured to define baseline as explained in Table 3. Improvement targets can be set up according to current baseline. Organizations can design metrics related to each of the above areas. These metrics will provide a maturity score for the entire supply chain. This score can be used for assessing supply chain performance, setting future goals and finally improving green operational practices. The same tool can be used as a research tool by SSCM researchers.

PERFORMANCE MEASUREMENT SYSTEM

An overview of entire framework of SSCM has been presented earlier in Figure 3. Measuring SSCM Performance is challenging due to participation of stakeholders from upstream and downstream processes. Level of deployment of Green practices across these members will determine operational maturity score for supply chain. This will determine the resultant green performance using different metrics defined in Table 3.

Nature of environmental impact is different in supply chains of different industries such as chemical, petro-chemical, mining, dairy, automobiles, healthcare, hospitality, furniture or retail etc.

Some industries produce GHGs during manufacturing process, whereas others pollute water bodies through hazardous chemicals. Automobiles running on fossil fuel produce GHGs during their running time. Also, it is important that supply chains are measured on all 3 parameters of TBL – Economic, Social and Environmental as defined in Table 4. This can be used as a reference to develop industry specific TBL metrics. This will provide insights into performance of the firm, industry sector on all the 3 key parameters.

Some additional frameworks being suggested in literature are GRI (Global Reporting Initiative), ISO 14031, Green SCOR (by SCC 2008), Sustainability Balanced Scorecard (SBSC). (Kafa et.al. 2013).

Table 3. Operational practices maturity score

Sr.	Sustainability Practices	Type	Supplier Process	Logisics Parnter	Focal Firm	Distributor/ Retailer	Customer	Environmental impact	Enironmental Score	Weighted Score
1	Green Engineering	Internal Process			X			1	10	10
2	Green Procurement	Internal and Upstream Process	X		X			5	2	10
3	Green Operations	Internal Process			X			10	2	20
4	Green Distribution	Downstream Process				X		5	4	20
5	Green Logistics	Upstream, Downstream process		X				5	5	25
6	Consumer Role	Internal Process					X	10	4	40
7	Green Marketing	Internal Process			X			1	8	8
			Weighted Score							15

Table 4. TBL performance measurement - references: (Azzone et. al., 1998)

Type	Economic	Social	Environmental	
Objectives	Gauge economic success and sustainability of an organization	Social sustainability - sustainable farming practices, stoppage of forest destruction	Measure ecological footprint of an organization	
Metrics for Focal firm and Suppliers	Profitability	Existence of social sustainability Policy	Natural resource use	Total Water usage
	EPS			Quantity of treated water
	ROI%			% of recycled water
	Growth%		Energy Use	MW Electricity Used
				Quantity of Fuel Consumption
				% of renewable energy sources
			GHG Emissions	Total GHG Emissions
			Wastage	Total Solid Wastes
Metrics for Logistics Firms	same as above	same as above	Energy use	Quantity of Fuel Consumption
			GHG Emissions	Total GHG Emissions

CONCLUSION

From literature review it is quite clear that global supply chains are having environmental impact leading to global warming. Firms need to strike a fine balance between economic, social and environmental sustainability, which are now commonly referred as triple bottom line. This paper primarily researches environmental sustainability. From historical perspective it is imperative that for long term survival of the firms need to practice social equity. For long term sustenance of the planet, supply chains need to adopt conservation of resources, alternate renewable resources and changes to different manufacturing, logistics and marketing strategies.

There are various forces that are driving firms to adopt SSCM across all the members of supply chain. Most effective among them seems to be the coercive pressures from regulatory bodies.

Literature presents various different ways in which each stage of supply chain can be made sustainable. Most common thread among them is replacing fossil fuels (Coal, Petroleum products) with renewable energy sources (such as solar, wind, geothermal etc.) at each stage of supply chain (at the supplier's operations, firm's own operations and for transportation) Secondly, conservation of resources such as water, material and energy through practices of Reduce-Reuse-Recycle, Remanufacture and the concept of closed loop supply chain. Idea is to reduce the use of resources by recovering them as much as practically possible. Third important aspect is addressing the design of products, processes and even industrial complexes to embed 12 sustainability principles. This approach will address the environmental issues at source by designing next generation products and processes which can bring revolution towards sustainability goals. Last but not the least, customer facing marketing folks also needs to create demand for greener products and develop a market segment of green consumers.

Commitment to 100% Renewable Energy Sources

In the increasingly globalized world, developed as well as developed countries' need for energy will continue to grow. Irrespective of the legislative requirements, focal firms in supply chains need to define aggressive targets towards using 100% renewable energy sources and developing next generation products which do not pollute the environment. Leading corporates in America are already using solar energy to power their operations. Walmart, Kohl's, Costco, Apple, Ikea, Macy's and many more have dedicated solar power plants. This will lower the carbon footprint of global supply chain on a massive scale.

REFERENCES

Abraham, M., & Nguyen, N. (2013). Sandestin Declaration: 9 Principles of Green Engineering. *Environment and Progress*, *2004*(22), 233–236. Retrieved from http://www.acs.org/content/acs/en/greenchemistry/what-is-green-chemistry/principles/sandestin-declaration.html doi:10.1002/ep.670220410

Adham, K. N., & Siwar, C. (2012). Empirical Investigation of Government Green Procurement (GGP) Practices In Malaysia. *OIDA International Journal of Sustainable Development*, *4*(4), 77–88.

Amaeshi, K. M., Osuji, O. K., & Nnodim, P. (2008). Corporate social responsibility in supply chains of global brands: A boundaryless responsibility? Clarifications, exceptions and implications. *Journal of Business Ethics*, *81*(1), 223–234. doi:10.1007/s10551-007-9490-5

Anastas, P. T., & Zimmerman, J. B. (2003). Peer reviewed: Design through the 12 principles of green engineering. *Environmental Science & Technology*, *37*(5), 94A–101A. doi:10.1021/es032373g PMID:12666905

Awaysheh, A., & Klassen, R. D. (2010). The impact of supply chain structure on the use of supplier socially responsible practices. *International Journal of Operations & Production Management, 30*(12), 1246–1268. doi:10.1108/01443571011094253

Azzone, G., & Noci, G. (1998). Identifying effective PMSs for the deployment of "green" manufacturing strategies. *International Journal of Operations & Production Management, 18*(4), 308–335. doi:10.1108/01443579810199711

Baker, M., & Hart, S. (2008). *The marketing book.* Routledge.

Banister, D., & Button, K. (1993). Environmental policy and transport: An overview. In *Transport, the Environment and Sustainable Development* (pp. 1-15).

Banker, S. (2013). Amazon Vs. Walmart: E-Commerce Vs. Omni-Channel Logistics. *Forbes Logistics and Transportation.* Retrieved from http://www.forbes.com/sites/stevebanker/2013/10/04/amazon-vs-walmart-e-commerce-vs-omni-channel-logistics/2/

Beer, S., & Lemmer, C. (2011). A critical review of "green" procurement: Life cycle analysis of food products within the supply chain. *Worldwide Hospitality and Tourism Themes, 3*(3), 229–244. doi:10.1108/17554211111142194

Borin, N., Lindsey-Mullikin, J., & Krishnan, R. (2013). An analysis of consumer reactions to green strategies. *Journal of Product and Brand Management, 22*(2), 118–128. doi:10.1108/10610421311320997

Brundtland, G., Khalid, M., Agnelli, S., Al-Athel, S., Chidzero, B., Fadika, L., ... & Okita, S. (1987). Our Common Future (Brundtland report).

Carter, C. R., & Rogers, D. S. (2008). A framework of sustainable supply chain management: Moving toward new theory. *International Journal of Physical Distribution & Logistics Management, 38*(5), 360–387. doi:10.1108/09600030810882816

Chen, Y. S., & Chang, C. H. (2012). Enhance green purchase intentions: The roles of green perceived value, green perceived risk, and green trust. *Management Decision, 50*(3), 502–520. doi:10.1108/00251741211216250

China, K. P. M. G. (2011). *China's 12th five-year plan: overview. KPMG Insight Series.* KPMG Adviso.

Choi, T., & Linton, T. (2011). Don't let your supply chain control your business.

Ciliberti, F., Pontrandolfo, P., & Scozzi, B. (2008). Investigating corporate social responsibility in supply chains: A SME perspective. *Journal of Cleaner Production, 16*(15), 1579–1588. doi:10.1016/j.jclepro.2008.04.016

Clark, W., J., C. Toms, L., & W. Green, K. (. (2014). Market-oriented sustainability: Moderating impact of stakeholder involvement. *Industrial Management & Data Systems, 114*(1), 21–36. doi:10.1108/IMDS-04-2013-0194

Conway, D. M. (2012). Sustainable procurement policies and practices at the state and local government level. In K. Hirokawa & P. Salkin (Eds.), *Greening Local Government.*

Côté, R. P., & Cohen-Rosenthal, E. (1998). Designing eco-industrial parks: A synthesis of some experiences. *Journal of Cleaner Production*, *6*(3), 181–188. doi:10.1016/S0959-6526(98)00029-8

Denend, L., & Plambeck, E. (2010). Walmart's Sustainability Strategy (B): 2010 Update.

Devi Juwaheer, T., Pudaruth, S., & Monique Emmanuelle Noyaux, M. (2012). Analysing the impact of green marketing strategies on consumer purchasing patterns in Mauritius. *World Journal of Entrepreneurship. Management and Sustainable Development*, *8*(1), 36–59.

Digalwar, K., A., R. Tagalpallewar, A., & K. Sunnapwar, V. (. (2013). Green manufacturing performance measures: An empirical investigation from Indian manufacturing industries. *Measuring Business Excellence*, *17*(4), 59–75. doi:10.1108/MBE-09-2012-0046

Dubey, R., & Bag, S. (2014). Antecedents of Green Manufacturing Practices: A Journey towards Manufacturing. In *Smart Manufacturing Innovation and Transformation: Interconnection and Intelligence: Interconnection and Intelligence* (p. 333).

Dubey, R., Bag, S., & Ali, S. S. (2014). Green supply chain practices and its impact on organisational performance: An insight from Indian rubber industry. *International Journal of Logistics Systems and Management*, *19*(1), 20–42. doi:10.1504/IJLSM.2014.064029

Dubey, R., Bag, S., Ali, S. S., & Venkatesh, V. G. (2013). Green purchasing is key to superior performance: An empirical study. *International Journal of Procurement Management*, *6*(2), 187–210. doi:10.1504/IJPM.2013.052469

Dubey, R., Gunasekaran, A., & Ali, S. S. (2015). Exploring the relationship between leadership, operational practices, institutional pressures and environmental performance: A framework for green supply chain. *International Journal of Production Economics*, *160*, 120–132. doi:10.1016/j.ijpe.2014.10.001

Elkington, J. (2004). Enter the triple bottom line. In A. Henriques, & J. Richardson (Eds.), *The triple bottom line: Does it all add up* (Ch. 1). London: Earthscan.

Ford, H., & Crowther, S. (1988). Today and tomorrow.

Friedman, T. L. (2006). The world is flat [updated and expanded]: A brief history of the twenty-first century. Macmillan.

Frizell, S. (2015). Toyota wants everyone to know how it made its hydrogen-powered car. *Time*. Retrieved from http://time.com/3654899/toyota-mirai-patents-hydrogen-power/

Glover, J. L., Champion, D., Daniels, K. J., & Dainty, A. J. D. (2014). An Institutional Theory perspective on sustainable practices across the dairy supply chain. *International Journal of Production Economics*, *152*, 102–111. doi:10.1016/j.ijpe.2013.12.027

Goldberg, R. A., & Yagan, J. D. (2007). McDonald's Corporation: managing a sustainable supply chain. Harvard Business School, 16.

Gore, A. (2006). *An inconvenient truth: The planetary emergency of global warming and what we can do about it*. Rodale.

Hassini, E., Surti, C., & Searcy, C. (2012). A literature review and a case study of sustainable supply chains with a focus on metrics. *International Journal of Production Economics*, *140*(1), 69–82. doi:10.1016/j.ijpe.2012.01.042

Henderson, R. M., & Nellemann, F. (2011). Sustainable Tea at Unilever. *Harvard Business School Marketing Unit Case*.

Herath, A. (2011).The Climate Change Debate: Man vs. Nature. *livescience*. Retrieved from http://www.livescience.com/16388-climate-change-debate-man-nature.html

Holt, D., & Ghobadian, A. (2009). An empirical study of green supply chain management practices amongst UK manufacturers. *Journal of Manufacturing Technology Management*, *20*(7), 933–956. doi:10.1108/17410380910984212

Hopkins, M. S., Haanaes, K., Balagopal, B., Velken, I., Kruschwitz, N., & Arthur, D. (2011). New sustainability study: The embracers seize advantage. *MIT Sloan Management Review*, *52*(3), 23.

Hsu, C. C., Choon Tan, K., Hanim Mohamad Zailani, S., & Jayaraman, V. (2013). Supply chain drivers that foster the development of green initiatives in an emerging economy. *International Journal of Operations & Production Management*, *33*(6), 656–688. doi:10.1108/IJOPM-10-2011-0401

Juran, J., & Godfrey, A. B. (1999). *Quality handbook*. Republished McGraw-Hill.

Kafa, N., Hani, Y., & El Mhamedi, A. (2013, September). Sustainability performance measurement for green supply chain management. In Management and Control of Production and Logistics (Vol. 6, No. 1, pp. 71-78).

Kaplan, R. S., & Norton, D. P. (1996). Using the balanced scorecard as a strategic management system.

Kaye Nijaki, L., & Worrel, G. (2012). Procurement for sustainable local economic development. *International Journal of Public Sector Management*, *25*(2), 133–153. doi:10.1108/09513551211223785

Kennedy, S., & Dowling, P. (2013) How Amazon.com Can Turn the Page on Supply Chain Sustainability. *supplychain247*. Retrieved from http://www.supplychain247.com/article/how_amazon_can_turn_the_page_on_supply_chain_sustainability

Kennedy, S., & Dowling, P. (2013). How Amazon.com Can Turn the Page on Supply Chain Sustainability. *GreenBiz*. Retrieved from http://www.greenbiz.com/blog/2013/08/07/amazon-turn-page-supply-chain-sustainability

Kotler, P. (2000). Marketing management: The millennium edition. *Englewood Cliffs, NJ: Prentice Hall. Maxham, JG Jr & Netemeyer, RG (2002) Modeling customer perceptions of complaint handling over time: the effects of perceived justice on satisfaction and intent. Journal of Retailing*, *78*(4), 239–252.

Laroche, M., Bergeron, J., & Barbaro-Forleo, G. (2001). Targeting consumers who are willing to pay more for environmentally friendly products. *Journal of Consumer Marketing*, *18*(6), 503–520. doi:10.1108/EUM0000000006155

Larson, A. (n. d.). Nike: Moving Down the Sustainability Track Through Chemical Substitution and Waste Reduction.

Lee, H., Duda, S., James, L., Mackwani, Z., Munoz, R., & Volk, D. (2007). Starbucks Corporation: Building a Sustainable Supply Chain.

Lee Weisstein, F., Asgari, M., & Siew, S. W. (2014). Price presentation effects on green purchase intentions. *Journal of Product and Brand Management, 23*(3), 230–239. doi:10.1108/JPBM-06-2013-0324

Leidreiter, A. (2014). 100% renewable energy as centerpiece of a climate action plan. *Energy Transition*. Retrieved from http://energytransition.de/2014/10/100-renewable-energy-as-centerpiece-of-a-climate-action-plan/

Lu, L., Bock, D., & Joseph, M. (2013). Green marketing: What the Millennials buy. *The Journal of Business Strategy, 34*(6), 3–10. doi:10.1108/JBS-05-2013-0036

Mahler, D. (2007). The sustainable supply chain. *Supply Chain Management Review, 11*(8), 59–60.

Marcacci, S. (2015). 7 Interesting global renewable energy trends from NREL (Charts Galore!). *Clean Technica*. Retrieved from http://cleantechnica.com/2015/01/26/7-interesting-global-renewable-energy-trends-from-nrel-charts-galore/

McDonough, W., Braungart, M., Anastas, P. T., & Zimmerman, J. B. (2003). Peer reviewed: Applying the principles of green engineering to cradle-to-cradle design. *Environmental Science & Technology, 37*(23), 434A–441A. doi:10.1021/es0326322 PMID:14700308

Mearian, L. (2015).The new facility is the largest such commercial solar deal in U.S. history. *COMPUTERWORLD*. Retrieved from http://www.computerworld.com/article/2882762/apples-850m-solar-plant-rockets-it-to-first-place-among-us-corporations.html

Mentzer, J. T., DeWitt, W., Keebler, J. S., Min, S., Nix, N. W., Smith, C. D., & Zacharia, Z. G. (2001). Defining supply chain management. *Journal of Business logistics, 22*(2), 1-25.

Muduli, K., & Barve, A. (2014). Impact of Soft Factors on GSCM Performance.*Proc. of the Intl. Conf. on Advances in Social Science, Management and Human Behaviour-- SMHB '13*

Nelson, D., Marsillac, E., & Rao, S. (2012). Antecedents and evolution of the green supply chain. *Journal of Operations and Supply Chain Management, 1*(1), 29–43.

Ōno, T. (1988). *Toyota production system: beyond large-scale production*. Productivity press.

Pazirandeh, A., & Jafari, H. (2013). Making sense of green logistics. *International Journal of Productivity and Performance Management, 62*(8), 889–904. doi:10.1108/IJPPM-03-2013-0059

Plambeck, E., Lee, H., & Yatsko, P. (2012). *Improving environmental performance in your Chinese supply chain*. MIT Sloan Management Review.

Polonsky, M.J. (1994). An introduction to green marketing. *Electronic Green Journal, 1*(2).

Polonsky, M. J. (1995). A stakeholder theory approach to designing environmental marketing strategy. *Journal of Business and Industrial Marketing, 10*(3), 29–46. doi:10.1108/08858629510096201

Rao, P. H. (2008). *Greening the supply chain: A guide for Asian managers*. SAGE Publications India.

Richardson, J. (2014). 100% Renewable Electricity Will Be Achieved In German State Soon. *Clean Technica*. Retrieved from http://cleantechnica.com/2014/06/23/100-renewable-electricity-will-achieved-german-state-soon/

Roberts, S. (2003). Supply chain specific? Understanding the patchy success of ethical sourcing initiatives. *Journal of Business Ethics*, *44*(2-3), 159–170. doi:10.1023/A:1023395631811

Rodrigue, J. P., Slack, B., & Comtois, C. (2001). Green logistics (the paradoxes of). The Handbook of Logistics and Supply-Chain Management, 2.

Roos, D., Womack, J. P., & Jones, D. (1990). *The Machine That Changed the World: The Story of Lean Production*. Harper Perennial, New York: Rawson.

Sanchez, X., A., Lehtiranta, L., D. Hampson, K., & Kenley, R. (. (2014). Evaluation framework for green procurement in road construction. *Smart and Sustainable Built Environment*, *3*(2), 153–169. doi:10.1108/SASBE-05-2013-0028

Sarkis, J., Zhu, Q., & Lai, K. H. (2011). An organizational theoretic review of green supply chain management literature. *International Journal of Production Economics*, *130*(1), 1–15. doi:10.1016/j.ijpe.2010.11.010

Shah, J. (2009). *Supply chain management: Text and Cases*. Pearson Education India.

Smyth, E. (2011), LED vs CFL — Which Light Bulb is More Efficient? *Clean Technica*. Retrieved from http://cleantechnica.com/2011/09/01/led-vs-cfl-which-light-bulb-is-more-efficient/

Solar Means Business 2015: Top U.S. Corporate Solar Users. (2015). *Solar Energy Industries Association*. Retrieved from http://www.seia.org/research-resources/solar-means-business-2015-top-us-corporate-solar-users

Srivastava, S. K. (2007). Green supply-chain management: A state-of-the-art literature review. *International Journal of Management Reviews*, *9*(1), 53–80. doi:10.1111/j.1468-2370.2007.00202.x

Stevenson, W. (2011). *Operations management*. McGraw-Hill/Irwin.

Toffel, M.W. (2014, March). Implementing Environmentally Sustainable Operations. *Harvard Business School Module Note 613-090.*

Topping, N. (2015).Transparency + a price on carbon can radically accelerate decarbonization. *The World Bank*. Retrieved From http://www.acs.org/content/acs/en/greenchemistry/what-is-green-chemistry/principles/sandestin-declaration.html

United States Environmental Protection Agency. (2015). Retrieved from http://www3.epa.gov/

Van Hoek, R. I. (1999). From reversed logistics to green supply chains. *Supply Chain Management: An International Journal*, *4*(3), 129–135. doi:10.1108/13598549910279576

Walmart Expands Online Sustainable Shopping. (2015). *Supplychain247*. Retrieved from http://www.supplychain247.com/article/walmart_expands_online_sustainable_shopping

Walton, S., & Huey, J. (1993). *Sam Walton, Made in America: My Story*. Bantam.

Whitelegg, J. (1993). *Transport for a sustainable future: the case for Europe*. Belhaven Pr.

Winston, A. (2014). Resilience in a hotter world. *Harvard Business Review*, *92*(4), 56–64. PMID:24830282

Wu, H. J., & Dunn, S. C. (1995). Environmentally responsible logistics systems. *International Journal of Physical Distribution & Logistics Management*, *25*(2), 20–38. doi:10.1108/09600039510083925

York, J., & Larson, A. (2006). *Nike: Moving down the sustainability track through chemical substitution and waste reduction. UVA-ENT-0098*. Charlottesville, VA: Darden Business Publishing.

APPENDIX

- **GSCM:** Green Supply Chain Management.
- **SSCM:** Sustainable Supply Chain Management.
- **GHGs:** Greenhouse Gases such as CO_2,CO etc.
- **7 Wastes Emphasized by Lean:** Overproduction, inventory, rework, over processing, waiting, unnecessary movement, transportation.
- **US Environmental Protection Agency's Definition of Scope 1, 2 and 3 Emissions**: see Table 5 for definitions.

Table 5. US Environmental Protection Agency's Definition of Scope 1,2 and 3 Emissions

Type of Emissions	Definition
Scope 1 Emissions	Direct GHG emissions from sources that are owned or controlled by the entity. Scope 1 can include emissions from fossil fuels burned on site, emissions from entity-owned or entity-leased vehicles, and other direct sources.
Scope 2 Emissions	Indirect GHG emissions resulting from the generation of electricity, heating and cooling, or steam generated off site but purchased by the entity, and the transmission and distribution (T&D) losses associated with some purchased utilities (e.g., chilled water, steam, and high temperature hot water).[1]
Scope 3 Emissions	Indirect GHG emissions from sources not owned or directly controlled by the entity but related to the entity's activities

Chapter 14
The Role of Sustainable Performance Measurement System in Global Supply Chain

Kijpokin Kasemsap
Suan Sunandha Rajabhat University, Thailand

ABSTRACT

This chapter summarizes the role of sustainable performance measurement system (PMS) in global supply chain, thus illustrating the theoretical and practical concepts of sustainable PMS and strategic supply chain management (SCM); and the significance of sustainable PMS in global supply chain. Therefore, it is essential for organizations to evaluate their sustainable PMS applications, establish a strategic plan to consistently explore their improvements, and rapidly respond to the sustainable PMS needs of customers in global supply chain. For practitioners and researchers, this chapter provides managerial guidelines for sustainable PMS applications utilized in global supply chain. Applying the sustainable PMS will significantly enhance organizational performance and reach strategic goals in global supply chain.

INTRODUCTION

In an era of globalization, supply chains are global (Rugman, Li, & Oh, 2009). Organizations are required to compete in globalized and turbulent markets (Cocca & Alberti, 2010). In order to survive in a dynamic environment, organizations need to be able to satisfy their stakeholders and excel at the same time along with all performance dimensions (Neely, Adams, & Kennerley, 2002). Issues associated with supply chain governance, standards for sustainable supply chain management (SCM), collaboration with suppliers, performance measurement, and accountability within supply chain are explored in global business (Morali & Searcy, 2013).

Supply chain sustainability has been of great interest in the last decade for academia and the industrial world because of various stakeholders' pressures to adopt a commitment to sustainability practices (Taticchi, Tonelli, & Pasqualino, 2013). Sustainable supply chain is a significant component of sustainable development in which the environmental and social criteria need to be fulfilled by supply chain

DOI: 10.4018/978-1-5225-0350-7.ch014

members to remain within supply chain, while it is expected that competitiveness would be maintained through meeting customer needs and related economic criteria (Taticchi et al., 2013).

The SCM program evaluation and monitoring is dependent on the development and application of performance measures (Bai & Sarkis, 2014). Existing performance measurement system (PMS) is rarely enough to improve organizational performance (Searcy, Karapetrovic, & McCartney, 2008). Performance measurement in supply chain application is universal (Cagnazzo, Taticchi, & Brun, 2009). Supply chain PMS is recognized due to the changing competitive nature from individual organizational competition to supply chain competing against each other (Bai, Sarkis, Wei, & Koh, 2012; Taticchi et al., 2012).

The increasing competition in the public and private sectors give rise to a growing interest in quality improvement and in designing and implementing PMS (Franceschini & Turina, 2013). There is an over-emphasis with supply chain efficiency improvements (i.e., time compression, cost cutting, quality improvement). Despite much theoretical discussion around improvement of supply chain competitiveness through delivering enhanced consumer value, there is little empirical research on how value proposition in the context of supply chain can be enhanced and aligned (Zokaei & Hines, 2007).

The strength of this chapter is on the thorough literature consolidation of sustainable PMS in global supply chain. The existing literature of sustainable PMS in global supply chain provides a contribution to practitioners and researchers by describing a comprehensive view of the functional applications of sustainable PMS in global supply chain to appeal to different segments of sustainable PMS in global supply chain in order to maximize the business impact of sustainable PMS in global supply chain.

BACKGROUND

Manufacturing industries are greatly affected by emerging technologies resulting in an influential increase in competition in local, regional, and global markets (Kasemsap, 2014). Integrity of supply chain activities is an influential factor to survive competitions in the global marketplace (Aliei, Sazvar, & Ashrafi, 2012). All organizations face the challenge of how to assess performance beyond current financial metrics (Meadows & Pike, 2010). Strategic alignment in the supply chain, assessed by the degree of matching between supply management and market requirements, is critical for the success of organizations in the global marketplace (Vachon, Halley, & Beaulieu, 2009). The competitive landscape has pushed organizations to compete not only on their own capabilities but with their entire supply chain (Hult, Ketchen, & Arrfelt, 2007). Increased customer requirements coupled with competitive pressure from globalization have forced managers to ensure that their organization's resources are well aligned not only across all functional areas but also throughout the entire supply chain (Vachon et al., 2009).

With an increasing pressure to act and report on sustainability strategies, an overwhelming number of principles, tools and reporting formats have emerged. Organizations must rely on many inter-organizational relationships to ensure productive and practical movements within their supply chains (Kotzab, Grant, Teller, & Halldorsson, 2009). Corporations adopt some of these to demonstrate their commitment to sustainable development (Beloff et al., 2004). Market opportunities are identified by understanding the competitive environment and translating this into a customer value proposition through the product and marketing strategy (Levitt, 1960; Porter, 1985). Competitive advantage is achieved if the supply chain can develop a strategy to deliver the value proposition at the lowest possible cost (Christopher, 2005).

PMS is considered as a means to gain competitive advantages and continuously react and adapt to external changes (Cocca & Alberti, 2010). PMS has expanded to the supply chain. Evaluating the supply chain and improving supply chain performance require the development of supply chain PMS (Bai et al., 2012). The expansion of PMS is due to the increased complexities of multiple processes and organizations that may be involved in product and service delivery and the role of supply chain relationships and processes in managing strategic advantages (Talluri & Sarkis, 2002; Gunasekaran, Patel, & McGaughey, 2004).

SUSTAINABLE PERFORMANCE MEASUREMENT SYSTEM IN GLOBAL SUPPLY CHAIN

This section explains the concepts of sustainable PMS and strategic SCM; and the significance of sustainable PMS in global supply chain.

Concept of Sustainable Performance Measurement System

Performance analysis tools have been available for decades (Schone & Molka, 2014). PMS has been acknowledged as a crucial element to increase the business performance (Sharma et al., 2005). PMS is a dynamic system that facilitates support of decision-making processes by gathering, elaborating and analyzing information (Neely et al., 2002). PMS can help an organization measure progress toward its goals and, in doing so, help the organization understand its current situation, the key issues it must address, and the options available (Searcy et al., 2008). Searcy et al. (2008) stated that the challenges and opportunities in organizational PMS can be grouped into three broad categories (i.e., PMS design, PMS implementation, and PMS evolution).

Performance measurement is essential for improvement (Emrouznejad & Dey, 2011). Performance measurement is an essential tool for implementing strategies aimed at achieving these goals (Kizer & Kirsh, 2012). Performance indicators play a key role in management practice (Strecker, Frank, Heise, & Kattenstroth, 2012). In order to perform a complete review of PMS, it is necessary to assess not only the effectiveness of the measures but also the effectiveness of the system as a whole (Cocca & Alberti, 2010).

Molla (2013) stated that performance has two dimensions (i.e., drivers and outcomes). The drivers, also known as leading performance indicators, refer to learning and innovation, processes, and customer value propositions. The outcomes, also known as lagging performance indicators, refer to financial results (Molla, 2013). The increasing importance of intangible assets in modern economies is driving organizations to include measures of intangible assets in managerial performance evaluations (Corona, 2009). Najmi, Rigas, and Fan (2005) described a structured framework for reviewing business performance and PMS simultaneously, in terms of strategic relevance of the measures as well as the efficiency and effectiveness of PMS itself. The framework employs a range of approaches and tools already available in literature (i.e., EFQM self-assessment process, affinity diagram, prioritization grid) in order to feature a high-level review card which identifies three review stages (i.e., ongoing, periodic, overall) of organizational performance (Cocca & Alberti, 2010).

PMS should be based not only on quantitative data but also qualitative data that helps in improving performance at all the levels of management (Soni & Kodali, 2010). The interest in performance evaluation has spawned a number of studies investigating criteria or evaluators for organizations' performance evaluation (Chen, Wang, & Lee, 2012). Performance evaluation takes industrial characteristics into account in order to form an accurate evaluation (Chang, Hung, & Huang, 2013). Sustainability performance evaluation can be eased for managers by using key performance indicators (KPIs), and use these KPIs to develop an easy and comparable performance measure (Bai & Sarkis, 2014). Sustainable development is generally interpreted to require improvement in the areas of economic, environmental, and social performance (Searcy et al., 2008). These three dimensions represent the triple bottom line (TBL) of organizational performance and are often referred to as the three pillars of sustainable development. Schwarz et al. (2002) stated that putting the concept of sustainable development into operation requires cost-effective ways to assess performance and measure the improvement.

The Supply Chain Operations Reference (SCOR) model, published in 1996 by Supply Chain Council (SCC), includes the patterns of business process reengineering, benchmarking, and process measurement (Jalalvand, Teimoury, Makui, Aryanezhad, & Jolai, 2011). The SCOR model is used to model the linkage of the strategic objectives and operational metrics in a hierarchical way (Kocaoglu, Gulsun, & Tanyas, 2013). The SCOR model has been used as a well-accepted SCM tool (Shoushtari, 2013). The use of SCOR model has been recommended as an important framework for supply chain PMS (Shepherd & Gunter, 2006; Gunasekaran & Kobu, 2007) and practical perspectives (Chae, 2009).

Moffett, Gilliespie, and McAdam (2008) stated that benchmarking is an emergent area of business improvement as a strategic initiative in many organizations. Benchmarking is a famous tool which is used as a tool for continuous improvement in organizations' performance and competitiveness in global business (Soni & Kodali, 2010). Benchmarking also provides a mechanism to make organizations more competitive, implement industry best practice, and develop measures of productivity (Soni & Kodali, 2010). The benchmarking process assists to establish effective goals and objectives improve business resources and identify the internal and external influences on modern organizations (Manning, Baines, & Chadd, 2008).

Concept of Strategic Supply Chain Management

Since its introduction as a concept in the 1980s, SCM has sustained significant modification and expansion (Stock, Boyer, & Harmon, 2010). Supply chains and SCM are the essential perspectives of business practice (Mentzer & Gundlach, 2010). SCM has moved up the corporate agenda to be cited as a reason for corporate success or failure (Godsell, Birtwistle, & van Hoek, 2010). Operations strategy can be defined as the set of infrastructural and structural decisions made by an organization that emerge from the corporate strategy and customers' requirements (Ritzman, Krejewski, & Klassen, 2004). SCM is defined as a shared objective of the whole supply chain (Zokaei & Hines, 2007). SCM exhibits the importance of consumer needs and attempts to capture the subtleties of consumer value as a source of differentiation and supply chain competitiveness (Christopher, 2005; Womack & Jones, 2005). It will be a specific advantage if supply chain succeeds in capturing the genuine consumer attributes, analyzing the value proposition at each step, identifying the misalignments with the consumer value, and transforming the products and processes to deliver those requirements (Zokaei & Hines, 2007).

To acquire a competitive advantage in the expanding market, manufacturing companies should be able to manage their supply chain as much effective as possible (Oztemel & Tekez, 2009). Business enterprises and their supply chains focus on economic activities (Jakhar, 2014). Building a global supply chain can be very challenging (Rugman et al., 2009). The concept of SCM has emerged in many different variations in a range of management streams such as relationship marketing (Gummesson, 1996), co-makership (Merli, 1991), value chain competitiveness (Porter, 1985), industrial networks (Hakansson & Snehota, 1995) and financial value chain analysis (Shank & Govindarajan, 1993). SCM must adopt different and more innovative strategies that support a better response to customer needs in an uncertain environment (Carvalho, Azevedo, & Cruz-Machado, 2013). Creating efficient supply chain is a major area of focus for enterprise architects (Attaran & Attaran, 2007). Globalization of trade, sophistication of technology and markets (La Londe & Masters, 1994), increased competition, and the dominance of the Japanese production philosophies (Womack & Jones, 1996) have potentially contributed to the evolution of SCM and its core concepts.

The strategic supply chain continues to be adopted by organizations as the medium for creating and sustaining a competitive advantage (Ireland & Webb, 2007). The ability to collaborate with supply chain partners becomes an essential core competence that is required from organizations when they are going to take up future challenges in global business (Westphal, Thoben, & Seifert, 2010). Competitive success for a strategic supply chain is contingent on management's ability to recognize changes in the competitive environment and then direct and coordinate action within and across organizations to effectively utilize resources and meet the demands of supply chain environment (Stonebraker & Afifi, 2004). Strategic supply chains seem to succeed or fail on the degree of organizational resources sharing among partners (i.e., information, knowledge, skills) and the supply chain partners' ability to practically use these organizational resources in the changing environments (Fawcett, Magnan, & McCarter, 2008).

Supply chain improvement literature remains focused on efficiency improvements such as time-based competition (Womack & Jones, 1996; Christopher, 2005), cash-to-cash time (Bowersox & Closs, 1996), quality-based competition (Womak, Jones, & Roos, 1990) and cost-based competition (Shank & Govindarajan, 1993). Many researchers emphasize the importance of supply chain effectiveness and enhanced consumer satisfaction in the context of supply chain (Zokaei & Simons, 2006). SCM directs all organizations to focus on developing innovative solutions to create individualized sources of consumer value (Mentzer, Dewitt, Keebler, Soonhong, Nancy, Smith, & Zacharia, 2001). Understanding consumers' attributes and jointly striving on augmentation of consumer satisfaction are imperative to successful SCM (Zokaei & Hines, 2007). Supply chain managers should consider the use of online auctions in order to achieve cost management objectives (Nadler & Kros, 2010).

The notion of aligning competitive priorities in supply chain results from observations of organizations or simulations (Vonderembse, Uppal, Huang, & Dismukes, 2006). Resource capitalization requires business process integration and joint teamwork to realize the value of supply chain resources (Shi & Liao, 2013). Business alignment should ensure congruence between product, marketing and supply chain strategy to gain sustainable competitive advantage. Strategic alignment in supply chain is measured by the difference between customer's requirements and the emphasis that the organization puts on these same requirements in dealing with its suppliers (Vachon et al., 2009). Supply chain alignment is achieved when there is congruence among supply chain strategy, infrastructure (the physical supply chain and its assets) and operating model (the way the physical supply chain will be managed) (Godsell & Harrison, 2006; Sun, Hsu, & Hwang, 2009).

SCM activities are widely implemented by Japanese automakers, particularly Toyota, who coordinate their supplier network and ensure that they contribute to fulfilling the needs of their end customers (Dyer & Hatch, 2004). Supply chain partners must possess a culture of viewing the supply chain as a whole and of recognizing the need for cooperative efforts (Defee & Fugate, 2010). The synergies produced by supply chain partners evolving to create new cross-organizational capabilities makes supply chain more competitive and less susceptible to rival supply chains that are not as proficient in continuously creating innovative capabilities (Defee & Fugate, 2010). Supply network encompasses an integrated network of key supply units, operating throughout supply chain (Srai & Gregory, 2008).

The benefits of economic integration are to offer economies of scale and cost savings. If economic integration is global then a global supply chain should be able to realize these benefits (Rugman et al., 2009). Supply chain structure, referred to Lean supply chain (Bruce, Daly, & Towers, 2004), builds on the principles of Lean management for process efficiency by systematically pursuing waste reduction and the elimination of non-value added operations (Hines, Holwe, & Rich, 2004). Inventories across the supply chain should be effectively managed through greater cooperation and better coordination (Saberi, Nookabadi, & Hejazi, 2012). Organizations need to build long-term relationships with their suppliers and customers by sharing information, risks, rewards, and goals (Mentzer et al., 2001). SCM requires the management of multiple relationships across supply chain (Lambert & Cooper, 2000).

Stakeholder pressure and sustainable SCM contribute to an organization's sustainability performance (Wolf, 2014). Managing buyer–supplier relationships has always been crucial to organizational success (Obal & Lancioni, 2013). Strategic relationships with critical suppliers must be understood in order to maximize the value creation in supply chain (Chen, Paulraj, & Lado, 2004). Successful management of buyer-supplier relationships contributes to organizational performance (Tan, Kannan, Handfield, & Ghosh, 1999). Kotabe, Martin, and Domoto (2003) exhibited the relation-specific assets developed between a supplier and a buyer as important factors for supply chains to generate competitive advantages for modern organizations.

The ongoing emergence of new technologies, such as Software-as-a-Service (SaaS) and social media, has made effective communications even more crucial as organizations decide which technologies are appropriate for them and their supply chain partners (Ganesan, 1994; Seppanen, Blomqvist, & Sundqvist, 2007). Communication between supply chain partners leads to the increased trust and commitment (Anderson & Weitz, 1992; Morgan & Hunt, 1994) that trust influences commitment (Ganesan, 1994; Morgan & Hunt, 1994) and that trust and commitment lead to the increased satisfaction and the buyer-supplier relationship success (Benton & Maloni, 2005; Palmatier, Dant, & Grewal, 2007). The buyer wants relationship-specific investments while the supplier only wants communication and information sharing to ensure uncertainty is minimized (Ambrose, Marshall, & Lynch, 2010).

Significance of Sustainable Performance Measurement System in Global Supply Chain

Over the past decade, the literature has clearly established the need for organizations to consider developing PMS for their supply chains (Gunasekaran et al., 2004; Shepherd & Gunter, 2006). Sustainability is becoming a mainstream issue for many organizations (Parisi, 2013). The importance of sustainability criteria to measure supply chain performance have increased in the competitive environment (Vasileiou & Morris, 2006). Effective PMS and metrics development are essential for making the business case for supply chain sustainability (Keating, Quazi, Kriz, & Coltman, 2008). The integration of tangible and intangible performance measures add to the complexity of managing supply chain performance (Giannakis, 2007). The ability of keeping PMS continuously updated is a challenge for modern organizations (Hudson, Smart, & Bourne, 2001). Developing a performance measurement model is a critical task for a supply chain and its members in order to examine their current status and identify improvement opportunities for steering their future direction (Lin & Li, 2010).

The evaluation and the set of PMS dimensions become more complex when considering supply chain sustainability is added (Ashby, Leat, & Hudson-Smith, 2012). Sustainable SCM performance measurement can be used for multiple purposes such as supplier selection, performance monitoring and development (Gimenez & Tachizawa, 2012; Hervani, Helms, & Sarkis, 2005). Investigation of PMS is needed for advancing SCM sustainability. A critical aspect of sustainable supply chain PMS is the identification of KPIs (Chae, 2009; Bai & Sarkis, 2012; Bai et al., 2012). Hundreds of measures for traditional business and operational supply chain evaluations may exist (Gunasekaran & Kobu, 2007). This number of performance measures increases greatly when additional environmental and social sustainability dimensions are included in supply chain evaluations. The need to identify KPI becomes more critical when such a large set of sustainable supply chain performance measures are used (Dotoli, Fanti, Meloni, & Zhou, 2006; Zhu, Dou, & Sarkis, 2010).

Sustainable supply chain PMS has its foundations in the emergent work on supply chain PMS (Bai & Sarkis, 2012). The competitive nature of supply chain has caused a shift from individual organizations competing against each other to supply chains competing against each other (Antai, 2011; Antai & Olson, 2013). Evaluating supply chain and improving supply chain performance requires the development of supply chain PMS and performance measures. Integrating sustainability into supply chain performance measurement is part of the adjustment, beginning with the integration of environmental, social and business performance measurement into supply chain (Bai & Sarkis, 2012; Olugu, Wong, & Shaharoun, 2011; Zhu et al., 2010). Traditional supply chain performance measures that focus on operational and business factors can add up to dozens or hundreds of measures, while adding sustainability dimensions increases the size and complexity of the performance measure set.

The balanced scorecard (Kaplan & Norton, 1996) is one of the more popular performance measurement tools and focuses on key business categories. Organizations place great emphasis on both supply chain and on balanced scorecard to develop effective measures to evaluate organizational performance (Chang, Hung, Wong, & Lee, 2013). The development of social and environmental balanced scorecard is a way to help managers identify the key performance factors to an organization's strategic sustainability objectives (Epstein & Wisner, 2001; Figge, Hahn, Schaltegger, & Wagner, 2002; Moller & Schaltegger, 2005). In addition to linking KPI to strategy and strategic planning, it can be integrated with managerial decision making, which in complex supply chain situations such as sustainable SCM is dependent on KPI identification (Cabral, Grilo, & Cruz-Machado, 2012; Grimm, Hofstetter, & Sarkis, 2013).

A major element within sustainable SCM is concerned with inter-organizational responsibility for various aspects of environmental and social performance (Dotoli et al., 2006). Within the environmental arena, several techniques exist to help managers map the environmental impacts along supply chains, such as the life-cycle assessment, product stewardship and design for environment (DFE) principles (Sarkis, 2006). Performance measurement, metrics and KPI are critical to these dimensions of life-cycle assessment (Bai & Sarkis, 2014). Developing tools for PMS to help and guide benchmarking for sustainable supply chains functionally improves the management and adoption of sustainable supply chain practices (Hervani et al., 2005).

Benchmarking is a valuable tool that provides an opportunity to learn from other organizations (Meybodi, 2009). Singh, Narain, and Yadav (2006) utilized benchmarking and PMS to investigate SCM practices at a number of Indian manufacturing organizations. Singh et al. (2006) found that Indian organizations use benchmarking as a continuous improvement tool. In a global marketplace, knowing how the best organizations conduct their business is an important element of successful competition. Internal benchmarking has been used as a tool for performance improvement in global supply chain (Soni & Kodali, 2010). Gunasekaran, Patel, and Tirtiroglu (2001) indicated that benchmarking is required for supply chain collaboration because supply chain members are concerned with both performance drivers and targets.

FUTURE RESEARCH DIRECTIONS

The strength of this chapter is on the thorough literature consolidation of sustainable PMS in global supply chain. The existing literature of sustainable PMS in global supply chain provides a contribution to practitioners and researchers by describing a comprehensive view of the functional applications of sustainable PMS in global supply chain to appeal to different segments of sustainable PMS in global supply chain in order to maximize the business impact of sustainable PMS in global supply chain. The classification of the current literature in the domains of sustainable PMS in global supply chain will provide the potential opportunities for future research. Future research direction should broaden the perspectives in the application of sustainable PMS in global supply chain in the knowledge-based organizations.

Practitioners and researchers should acknowledge the applicability of a more multidisciplinary approach toward research activities in promoting sustainable PMS in global supply chain with knowledge management-related variables (i.e., knowledge-sharing behavior, knowledge creation, organizational learning, learning orientation, and motivation to learn). It will be useful to bring additional disciplines together (i.e., strategic management, information technology, organizational development, and human resources) to support a more holistic examination of sustainable PMS in global supply chain in order to combine or transfer existing theories and approaches to inquiry in this area. An examination of linkages between sustainable PMS in global supply chain and cloud computing would seem to be viable for future research efforts.

CONCLUSION

This chapter summarized the role of sustainable PMS in global supply chain, thus illustrating the theoretical and practical concepts of sustainable PMS and strategic SCM; and the significance of sustainable PMS in global supply chain. Therefore, it is essential for organizations to evaluate their sustainable PMS applications, establish a strategic plan to consistently explore their improvements, and rapidly respond to the sustainable PMS needs of customers in global supply chain. For practitioners and researchers, this chapter provides managerial guidelines for sustainable PMS applications utilized in global supply chain. Applying the sustainable PMS will greatly improve organizational performance and gain sustainable competitive advantage in global supply chain.

REFERENCES

Aliei, M., Sazvar, A., & Ashrafi, B. (2012). Assessment of information technology effects on management of supply chain based on fuzzy logic in Iran tail industries. *International Journal of Advanced Manufacturing Technology*, *63*(1-4), 215–223. doi:10.1007/s00170-012-3900-2

Ambrose, E., Marshall, D., & Lynch, D. (2010). Buyer supplier perspectives on supply chain relationships. *International Journal of Operations & Production Management*, *30*(12), 1269–1290. doi:10.1108/01443571011094262

Anderson, E., & Weitz, B. A. (1992). The use of pledges to build and sustain commitment in distribution channels. *JMR, Journal of Marketing Research*, *29*(1), 18–34. doi:10.2307/3172490

Antai, I. (2011). Supply chain vs. supply chain competition: A niche-based approach. *Management Research Review*, *34*(10), 1107–1124. doi:10.1108/01409171111171500

Antai, I., & Olson, H. (2013). Interaction: A new focus for supply chain vs. supply chain competition. *International Journal of Physical Distribution & Logistics Management*, *43*(7), 511–528. doi:10.1108/IJPDLM-06-2012-0195

Ashby, A., Leat, M., & Hudson-Smith, M. (2012). Making connections: A review of supply chain management and sustainability literature. *Supply Chain Management: An International Journal*, *17*(5), 497–516. doi:10.1108/13598541211258573

Attaran, M., & Attaran, S. (2007). Collaborative supply chain management: The most promising practice for building efficient and sustainable supply chains. *Business Process Management Journal*, *13*(3), 390–404. doi:10.1108/14637150710752308

Bai, C., & Sarkis, J. (2012). Supply-chain performance-measurement system management using neighbourhood rough sets. *International Journal of Production Research*, *50*(9), 2484–2500. doi:10.1080/00207543.2011.581010

Bai, C., & Sarkis, J. (2014). Determining and applying sustainable supplier key performance indicators. *Supply Chain Management: An International Journal*, *19*(3), 275–291. doi:10.1108/SCM-12-2013-0441

Bai, C., Sarkis, J., Wei, X., & Koh, L. (2012). Evaluating ecological sustainable performance measures for supply chain management. *Supply Chain Management: An International Journal*, *17*(1), 78–92. doi:10.1108/13598541211212221

Beloff, B., Tanzil, D., & Lines, M. (2004). Sustainable development performance assessment. *Environment and Progress*, *23*(4), 271–276. doi:10.1002/ep.10045

Benton, W. C., & Maloni, M. (2005). The influence of power driven buyer/seller relationships on supply chain satisfaction. *Journal of Operations Management*, *23*(1), 1–22. doi:10.1016/j.jom.2004.09.002

Bowersox, D. J., & Closs, D. C. (1996). *Logistical management: The integrated supply chain process*. New York, NY: McGraw-Hill.

Bruce, M., Daly, L., & Towers, N. (2004). Lean or agile: A solution for supply chain management in the textiles and clothing industry. *International Journal of Operations & Production Management*, *24*(1-2), 151–170. doi:10.1108/01443570410514867

Cabral, I., Grilo, A., & Cruz-Machado, V. (2012). A decision-making model for lean, agile, resilient and green supply chain management. *International Journal of Production Research*, *50*(17), 4830–4845. doi:10.1080/00207543.2012.657970

Cagnazzo, L., Taticchi, P., & Brun, A. (2009). The role of performance measurement systems to support quality improvement initiatives at supply chain level. *International Journal of Productivity and Performance Management*, *59*(2), 163–185.

Carvalho, H., Azevedo, S. G., & Cruz-Machado, V. (2013). Agile and resilient approaches to supply chain management: Influence on performance and competitiveness. *Logistics Research*, *4*(1-2), 49–62. doi:10.1007/s12159-012-0064-2

Chae, B. (2009). Developing key performance indicators for supply chain: An industry perspective. *Supply Chain Management: An International Journal, 14*(6), 422–428. doi:10.1108/13598540910995192

Chang, C. C., Hung, S. W., & Huang, S. Y. (2013). Evaluating the operational performance of knowledge-based industries: The perspective of intellectual capital. *Quality & Quantity, 47*(3), 1367–1383. doi:10.1007/s11135-011-9595-x

Chang, H. H., Hung, C. J., Wong, K. H., & Lee, C. H. (2013). Using the balanced scorecard on supply chain integration performance—a case study of service businesses. *Service Business, 7*(4), 539–561. doi:10.1007/s11628-012-0175-5

Chen, I. J., Paulraj, A., & Lado, A. A. (2004). Strategic purchasing, supply management, and firm performance. *Journal of Operations Management, 22*(5), 505–523. doi:10.1016/j.jom.2004.06.002

Chen, S. H., Wang, P. W., & Lee, H. T. (2012). Performance measurement considering the interdependence of evaluators and criteria. *Quality & Quantity, 46*(1), 103–115. doi:10.1007/s11135-010-9330-z

Christopher, M. (2005). *Logistics and supply chain management: Creating value-adding networks.* London, UK: Pitman.

Cocca, P., & Alberti, M. (2010). A framework to assess performance measurement systems in SMEs. *International Journal of Productivity and Performance Management, 59*(2), 186–200. doi:10.1108/17410401011014258

Corona, C. (2009). Dynamic performance measurement with intangible assets. *Review of Accounting Studies, 14*(2-3), 314–348. doi:10.1007/s11142-009-9095-6

Defee, C. C., & Fugate, B. S. (2010). Changing perspective of capabilities in the dynamic supply chain era. *International Journal of Logistics Management, 21*(2), 180–206. doi:10.1108/09574091011071915

Dotoli, M., Fanti, M. P., Meloni, C., & Zhou, M. (2006). Design and optimization of integrated e-supply chain for agile and environmentally conscious manufacturing. *IEEE Transactions on Systems, Man, and Cybernetics, 36*(1), 62–75. doi:10.1109/TSMCA.2005.859189

Dyer, J. H., & Hatch, N. W. (2004). Using supplier network to learn faster. *MIT Sloan Management Review, 45*(3), 57–63.

Emrouznejad, A., & Dey, P. K. (2011). Performance measurement in the health sector: Uses of frontier efficiency methodologies and multi-criteria decision making. *Journal of Medical Systems, 35*(5), 977–979. doi:10.1007/s10916-010-9622-9 PMID:21072574

Epstein, M. J., & Wisner, P. S. (2001). Using a balanced scorecard to implement sustainability. *Environmental Quality Management, 11*(2), 1–10. doi:10.1002/tqem.1300

Fawcett, S. E., Magnan, G. M., & McCarter, M. W. (2008). Benefits, barriers, and bridges to effective supply chain management. *Supply Chain Management: An International Journal, 13*(1), 35–48. doi:10.1108/13598540810850300

Figge, F., Hahn, T., Schaltegger, S., & Wagner, M. (2002). The sustainability balanced scorecard: Linking sustainability management to business strategy. *Business Strategy and the Environment*, *11*(5), 269–284. doi:10.1002/bse.339

Franceschini, F., & Turina, E. (2013). Quality improvement and redesign of performance measurement systems: An application to the academic field. *Quality & Quantity*, *47*(1), 465–483. doi:10.1007/s11135-011-9530-1

Ganesan, S. (1994). Determinants of long-term orientation in buyer-seller relationships. *Journal of Marketing*, *58*(2), 1–19. doi:10.2307/1252265

Giannakis, M. (2007). Performance measurement of supplier relationships. *Supply Chain Management: An International Journal*, *12*(6), 400–411. doi:10.1108/13598540710826335

Gimenez, C., & Tachizawa, E. M. (2012). Extending sustainability to suppliers: A systematic literature review. *Supply Chain Management: An International Journal*, *17*(5), 531–543. doi:10.1108/13598541211258591

Godsell, J., Birtwistle, A., & van Hoek, R. (2010). Building the supply chain to enable business alignment: Lessons from British American Tobacco (BAT). *Supply Chain Management: An International Journal*, *15*(1), 10–15. doi:10.1108/13598541011018085

Godsell, J., Harrison, A., Emberson, C., & Storey, J. (2006). Customer responsive supply chain strategy: An unnatural act? *International Journal of Logistics: Research and Applications*, *9*(1), 47–56. doi:10.1080/13675560500534664

Grimm, J. H., Hofstetter, J. S., & Sarkis, J. (2013). Critical factors for sub-supplier management: A sustainable food supply chains perspective. *International Journal of Production Economics*, *152*, 159–173. doi:10.1016/j.ijpe.2013.12.011

Gummesson, E. (1996). Relationship marketing and the imaginary organisation: A synthesis. *European Journal of Marketing*, *30*(2), 31–44. doi:10.1108/03090569610106635

Gunasekaran, A., & Kobu, B. (2007). Performance measures and metrics in logistics and supply chain management: A review of recent literature (1995-2004) for research and applications. *International Journal of Production Research*, *45*(12), 2819–2840. doi:10.1080/00207540600806513

Gunasekaran, A., Patel, C., & McGaughey, R. E. (2004). A framework for supply chain performance measurement. *International Journal of Production Economics*, *87*(3), 333–347. doi:10.1016/j.ijpe.2003.08.003

Gunasekaran, A., Patel, C., & Tirtiroglu, E. (2001). Performance measurement and metrics in a supply chain environment. *International Journal of Operations & Production Management*, *21*(1), 71–87. doi:10.1108/01443570110358468

Hakansson, H., & Snehota, I. (1995). *Developing business relationships*. London, UK: Routledge.

Hervani, A. A., Helms, M. M., & Sarkis, J. (2005). Performance measurement for green supply chain management. *Benchmarking: An International Journal*, *12*(4), 330–353. doi:10.1108/14635770510609015

Hines, P., Holwe, M., & Rich, N. (2004). Learning to evolve: A review of contemporary lean thinking. *International Journal of Operations & Production Management*, *24*(10), 994–1011. doi:10.1108/01443570410558049

Hudson, M., Smart, A., & Bourne, M. (2001). Theory and practice in SME performance measurement system. *International Journal of Operations & Production Management*, *21*(8), 1096–1115. doi:10.1108/EUM0000000005587

Hult, G. T. M., Ketchen, D. J., & Arrfelt, M. (2007). Strategic supply chain management: Improving performance through a culture of competitiveness and knowledge development. *Strategic Management Journal*, *28*(10), 1035–1052. doi:10.1002/smj.627

Ireland, R. D., & Webb, J. W. (2007). A multi-theoretic perspective on trust and power in strategic supply chains. *Journal of Operations Management*, *25*(2), 482–497. doi:10.1016/j.jom.2006.05.004

Jakhar, S. K. (2014). Designing the green supply chain performance optimisation model. *Global Journal of Flexible Systems Management*, *15*(3), 235–259. doi:10.1007/s40171-014-0069-6

Jalalvand, F., Teimoury, E., Makui, A., Aryanezhad, M. B., & Jolai, F. (2011). A method to compare supply chains of an industry. *Supply Chain Management: An International Journal*, *16*(2), 82–97. doi:10.1108/13598541111115347

Kaplan, R. S., & Norton, D. P. (1996). *The balanced scorecard: Translating strategy into action*. Boston, MA: Harvard Business School Press.

Kasemsap, K. (2014). The role of lean production on organizational performance. In V. Modrak & P. Semanco (Eds.), *Handbook of research on design and management of lean production systems* (pp. 358–388). Hershey, PA: IGI Global. doi:10.4018/978-1-4666-5039-8.ch017

Keating, B., Quazi, A., Kriz, A., & Coltman, T. (2008). In pursuit of a sustainable supply chain: Insights from Westpac Banking Corporation. *Supply Chain Management: An International Journal*, *13*(3), 175–179. doi:10.1108/13598540810871217

Kizer, K. W., & Kirsh, S. R. (2012). The double edged sword of performance measurement. *Journal of General Internal Medicine*, *27*(4), 395–397. doi:10.1007/s11606-011-1981-5 PMID:22271270

Kocaoglu, B., Gulsun, B., & Tanyas, M. (2013). A SCOR based approach for measuring a benchmarkable supply chain performance. *Journal of Intelligent Manufacturing*, *24*(1), 113–132. doi:10.1007/s10845-011-0547-z

Kotabe, M., Martin, X., & Domoto, H. (2003). Gaining from vertical partnerships: Knowledge transfer, relationship duration and supplier performance improvements in the U.S. and Japanese automotive industries. *Strategic Management Journal*, *24*(4), 293–316. doi:10.1002/smj.297

Kotzab, H., Grant, D. B., Teller, C., & Halldorsson, A. (2009). Supply chain management and hypercompetition. *Logistics Research*, *1*(1), 5–13. doi:10.1007/s12159-008-0002-5

La Londe, B. J., & Masters, J. M. (1994). Emerging logistics strategies: Blueprints for the next century. *International Journal of Physical Distribution & Logistics Management*, *24*(7), 35–47. doi:10.1108/09600039410070975

Lambert, D. M., & Cooper, M. C. (2000). Issues in supply chain management. *Industrial Marketing Management, 29*(1), 65–83. doi:10.1016/S0019-8501(99)00113-3

Levitt, T. (1960). Marketing myopia. *Harvard Business Review, 38*(4), 45–56. PMID:15252891

Lin, L. C., & Li, T. S. (2010). An integrated framework for supply chain performance measurement using six-sigma metrics. *Software Quality Journal, 18*(3), 387–406. doi:10.1007/s11219-010-9099-2

Manning, L., Baines, R., & Chadd, S. (2008). Benchmarking the poultry meat supply chain. *Benchmarking: An International Journal, 15*(2), 148–165. doi:10.1108/14635770810864866

Meadows, M., & Pike, M. (2010). Performance management for social enterprises. *Systemic Practice and Action Research, 23*(2), 127–141. doi:10.1007/s11213-009-9149-5

Mentzer, J. T., DeWitt, W., Keebler, J. S., Min, S., Nix, N. W., Smith, C. D., & Zacharia, Z. G.. (2001). Defining supply chain management. *Journal of Business Logistics, 22*(2), 1–25. doi:10.1002/j.2158-1592.2001.tb00001.x

Mentzer, J. T., & Gundlach, G. (2010). Exploring the relationship between marketing and supply chain management: Introduction to the special issue. *Journal of the Academy of Marketing Science, 38*(1), 1–4. doi:10.1007/s11747-009-0150-4

Merli, G. (1991). *Co-makership: The new supply strategy for manufacturers*. Cambridge, MA: Productivity Press.

Meybodi, M. Z. (2009). Benchmarking performance measures in traditional and just-in-time companies. *Benchmarking: An International Journal, 16*(1), 88–102. doi:10.1108/14635770910936531

Moffett, S., Anderson-Gilliespie, K., & McAdam, R. (2008). Benchmarking and performance measurement: A statistical analysis. *Benchmarking: An International Journal, 15*(4), 368–381. doi:10.1108/14635770810887203

Molla, A. (2013). Identifying IT sustainability performance drivers: Instrument development and validation. *Information Systems Frontiers, 15*(5), 705–723. doi:10.1007/s10796-013-9415-z

Möller, A., & Schaltegger, S. (2005). The sustainability balanced scorecard as a framework for eco-efficiency analysis. *Journal of Industrial Ecology, 9*(4), 73–83. doi:10.1162/108819805775247927

Morali, O., & Searcy, C. (2013). A review of sustainable supply chain management practices in Canada. *Journal of Business Ethics, 117*(3), 635–658. doi:10.1007/s10551-012-1539-4

Morgan, R. M., & Hunt, S. D. (1994). The commitment-trust theory of relationship marketing. *Journal of Marketing, 58*(3), 20–38. doi:10.2307/1252308

Nadler, S. S., & Kros, J. F. (2010). An assessment of supply chain managers' trust in online auctions. *Industrial Management & Data Systems, 110*(6), 805–822. doi:10.1108/02635571011055063

Najmi, M., Rigas, J., & Fan, I. S. (2005). A framework to review performance measurement systems. *Business Process Management Journal, 11*(2), 109–122. doi:10.1108/14637150510591129

Neely, A., Adams, C., & Kennerley, M. (2002). *The performance prism: The scorecard for measuring and managing stakeholder relationship*. London, UK: Financial Times-Prentice Hall.

Obal, M., & Lancioni, R. A. (2013). Maximizing buyer–supplier relationships in the digital era: Concept and research agenda. *Industrial Marketing Management, 42*(6), 851–854. doi:10.1016/j.indmarman.2013.06.002

Olugu, E. U., Wong, K. Y., & Shaharoun, A. M. (2011). Development of key performance measures for the automobile green supply chain. *Resources, Conservation and Recycling, 55*(6), 567–579. doi:10.1016/j.resconrec.2010.06.003

Oztemel, E., & Tekez, E. K. (2009). Interactions of agents in performance based supply chain management. *Journal of Intelligent Manufacturing, 20*(2), 159–167. doi:10.1007/s10845-008-0229-7

Palmatier, R. W., Dant, R. P., & Grewal, D. (2007). A comparative longitudinal analysis of theoretical perspectives of interorganizational relationship performance. *Journal of Marketing, 71*(4), 172–194. doi:10.1509/jmkg.71.4.172

Parisi, C. (2013). The impact of organisational alignment on the effectiveness of firms' sustainability strategic performance measurement systems: An empirical analysis. *Journal of Management & Governance, 17*(1), 71–97. doi:10.1007/s10997-012-9219-4

Porter, M. E. (1985). *Competitive advantage: Creating and sustaining superior performance*. New York, NY: Free Press.

Ritzman, L. P., Krejewski, L. J., & Klassen, R. D. (2004). *Foundations of operations management*. Upper Saddle River, NJ: Prentice-Hall.

Rugman, A. M., Li, J., & Oh, C. H. (2009). Are supply chains global or regional? *International Marketing Review, 26*(4-5), 384–395.

Saberi, S., Nookabadi, A. S., & Hejazi, S. R. (2012). Applying agent-based system and negotiation mechanism in improvement of inventory management and customer order fulfilment in multi echelon supply chain. *Arabian Journal for Science and Engineering, 37*(3), 851–861. doi:10.1007/s13369-012-0197-2

Sarkis, J. (2006). *Greening the supply chain*. Berlin, Germany: Springer-Verlag. doi:10.1007/1-84628-299-3

Schone, R., & Molka, D. (2014). Integrating performance analysis and energy efficiency optimizations in a unified environment. *Computer Science - Research for Development, 29*(3-4), 231–239. doi:10.1007/s00450-013-0243-7

Schwarz, J., Beloff, B., & Beaver, E. (2002). Use sustainability metrics to guide decision-making. *Chemical Engineering Progress, 98*(7), 58–63.

Searcy, C., Karapetrovic, S., & McCartney, D. (2008). Application of a systems approach to sustainable development performance measurement. *International Journal of Productivity and Performance Management, 57*(2), 182–197. doi:10.1108/17410400810847429

Seppanen, R., Blomqvist, K., & Sundqvist, S. (2007). Measuring inter-organizational trust: A critical review of the empirical research in 1990–2003. *Industrial Marketing Management*, *36*(2), 249–265. doi:10.1016/j.indmarman.2005.09.003

Shank, J. K., & Govindarajan, V. (1993). *Strategic cost management*. New York, NY: Free Press.

Sharma, M. K., Bhagwat, R., & Dangayach, G. S. (2005). Practice of performance measurement: Experience from Indian SMEs. *International Journal of Globalisation and Small Business*, *1*(2), 183–213. doi:10.1504/IJGSB.2005.008014

Shepherd, C., & Gunter, H. (2006). Measuring supply chain performance: Current research and future directions. *International Journal of Productivity and Performance Management*, *55*(3-4), 242–258. doi:10.1108/17410400610653219

Shi, X., & Liao, Z. (2013). The mediating effects of interfirm business process integration and joint teamwork on firm performance in supply chains. *Asia Pacific Journal of Management*, *30*(4), 1243–1264. doi:10.1007/s10490-012-9308-6

Shoushtari, K. D. (2013). Redesigning a large supply chain management system to reduce the government administration: A socio-functional systems approach. *Systemic Practice and Action Research*, *26*(2), 195–216. doi:10.1007/s11213-012-9244-x

Singh, A., Narain, R., & Yadav, R. C. (2006). Benchmarking and performance measurement of supply chain management practices: A survey of Indian organizations. *International Journal of Services and Operations Management*, *2*(4), 313–334. doi:10.1504/IJSOM.2006.010250

Soni, G., & Kodali, R. (2010). Internal benchmarking for assessment of supply chain performance. *Benchmarking: An International Journal*, *17*(1), 44–76. doi:10.1108/14635771011022316

Srai, J. S., & Gregory, M. (2008). A supply network configuration perspective on international supply chain development. *International Journal of Operations & Production Management*, *28*(5), 386–411. doi:10.1108/01443570810867178

Stock, J. R., Boyer, S. L., & Harmon, T. (2010). Research opportunities in supply chain management. *Journal of the Academy of Marketing Science*, *38*(1), 32–41. doi:10.1007/s11747-009-0136-2

Stonebraker, P. W., & Afifi, R. (2004). Toward a contingency theory of supply chains. *Management Decision*, *42*(9), 1131–1144. doi:10.1108/00251740410565163

Strecker, S., Frank, U., Heise, D., & Kattenstroth, H. (2012). MetricM: A modeling method in support of the reflective design and use of performance measurement systems. *Information Systems and e-Business Management, 10*(2), 241-276.

Sun, S., Hsu, M., & Hwang, W.-J. (2009). The impact of alignment between supply chain strategy and environmental uncertainty on SCM performance. *Supply Chain Management: An International Journal*, *14*(3), 201–244. doi:10.1108/13598540910954548

Talluri, S., & Sarkis, J. (2002). A model for performance monitoring of suppliers. *International Journal of Production Research*, *40*(16), 4257–4269. doi:10.1080/00207540210152894

Tan, K. C., Kannan, V. R., Handfield, R. B., & Ghosh, S. (1999). Supply chain management: An empirical study of its impact on performance. *International Journal of Operations & Production Management*, *19*(10), 1034–1052. doi:10.1108/01443579910287064

Taticchi, P., Balachandran, K., & Tonelli, F. (2012). Performance measurement and management systems: State of the art, guidelines for design and challenges. *Measuring Business Excellence*, *16*(2), 41–54. doi:10.1108/13683041211230311

Taticchi, P., Tonelli, F., & Cagnazzo, L. (2009). A decomposition and hierarchical approach for business performance measurement and management. *Measuring Business Excellence*, *13*(4), 47–57. doi:10.1108/13683040911006783

Taticchi, P., Tonelli, F., & Pasqualino, R. (2013). Performance measurement of sustainable supply chains: A literature review and a research agenda. *International Journal of Productivity and Performance Management*, *62*(8), 782–804. doi:10.1108/IJPPM-03-2013-0037

Vachon, S., Halley, A., & Beaulieu, M. (2009). Aligning competitive priorities in the supply chain: The role of interactions with suppliers. *International Journal of Operations & Production Management*, *29*(4), 322–340. doi:10.1108/01443570910945800

Vasileiou, K., & Morris, J. (2006). The sustainability of the supply chain for fresh potatoes in Britain. *Supply Chain Management: An International Journal*, *11*(4), 317–327. doi:10.1108/13598540610671761

Vonderembse, M. A., Uppal, M., Huang, S. H., & Dismukes, J. P. (2006). Designing supply chains: Towards theory development. *International Journal of Production Economics*, *100*(2), 223–238. doi:10.1016/j.ijpe.2004.11.014

Westphal, I., Thoben, K. D., & Seifert, M. (2010). Managing collaboration performance to govern virtual organizations. *Journal of Intelligent Manufacturing*, *21*(3), 311–320. doi:10.1007/s10845-008-0182-5

Wolf, J. (2014). The relationship between sustainable supply chain management, stakeholder pressure and corporate sustainability performance. *Journal of Business Ethics*, *119*(3), 317–328. doi:10.1007/s10551-012-1603-0

Womack, J., & Jones, D. T. (1996). *Lean thinking: Banish waste and create wealth in your corporation*. London, UK: Simon and Schuster.

Womack, J., & Jones, D. T. (2005). *Lean solutions: How companies and customers can create value and wealth together*. London, UK: Simon and Schuster.

Womack, J., Jones, D. T., & Roos, D. (1990). *The machine that changed the world*. New York, NY: Rawson Associates.

Zhu, Q., Dou, Y., & Sarkis, J. (2010). A portfolio-based analysis for green supplier management using the analytical network process. *Supply Chain Management: An International Journal*, *15*(4), 306–319. doi:10.1108/13598541080001414

Zokaei, K., & Hines, P. (2007). Achieving consumer focus in supply chains. *International Journal of Physical Distribution & Logistics Management*, *37*(3), 223–247. doi:10.1108/09600030710742434

Zokaei, K., & Simons, D. W. (2006). Value chain analysis in improvement of customer focus: A case study of UK red meat industry. *International Journal of Logistics Management*, *17*(2), 141–162. doi:10.1108/09574090610689934

ADDITIONAL READINGS

Bai, C., & Sarkis, J. (2010). Integrating sustainability into supplier selection with grey system and rough set methodologies. *International Journal of Production Economics*, *124*(1), 252–264. doi:10.1016/j.ijpe.2009.11.023

Bai, C., & Sarkis, J. (2011). Evaluating supplier development programs with a grey based rough set methodology. *Expert Systems with Applications*, *38*(11), 13505–13517. doi:10.1016/j.eswa.2011.02.137

Bentes, A. V., Carneiro, J., da Silva, J. F., & Kimura, H. (2012). Multidimensional assessment of organizational performance: Integrating BSC and AHP. *Journal of Business Research*, *65*(12), 1790–1799. doi:10.1016/j.jbusres.2011.10.039

Bjorklund, M., Martinsen, U., & Abrahamsson, M. (2012). Performance measurements in the greening of supply chains. *Supply Chain Management: An International Journal*, *17*(1), 29–39. doi:10.1108/13598541211212186

Bruzzone, A. A. G., Anghinolfi, D., Paolucci, M., & Tonelli, F. (2012). Energy-aware scheduling for improving manufacturing process sustainability: A mathematical model for flexible flow shops. *CIRP Annals – Manufacturing Technology*, *61*(1), 459–462. doi:10.1016/j.cirp.2012.03.084

Carbone, V., Moatti, V., & Esposito Vinzi, V. (2012). Mapping corporate responsibility and sustainable supply chains: An exploratory perspective. *Business Strategy and the Environment*, *21*(7), 475–494. doi:10.1002/bse.1736

Colicchia, C., Melacini, M., & Perotti, S. (2011). Benchmarking supply chain sustainability: Insights from a field study. *Benchmarking: An International Journal*, *18*(5), 705–732. doi:10.1108/14635771111166839

Dinter, B. (2013). Success factors for information logistics strategy: An empirical investigation. *Decision Support Systems*, *54*(3), 1207–1218. doi:10.1016/j.dss.2012.09.001

Erol, I., Sencer, S., & Sari, R. (2011). A new fuzzy multi-criteria framework for measuring sustainability performance of a supply chain. *Ecological Economics*, *70*(6), 1088–1100. doi:10.1016/j.ecolecon.2011.01.001

Hassini, E., Surti, C., & Searcy, C. (2012). A literature review and a case study of sustainable supply chains with a focus on metrics. *International Journal of Production Economics*, *140*(1), 69–82. doi:10.1016/j.ijpe.2012.01.042

La Rovere, E. L., Soares, J. B., Oliveira, L. B., & Lauria, T. (2010). Sustainable expansion of electricity sector: Sustainability indicators as an instrument to support decision-making. *Renewable & Sustainable Energy Reviews*, *14*(1), 422–429. doi:10.1016/j.rser.2009.07.033

Lee, K. H., & Saen, R. F. (2012). Measuring corporate sustainability management: A data envelopment analysis approach. *International Journal of Production Economics*, *140*(1), 219–226. doi:10.1016/j.ijpe.2011.08.024

Lehtinen, J., & Ahola, T. (2010). Is performance measurement suitable for an extended enterprise? *International Journal of Operations & Production Management*, *30*(2), 181–204. doi:10.1108/01443571011018707

Manning, L. (2013). Corporate and consumer social responsibility in the food supply chain. *British Food Journal*, *115*(1), 9–29. doi:10.1108/00070701311289858

Mirhedayatian, S. M., Azadi, M., & Saen, R. F. (2014). A novel network data envelopment analysis model for evaluating green supply chain management. *International Journal of Production Economics*, *147*, 544–554. doi:10.1016/j.ijpe.2013.02.009

Seuring, S. (2013). A review of modeling approaches for sustainable supply chain management. *Decision Support Systems*, *54*(4), 1513–1520. doi:10.1016/j.dss.2012.05.053

Tonelli, F., Evans, S., & Taticchi, P. (2013). Industrial sustainability: Challenges, perspectives, actions. *International Journal of Business Innovation Research*, *7*(2), 143–163. doi:10.1504/IJBIR.2013.052576

Wang, H. F., & Hsu, H. W. (2010). A closed-loop logistic model with a spanning-tree based genetic algorithm. *Computers & Operations Research*, *37*(2), 376–389. doi:10.1016/j.cor.2009.06.001

Yakovleva, N., Sarkis, J., & Sloan, T. (2012). Sustainable benchmarking of supply chains: The case of the food industry. *International Journal of Production Research*, *50*(5), 1297–1317. doi:10.1080/00207543.2011.571926

KEY TERMS AND DEFINITIONS

Competitive Advantage: A superiority gained by an organization when it can provide the same value as its competitors but at a lower price, or can charge higher prices by providing greater value through differentiation.

Management: The organization and coordination of the activities of a business in order to achieve defined objectives.

Performance: The accomplishment of a given task measured against preset known standards of accuracy, completeness, cost, and speed.

Performance Measure: The application of quantitative measurements to determine if the activities being pursued by an organization are effective in obtaining the goals of the organization.

Strategic Alignment: The process of bringing the actions of an organization's business divisions and staff members into line with the planned objectives.

Supply Chain: The network of retailers, distributors, transporters, storage facilities and suppliers that participate in the sale, delivery and production of a particular product.

Supply Chain Management: The management of material and information flow in a supply chain to provide the highest degree of customer satisfaction at the lowest possible cost.

Value Chain: The interlinked value-adding activities that convert inputs into outputs which, in turn, add to the bottom line and help create competitive advantage.

Compilation of References

Abraham, M., & Nguyen, N. (2013). Sandestin Declaration: 9 Principles of Green Engineering. *Environment and Progress, 2004*(22), 233–236. Retrieved from http://www.acs.org/content/acs/en/greenchemistry/what-is-green-chemistry/principles/sandestin-declaration.html doi:10.1002/ep.670220410

Abu-Baker, N., & Naser, K. (2000). Empirical evidence on corporate social disclosure (CSD) practices in Jordan. *International Journal of Commerce and Management, 10*(3/4), 18–34. doi:10.1108/eb047406

Acharya, B. G., Mohanty, C. P., & Mahapatra, S. S. (2013). Multi-objective Optimization Machining of Hardened Steel Using NSGA- II. *Procedia Engineering, 51*, 554–560. doi:10.1016/j.proeng.2013.01.078

Acosta, P., Acquier, A., & Delbard, O. (2013). Adoption of sustainable supply chain management programs from a supplier perspective. Proceedings of theXXII Conférence Internationale de Management Stratégique.

Adams, C. A., Muir, S., & Hoque, Z. (2014). Measurement of sustainability performance in the public sector. *Sustainability Accounting. Management and Policy Journal, 5*(1), 46–67.

Adham, K. N., & Siwar, C. (2012). Empirical Investigation of Government Green Procurement (GGP) Practices In Malaysia. *OIDA International Journal of Sustainable Development, 4*(4), 77–88.

After Two Large Gains, Rate of Atmospheric CO2 increase Returns to Average, NOAA Reports. (2005, March 31). *National Oceanic and Atmospheric Administration*. Retrieved from http://www.noaanews.noaa.gov/stories2005/s2412.htm

Ageron, B., Gunasekaran, A., & Spalanzani, A. (2012). Sustainable supply management: An empirical study. *International Journal of Production Economics, 140*(1), 168–182. doi:10.1016/j.ijpe.2011.04.007

Aguilera, R. V., Rupp, D. E., Williams, C. A., & Ganapathi, J. (2007). Putting the S back in corporate social responsibility: A multilevel theory of social change in organizations. *Academy of Management Review, 32*(3), 836–863. doi:10.5465/AMR.2007.25275678

Al Khidir, T., & Zailani, S. (2009). Going green in supply chain towards environmental sustainability. *Global Journal of Environmental Research, 3*(3), 246–251.

Aliei, M., Sazvar, A., & Ashrafi, B. (2012). Assessment of information technology effects on management of supply chain based on fuzzy logic in Iran tail industries. *International Journal of Advanced Manufacturing Technology, 63*(1-4), 215–223. doi:10.1007/s00170-012-3900-2

Allen, R. (1980). *How to save the world.* London: Kogan Page.

Allen, T. D. (2001). Family-Supportive Work Environments: The Role of Organizational Perceptions. *Journal of Vocational Behavior, 58*(3), 414–435. doi:10.1006/jvbe.2000.1774

Almeida, C. M. V. B., Santos, A. P. Z., Bonilla, S. H., Giannetti, B. F., & Huisingh, D. (2013). The roles, perspectives and limitations of environmental accounting in higher educational institutions: An energy synthesis study of the engineering programme at the Paulista University in Brazil. *Journal of Cleaner Production*, *52*, 380–391. doi:10.1016/j.jclepro.2013.03.008

Amaeshi, K. M., Osuji, O. K., & Nnodim, P. (2008). Corporate social responsibility in supply chains of global brands: A boundaryless responsibility? Clarifications, exceptions and implications. *Journal of Business Ethics*, *81*(1), 223–234. doi:10.1007/s10551-007-9490-5

Amar, A. D. (2004). Motivating knowledge workers to innovate: A model integrating Motivation dynamics and antecedents. *European Journal of Innovation Management*, *7*(2), 89–101. doi:10.1108/14601060410534366

Ambrose, E., Marshall, D., & Lynch, D. (2010). Buyer supplier perspectives on supply chain relationships. *International Journal of Operations & Production Management*, *30*(12), 1269–1290. doi:10.1108/01443571011094262

Amran, A., & Devi, S. S. (2008). The Impact of Government and Foreign Affiliate Influence on Corporate Social Reporting: The Case of Malaysia. *Managerial Auditing Journal*, *23*(4), 386–404. doi:10.1108/02686900810864327

Anant, T. C. A., Basu, K., Chopra, K., Dasgupta, P., Desai, N., Gundimeda, H. P., ... & Sundaram, K. (2013). Green National Accounts in India: A Framework.

Anastas, P. T., & Zimmerman, J. B. (2003). Peer reviewed: Design through the 12 principles of green engineering. *Environmental Science & Technology*, *37*(5), 94A–101A. doi:10.1021/es032373g PMID:12666905

Andersen, M., & Skjoett-Larsen, T. (2009). Corporate social responsibility in global supply chains. *Supply Chain Management: An International Journal*, *142*, 75–86.

Anderson, A. R. (1998). Cultivating the Garden of Eden: Environmental entrepreneuring. *Journal of Organizational Change Management*, *11*(2), 135–144. doi:10.1108/09534819810212124

Anderson, E., & Weitz, B. A. (1992). The use of pledges to build and sustain commitment in distribution channels. *JMR, Journal of Marketing Research*, *29*(1), 18–34. doi:10.2307/3172490

Anderson, J. C., & Gerbing, D. W. (1988). Structural equation modeling in practice: A review and recommended two-step approach. *Psychological Bulletin*, *103*(3), 411–423. doi:10.1037/0033-2909.103.3.411

Anderson, S. W., Daly, J. D., & Johnson, M. F. (1999). 'Why firms seek ISO 9000 certification: Regulatory compliance advantage? *Production and Operations Management*, *8*(1), 28–43. doi:10.1111/j.1937-5956.1999.tb00059.x

Andersson, B., Bider, I., Johannesson, P., & Perjons, E. (2005). Towards a formal definition of goal-oriented business process patterns. *Business Process Management Journal*, *11*(6), 650–662. doi:10.1108/14637150510630846

Antai, I. (2011). Supply chain vs. supply chain competition: A niche-based approach. *Management Research Review*, *34*(10), 1107–1124. doi:10.1108/01409171111171500

Antai, I., & Olson, H. (2013). Interaction: A new focus for supply chain vs. supply chain competition. *International Journal of Physical Distribution & Logistics Management*, *43*(7), 511–528. doi:10.1108/IJPDLM-06-2012-0195

Arifin, K., Aiyub, K., Awang, A., Jahi, J. m., & Iteng, R. (2009). Implementation of Integrated Management System in Malaysia: The level of organization's understanding and awareness. *European Journal of Scientific Research*, *31*(2), 188–195.

Arthur, J. (1994). Effects of human resource systems on manufacturing performance and turnover. *Academy of Management Journal*, *37*(3), 670–687. doi:10.2307/256705

AS9100: Quality Management Systems: requirements for aviation, space and defense organizations – revision c. (2009). SAE International, London.

Ashby, A., Leat, M., & Hudson-Smith, M. (2012). Making connections: A review of supply chain management and sustainability literature. *Supply Chain Management: An International Journal*, *175*, 497–516.

Asheim, G. B., Buchholz, W., & Tungodden, B. (2007). *Justifying sustainability* (pp. 33–51). Springer Netherlands.

Ashton, C., & Morton, L. (2005). Managing talent for competitive advantage. *Strategic Human Resources Review*, *4*(5), 28–31. doi:10.1108/14754390580000819

Asokan, P., Kumar, R. R., Jeyapaul, R., & Santhi, M. (2008). Development of Multi-objective Optimization Models for Electrochemical Machining Process. *International Journal of Advanced Manufacturing Technology*, *39*(1-2), 55–63. doi:10.1007/s00170-007-1204-8

Attaran, M., & Attaran, S. (2007). Collaborative supply chain management: The most promising practice for building efficient and sustainable supply chains. *Business Process Management Journal*, *13*(3), 390–404. doi:10.1108/14637150710752308

Attri, R., Dev, N., & Sharma, V. (2013). Interpretive Structural Modeling (ISM) approach: An overview. *Research Journal of Management Sciences*, *2*(2), 3–8.

Awaysheh, A., & Klassen, R. D. (2010). The impact of supply chain structure on the use of supplier socially responsible practices. *International Journal of Operations & Production Management*, *30*(12), 1246–1268. doi:10.1108/01443571011094253

Axelrod, R., Riolo, R., & Cohen, D. (2002). Beyond Geography: Cooperation with Persistent Links in the Absence of Clustered Neighborhoods. *Personality and Social Psychology Review*, *6*(4), 341–346. doi:10.1207/S15327957PSPR0604_08

Azevedo, S. G., & Carvalho, H., & CruzMachado, V. (2011). The influence of green practices on supply chain performance: A case study approach. *Transportation Research Part E, Logistics and Transportation Review*, *47*(6), 850–871. doi:10.1016/j.tre.2011.05.017

Azzone, G., & Noci, G. (1998). Identifying effective PMSs for the deployment of 'green' manufacturing strategies. *International Journal of Operations & Production Management*, *18*(4), 308–335. doi:10.1108/01443579810199711

Baas, L. (2007). To make zero emissions technologies and strategies become a reality, the lessons learned of cleaner production dissemination have to be known. *Journal of Cleaner Production*, *15*(13), 1205–1216. doi:10.1016/j.jclepro.2006.07.017

Bag, S., & Anand, N. (2015). Modelling barriers of sustainable supply chain network design using interpretive structural modelling: An insight from food processing sector in India. *International Journal of Automation and Logistics*, *1*(3), 234–255. doi:10.1504/IJAL.2015.071722

Bai, C. A., & Sarkis, J. (2014). Determining and Applying Sustainable Supplier Key Performance Indicators. *Supply Chain Management: An International Journal*, *19*(3), 5–5.

Bai, C., & Sarkis, J. (2010). Greener supplier development: Analytical evaluation using rough Set theory. *Journal of Cleaner Production*, *18*(12), 1200–1210. doi:10.1016/j.jclepro.2010.01.016

Bai, C., & Sarkis, J. (2010). Integrating sustainability into supplier selection with grey system and rough set methodologies. *International Journal of Production Economics*, *124*(1), 252–264. doi:10.1016/j.ijpe.2009.11.023

Bai, C., & Sarkis, J. (2012). Supply-chain performance-measurement system management using neighbourhood rough sets. *International Journal of Production Research*, *50*(9), 2484–2500. doi:10.1080/00207543.2011.581010

Bai, C., & Sarkis, J. (2014). Determining and applying sustainable supplier key performance indicators. *Supply Chain Management: An International Journal*, *19*(3), 275–291. doi:10.1108/SCM-12-2013-0441

Bai, C., Sarkis, J., Wei, X., & Koh, L. (2012). 'Evaluating ecological sustainable performance measures for supply chain management', *Supply Chain Management. International Journal (Toronto, Ont.)*, *17*(1), 78–92.

Bai, C., Sarkis, J., Wei, X., & Koh, L. (2012). Evaluating ecological sustainable performance measures for supply chain management. *Supply Chain Management: An International Journal*, *17*(1), 78–92. doi:10.1108/13598541211212221

Baker, M., & Hart, S. (2008). *The marketing book*. Routledge.

Ball, A. (2005). Environmental accounting and change in UK local government. *Accounting, Auditing & Accountability Journal*, *18*(3), 346–373. doi:10.1108/09513570510600738

Bamber, C. J., Sharp, J. M., & Castka, P. (2004). Third party assessment: The role of the maintenance function in an integrated management system. *Journal of Quality in Maintenance Engineering*, *10*(1), 26–36. doi:10.1108/13552510410526848

Banerjee, S. B. (2003). Who sustains whose development? Sustainable development and the reinvention of nature. *Organization Studies*, *24*(1), 143–180. doi:10.1177/0170840603024001341

Banister, D., & Button, K. (1993). Environmental policy and transport: An overview. In *Transport, the Environment and Sustainable Development* (pp. 1-15).

Banker, S. (2013). Amazon Vs. Walmart: E-Commerce Vs. Omni-Channel Logistics. *Forbes Logistics and Transportation*. Retrieved from http://www.forbes.com/sites/stevebanker/2013/10/04/amazon-vs-walmart-e-commerce-vs-omni-channel-logistics/2/

Bansal, P. (2002). The corporate challenges of sustainable development. *The Academy of Management Executive*, *16*(2), 122–131. doi:10.5465/AME.2002.7173572

Bansal, P. (2005). Evolving sustainably: A longitudinal study of corporate sustainable development. *Strategic Management Journal*, *26*(3), 197–218. doi:10.1002/smj.441

Barari, S., Agarwal, G., Zhang, W. J., Mahanty, B., & Tiwari, M. K. (2012). A decision framework for the analysis of green supply chain contracts: An evolutionary game approach. *Expert Systems with Applications*, *39*(3), 2965–2976. doi:10.1016/j.eswa.2011.08.158

Barbera, A. J., & McConnell, V. D. (1990). The impact of environmental regulations on industry productivity: Direct and indirect effects. *Journal of Environmental Economics and Management*, *18*(1), 50–65. doi:10.1016/0095-0696(90)90051-Y

Barber, J. (2007). Mapping the movement to achieve sustainable production and consumption in North America. *Journal of Cleaner Production*, *15*(6), 499–512. doi:10.1016/j.jclepro.2006.05.010

Barney, J. B. (1986). Organizational culture: Can it be a source of sustained competitive advantage? *Academy of Management Review*, *11*(3), 656–665. doi:10.5465/AMR.1986.4306261

Barney, J. B. (1991). Firm resources and sustained competitive advantage. *Journal of Management*, *17*(1), 99–120. doi:10.1177/014920639101700108

Barney, J. B. (2012). Purchasing, supply chain management and sustained competitive advantage: The relevance of resource-based theory. *Journal of Supply Chain Management*, *48*(2), 3–6. doi:10.1111/j.1745-493X.2012.03265.x

Baron, R. M., & Kenny, D. A. (1986). The Moderator-Mediator Variable Distinction in Social Psychological Research: Conceptual, Strategic, and Statistical Considerations. *Journal of Personality and Social Psychology*, *51*(6), 1173–1182. doi:10.1037/0022-3514.51.6.1173 PMID:3806354

Barral, M. (2007). La gestión integrada y la dinámica de sistemas: Criterios a aplicar en los laboratorios clínicos. *Acta Bioquímica Clínica Latinoamericana*, *41*(3), 407–418.

Barratt, M. (2004). Understanding the meaning of collaboration in the supply chain. *Supply Chain Management: an international journal, 9*(1), 30-42.

Barron, L., & Gauntlet, E. (2002, March 3–6). WACOSS 'Housing and Sustainable Communities Indicators Project.' Paper presented at 'Sustaining our Communities' International Local Agenda 21 Conference, Adelaide (p 4). Retrieved from http://www.adelaide.sa.gov.au/soc/pdf/barron_gauntlett.pdf

Barron, V. (2010). How standards contribute to sustainable development. *BSI White Paper.* Retrieved from http://www.bsigroup.co.uk/en/Forms/Document-Downloads/How-standards-contribute-to-sustainable-development/

Bartley, T. (2003). Certifying forests and factories: States, social movements, and the rise of private regulation in the apparel and forest products fields. *Politics & Society*, *31*(3), 433–464. doi:10.1177/0032329203254863

Baskaran, V., Nachiappan, S., & Rahman, S. (2012). 'Indian textile suppliers' sustainability evaluation using the grey approach'. *International Journal of Production Economics*, *135*(2), 647–658. doi:10.1016/j.ijpe.2011.06.012

Bassi, L. J., & Van Buren, M. E. (1999). Sharpening the leading edge. *Training & Development*, *53*(l), 23–32.

Bayati, A., & Taghavi, A. (2007). The impacts of acquiring ISO 9000 certification on the performance of SMEs in Tehran. *The TQM Magazine*, *19*(2), 140–149. doi:10.1108/09544780710729980

Beachy, R. N. (2010). Science and sustainability: The emerging consensus. *Bioscience*, *60*(6), 406–407. doi:10.1525/bio.2010.60.6.2

Beamon, B. M. (1999). Designing the green supply chain. *Logistics information management, 12*(4), 332-342.

Beamon, B. M. (1999). Measuring supply chain performance. *International Journal of Operations & Production Management*, *19*(3), 275–292. doi:10.1108/01443579910249714

Bebbington, J. (2001, June). Sustainable development: A review of the international development, business and accounting literature. *Accounting Forum*, *25*(2), 128–157. doi:10.1111/1467-6303.00059

Beckmerhagen, I. A., Berg, H. P., Karapetrovic, S. V., & Willborn, W. O. (2003). Integration of management systems: Focus on safety in the nuclear industry. *International Journal of Quality & Reliability Management*, *20*(2), 209–228. doi:10.1108/02656710310456626

Beer, S., & Lemmer, C. (2011). A critical review of "green" procurement: Life cycle analysis of food products within the supply chain. *Worldwide Hospitality and Tourism Themes*, *3*(3), 229–244. doi:10.1108/17554211111142194

Beloff, B., Tanzil, D., & Lines, M. (2004). Sustainable development performance assessment. *Environment and Progress*, *23*(4), 271–276. doi:10.1002/ep.10045

Benardos, P. G., & Vosniakos, G. C. (2003). Predicting surface roughness in machining: A review. *International Journal of Machine Tools & Manufacture*, *43*(8), 833–844. doi:10.1016/S0890-6955(03)00059-2

Benjaafar, S., Li, Y., & Daskin, M. (2013). Carbon footprint and the management of supply chains: Insights from simple models. *IEEE Transactions on* Automation Science and Engineering, *10*(1), 99–116.

Bennett, M., & James, P. (Eds.). (1998). *The Green bottom line: environmental accounting for management: current practice and future trends*. Greenleaf Publishing.

Benton, W. C., & Maloni, M. (2005). The influence of power driven buyer/seller relationships on supply chain satisfaction. *Journal of Operations Management*, *23*(1), 1–22. doi:10.1016/j.jom.2004.09.002

Berkhout, F., Hertin, J., & Gann, D. H. (2006). Learning To Adapt: Organizational Adaptations to Climate Change Impact. *Climatic Change*, *78*(1), 135–156. doi:10.1007/s10584-006-9089-3

Bernardo, M., Casadesus, M., Karapetrovic, S., & Heras, I. (2009). How integrated are environmental, quality and other standardized management systems? An empirical study. *Journal of Cleaner Production*, *17*(8), 742–750. doi:10.1016/j.jclepro.2008.11.003

Bernardo, M., Casadesus, M., Karapetrovic, S., & Heras, I. (2010). An empirical study on the integration of management system audits. *Journal of Cleaner Production*, *18*(5), 486–495. doi:10.1016/j.jclepro.2009.12.001

Berns, M., Townend, A., & Khayat, Z. et al.. (2009). Sustainability and Competitive Advantage. *MIT Sloan Management Review*, *51*(1), 19–26.

Berns, M., Townend, A., Khayat, Z., Balagopal, B., Reeves, M., Hopkins, M. S., & Kruschwitz, N. (2009). Sustainability and competitive advantage. *MIT Sloan Management Review*, *51*(2), 19–26.

Berns, M., Townend, A., Khayat, Z., Balagopal, B., Reeves, M., Hopkins, M. S., & Kruschwitz, N. (2009). The business of sustainability: What it means to managers now. *MIT Sloan Management Review*, *51*(1), 20–26.

Bhattacharya, C. B., Sen, S., & Korschun, D. (2012). Using corporate social responsibility to win the war for talent. *MIT. Sloan Management Review*, 49(2).

Bhattacharyya, B., & Sorkhel, S. K. (1999). Investigation for controlled electrochemical machining through response surface methodology-based approach. *Journal of Materials Processing Technology*, *86*(1-3), 200–207. doi:10.1016/S0924-0136(98)00311-2

Bierma, T. J., & Waterstraat, F. L. (1999). Cleaner production from chemical suppliers: Understanding shared savings contracts. *Journal of Cleaner Production*, *7*(2), 145–158. doi:10.1016/S0959-6526(98)00073-0

Biggemann, S., Williams, M., & Kro, G. (2014). Building in sustainability, social responsibility and value co-creation. *Journal of Business and Industrial Marketing*, *29*(4), 304–312. doi:10.1108/JBIM-08-2013-0161

Binder, M., & Edwards, J. S. (2010). Using grounded theory method for theory building in operations management research: A study on inter-firm relationship governance. *International Journal of Operations & Production Management*, *30*(3), 232–259. doi:10.1108/01443571011024610

Bishop, J. W. (1998, February 14). The relationship between quality and support in the workplace. *Paper presented at the conference on Quality and Management: Quality now and Direction for the 21st Century*, Arizona State University, Tempe, Arizona, USA.

Bjorklund, M. (2011). Influence from the business environment to environmental purchasing – drivers and hinders of purchasing green transportation services. *Journal of Purchasing and Supply Management*, *17*(1), 11–22. doi:10.1016/j.pursup.2010.04.002

Bjorklund, M., Martinsen, U., & Abrahamsson, M. (2012). 'Performance measurements in the greening of the supply chains', *Supply Chain Management. International Journal (Toronto, Ont.)*, *17*(1), 29–39.

Blacconiere, W. G., & Patten, D. M. (1994). Environmental disclosures, regulatory costs, and changes in firm value. *Journal of Accounting and Economics*, *18*(3), 357–377. doi:10.1016/0165-4101(94)90026-4

Black, S. E., & Lynch, L. M. (1996). Human capital investments and productivity. *The American Economic Review*, *86*(2), 68.

Blengini, G. A., & Shields, D. J. (2010). Green labels and sustainability reporting: Overview of the building products supply chain in Italy. *Management of Environmental Quality: An International Journal*, *21*(4), 477–493. doi:10.1108/14777831011049115

Blome, C., Paulraj, A., & Schuetz, K. (2014). Supply chain collaboration and sustainability: A profile deviation analysis. *International Journal of Operations & Production Management*, *34*(5), 639–663. doi:10.1108/IJOPM-11-2012-0515

Bockerman, P., & Ilmakunnas, P. (2004). Job dimensions, job satisfaction, and on-the-job search: is there a nexus? *HECER Discussion Paper* No. 36, University of Helsinki.

Boltic, Z., Ruzic, N., Jovanovic, M., Savic, M., Jovanovic, J., & Petrovic, S. (2013). Cleaner production aspects of tablet coating process in pharmaceutical industry: Problem of VOCs emission. *Journal of Cleaner Production*, *44*(1), 123–132. doi:10.1016/j.jclepro.2013.01.004

Borin, N., Lindsey-Mullikin, J., & Krishnan, R. (2013). An analysis of consumer reactions to green strategies. *Journal of Product and Brand Management*, *22*(2), 118–128. doi:10.1108/10610421311320997

Bose, I., & Pal, R. (2012). Do green supply chain management initiatives impact stock prices of firms? *Decision Support Systems*, *52*(3), 624–634. doi:10.1016/j.dss.2011.10.020

Boudreau, J. W., & Ramstad, P. M. (2005). Talentship and the new paradigm for human resource management: From professional practice to strategic talent decision science. *Human Resource Planning*, *28*(2), 17–26.

Bouma, J. J., & van der Veen, M. (2002). Wanted: A theory for environmental management accounting. In Environmental management accounting: Informational and institutional developments (pp. 279-290). Springer Netherlands. doi:10.1007/0-306-48022-0_22

Bowen, F. E., Cousins, P. D., Lamming, R. C., & Farukt, A. C. (2001). The role of supply management capabilities in green supply. *Production and Operations Management*, *10*(2), 174–189. doi:10.1111/j.1937-5956.2001.tb00077.x

Bowersox, D. J., & Closs, D. C. (1996). *Logistical management: The integrated supply chain process*. New York, NY: McGraw-Hill.

Bracci, E., & Maran, L. (2013). Environmental management and regulation: Pitfalls of environmental accounting? *Management of Environmental Quality: An International Journal*, *24*(4), 538–554. doi:10.1108/MEQ-04-2012-0027

Bramley, G., & Power, S. (2009). Urban form and social sustainability: The role of density and housing type. *Environment and Planning. B, Planning & Design*, *361*(1), 30–48. doi:10.1068/b33129

Brandenburg, M., Govindan, K., Sarkis, J., & Seuring, S. (2014). Quantitative models for supply chain management: Developments and directions. *European Journal of Operational Research*, *233*(2), 299–312. doi:10.1016/j.ejor.2013.09.032

Brannick, J. (1999). Employees; Job satisfaction; Labor turnover; Recruiting. *AMACOM (New York)*. Retrieved from http://64.17.184.140/wp-content/uploads/2012/12/V4I1-7.pdf

Brown, G., & Stone, L. (2007). Cleaner production in New Zealand: Taking stock. *Journal of Cleaner Production*, *15*(8-9), 716–728. doi:10.1016/j.jclepro.2006.06.025

Brown, S., Squire, B., & Blackmon, K. (2007). The contribution of manufacturing strategy involvement and alignment to world-class manufacturing performance. *International Journal of Operations & Production Management*, *27*(3), 282–302. doi:10.1108/01443570710725554

Bruce, M., Daly, L., & Towers, N. (2004). Lean or agile: A solution for supply chain management in the textiles and clothing industry. *International Journal of Operations & Production Management*, *24*(1-2), 151–170. doi:10.1108/01443570410514867

Brundtland, G., Khalid, M., Agnelli, S., Al-Athel, S., Chidzero, B., Fadika, L., ... & Okita, S. (1987). Our Common Future (Brundtland report).

Brundtland, G. H. (1987). *Report of the World Commission on environment and development:" our common future*. United Nations.

Bruntland, G. 1987. Our common future: The world commission on environment and development.

BS OHSAS 18001: Occupational Health and Safety. (2008). British Standards Institution.

Buchholz, R. A. (1991). Corporate responsibility and the good society: From economics to ecology. *Business Horizons*, *34*(4), 19–31. doi:10.1016/0007-6813(91)90003-E

Buchholz, R. A., & Rosenthal, S. B. (2005). Toward a contemporary conceptual framework for stakeholder theory. *Journal of Business Ethics*, *58*(1-3), 137–148. doi:10.1007/s10551-005-1393-8

Buckingham, M., & Clifton, D. O. (2001). *Now, discover your strengths*. New York: Free Press.

Buckley, R. (2002). Tourism ecolabels. *Annals of Tourism Research*, *29*(1), 183–208. doi:10.1016/S0160-7383(01)00035-4

Burritt, R. L., & Saka, C. (2006). Environmental management accounting applications and eco- efficiency: Case studies from Japan. *Journal of Cleaner Production*, *14*(14), 1262–1275. doi:10.1016/j.jclepro.2005.08.012

Burritt, R., & Saka, C. (2006). *Quality of physical environmental management accounting information. Lessons from pollutant release and transfer registers* (pp. 373–407). Springer Netherlands; doi:10.1007/978-1-4020-4974-3_17

Burton, I. (1987). Report on Reports: Our Common Future: The World Commission on Environment a-nd Development. *Environment: Science and Policy for Sustainable Development*, *29*(5), 25–29. doi:10.1080/00139157.1987.9928891

Business strategy for sustainable development: leadership and accountability for the'90s. (1992Deloitte & Touche, International Institute for Sustainable Development, and Business Council for Sustainable Development.

Butler, T., & Waldroop, J. (2001). *Job sculpting: The art of retaining your best people. Harvard Business Review on finding and keeping the best people* (pp. 179–203). Boston: Harvard Business School Press.

Buy now don't pay later. (2004). *Welsh Procurement Initiative*. Retrieved from http://prp.wales.gov.uk/docs/prp/generalgoodsservices/130612buynowdontpaylater.pdf

Cable, D. M., & Judge, T. A. (1997). Interviewer's perception of person organization fit and organizational selection decisions. *The Journal of Applied Psychology*, *82*(4), 61. doi:10.1037/0021-9010.82.4.546 PMID:9378683

Cabral, I., Grilo, A., & Cruz-Machado, V. (2012). A decision-making model for lean, agile, resilient and green supply chain management. *International Journal of Production Research*, *50*(17), 4830–4845. doi:10.1080/00207543.2012.657970

Cagnazzo, L., Taticchi, P., & Brun, A. (2009). The role of performance measurement systems to support quality improvement initiatives at supply chain level. *International Journal of Productivity and Performance Management*, *59*(2), 163–185.

Cairns, R. D. (2000). Sustainability accounting and green accounting. *Environment and Development Economics*, *5*(1), 49–54. doi:10.1017/S1355770X00000048

Campbell, J. L. (2007). Why would corporations behave in socially responsible ways? An institutional theory of corporate social responsibility. *Academy of Management Review*, *32*(3), 946–967. doi:10.5465/AMR.2007.25275684

Caniato, F., Caridi, M., Crippa, L., & Moretto, A. (2012). Environmental sustainability in fashion supply chains. *International Journal of Production Economics*, *135*(2), 559–570. doi:10.1016/j.ijpe.2011.06.001

Caniels, M. C. J., Gehrsitz, M. H., & Semeijn, J. (2013). Participation of suppliers in greening the supply chains: An empirical analysis of German automotive suppliers. *Journal of Purchasing and Supply Management*, *19*(3), 134–143. doi:10.1016/j.pursup.2013.02.005

Carter, C. R., & Jennings, M. M. (2004). The role of purchasing in corporate social responsibility: a structural equation analysis. *Journal of business Logistics*, *25*(1), 145-186.

Carter, C.R., Jennings, M.M. (2002a). Logistics social responsibility: an integrative framework. *Journal of business logistics*, 23(1), 145-180.

Carter, C. R. (2005). Purchasing social responsibility and firm performance: The key mediating roles of organizational learning and supplier performance. *International Journal of Physical Distribution & Logistics Management*, *35*(3), 177–194. doi:10.1108/09600030510594567

Carter, C. R., Auskalnis, R., & Ketchum, C. (1999). 'Purchasing from minority business Enterprises: A cross-industry comparison of best practices. *Journal of Supply Chain Management*, *35*(1), 28–32. doi:10.1111/j.1745-493X.1999.tb00053.x

Carter, C. R., & Dresner, M. (2001). Purchasing's Role in Environmental Management: Cross-Functional Development of Grounded Theory. *Journal of Supply Chain Management*, *37*(3), 12–27. doi:10.1111/j.1745-493X.2001.tb00102.x

Carter, C. R., & Easton, P. L. (2011). Sustainable supply chain management: Evolution and future directions. *International Journal of Physical Distribution & Logistics Management*, *41*(1), 46–62. doi:10.1108/09600031111101420

Carter, C. R., & Jennings, M. M. (2002b). Social responsibility and supply chain relationships. *Transportation Research Part E, Logistics and Transportation Review*, *38*(1), 37–52. doi:10.1016/S1366-5545(01)00008-4

Carter, C. R., Kale, R., & Grimm, C. M. (2000). Environmental purchasing and firm performance: An empirical investigation. *Transportation Research Part E, Logistics and Transportation Review*, *36*(3), 219–228. doi:10.1016/S1366-5545(99)00034-4

Carter, C. R., & Rogers, D. S. (2008). A framework for sustainable supply chain management: Moving toward new theory. *International Journal of Physical Distribution & Logistics Management*, *38*(5), 360–387. doi:10.1108/09600030810882816

Carter, P. L., Carter, J. R., Monczka, R. M., Slaight, T. H., & Swan, A. J. (2000). The Future of Purchasing and Supply: A Ten-Year Forecast1. *Journal of Supply Chain Management*, *36*(1), 14–26. doi:10.1111/j.1745-493X.2000.tb00066.x

Carvalho, H., Azevedo, S. G., & Cruz-Machado, V. (2013). Agile and resilient approaches to supply chain management: Influence on performance and competitiveness. *Logistics Research*, *4*(1-2), 49–62. doi:10.1007/s12159-012-0064-2

Cascio, W. F. (1987). *Applied psychology in personnel management*. Englewood Cliffs, NJ: Prentice-Hall.

Castka, P., & Balzarova, M.A. (2008). ISO 26000 and supply chains - On the diffusion of the social responsibility standard. *International Journal of Production Economics*, *111*(2), 274–286. doi:10.1016/j.ijpe.2006.10.017

Celik, M. (2009). Establishing an Integrated Process Management System (IPMS) in ship management companies. *Expert Systems with Applications*, *36*(4), 8152–8171. doi:10.1016/j.eswa.2008.10.022

Cerin, P., & Karlson, L. (2002). Business incentives for sustainability: A property rights approach. *Ecological Economics*, *40*(1), 13–22. doi:10.1016/S0921-8009(01)00275-0

Cerin, P., & Scholtens, B. (2011). Linking responsible investments to societal influence: Motives, assessments and risks(editorial). *Sustainable Development*, *19*(5), 289–358.

Cetinkaya, B., Cuthbertson, R., Ewer, G., Klaas-Wissing, T., Piotrowicz, W., & Tyssen, C. (2011). *Sustainable supply chain management: practical ideas for moving towards best practice*. Springer.

Chaabane, A., Ramudhin, A., & Paquet, M. (2012). Design of sustainable supply chains under the emission trading scheme. *International Journal of Production Economics*, *135*(1), 37–49. doi:10.1016/j.ijpe.2010.10.025

Chae, B. (2009). Developing key performance indicators for supply chain: An industry perspective. *Supply Chain Management: An International Journal*, *14*(6), 422–428. doi:10.1108/13598540910995192

Chang, C. C., Hung, S. W., & Huang, S. Y. (2013). Evaluating the operational performance of knowledge-based industries: The perspective of intellectual capital. *Quality & Quantity*, *47*(3), 1367–1383. doi:10.1007/s11135-011-9595-x

Chang, H. H., Hung, C. J., Wong, K. H., & Lee, C. H. (2013). Using the balanced scorecard on supply chain integration performance—a case study of service businesses. *Service Business*, *7*(4), 539–561. doi:10.1007/s11628-012-0175-5

Chapman, J. A., & Jehn, K. A. (1994). Assessing the relationship between industry characteristics and organizational culture: How different can you be? *Academy of Management Journal*, *37*(3), 522–553. doi:10.2307/256699

Chari, F., & Chiriseri, L. (2014). Barriers to sustainable procurement in Zimbabwe. *Greener Journal of Business and Management Studies*, *4*(1), 14–18.

Chatterjee, B., & Zaman Mir, M. (2008). The current status of environmental reporting by Indian companies. *Managerial Auditing Journal*, *23*(6), 609–629. doi:10.1108/02686900810882138

Chatterjee, P., & Finger, M. (1994). *The earth brokers: Power, politics, and world development*. US: Taylor & Francis.

Cheese, P., Thormas, R. J., & Craig, E. (2008). *The talent powered organization: strategies for globalization, talent management and high performance*. London, Philadelphia: Kogan.

Cheloha, R., & Swain, J. (2005). Talent management system key to effective succession planning. *Canadian HR Reporter*, *18*(17), 5–7.

Chen, C. C. (2005). Incorporating green purchasing into the frame of ISO 14000. *Journal of Cleaner Production*, *13*(9), 927–933. doi:10.1016/j.jclepro.2004.04.005

Chen, I. J., & Paulraj, A. (2004). Towards a theory of supply chain management: The constructs and measurements. *Journal of Operations Management*, *22*(2), 119–150. doi:10.1016/j.jom.2003.12.007

Chen, I. J., Paulraj, A., & Lado, A. A. (2004). Strategic purchasing, supply management, and firm performance. *Journal of Operations Management*, *22*(5), 505–523. doi:10.1016/j.jom.2004.06.002

Chen, M.-K., Tai, T.-W., & Hung, T.-Y. (2012). Component selection system for green supply chain. *Expert Systems with Applications*, *39*(5), 5687–5701. doi:10.1016/j.eswa.2011.11.102

Chen, S. H., Wang, P. W., & Lee, H. T. (2012). Performance measurement considering the interdependence of evaluators and criteria. *Quality & Quantity*, *46*(1), 103–115. doi:10.1007/s11135-010-9330-z

Chen, Y. S., & Chang, C. H. (2012). Enhance green purchase intentions: The roles of green perceived value, green perceived risk, and green trust. *Management Decision*, *50*(3), 502–520. doi:10.1108/00251741211216250

Chien, M. K., & Shih, L. H. (2007). An empirical study of the implementation of green supply chain management practices in the electrical and electronic industry and their relation to organizational performances.

China, K. P. M. G. (2011). *China's 12th five-year plan: overview. KPMG Insight Series*. KPMG Adviso.

Choi, T., & Linton, T. (2011). Don't let your supply chain control your business.

Choi, T. M., Lo, C. K., Wong, C. W., & Yee, R. W. (2012). Green manufacturing and distribution in the fashion and apparel industries. *International Journal of Production Economics*, *135*(2), 531. doi:10.1016/j.ijpe.2011.07.012

Choi, T. Y., & Hartley, J. L. (1996). An exploration of supplier selection practices across the supply chain. *Journal of Operations Management*, *14*(4), 333–343. doi:10.1016/S0272-6963(96)00091-5

Cholette, S., & Venkat, K. (2009). The energy and carbon intensity of wine distribution: A study of logistical options for delivering wine to consumers. *Journal of Cleaner Production*, *17*(16), 1401–1413. doi:10.1016/j.jclepro.2009.05.011

Choudhary, S. G. (2010). Planet first. *Quest. TQMS*, *5*, 20–26.

Chouinard, Y., Ellison, J., & Ridgeway, R. (2011).The sustainable economy. *Harvard Business review, 89*(10), 5-2-62.

Christmann, P., & Taylor, G. (2001). Globalization and the environment: Determinants of firm self-regulation in China. *Journal of international business studies*, 439-458. Retrieved from http://www.gurn.info

Christmann, P. (2004). Multinational companies and the natural environment: Determinants of global environmental policy. *Academy of Management Journal*, *47*(5), 747–760. doi:10.2307/20159616

Christopher, M. (2005). *Logistics and supply chain management: Creating value-adding networks*. London, UK: Pitman.

Christopher, M., & Lee, H. (2004). Mitigating supply chain risk through improved confidence. *International Journal of Physical Distribution & Logistics Management*, *34*(5), 388–396. doi:10.1108/09600030410545436

Churchill, G. A. (1979). A paradigm for developing better measures for marketing constructs. *JMR, Journal of Marketing Research*, *16*(1), 64–73. doi:10.2307/3150876

Ciceri, N. D., Garetti, M., & Sperandio, S. (2010). From product end-of-life sustainable considerations to design management. In Advances in Production Management Systems. New Challenges, New Approaches (pp. 152-159). Springer Berlin Heidelberg.

Ciliberti, F., Pontrandolfo, P., & Scozzi, B. (2008). Investigating corporate social responsibility in supply chains: A SME perspective. *Journal of Cleaner Production*, *16*(15), 1579–1588. doi:10.1016/j.jclepro.2008.04.016

Ciliberti, F., Pontrandolfo, P., & Scozzi, B. (2008). Logistics social responsibility: Standard adoption and practices in Italian companies. *International Journal of Production Economics*, *113*(1), 88–106. doi:10.1016/j.ijpe.2007.02.049

Clarke, K. F. (2001). What businesses are doing to attract and retain employee- becoming an employer of choice. *Employee Benefits Journal*, 2001, 34–37. PMID:11272511

Clark, J. W., Toms, L. C., & Green, K. W. (2014). Market-oriented sustainability: Moderating impact of stakeholder involvement. *Industrial Management & Data Systems*, *114*(1), 21–36. doi:10.1108/IMDS-04-2013-0194

Clarkson, M. E. (1995). A stakeholder framework for analyzing and evaluating corporate social performance. *Academy of Management Review*, *20*(1), 92–117.

Clelland, I. J., Dean, T. J., & Douglas, T. J. (2000). Stepping towards sustainable business: An evaluation of waste minimization practices in US manufacturing. *Interfaces*, *30*(3), 107–124. doi:10.1287/inte.30.3.107.11661

Clemens, B., & Douglas, T. J. (2005). Understanding strategic responses to institutional pressures. *Journal of Business Research*, *58*(9), 1205–1213. doi:10.1016/j.jbusres.2004.04.002

Clemens, B., & Douglas, T. J. (2006). 'Does coercion drive firms to adopt 'voluntary' green initiatives? Relationships among coercion, superior firm resources, and voluntary green initiatives'. *Journal of Business Research*, *59*(4), 483–491. doi:10.1016/j.jbusres.2005.09.016

Clemo, K. (2008). Preparing for climate change: Insurance and small Business. *The Geneva Papers on Risk and Insurance. Issues and Practice*, *33*(1), 110–116. doi:10.1057/palgrave.gpp.2510160

Climate Change and Business: from Supply Chain Carbon Emissions to a Wider Risk Set2008*Business and the Environment*, 19(7), 1-3.

Closs, D. J., Speier, C., & Meacham, N. (2011). Sustainability to support end-to-end value chains: The role of supply chain management. *Journal of the Academy of Marketing Science*, *39*(1), 101–116. doi:10.1007/s11747-010-0207-4

Coates, T. J., & Thoresen, C. E. (1976). Teacher anxiety: A review with recommendations. *Review of Educational Research*, *46*(2), 159–184. doi:10.3102/00346543046002159

Cocca, P., & Alberti, M. (2010). A framework to assess performance measurement systems in SMEs. *International Journal of Productivity and Performance Management*, *59*(2), 186–200. doi:10.1108/17410401011014258

Connelly, B. L., Ketchen, D. J. Jr, & Slater, S. F. (2011). Toward a "theoretical toolbox" for sustainability research in marketing. *Journal of the Academy of Marketing Science*, *39*(1), 86–100. doi:10.1007/s11747-010-0199-0

Consumer perceptions on green products. (2015). *Green Purchasing Network India-GPNI*. Retrieved from http://gpnindia.org/green-products/

Conway, D. M. (2012). Sustainable procurement policies and practices at the state and local government level. In K. Hirokawa & P. Salkin (Eds.), *Greening Local Government*.

Coomer, J. (1979). *Quest for a sustainable society*. Oxford: Pergamon.

Corbett, C. J., & Klassen, R. D. (2006). Extending the horizons: Environmental excellence as key to improving operations. *Manufacturing & Service Operations Management*, *8*(1), 5–22. doi:10.1287/msom.1060.0095

Corbett, C. J., & Kleindorfer, P. R. (2003). Environmental management and operations management: Introduction to the third special issue. *Production and Operations Management*, *12*(3), 287–289. doi:10.1111/j.1937-5956.2003.tb00203.x

Corley, K. G., & Gioia, D. A. (2011). Building theory about theory building: What constitutes a theoretical contribution? *Academy of Management Review*, *36*(1), 12–32. doi:10.5465/AMR.2011.55662499

Corona, C. (2009). Dynamic performance measurement with intangible assets. *Review of Accounting Studies*, *14*(2-3), 314–348. doi:10.1007/s11142-009-9095-6

Côté, R. P., & Cohen-Rosenthal, E. (1998). Designing eco-industrial parks: A synthesis of some experiences. *Journal of Cleaner Production*, *6*(3), 181–188. doi:10.1016/S0959-6526(98)00029-8

Coughlan, P., & Coghlan, D. (2002). Action research for operations management. *International Journal of Operations & Production Management*, *22*(2), 220–240. doi:10.1108/01443570210417515

Cranny, C. J., Smith, P. C., & Stone, E. F. (1992). *Job satisfaction: How people feel about their jobs and how it affects their performance. Lexington Books*. New York: Lexington Books.

Crespin-Mazet, F., & Dontenwill, E. (2012). Sustainable procurement: Building legitimacy in the supply network. *Journal of Purchasing and Supply Management*, *18*(4), 207–217. doi:10.1016/j.pursup.2012.01.002

Cronbach, L. J. (1951). Coefficient alpha and the internal structure of tests. *psychometrika*, 16(3), 297-334.

Croom, S., Romano, P., & Giannakis, M. (2000). Supply chain management: an analytical framework for critical literature review. *European journal of purchasing & supply management, 6*(1), 67-83.

Crozier, M., & Thoenig, J.-C. (1976). The regulation of complex organized systems. *Administrative Science Quarterly*, *21*(4), 547–570. doi:10.2307/2391716

Daly, H. E. (1990). Toward some operational principles of sustainable development. *Ecological Economics*, *2*(1), 1–6. doi:10.1016/0921-8009(90)90010-R

Dam, L., & Petkova, B. N. (2014). The impact of environmental supply chain sustainability programs on shareholder wealth. *International Journal of Operations & Production Management*, *34*(5), 586–609. doi:10.1108/IJOPM-10-2012-0482

Dansk Standard. (2005). *Management systems*: guidelines for integrated management system. DS 8001.

Darnall, N., Henriques, I., & Sadorsky, P. (2008). Do environmental management systems improve business performance in an international setting? *Journal of International Management*, *14*(4), 364–376. doi:10.1016/j.intman.2007.09.006

Darnall, N., Jolley, G. J., & Handfield, R. (2008). Environmental management systems and green supply chain management: Complements for sustainability? *Business Strategy and the Environment*, *17*(1), 30–45. doi:10.1002/bse.557

Dash, J. (2001). Schools push soft skills for info security majors. *Computerworld*, *35*(6), 24.

Das, M. K., Kumar, K., Barman, T. K., & Sahoo, P. (2014). Optimization of Surface Roughness and MRR in Electrochemical Machining of EN31 Tool Steel using Grey-Taguchi Approach. *Procedia Materials Science*, *6*, 729–740. doi:10.1016/j.mspro.2014.07.089

Davidson, C. I., Hendrickson, C. T., Matthews, H. S., Bridges, M. W., Allen, D. T., Murphy, C. F., & Austin, S. et al. (2010). Preparing future engineers for challenges of the 21st century: Sustainable engineering. *Journal of Cleaner Production*, *18*(7), 698–701. doi:10.1016/j.jclepro.2009.12.021

Davies, R. (2001). How to boost staff retention. *People management*, 7(8), 54-56.

Davies, A. J., & Kochar, A. K. (2002). Manufacturing Best Practice and performance studies: A critique. *International Journal of Operations & Production Management*, *22*(3), 289–305. doi:10.1108/01443570210417597

Davies, A. R., & Mullin, S. J. (2011). Greening the economy: Interrogating sustainability innovations beyond the mainstream. *Journal of Economic Geography*, *11*(5), 793–816. doi:10.1093/jeg/lbq050

De Cicco, F. (2004). *Sistemas Integrados de Gestão*: agregando valor aos sistemas ISO 9000 (in Portuguese). Retrieved from http://www.qsp.com.br/artigo.shtml

De Giovanni, P. (2012). Do internal and external environmental management contribute to the triple bottom line? *International Journal of Operations & Production Management*, *32*(3), 265–290. doi:10.1108/01443571211212574

Deb, K. (2002). *Multi-Objective Optimization using Evolutionary Algorithms*. Singapore: John Wiley & Sons.

Deb, K., Pratap, A., Agarwal, S., & Meyarivan, T. (2002). A fast and elitist multi-objective genetic algorithm: NSGA-II. *IEEE Transactions on Evolutionary Computation*, *6*(2), 182–197. doi:10.1109/4235.996017

Defee, C. C., & Fugate, B. S. (2010). Changing perspective of capabilities in the dynamic supply chain era. *International Journal of Logistics Management*, *21*(2), 180–206. doi:10.1108/09574091011071915

Deif, A. M. (2011). A system model for green manufacturing. *Journal of Cleaner Production*, *19*(14), 1553–1559. doi:10.1016/j.jclepro.2011.05.022

Dekker, R., Bloemhof, J., & Mallidis, I. (2012). Operations research for green logistics – an overview of aspects, issues, contributions and challenges. *European Journal of Operational Research*, *219*(3), 671–679. doi:10.1016/j.ejor.2011.11.010

Deloitte report (Energy Resources). (n. d.). Retrieved from www.deloitte.com

Denend, L., & Plambeck, E. (2010). Walmart's Sustainability Strategy (B): 2010 Update.

Deng, J. (1989). Introduction to Grey System. *Journal of Grey System*, *1*, 1–24.

Despeisse, M., Ball, P. D., & Evans, S. (2012). Modelling and Tactics for Sustainable Manufacturing: An Improvement Methodology. In Sustainable Manufacturing (pp. 9-16). Berlin Heidelberg: Springer.

Despeisse, M., Ball, P. D., Evans, S., & Levers, A. (2012). Industrial Ecology at Factory Level – a Conceptual Model. *Journal of Cleaner Production*, *31*(3–4), 30–39. doi:10.1016/j.jclepro.2012.02.027

Despeisse, M., Mbaye, F., Ball, P. D., & Levers, A. (2012). The emergence of sustainable manufacturing practices. *Production Planning and Control*, *23*(5), 354–376. doi:10.1080/09537287.2011.555425

Despeisse, M., Oates, M. R., & Ball, P. D. (2013). Sustainable manufacturing tactics and cross-functional factory modeling. *Journal of Cleaner Production*, *42*, 31–41. doi:10.1016/j.jclepro.2012.11.008

Devi Juwaheer, T., Pudaruth, S., & Monique Emmanuelle Noyaux, M. (2012). Analysing the impact of green marketing strategies on consumer purchasing patterns in Mauritius. *World Journal of Entrepreneurship. Management and Sustainable Development*, *8*(1), 36–59.

Diabat, A., & Govindan, K. (2011). An analysis of the drivers affecting the implementation of green supply chain management. *Resources, Conservation and Recycling*, *55*(6), 659–667. doi:10.1016/j.resconrec.2010.12.002

Digalwar, A.K., Tagalpallewar, A.R. & Sunnapwar, V.K. (2013) Green manufacturing performance measures: an empirical investigation from Indian manufacturing industries. *Measuring Business Excellence*, 17(4), 59-75.

Digalwar, K., A., R. Tagalpallewar, A., & K. Sunnapwar, V. (. (2013). Green manufacturing performance measures: An empirical investigation from Indian manufacturing industries. *Measuring Business Excellence*, *17*(4), 59–75. doi:10.1108/MBE-09-2012-0046

Dillard, J., Dujon, V., & King, M. C. (2009). *Understanding the social dimension of sustainability. Unitec Kington.* Routledge.

Dillard, J., Dujon, V., & King, M. C. (Eds.). (2008). *Understanding the social dimension of sustainability*. Routledge.

DiMaggio, P. J., & Powell, W. W. (1983). The iron cage revisited: Industrial isomorphism and collective rationality in organizational fields. *American Sociological Review*, *48*(2), 147–160. doi:10.2307/2095101

Dixon, S. E. A., & Clifford, A. (2007). Ecopreneurship – a new approach to managing the triple bottom line. *Journal of Organizational Change Management*. doi:10.1108/09534810710740164

Djalali, A., & Vollaard, P. (2008). The complex history of sustainability: an index of trends, authors, projects and fiction. After Zero, 18(1), 160p.

Dobbs, R., Remes, J., Manyika, J., Roxburgh, C., Smit, S., & Schaer, F. (2012). *Urban world: Cities and the rise of the consuming class*. McKinsey Global Institute.

Dorweiler, V. P., & Yakhou, M. (2004). Environmental Accounting: An Essential Component of Business Strategy. *Asian Institute of Technology Thailand Online*, *13*(2), 65–77.

Dos Santos, M. A., Svensson, G., & Padin, C. (2013). Indicators of sustainable business practices: Woolworths in South Africa. *Supply Chain Management: An International Journal*, *18*(1), 104–108. doi:10.1108/13598541311293212

Dos Santos, M. A., Svensson, G., & Padin, C. (2014). Implementation, monitoring and evaluation of sustainable business practices: Framework and empirical illustration. *Corporate Governance. The International Journal of Business in Society*, *14*(4).

Dotoli, M., Fanti, M. P., Meloni, C., & Zhou, M. (2006). Design and optimization of integrated e-supply chain for agile and environmentally conscious manufacturing. *IEEE Transactions on Systems, Man, and Cybernetics*, *36*(1), 62–75. doi:10.1109/TSMCA.2005.859189

Dowlatshahi, S. (2000). Developing a theory of reverse logistics. *Interfaces*, *30*(3), 143–155. doi:10.1287/inte.30.3.143.11670

Drumwright, M. E. (1994). Socially responsible organizational buying: Environmental concern as a noneconomic buying criterion. *Journal of Marketing*, *58*(3), 1–19. doi:10.2307/1252307

D'Souza, C., & Peretiatko, R. (2002). The nexus between industrialization and environment: A case study of Indian enterprises. *Environmental Management and Health*, *13*(1), 80–97. doi:10.1108/09566160210417859

Du Pisani, J. A. (2006). Sustainable development: Historical roots of the concept. *Environmental Sciences*, *3*(2), 83–96. doi:10.1080/15693430600688831

Dubey, R., & Bag, S. (2014). Antecedents of Green Manufacturing Practices: A Journey towards Manufacturing. In *Smart Manufacturing Innovation and Transformation: Interconnection and Intelligence: Interconnection and Intelligence* (p. 333).

Dubey, R., & Bag, S. (2014). Antecedents of Green Manufacturing Practices: A Journey towards Manufacturing. *Smart Manufacturing Innovation and Transformation: Interconnection and Intelligence*, 333.

Dubey, R., & Bag, T. (2014) Antecedents of green manufacturing practices: a journey towards manufacturing sustainability. In Smart Manufacturing Innovation and Transformation: Interconnection and Intelligence (Ch. 13, pp. 333-354). doi:10.4018/978-1-4666-5836-3.ch013

Dubey, R., Gunasekaran, A., Singh, S., and Singh, T., (2015b). Building theory of sustainable manufacturing using total interpretive structural modeling. *International Journal of Systems Science Operations & Logistics*. DOI:.10.1080/23302674.2015.1025890

Dubey, R., & Ali, S. S. (2014). Identification of flexible manufacturing system dimensions and their interrelationship using total interpretive structural modeling and fuzzy MICMAC analysis. *Global Journal of Flexible Systems Management*, *15*(2), 131–143. doi:10.1007/s40171-014-0058-9

Dubey, R., & Bag, S. (2013). Exploring the dimensions of sustainable practices: An empirical study on Indian manufacturing firms. *International Journal of Operations and Quantitative Management*, *19*(2), 123–146.

Dubey, R., Bag, S., & Ali, S. S. (2014). Green supply chain practices and its impact on organizational performance: An insight from Indian rubber industry. *International Journal of Logistics Systems and Management*, *19*(1), 20–42. doi:10.1504/IJLSM.2014.064029

Dubey, R., Bag, S., Ali, S. S., & Venkatesh, V. G. (2013). Green purchasing is key to superior performance: An empirical study. *International Journal of Procurement Management*, *6*(2), 187–210. doi:10.1504/IJPM.2013.052469

Dubey, R., & Gunasekaran, A. (2015). Shortage of sustainable supply chain talent: An industrial training framework. *Industrial and Commercial Training*, *47*(2), 86–94.

Dubey, R., Gunasekaran, A., & Ali, S. S. (2014). Exploring the Relationship between Leadership, Operational Practices, Institutional Pressures and Environmental Performance: A Framework for Green Supply Chain. *International Journal of Production Economics*.

Dubey, R., Gunasekaran, A., & Ali, S. S. (2015). Exploring the relationship between leadership, operational practices, institutional pressures and environmental pressure: A framework for green supply chain. *International Journal of Production Economics*, *160*(1), 120–132. doi:10.1016/j.ijpe.2014.10.001

Dubey, R., Gunasekaran, A., & Chakrabarty, A. (2015a). World-class sustainable manufacturing: Framework and a performance measurement system. *International Journal of Production Research*, *53*(17), 5207–5223. doi:10.1080/00207543.2015.1012603

Dubey, R., Gunasekaran, A., Childe, S. J., Wamba, S. F., & Papadopoulos, T. (2015). The impact of big data on world-class sustainable manufacturing. *International Journal of Advanced Manufacturing Technology*, *2015*, 1–15.

Dubey, R., Gunasekaran, A., Singh, S., & Singh, T. (2015). Building theory of sustainable manufacturing using total interpretive structural modelling. *International Journal of Production Research*.

Dubey, R., Gunasekaran, A., & Stephen, J. C. (2015c). The design of a responsive sustainable supply chain network under uncertainty. *International Journal of Advanced Manufacturing Technology*, *80*(1-4), 427–445. doi:10.1007/s00170-015-6967-8

Duijm, N. J., Fiévez, C., Gerbec, M., Hauptmanns, U., & Konstandinidou, M. (2008). Management of health, safety and environment in process industry. *Safety Science*, *46*(6), 908–920. doi:10.1016/j.ssci.2007.11.003

Duperrin, J. C., & Godet, M. (1973). *The method for hierarchical system elements* (pp. 45–51). Paris: Rapp. Econ. De CEA.

Dyer, J. H., & Hatch, N. W. (2004). Using supplier network to learn faster. *MIT Sloan Management Review*, *45*(3), 57–63.

Dyllick, T., & Hockerts, K. (2002). Beyond the business case for corporate sustainability. *Business Strategy and the Environment*, *11*(2), 130–141. doi:10.1002/bse.323

Ebeid, S. J., Hewidy, M. S., El-Taweel, T. A., & Youssef, A. H. (2004). Towards higher accuracy for ECM hybridized with low-frequency vibrations using the response surface methodology. *Journal of Materials Processing Technology*, *149*(1-3), 432–438. doi:10.1016/j.jmatprotec.2003.10.046

Eccles, R. G., Perkins, K. M., & Serafeim, G. (2012). How to become a sustainable company. *MIT Sloan Management Review*, *53*(4), 43–50.

Eccles, R. G., Perkins, K. M., & Serafeim, G. (2012). How to become a sustainable company. *MIT Sloan Management Review*, *53*(4), 43–50.

Edward, J. R. (1999). Person job fit: a conceptual integration, literature review, and methodological critique. In C.L. Cooper & I.T. Robertson (Ed.), International review of industrial and organizational psychology (Vol. 6, pp. 283-357). John Wiley and Sons Ltd.

Edwards, J. B., McKinnon, A. C., & Cullinane, S. L. (2010). Comparative analysis of the carbon footprints of conventional and online retailing: A "last mile" perspective. *International Journal of Physical Distribution & Logistics Management*, *40*(1/2), 103–123. doi:10.1108/09600031011018055

Ehrgott, M., Reimann, F., Kaufmann, L., & Carter, C. R. (2011). Social sustainability in selecting emerging economy suppliers. *Journal of Business Ethics*, *98*(1), 99–119. doi:10.1007/s10551-010-0537-7

Eiadat, Y., Kelly, A., Roche, F., & Eyadat, H. (2008). Green and competitive? An empirical test of the mediating role of environmental innovation strategy. *Journal of World Business*, *43*(2), 131–145. doi:10.1016/j.jwb.2007.11.012

Eisenberger, R., Fasolo, P., & Davis-LaMastro, V. (1990). Perceived organizational support and employee diligence, commitment, and innovation. *The Journal of Applied Psychology*, *75*(1), 51–59. doi:10.1037/0021-9010.75.1.51

Eisenhardt, K. M. (1989). Agency theory: An assessment and review. *Academy of Management Review*, *14*(1), 57–74.

El Serafy, S. (1997). Green accounting and economic policy. *Ecological Economics*, *21*(3), 217–229. doi:10.1016/S0921-8009(96)00107-3

Elkington, J. (1997). Cannibals with forks. *The triple bottom line of 21st century.*

Elkington, J. (2004). Enter the triple bottom line. In A. Henriques, & J. Richardson (Eds.), *The triple bottom line: Does it all add up* (Ch. 1). London: Earthscan.

Elkington, J. (1998). *Cannibals with Forks: The Triple Bottom Line of 21st Century Business*. Canada: New Society Publishers.

Elkin, T., McLaren, D., & Hillman, M. (1991). *Reviving the City*. London: Friends of the Earth.

Elliott, J. (2012). *An Introduction to Sustainable Development*. USA: Routledge.

Ellis, M., Kisling, E., & Hackworth, R. G. (2014). Teaching Soft Skills Employers Need. *Community College Journal of Research and Practice*, *38*(5), 433–453. doi:10.1080/10668926.2011.567143

Ellram, L. M., & Cooper, M. C. (2014). Supply Chain Management: It's All About the Journey, Not the Destination. *Journal of Supply Chain Management*, *50*(1), 8–20. doi:10.1111/jscm.12043

Emmelhainz, M. A., & Adams, R. J. (1999). The apparel industry response to "sweatshop" concerns: A review and analysis of codes of conduct. *Journal of Supply Chain Management*, *35*(2), 51–57. doi:10.1111/j.1745-493X.1999.tb00062.x

Emrouznejad, A., & Dey, P. K. (2011). Performance measurement in the health sector: Uses of frontier efficiency methodologies and multi-criteria decision making. *Journal of Medical Systems*, *35*(5), 977–979. doi:10.1007/s10916-010-9622-9 PMID:21072574

Epstein, M. J., & Wisner, P. S. (2001). Using a balanced scorecard to implement sustainability. *Environmental Quality Management*, *11*(2), 1–10. doi:10.1002/tqem.1300

Eresi, K. (1996). Environment-Information Disclosure in Annual Reports. *Chartered Accountant-New Delhi*, *44*, 45–48.

Eskildesen, J., & Hahlgaard, J. (2000). A meta-analysis of antecedents and correlates of employee turnover: Update, Moderator test, and research implications for the next millennium. *Journal of Management*, *26*(3), 463–488. doi:10.1177/014920630002600305

Faber, N., Jorna, R., & Van Engelen, J. (2005). The Sustainability of "Sustainability"- A Study into The Conceptual Foundations of the Notion of Sustainability. *Journal of Environmental Assessment Policy and Management*, *7*(1), 1–33. doi:10.1142/S1464333205001955

Fan, L. T., & Tengyan, Z. (2012). Estimation of energy dissipation and cost: the foundation for sustainability assessment in process design. In D.C.Y. Foo, M.M. El-Halwagi, & R.R. Tan (Eds.), Recent Advances in Sustainable Process Design and Optimization. Singapore: World Scientific Publishing Co. Pte Ltd.

Fang, K., Uhan, N., Zhao, F., & Sutherland, J. W. (2011). A new shop scheduling approach in support of sustainable manufacturing. In *Glocalized Solutions for Sustainability in Manufacturing* (pp. 305–310). Springer Berlin Heidelberg. doi:10.1007/978-3-642-19692-8_53

Fang, X., Qu, N., Zhang, Y., Xu, Z., & Zhu, D. (2013). Effects of pulsating electrolyte flow in electrochemical machining. *Journal of Materials Processing Technology*, *214*(1), 36–43. doi:10.1016/j.jmatprotec.2013.07.012

Farber, B. (1991). *Crisis in education: Stress and burnout in the American teacher*. San Francisco, CA: Jossey-Bass.

Farris, D. R., & Sage, A. P. (1975). On the use of interpretive structural modeling for worth assessment. *Computers & Electrical Engineering*, *2*(2), 149–174. doi:10.1016/0045-7906(75)90004-X

Fawcett, S. E., Magnan, G. M., & McCarter, M. W. (2008). Benefits, barriers, and bridges to effective supply chain management. *Supply Chain Management: An International Journal*, *13*(1), 35–48. doi:10.1108/13598540810850300

Fearne, A., Martinez, M. G., & Dent, B. (2012). Dimensions of sustainable value chains: Implications for value chain analysis. *Supply Chain Management: An International Journal*, *17*(6), 575–581. doi:10.1108/13598541211269193

Feisel, H., Hartmann, E., & Giunipero, L. (2011). The importance of human aspect in the supply function: Strategies for developing PSM proficiency. *Journal of Purchasing and Supply Management*, *17*(2), 54–67. doi:10.1016/j.pursup.2010.07.001

Feng, S. C., & Joung, C. B. (2009, December). An overview of a proposed measurement infrastructure for sustainable manufacturing. *Proceedings of the 7th Global Conference on Sustainable Manufacturing*. Retrieved from http://www.nist.gov/customcf/get_pdf.cfm

Ferreira, S.S. (2009) Modelo de sistema integrado de gestão em eficiência energética e segurança e saúde no trabalho. Brasília: PPGENE, DM.

Ferroli, P.C.M. & Librelotto, L.I. (2011). Integração da sustentabilidade em ferramenta projetual: feap-sus. *Produção online*, 11(2), 447-475.

Field, A. (2009). *Discovering statistics using SPSS*. Sage publications.

Figge, F., Hahn, T., Schaltegger, S., & Wagner, M. (2002). The sustainability balanced scorecard: Linking sustainability management to business strategy. *Business Strategy and the Environment*, *11*(5), 269–284. doi:10.1002/bse.339

Flammer, C. (2013). Corporate Social Responsibility and Shareholder Reaction: The Environmental Awareness of Investors. *Academy of Management Journal*, *56*(3), 758–781. doi:10.5465/amj.2011.0744

Flint, D. J., & Golicic, S. L. (2009). Searching for competitive advantage through sustainability: A qualitative study in the New Zealand wine industry. *International Journal of Physical Distribution & Logistics Management*, *39*(10), 841–860. doi:10.1108/09600030911011441

Florida, R. (1996). Lean and green: The move to environmentally conscious manufacturing. *California Management Review*, *39*(1), 80–105. doi:10.2307/41165877

Flynn, B. B., Schroeder, R. G., & Flynn, E. J. (1999). World class manufacturing: An investigation of Hayes and Wheelwright's foundation. *Journal of Operations Management*, *17*(3), 249–269. doi:10.1016/S0272-6963(98)00050-3

Flyvbjerg, B. (2010). Five misunderstandings about case-study research. In C. Seale, G. Gobo, J. F. Gubrium, & D. Silverman (Eds.), *Qualitative Research Proactive*. Los Angeles/London/New Delhi/Singapore/Washington D.C.: Sage.

FoFdation: Foundation for the Sustainable Factory of the Future. (2012).

Ford, H., & Crowther, S. (1988). Today and tomorrow.

Fornell, C., & Larcker, D. F. (1981). Evaluating Structural Equation Models with Unobservable Variables and Measurement Error. *JMR, Journal of Marketing Research*, *18*(1), 39–50. doi:10.2307/3151312

Foster, J. (2001). Education as sustainability. *Environmental Education Research*, *2001*, 153–165. doi:10.1108/ICT-08-2014-0052

Franceschini, F., & Turina, E. (2013). Quality improvement and redesign of performance measurement systems: An application to the academic field. *Quality & Quantity*, *47*(1), 465–483. doi:10.1007/s11135-011-9530-1

Frazis, H., Gittleman, M., Horrigan, M., & Joyce, M. (1998). Results from the 1995 survey of employer provided training. *Monthly Labor Review*, *21*(6), 3–14.

Frederick, W. C. (1994). From CSR1 to CSR2 The Maturing of Business-and-Society Thought. *Business & Society*, *33*(2), 150–164. doi:10.1177/000765039403300202

Freeman, R. E. (1984). *Strategic Management: A Stakeholder Approach*. Boston: Pitman.

Fresner, J., & Engelhardt, G. (2004). Experiences with integrated management systems for two small companies in Austria. *Journal of Cleaner Production*, *12*(6), 623–631. doi:10.1016/j.jclepro.2003.09.013

Freyermuth. (2007). Retaining Employees in a Tightening Labor Market. RSM McGladrey.

Friedman, T. L. (2006). The world is flat [updated and expanded]: A brief history of the twenty-first century. Macmillan.

Friesen, D. (1986, April). Overall stress and job satisfaction as predictors of burnout. Paper presented at the annual meeting of the American Educational Research Association, San Francisco, CA, USA.

Frizell, S. (2015). Toyota wants everyone to know how it made its hydrogen-powered car. *Time*. Retrieved from http://time.com/3654899/toyota-mirai-patents-hydrogen-power/

Frondel, M., Horbach, J., & Rennings, K. (2008). What triggers environmental management and innovation? Empirical evidence from Germany. *Ecological Economics*, *66*(1), 153–160. doi:10.1016/j.ecolecon.2007.08.016

Gaan, N. S. (2008). Social Support, Job Attitudes and Job Outcome across Gender. *The Icfai University Journal of Organizational Behavior*, *52*, 34–44.

Gaiduk, R., Gaiduk, J., & Fields, D. (2009). Limiting the Brain Drain. Determinants of Employee Organizational attachment in Lithuania. *Baltic Journal of Management*, *4*(2), 149–168. doi:10.1108/17465260910958782

Galpin, T., Whitttington, J. L., & Bell, G. (2015). Is your sustainability strategy sustainable? Creating a culture of sustainability. *Corporate Governance: The International Journal of Business in Society*, *15*(1), 1–17. doi:10.1108/CG-01-2013-0004

Ganesan, S. (1994). Determinants of long-term orientation in buyer-seller relationships. *Journal of Marketing*, *58*(2), 1–19. doi:10.2307/1252265

Garbie, I. H. (2014). An analytical technique to model and assess sustainable development index in manufacturing enterprises. *International Journal of Production Research.*

Garbie, I. H. (2013). DFSME: Design for sustainable manufacturing enterprises (an economic viewpoint). *International Journal of Production Research*, *51*(2), 479–503. doi:10.1080/00207543.2011.652746

Garbie, I. H. (2014). An analytical technique to model and assess sustainable development index in manufacturing enterprises. *International Journal of Production Research*, *52*(16), 4876–4915. doi:10.1080/00207543.2014.893066

Garbie, I. H., & Shikdar, A. (2011). Analysis and estimation of complexity level in industrial firms. *International Journal of Industrial and Systems Engineering*, *8*(2), 175–197. doi:10.1504/IJISE.2011.041368

Garetti, M., Mummolo, G., & Taisch, M. (2012). (Ed.). Special issue on-sustainable manufacturing. Production Planning & Control, 23(2-3), 79-82

Garetti, M., Mummolo, G., & Taisch, M. (2012). Special issue on "sustainable manufacturing". *Production Planning and Control*, *23*(2-3), 79–82. doi:10.1080/09537287.2011.591617

Garetti, M., & Taisch, M. (2011). Sustainable manufacturing: Trends and research challenges. *Production Planning and Control*, *23*(2-3), 83–104. doi:10.1080/09537287.2011.591619

Garg, P., & Rastongi, R. (2006). New model of job design motivation employees Performance. *Journal of Management Development*, *25*(6), 572–587. doi:10.1108/02621710610670137

Garg, R. B. L. (1985). Management Education in India. *Management Accountant*, *20*(1), 19.

Gavronski, I., Klassen, R. D., Vachon, S., & do Nascimento, L. S. M. (2011). A resource-based view of green supply management. *Transportation Research Part E, Logistics and Transportation Review*, *47*(6), 872–885. doi:10.1016/j.tre.2011.05.018

Getabstract. (2014). *The World's Largest Library of Business Book Summaries*. Retrieved from http://www.getabstract.com/

Ghadimi, P., Azadnia, A. H., Mohd Yusof, N., & Mat Saman, M. Z. (2012). A weighted fuzzy approach for product sustainability assessment: A case study in automotive industry. *Journal of Cleaner Production*, *33*, 10–21. doi:10.1016/j.jclepro.2012.05.010

Ghosh, D., & Shah, J. (2012). A comparative analysis of greening polices across supply chain structures. *International Journal of Production Economics*, *135*(2), 568–583. doi:10.1016/j.ijpe.2011.05.027

Giannakis, M. (2007). Performance measurement of supplier relationships. *Supply Chain Management: An International Journal*, *12*(6), 400–411. doi:10.1108/13598540710826335

Gibb, S. (2014). Soft skills assessment: Theory development and the research agenda. *International Journal of Lifelong Education*, *33*(4), 455–471. doi:10.1080/02601370.2013.867546

Gibson, R. B. (2006). Sustainability assessment: Basic components of a practical approach. *Impact Assessment and Project Appraisal*, *24*(3), 170–182. doi:10.3152/147154606781765147

Giddens, A. (1984). *The Constitution of Society: Outline of the Theory of Structuration*. Berkeley, CA: University of California Press.

Gilbert, S. (2000). Integrated summary of the conference on "Greening Supply Chain".

Gimenez, C., Sierra, V., & Rodon, J. (2012). Sustainable operations: Their impact on the triple bottom line. *International Journal of Production Economics*, *140*(1), 149–159. doi:10.1016/j.ijpe.2012.01.035

Gimenez, C., & Tachizawa, E. M. (2012). Extending sustainability to suppliers: A systematic literature review. *Supply Chain Management: An International Journal*, *17*(5), 531–543. doi:10.1108/13598541211258591

Giunipero, L. C. (2000). *A skills-based analysis of the world class purchaser*. Center for Advanced Purchasing Studies.

Giunipero, L. C., & Eltantawy, R. A. (2004). Securing the upstream supply chain: A risk management approach. *International Journal of Physical Distribution & Logistics Management*, *34*(9), 698–713. doi:10.1108/09600030410567478

Giunipero, L. C., Hooker, R. E., & Denslow, D. (2012). Purchasing and supply management sustainability. *Journal of Purchasing and Supply Management*, *18*(4), 258–269. doi:10.1016/j.pursup.2012.06.003

Gladwin, T. N., Krause, T. S., & Kennelly, J. J. (1995). Beyond eco efficiency: Towards socially sustainable business. *Sustainable Development*, *31*(1), 35–43. doi:10.1002/sd.3460030105

Global Reporting Initiative. (2014). Retrieved from http://www.globalreporting.org

Glover, J. L., Champion, D., Daniels, K. J., & Dainty, A. D. J. (2014). An institutional theory perspective on sustainable practices across the diary supply chain. *International Journal of Production Economics*, *152*(1), 102–111. doi:10.1016/j.ijpe.2013.12.027

Godfrey, R. (1998). Ethical purchasing: developing the supply chain beyond the environment. In *Greener purchasing: Opportunities and innovations* (pp. 244-251).

Godschalk, D. R. (2004). Land use planning challenges: Coping with conflicts in visions of sustainable development and livable communities. *Journal of the American Planning Association*, *70*(1), 5–13. doi:10.1080/01944360408976334

Godschalk, S. K. (2009). Does Corporate environmental accounting make business sense? In *Environmental management accounting for cleaner production* (pp. 249–265). Springer Netherlands. Doi:10.1007/978-1-4020-8913-8_13

Godsell, J., Birtwistle, A., & van Hoek, R. (2010). Building the supply chain to enable business alignment: Lessons from British American Tobacco (BAT). *Supply Chain Management: An International Journal*, *15*(1), 10–15. doi:10.1108/13598541011018085

Godsell, J., Harrison, A., Emberson, C., & Storey, J. (2006). Customer responsive supply chain strategy: An unnatural act? *International Journal of Logistics: Research and Applications*, *9*(1), 47–56. doi:10.1080/13675560500534664

Goebel, P., Reuter, C., Pibernik, R., & Sichtmann, C. (2012). The influence of ethical culture on supplier selection in the context of sustainable sourcing. *International Journal of Production Economics*, *140*(1), 7–17. doi:10.1016/j.ijpe.2012.02.020

Gold, M. (2001). Breaking all the rules for recruitment and retention. *Journal of career planning and employment*, 61(3), 6-8.

Goldberg, R. A., & Yagan, J. D. (2007). McDonald's Corporation: managing a sustainable supply chain. Harvard Business School, 16.

Gold, S., Seuring, S., & Beske, P. (2010). Sustainable supply chain management and inter-organizational resources: A literature review. *Corporate Social Responsibility and Environmental Management*, *17*(4), 230–245.

Goldsmith, E., Allen, R., Allaby, M., Davoll, J., & Lawrence, S. (1972). *A blueprint for survival*. Harmondsworth: Penguin.

Goldstein, I. (1991). Training in organizations. In M.D. Dunnette, & L.M. Hough (Ed.), Handbook of Industrial and Organizational Psychology (2nd ed.,Vol. 2, pp. 507-619).

Gomez-Mmejia, L. R., Balkin, D. B., & Cardy, R. L. (1995). *Managing human resources*. Englewood cliffs, NJ, USA: Prentice-Hall, Inc.

Gonzalez, P., Sarkis, J., & Adenso-Diaz, B. (2008). Environmental management system certification and its influence on corporate practices: Evidence from the automotive industry. *International Journal of Operations & Production Management*, *28*(11), 1021–1041. doi:10.1108/01443570810910179

Gonzalez-Torre, P. L., Adenso-Diaz, B., & Artiba, H. (2004). Environmental and reverse logistics policies in European bottling and packaging firms. *International Journal of Production Economics*, *88*(1), 95–104. doi:10.1016/S0925-5273(03)00181-6

Gopalakrishnan, K., Yusuf, Y. Y., Musa, A., Abubakar, T., & Ambursa, H. M. (2012). Sustainable supply chain management: A case study of British Aerospace (BAe) Systems. *International Journal of Production Economics*, *140*(1), 193–203. doi:10.1016/j.ijpe.2012.01.003

Gore, A. (2006). *An inconvenient truth: The planetary emergency of global warming and what we can do about it*. Rodale.

Gorman, H. (2000). Winning hearts and minds? - Emotional labour and learning for care management. *Journal of Social Work Practice*, *14*(2), 149–158. doi:10.1080/02650530020020546

Gotschol, A., De Giovanni, P., & Vinzi, V. E. (2014). Is environmental management an economically sustainable business? *Journal of Environmental Management*, *144*, 73–82. doi:10.1016/j.jenvman.2014.05.001 PMID:24921964

Govindan, K., Kaliyan, M., Kannan, D., & Haq, A. N. (2014). Barriers analysis for green supply chain management implementation in Indian industries using analytic hierarchy process. *International Journal of Production Economics*, *147*, 555–568. doi:10.1016/j.ijpe.2013.08.018

Goworek, H. (2011). Social and environmental sustainability in the clothing industry: A case study of a fair trade retailer. *Social Responsibility Journal*, *7*(1), 74–86. doi:10.1108/17471111111114558

Graddick, M. N. (1988). *Corporate Philosophies of Employee Development, Career Growth and Human Resource Strategies: The Role of the Human Resource Professional in Employee Development* (M. London & E. M. Mone, Eds.). Westport, CT, USA: Quorum Books.

Graedel, T. E., Allwood, J., Birat, J. P., Reck, B. K., Sibley, S. F., Sonnemann, G., & Hagelüken, C. (2011). *Recycling Rates of Metals-A Status Report. A Report of the Working Group on the Global Metal Flows to the International Resource Panel*. United Nations Environment Programme.

Granovetter, M. (1985). Economic action and social structure: The problem of embeddedness. *American Journal of Sociology*, *91*(3), 481–510. doi:10.1086/228311

Gray, R., & Bebbington, J. (2007). Corporate sustainability, accountability and the pursuit of the impossible dream. Handbook of sustainable development, 376-394.

Gray, R. (2001). Thirty years of social accounting, reporting and auditing: What (if anything) have we learnt? *Business Ethics (Oxford, England)*, *10*(1), 9–15. doi:10.1111/1467-8608.00207

Gray, R. (2010). Is accounting for sustainability actually accounting for sustainability… and how would we know? An exploration of narratives of organisations and the planet. *Accounting, Organizations and Society*, *35*(1), 47–62. doi:10.1016/j.aos.2009.04.006

Gray, R., Owen, D., & Maunders, K. (1988). Corporate social reporting: Emerging trends in accountability and the social contract. *Accounting, Auditing & Accountability Journal*, *1*(1), 6–20. doi:10.1108/EUM0000000004617

Green Purchasing Australia report. (2009). Retrieved from http://www.ecobuy.org.au

Greenhaus, J. H., & Callanan, G. A. (1994). *Career Management*. Fort Worth, Texas: The Dryden Press.

Green, K. W. Jr, Whitten, D., & Inman, R. A. (2008). The impact of logistics performance on organizational performance in a supply chain context. *Supply Chain Management: An International Journal*, *13*(4), 317–327. doi:10.1108/13598540810882206

Green, K. W. Jr, Zelbst, P. J., Meacham, J., & Bhadauria, V. S. (2012). Green supply chain management practices: Impact on performance. *Supply Chain Management: An International Journal*, *17*(3), 290–305. doi:10.1108/13598541211227126

Green, K., Morton, B., & New, S. (1996). Purchasing and environmental management: Interactions, policies and opportunities. *Business Strategy and the Environment*, *5*(3), 188–197. doi:10.1002/(SICI)1099-0836(199609)5:3<188::AID-BSE60>3.0.CO;2-P

Green, K., Morton, B., & New, S. (1998). Green purchasing and supply policies: Do they improve companies' environmental performance? *Supply Chain Management: An International Journal*, *3*(2), 89–95. doi:10.1108/13598549810215405

Griffin, A., & Brimacombe, L. G. (2010). Steely resolve'. *Quest. TQMS*, *5*, 27–32.

Grimm, J. H., Hofstetter, J. S., & Sarkis, J. (2013). Critical factors for sub-supplier management: A sustainable food supply chains perspective. *International Journal of Production Economics*, *152*, 159–173. doi:10.1016/j.ijpe.2013.12.011

Grober, U. (2007). *Deep roots: a conceptual history of "sustainable development" (Nachhaltigkeit)*. Berlin: WZB.

Grosvold, J., Hoejmose, S., & Roehrich, J. (2014). Squaring the Circle: Management, Measurement and Performance of Sustainability in Supply Chains. *Supply Chain Management: An International Journal*, *19*(3), 6–6.

Guang Shi, V., Lenny Koh, S. C., Baldwin, J., & Cucchiella, F. (2012). Natural resource based green supply chain management. *Supply Chain Management: An International Journal*, *17*(1), 54–67. doi:10.1108/13598541211212203

Gudmundsson, E., & Roheim, C. A. (2000). Ecolabeling seafood for sustainable production: Implications for fisheries management. *Marine Resource Economics*, *15*(2), 97–113. doi:10.1086/mre.15.2.42629294

Guide, V. D. R., Harrison, T. P., & Van Wassenhove, L. N. (2003). The challenge of closed-loop supply chains. *Interfaces*, *33*(6), 3–6. doi:10.1287/inte.33.6.3.25182

Guillén, O. P., & Carmenates, Y. P. (2009). Sistema de Gestión Integrado: Calidad, Medio Ambiente y Seguridad y Salud en el Trabajo como herramienta del Sistema de Dirección y Gestión en la Empresa de Investigaciones y Proyectos Hidráulicos de Ciego de Ávila. *Enlace. Inter-American Commission of Women*, *87*(5), 1–20.

Gummesson, E. (1996). Relationship marketing and the imaginary organisation: A synthesis. *European Journal of Marketing*, *30*(2), 31–44. doi:10.1108/03090569610106635

Gunasekaran, A., & Gallear, D. (2012). Special issue on sustainable development of manufacturing and services. *International Journal of Production Economics*, *140*(1), 1–6. doi:10.1016/j.ijpe.2012.07.005

Gunasekaran, A., Irani, Z., & Papadopoulos, T. (2013). Modelling and analysis of sustainable operations management: Certain investigations for research and applications. *The Journal of the Operational Research Society*, *65*(6), 806–823. doi:10.1057/jors.2013.171

Gunasekaran, A., & Kobu, B. (2007). Performance measures and metrics in logistics and supply chain management: A review of recent literature (1995-2004) for research and applications. *International Journal of Production Research*, *45*(12), 2819–2840. doi:10.1080/00207540600806513

Gunasekaran, A., Patel, C., & McGaughey, R. E. (2004). A framework for supply chain performance measurement. *International Journal of Production Economics*, *87*(3), 333–347. doi:10.1016/j.ijpe.2003.08.003

Gunasekaran, A., Patel, C., & Tirtiroglu, E. (2001). Performance measurement and metrics in a supply chain environment. *International Journal of Operations & Production Management*, *21*(1), 71–87. doi:10.1108/01443570110358468

Gunasekaran, A., & Spalanzani, A. (2012). Sustainability of manufacturing and services: Investigations for research and applications. *International Journal of Production Economics*, *140*(1), 35–47. doi:10.1016/j.ijpe.2011.05.011

Gungor, A., & Gupta, S. M. (1999). Issues in environmentally conscious manufacturing and product recovery: A survey. *Computers & Industrial Engineering*, *36*(4), 811–853. doi:10.1016/S0360-8352(99)00167-9

Gupta, S., & Pasule-Desai, O. D. (2011). Sustainable supply chain management: Review and research opportunities. *IIMB Management Review*, *23*(4), 234–245. doi:10.1016/j.iimb.2011.09.002

Hacking, T., & Gurthrie, P. (2008, February). A framework for clarifying the meaning of Triple Bottom-Line, Integrated, and Sustainability Assessment. *Environmental Impact Assessment Review*, *28*(2-3), 73–89. doi:10.1016/j.eiar.2007.03.002

Hair, J. F. Jr, Black, W. C., Babin, B. J., Anderson, R. E., & Tantham, R. L. (2011). *Multivariate Data Analysis* (6th ed.). India: Person Education Inc.

Hakansson, H., & Snehota, I. (1995). *Developing business relationships*. London, UK: Routledge.

Hall, D. (1996). *The Career Is Dead, Long Live the Career: A Relational Approach to Careers*. San Francisco, CA: Jossey-Bass Publishers.

Hall, D. T., & Moss, J. E. (1998). The new protean career contract: Helping organizations and employees adapt. *Organizational Dynamics*, *26*(3), 22–37. doi:10.1016/S0090-2616(98)90012-2

Halldorsson, A., & Kovacs, G. (2010). The sustainable agenda and energy efficiency: Logistics solutions and supply chains in times of climate change. *International Journal of Physical Distribution & Logistics Management*, *40*(1/2), 5–13. doi:10.1108/09600031011018019

Hall, J. (2000). Environmental supply chain dynamics. *Journal of Cleaner Production*, *8*(6), 455–471. doi:10.1016/S0959-6526(00)00013-5

Hamidi, N., Omidvari, M., & Meftahi, M. (2012). The effect of integrated management system on safety and productivity indices: Case study, Iranian cement industries. *Safety Science*, *50*(5), 1180–1189. doi:10.1016/j.ssci.2012.01.004

Hamner, B. (2006). Effects of green purchasing strategies on supplier behaviour. In *Greening the supply chain* (pp. 25–37). Springer London. doi:10.1007/1-84628-299-3_2

Handfield, R. B., Krause, D. R., Scannell, T. V., & Monczka, R. M. (2000). Avoid the pitfalls in supplier development. *Mit Sloan Management Review*, 41(2).

Handfield, R., Sroufe, R., & Walton, S. (2005). Integrating environmental management and supply chain strategies. *Business Strategy and the Environment*, *14*(1), 1–19. doi:10.1002/bse.422

Handfield, R., Walton, S. V., Sroufe, R., & Melnyk, S. A. (2002). Applying environmental criteria to supplier assessment: A study in the application of the Analytical Hierarchy Process. *European Journal of Operational Research*, *141*(1), 70–87. doi:10.1016/S0377-2217(01)00261-2

Hansson, J., Backlund, F., & Lycke, L. (2003). Managing commitment: Increasing the odds for successful implementation of TQM, TPM or RCM. *International Journal of Quality & Reliability Management*, *20*(9), 993–1008. doi:10.1108/02656710310500815

Harte, G., & Owen, D. (1991). Environmental disclosure in the annual reports of British companies: A research note. *Accounting, Auditing & Accountability Journal*, *4*(3). doi:10.1108/09513579110144442

Hart, S. L. (1995). A natural resource-based view of the firm. *Academy of Management Review*, *20*(4), 986–1014.

Hart, S. L. (1997). 'Beyond greening: Strategies for a sustainable world. *Harvard Business Review*, *75*, 66–77. PMID:10174798

Hasan, M. (2013). 'Sustainable supply chain management practices and operational performance. *American Journal of Industrial and Business Management*, *3*(1), 42–48. doi:10.4236/ajibm.2013.31006

Hassini, E., Surti, C., & Searcy, C. (2012). A literature review and a case study of sustainable supply chains with a focus on metrics. *International Journal of Production Economics*, *140*(1), 69–82. doi:10.1016/j.ijpe.2012.01.042

Haugh, H. M., & Talwar, A. (2010). How do corporations embed sustainability across the organization? *Academy of Management Learning & Education*, *9*(3), 384–396. doi:10.5465/AMLE.2010.53791822

Hayes, R. H., & Wheelwright, S. C. (1985), Restoring our Competitive Edge: Competing through Manufacturing, John Wiley & Sons, New York, NY.

Hayes, R. H., & Wheelwright, S. C. (1984). *Restoring Our Competitive Edge: Competing Through Manufacturing*. New York: John Wiley.

Heilala, J., Vatanen, S., Tonteri, H., Montonen, J., Lind, S., Johansson, B., & Stahre, J. (2008, December). Simulation-based sustainable manufacturing system design. *Proceedings of theSimulation Conference WSC '08* (pp. 1922-1930). IEEE. doi:10.1109/WSC.2008.4736284

Henderson, R. M., & Nellemann, F. (2011). Sustainable Tea at Unilever. *Harvard Business School Marketing Unit Case*.

Henderson, S. (2011). The development of competitive advantage through sustainable event management. *Worldwide Hospitality and Tourism Themes*, *3*(3), 245–257. doi:10.1108/17554211111142202

Henriques, A. (2004). CSR, Sustainability and Triple Bottom Line. In A. Henriques & J. Richardson (Eds.), *The Triple Bottom Line: does it all add up?* (pp. 17–25). London: Earthcan.

Hens, L., & Nath, B. (2005). *'The world summit on sustainable development*. Dordrecht: Springer. doi:10.1007/1-4020-3653-1

Herath, A. (2011).The Climate Change Debate: Man vs. Nature. *livescience*. Retrieved from http://www.livescience.com/16388-climate-change-debate-man-nature.html

Hervani, A. A., Helms, M. M., & Sarkis, J. (2005). Performance measurement for green supply chain management. *Benchmarking International Journal (Toronto, Ont.)*, *12*(4), 330–353.

Hervani, A. A., Helms, M. M., & Sarkis, J. (2005). Performance measurement for green supply chain management. *Benchmarking: An International Journal*, *12*(4), 330–353. doi:10.1108/14635770510609015

Herzig, C., Viere, T., Schaltegger, S., & Burritt, R. L. (2012). *Environmental Management Accounting: Case Studies of South-East Asian Companies*. Routledge.

Highhouse, S., Stierwalt, S. L., Bachiochi, P., Elder, A. E., & Fisher, G. (1999). Effects of advertised human resource management practices on attraction of African American applicants. *Personnel Psychology*, *52*(2), 425–442. doi:10.1111/j.1744-6570.1999.tb00167.x

Hilson, G., & Nayee, V. (2002). Environmental management system implementation in the mining industry: A key to achieving cleaner production. *International Journal of Mineral Processing*, *64*(1), 19–41. doi:10.1016/S0301-7516(01)00071-0

Hines, P., Holwe, M., & Rich, N. (2004). Learning to evolve: A review of contemporary lean thinking. *International Journal of Operations & Production Management*, *24*(10), 994–1011. doi:10.1108/01443570410558049

Hirsch, P. M. (1975). Organizational effectiveness and the institutional environment. *Administrative Science Quarterly*, *20*(3), 327–344. doi:10.2307/2391994

Hitchcock, T. (2012). Low carbon and green supply chains: The legal drivers and commercial pressures. *Supply Chain Management, International Journal (Toronto, Ont.)*, *17*(1), 98–101.

Hitchcock, T. (2012). Low carbon and green supply chains: The legal drivers and commercial pressures. *Supply Chain Management: An International Journal*, *17*(1), 98–101. doi:10.1108/13598541211212249

Hoejmose, S. U., & Adrien-Kirby, A. J. (2012). Socially and environmentally responsible procurement: A literature review and future research agenda of a managerial issue in the 21st century. *Journal of Purchasing and Supply Management*, *18*(4), 232–242. doi:10.1016/j.pursup.2012.06.002

Hoejmose, S., Brammer, S., & Millington, A. (2012). Green supply chain management: The role of trust and top management in B2B and B2C markets. *Industrial Marketing Management*, *12*(4), 609–620. doi:10.1016/j.indmarman.2012.04.008

Hoejmose, S., Brammer, S., & Millington, A. (2013). An empirical examination of the relationship between business strategy and socially responsible supply chain management. *International Journal of Operations & Production Management*, *33*(5), 589–621. doi:10.1108/01443571311322733

Ho, J. C., Shalishali, M. K., Tseng, T., & Ang, D. S. (2009). Opportunities in green supply chain management. *The Coastal Business Journal*, *8*(1), 18–31.

Ho, L. W., Dickinson, N. M., & Chan, G. (2010, February). Green procurement in the Asian public sector and the Hong Kong private sector. In *Natural resources forum*. UK: Blackwell Publishing Ltd. doi:10.1111/j.1477-8947.2010.01274.x

Holdsworth, R. (2003). Development and implementation of an integrated management system: Applications approach to design. *Journal of Hazardous Materials Practical*, *104*(1), 193–205. doi:10.1016/j.jhazmat.2003.08.001 PMID:14602409

Holt, D., & Ghobadian, A. (2009). An empirical study of green amongst UK manufacturers. *Journal of Manufacturing Technology Management*, *20*(7), 933–956. doi:10.1108/17410380910984212

Hong, P., Roh, J. J., & Rawski, G. (2012). Benchmarking sustainability practices: Evidence from manufacturing firms. *Benchmarking: An International Journal*, *19*(4/5), 634–648. doi:10.1108/14635771211258052

Hopkins, M. S., Haanaes, K., Balagopal, B., Velken, I., Kruschwitz, N., & Arthur, D. (2011). New sustainability study: The embracers seize advantage. *MIT Sloan Management Review*, *52*(3), 23.

How climate change affects India. (2011). *India climate Portal*. Retrieved from http://www.indiaclimateportal.org

Howard, S., & Bristow, S. (2007). Business and Climate Change: Rising Public Awareness Creates Significant Opportunity. *UN Chronicle*, *44*(2), 56.

Hsu, C. C., Choon Tan, K., Hanim Mohamad Zailani, S., & Jayaraman, V. (2013). Supply chain drivers that foster the development of green initiatives in an emerging economy. *International Journal of Operations & Production Management*, *33*(6), 656–688. doi:10.1108/IJOPM-10-2011-0401

Hsu, C. W., & Hu, A. H. (2008). Green supply chain management in the electronic industry. *International Journal of Environmental Science and Technology*, *5*(2), 205–216. doi:10.1007/BF03326014

Hsu, C. W., & Hu, A. H. (2009). Applying hazardous substance management to supplier selection using analytic network process. *Journal of Cleaner Production*, *17*(2), 255–264. doi:10.1016/j.jclepro.2008.05.004

Hsu, C. W., Kuo, T. C., Chen, S. H., & Hu, A. H. (2013). Using DEMATEL to develop a carbon management model of supplier selection in green supply chain management. *Journal of Cleaner Production*, *56*, 164–172. doi:10.1016/j.jclepro.2011.09.012

Hsu, M. K., Jiang, J. J., Klein, G., & Tang, Z. (2003). Perceived career incentives and intent to leave. *Information & Management*, *40*(5), 361–369. doi:10.1016/S0378-7206(02)00018-6

Hu, A. H., & Hsu, C. (2010). Critical factors for implementing green supply chain management practice: An empirical study of electrical and electronics industries in Taiwan. *Management Research Review*, *33*(6), 586–608. doi:10.1108/01409171011050208

Huberman, M. (1989). The professional life cycle of teachers. *Teachers College Record*, *9*(1), 31–57.

Hudson, M., Smart, A., & Bourne, M. (2001). Theory and practice in SME performance measurement system. *International Journal of Operations & Production Management*, *21*(8), 1096–1115. doi:10.1108/EUM0000000005587

Hult, G. T. M., Ketchen, D. J., & Arrfelt, M. (2007). Strategic supply chain management: Improving performance through a culture of competitiveness and knowledge development. *Strategic Management Journal*, *28*(10), 1035–1052. doi:10.1002/smj.627

Hult, G.T.M. (2011). Market –focused sustainability: Market orientation plus! *Journal of the Academy of Marketing Science*, *39*(1), 1–6. doi:10.1007/s11747-010-0223-4

Humphreys, P. K., Shiu, W. K., & Chan, F. T. S. (2001). Collaborative buyer-supplier relationships in Hong Kong manufacturing firms. *Supply Chain Management: an international journal*, *6*(4), 152-162.

Huq, F., Stevenson, M., Zorzini, M., & Hendry, L. (2013). Social sustainability in the Apparel Supply Chain: an Institutional Theory perspective. *Proceedings of the 20th European Operations Management Association (EurOMA)*.

Huq, F. A., Stevenson, M., & Zorzini, M. (2014). Social sustainability in developing country suppliers: An exploratory study in the ready made garments industry of Bangladesh. *International Journal of Operations & Production Management*, *34*(5), 610–638. doi:10.1108/IJOPM-10-2012-0467

Huselid, M. A. (1995). The impact of human resource management practices on turnover productivity, and corporate financial performance. *Academy of Management Journal*, *38*(3), 635–672. doi:10.2307/256741

Husgafvel, R., Pajunen, N., Virtanen, K., Paavola, I. L., Päällysaho, M., Inkinen, V., ... & Ekroos, A. (2014). Social sustainability performance indicators–experiences from process industry. *International Journal of Sustainable Engineering*.

Hutchins, M. J., & Sutherland, J. W. (2008). An exploration of measures of social sustainability and their application to supply chain decisions. *Journal of Cleaner Production*, *16*(15), 1688–1698. doi:10.1016/j.jclepro.2008.06.001

Hutton, R. B., Cox, D. B., Clouse, M. L., Gaensbauer, J., & Banks, B. D. (2007). The role of sustainable development in risk assessment and management for multinational corporations. *Multinational Business Review*, *15*(1), 89.

Hytter, A. (2008). *Dark Side Leaders, Work Environment and Employee Health*. Retrieved from hvxu.se/ehv/forskning/hofreseminarier/2008/080514%20DarkSide%20Final%20version.pdf

Ihsan Malik & Basharat Naeem. (2009), Research methods in the social Sciences. Melbourne: Martin Press, Inc.

Ijomah, W. L., McMahon, C. A., Hammond, G. P., & Newman, S. T. (2007). Development of design for remanufacturing guidelines to support sustainable manufacturing. *Robotics and Computer-integrated Manufacturing*, *23*(6), 712–719. doi:10.1016/j.rcim.2007.02.017

Iles, P., Chuai, X., & Preece, D. A. (2010). Talent management and HRM in multinational companies in Beijing: Definitions, differences and drivers. *Journal of World Business*, *45*(2), 179–189. doi:10.1016/j.jwb.2009.09.014

Indicators of sustainable development: guidelines and methodologies. (2001). UNDSD. Retrieved from http://www.un.org/esa/sustdev

Intergovernmental Panel on Climate Change, Copenhagen, Denmark. (2014, October 31). Retrieved from http://www.ipcc.ch/scripts/_session_template.php?page=_40ipcc.htm

International Trade Administration. (2013). Retrieved from http://trade.gov/competitiveness/sustainablemanufacturing/how_doc_defines_SM.asp

Investing in Education: Analysis of the 1999 World Education Indicators. (2000). OECD, Paris.

Ireland, R. D., & Webb, J. W. (2007). A multi-theoretic perspective on trust and power in strategic supply chains. *Journal of Operations Management*, *25*(2), 482–497. doi:10.1016/j.jom.2006.05.004

Irshad, M. (2012). Factors affecting Employee Retention Evidence from Literature Review. *Abasyn Journal of Social Sciences*, 4(1), 84-102. Retrieved from http://64.17.184.140/wp-content/uploads/2012/12/V4I1-7.pdf

ISO. (1998). *TS 9000 – Quality management and quality assurance standards-part 3: guidelines for the application of ISO 9001 to the development, supply and maintenance of software*. Switzerland: International Organization for Standardization.

ISO. (2002). *TS 16949:2002 – Quality Management System – Particular requirements for the application of ISO9001:2000 for automotive production and relevant service part organizations*. Switzerland: International Organization for Standardization.

ISO. (2006). *ISO 22000: Food safety management systems - Requirements for any organization in the food chain*. Switzerland: International Organization for Standardization.

ISO. (2007a). *ISO 27001: Information technology - Security techniques-Information Security management systems-Requirements*. Switzerland: International Organization for Standardization.

ISO. (2007b). *ISO 28000: Specifications for security risk management systems for the supply chain*. Switzerland: International Organization for Standardization.

ISO. (2008). *ISO 9001: Quality management systems — Requirements*. Switzerland: International Organization for Standardization.

ISO. (2008b). *ISO 14001: Environmental management systems—Requirements with guidance for use*. Switzerland: International Organization for Standardization.

ISO. (2008c). *The Integrated Use of Management System Standards*. Switzerland: International Organization for Standardization.

ISO. (2009). *ISO 26000: Guidance on social responsibility*. Switzerland: International Organization for Standardization.

ISO. (2013). *The ISO survey of management system standard certifications – 2012*: Executive summary. Retrieved from http://www.iso.org/iso/iso_survey_executive-summary.pdf

J., Elkington, (1999). Triple bottom line revolution: reporting for the third millennium. *Australian CPA*, 16(9), 75.

Jabbaripour, B., Sadeghi, M. H., Faridvand, S., & Shabgard, M. R. (2012). Investigating the effects of EDM parameters on surface integrity-MRR and TWR in machining of Ti-6Al-4V. *Mining Science and Technology*, *16*, 419–444.

Jagadeesa, T. (2013). *Non Traditional Manufacturing processes-an introduction. Lecture notes, Sept., 4th*. India: NIT Calicut.

Jain, N. K., & Jain, V. K. (2007). Optimization of electro-chemical machining process parameter using genetic algorithms. *Mining Science and Technology*, *11*, 235–258.

Jakhar, S. K. (2014). Designing the green supply chain performance optimisation model. *Global Journal of Flexible Systems Management*, *15*(3), 235–259. doi:10.1007/s40171-014-0069-6

Jalalvand, F., Teimoury, E., Makui, A., Aryanezhad, M. B., & Jolai, F. (2011). A method to compare supply chains of an industry. *Supply Chain Management: An International Journal*, *16*(2), 82–97. doi:10.1108/13598541111115347

Jamali, D. (2006). Insights into triple bottom line integration from a learning organization perspective. *Business Process Management Journal*, *12*(6), 809–821. doi:10.1108/14637150610710945

Janicke, M. (2008). Ecological modernization: New perspectives. *Journal of Cleaner Production*, *16*(5), 557–565. doi:10.1016/j.jclepro.2007.02.011

Janković, S., & Krivačić, D. (2014). Environmental accounting as perspective for hotel sustainability: Literature review. *Tourism and Hospitality Management*, *20*(1), 103–120.

Jasch, C. (2001). Environmental Management Accounting Metrics, Procedures and Principles. *UN Division for Sustainable Development*. DOI:10.1007/0-306-48022-0_3

Jayal, A. D., Badurdeen, F., Dillon, O. W. Jr, & Jawahir, I. S. (2010). Sustainable manufacturing: Modeling and optimization challenges at the product, process and system levels. *CIRP Journal of Manufacturing Science and Technology*, *2*(3), 144–152. doi:10.1016/j.cirpj.2010.03.006

Jenkins, H., & Yakovleva, N. (2006). Corporate social responsibility in the mining industry: Exploring trends in social and environmental disclosure. *Journal of Cleaner Production*, *14*(3), 271–284. doi:10.1016/j.jclepro.2004.10.004

Jennings, P. D., & Zandbergen, P. A. (1995). Ecologically sustainable organizations: An institutional approach. *Academy of Management Review*, *20*(4), 1015–1052.

Jharkharia, S., & Shankar, R. (2005). IT-enablement of supply chains: Understanding the barriers. *Journal of Enterprise Information Management*, *18*(1), 11–27. doi:10.1108/17410390510571466

Jimenez-Gonzalez, C., Poechlauer, P., Broxterman, Q. B., Yang, B. S., am Ende, D., Baird, J., & Manley, J. et al. (2011). Key green engineering research areas for sustainable manufacturing: A perspective from pharmaceutical and fine chemicals manufacturers. *Organic Process Research & Development*, *15*(4), 900–911. doi:10.1021/op100327d

Jin-Hai, L., Anderson, A. R., & Harrison, R. T. (2003). The evolution of agile manufacturing. *Business Process Management Journal*, *9*(2), 170–189. doi:10.1108/14637150310468380

John, B. (2007). The Business of Climate change. *CMA Management*, *81*(1), 48–49.

Johnson, S. (2004). *Environmental management accounting*. Association of Chartered Accountants.

Johnstone, N., & Labonne, J. (2009). Why do manufacturing facilities introduce environmental management systems? Improving and/or signaling performance. *Ecological Economics*, *68*(3), 719–730. doi:10.1016/j.ecolecon.2008.06.003

Jones, C., Hesterly, W. S., & Borgatti, S. P. (1997). A general theory of network governance: Exchange conditions and social mechanisms. *Academy of Management Review*, *22*(4), 911–945.

Jones, R. (2005). Finding sources of brand value: Developing a stakeholder model of brand equity. *The Journal of Brand Management*, *13*(1), 10–32. doi:10.1057/palgrave.bm.2540243

Jones, T. (1995). Instrumental stakeholder theory: A synthesis of ethics and economics. *Academy of Management Review*, *20*(2), 404–437.

Jorgensen, M. S., Jorgensen, U., Hendriksen, K., Hirsbak, S., Thomsen, H. H., & Thorsen, N. (2010). Environmental management in Danish transnational textile product chains. *Management Research Review*, *33*(4), 357–379. doi:10.1108/01409171011030462

Jørgensen, T. H. (2008). Towards more sustainable management systems: Through life cycle management and integration. *Journal of Cleaner Production*, *16*(10), 1071–1080. doi:10.1016/j.jclepro.2007.06.006

Jørgensen, T. H., Remmen, A., & Mellado, M. D. (2006). Integrated management systems: Three different levels of integration. *Journal of Cleaner Production*, *14*(8), 713–722. doi:10.1016/j.jclepro.2005.04.005

Jorgensen, T. H., & Simonsen, G. (2002). Prospects of a unified management system. *Corporate Social Responsibility and Environmental Management*, *9*(2), 91–98. doi:10.1002/csr.11

Joung, C. B., Carrell, J., Sarkar, P., & Feng, S. C. (2013). Categorization of indicators for sustainable manufacturing. *Ecological Indicators*, *24*, 148–157. doi:10.1016/j.ecolind.2012.05.030

Jovane, F., Yoshikawa, H., Alting, L., Boër, C. R., Westkamper, E., Williams, D., & Paci, A. M. et al. (2008). The incoming global technological and industrial revolution towards competitive sustainable manufacturing. *CIRP Annals-Manufacturing Technology*, *57*(2), 641–659. doi:10.1016/j.cirp.2008.09.010

Judge, T. A., Hanisch, K. A., & Drankoski, R. D. (1995). Human resource management and employee attitudes. In G. R. Ferris, S. D. Rosen, & D. T. Barnum (Eds.), *Handbook of human resources management*. United Kingdom: Blackwell Publishers, Ltd.

Juran, J., & Godfrey, A. B. (1999). *Quality handbook*. Republished McGraw-Hill.

Kaebernick, H., Kara, S., & Sun, M. (2003). Sustainable product development and manufacturing by considering environmental requirements. *Robotics and Computer-integrated Manufacturing*, *19*(6), 461–468. doi:10.1016/S0736-5845(03)00056-5

Kafa, N., Hani, Y., & El Mhamedi, A. (2013, September). Sustainability performance measurement for green supply chain management. In Management and Control of Production and Logistics (Vol. 6, No. 1, pp. 71-78).

Kaighobadi, M., & Venkatesh, K. (1994). Flexible manufacturing systems: An overview. *International Journal of Operations & Production Management*, *14*(4), 26–49. doi:10.1108/01443579410056029

Kaiser, H. F. (1974). An index of factorial simplicity. *Psychometrika*, *39*(1), 31–36. doi:10.1007/BF02291575

Källström, H.N., Ljung, M. (2005). Social sustainability and collaborative learning. *AMBIO: A Journal of the Human Environment*; 34(4), 376-382.

Kang, J., Bennett, M. W., Carbado, D. W., Casey, P., Dasgupta, N., Faigman, D. L., & Mnookin, J. (2012). Implicit bias in the courtroom. *UCLA Law Review*, *59*(5).

Kang, S.-H., Kang, B., Shim, K., Kim, D., & Han, J. (2012). A theoretical framework for strategy development to introduce sustainable supply chain management. *Procedia: Social and Behavioral Sciences*, *40*(1), 631–635. doi:10.1016/j.sbspro.2012.03.241

Kannegiesser, M., & Günther, H. O. (2014). Sustainable development of global supply chains-part 1: Sustainability optimization framework. *Flexible Services and Manufacturing Journal*, *26*(1-2), 24–47. doi:10.1007/s10696-013-9176-5

Kaplan, R. S., & Norton, D. P. (1996). Using the balanced scorecard as a strategic management system.

Kaplan, R. S., & Norton, D. P. (1996). *The balanced scorecard: Translating strategy into action*. Boston, MA: Harvard Business School Press.

Karapetrovic, S. (2002). Strategies for the integration of management systems and standards. *The TQM Magazine*, *14*(1), 61–68. doi:10.1108/09544780210414254

Karapetrovic, S. (2003). Musings on integrated management systems. *Measuring Business Excellence*, *7*(1), 4–13. doi:10.1108/13683040310466681

Karapetrović, S., Casadesus, M., & Heras, I. (2010). Analysis of integration within the standards-based Integrated Management Systems.*Proceedings of International Quality Conference* (pp. 1-12).

Kara, S., Ibbotson, S., & Kayis, B. (2014). Sustainable product development in practice: An international survey. *Journal of Manufacturing Technology Management*, *25*(6), 848–872. doi:10.1108/JMTM-09-2012-0082

Kasemsap, K. (2014). The role of lean production on organizational performance. In V. Modrak & P. Semanco (Eds.), *Handbook of research on design and management of lean production systems* (pp. 358–388). Hershey, PA: IGI Global. doi:10.4018/978-1-4666-5039-8.ch017

Kaye Nijaki, L., & Worrel, G. (2012). Procurement for sustainable local economic development. *International Journal of Public Sector Management*, *25*(2), 133–153. doi:10.1108/09513551211223785

Keating, B., Quazi, A., Kriz, A., & Coltman, T. (2008). In pursuit of a sustainable supply chain: Insights from Westpac Banking Corporation. *Supply Chain Management: An International Journal*, *13*(3), 175–179. doi:10.1108/13598540810871217

Kennedy, S., & Dowling, P. (2013) How Amazon.com Can Turn the Page on Supply Chain Sustainability. *supplychain247*. Retrieved from http://www.supplychain247.com/article/how_amazon_can_turn_the_page_on_supply_chain_sustainability

Kennedy, S., & Dowling, P. (2013). How Amazon.com Can Turn the Page on Supply Chain Sustainability. *GreenBiz*. Retrieved from http://www.greenbiz.com/blog/2013/08/07/amazon-turn-page-supply-chain-sustainability

Ke, W., Liu, H., Wei, K. K., Gu, J., & Chen, H. (2009). How do mediated and non-mediated power affect electronic supply chain management system adoption? The mediating effects of trust and institutional pressures. *Decision Support Systems*, *46*(4), 839–851. doi:10.1016/j.dss.2008.11.008

Khan, S. A., Kumar, S., Hussain, M. Z., & Kalra, N. (2009). Climate Change, Climate Variability and Indian Agriculture: Impacts, Vulnerability and Adaptation Strategies. In S. N. Singh (Ed.), *Climate Change and Crops. Environmental Science and Engineering*. Heidelberg: Springer-Verlag Berlin. doi:10.1007/978-3-540-88246-6_2

Kim, J. K. (2000, February). Does attitude towards money moderate the relationship between intrinsic job satisfaction and voluntary turnover? *Human Relations*, *53*, 213–245.

Kim, J., Park, K., Hwang, Y., & Park, I. (2010). Sustainable manufacturing: A case study of the forklift painting process. *International Journal of Production Research*, *48*(10), 3061–3078. doi:10.1080/00207540902791785

Kim, Y., & Choi, S. M. (2005). Antecedents of green purchase behavior: An examination of collectivism, environmental concern, and PCE. *Advances in Consumer Research. Association for Consumer Research (U. S.)*, *32*, 592.

King, A. A., & Lenox, M. J. (2001). Lean and green? And empirical examination of the relationship between lean production and environmental performance. *Production and Operations Management*, *10*(3), 244–256. doi:10.1111/j.1937-5956.2001.tb00373.x

Kirby, S., & Richard, O. C. (2000). Impact of marketing work-place diversity on employee job involvement and organizational commitment. *The Journal of Social Psychology*, *140*(3), 367–378. doi:10.1080/00224540009600477 PMID:10902377

Kiron, D., Kruschwitz, N., Rubel, H., Reeves, M., & Fuisz-Kehrbach, S. (2013). Sustainability's next frontier: Walking the talk on the sustainability issues that matter most. *MIT Sloan Management Review Research Report*.

Kiron, D., Kruschwitz, N., Haanaes, K., & von Streng Velken, I. (2012). Sustainability nears a tipping point. *MIT Sloan Management Review*, *53*(2), 69–74.

Kiron, D., Kruschwitz, N., Haanes, K., & Kehrbach, S. K. (2013). How serious is climate change to Business? *MIT Sloan Management Review*, *55*(1), 75–77.

Kizer, K. W., & Kirsh, S. R. (2012). The double edged sword of performance measurement. *Journal of General Internal Medicine*, *27*(4), 395–397. doi:10.1007/s11606-011-1981-5 PMID:22271270

Klassen, R. D., & McLaughlin, C. (1996). The impact of environmental management on firm performance. *Management Science*, *42*(8), 1199–1214. doi:10.1287/mnsc.42.8.1199

Klassen, R. D., & Vachon, S. (2003). Collaboration and evaluation in the supply chain: Their impact on plant-level environmental investment. *Production and Operations Management*, *12*(3), 336–352. doi:10.1111/j.1937-5956.2003.tb00207.x

Klassen, R. D., & Whybark, D. C. (1999). The impact of environmental technologies on manufacturing performance. *Academy of Management Journal*, *42*(6), 599–615. doi:10.2307/256982

Kleindorfer, P. R., Singhal, K., & Wassenhove, L. N. (2005). Sustainable operations management. *Production and Operations Management*, *14*(4), 482–492. doi:10.1111/j.1937-5956.2005.tb00235.x

Klocke, F., Zeis, M., Klink, A., & Veselovac, D. (2013). Experimental research on the electrochemical machining of modern titanium and nickel-based alloys for aero engine components. *Procedia CIRP*, *6*, 368–372. doi:10.1016/j.procir.2013.03.040

Kocaoglu, B., Gulsun, B., & Tanyas, M. (2013). A SCOR based approach for measuring a benchmarkable supply chain performance. *Journal of Intelligent Manufacturing*, *24*(1), 113–132. doi:10.1007/s10845-011-0547-z

Kogg, B. (2003). Greening a cotton-textile supply chain. *Greener Management International, 2003*(43), 52-64.

Kogg, B., & Mont, O. (2012). Environmental and social responsibility in supply chains: The practice of choice and inter-organisational management. *Ecological Economics*, *83*, 154–163. doi:10.1016/j.ecolecon.2011.08.023

Konovsky, M. A. (2000). Understanding procedural justice and its impact on business organizations. *Journal of Management*, *26*(3), 489–511. doi:10.1177/014920630002600306

Kopnina, H. (2011). Consumption in environmental education: Developing curriculum that addresses cradle to cradle principles. *Factis Pax*, *5*(3), 374–388.

Kordestani, A., Peighambari, K., & Foster, T. (2015). Emerging trends in sustainability research: A look back as we begin to look forward. *International Journal of Environment and Sustainable Development*, *14*(2), 154–169. doi:10.1504/IJESD.2015.068602

Koren, Y., Heisel, U., Jovane, F., Moriwaki, T., Pritschow, G., Ulsoy, G., & Van Brussel, H. (1999). Reconfigurable manufacturing systems. *CIRP Annals-Manufacturing Technology*, *48*(2), 527–540. doi:10.1016/S0007-8506(07)63232-6

Kortelainen, K. (2008b). 'Global supply chains and social requirements: Case studies of labour condition auditing in the People's Republic of China'. *Business Strategy and the Environment*, *17*(7), 431–443. doi:10.1002/bse.634

Kortelainen, M. (2008a). Dynamic environmental performance analysis: A Malmquist index approach. *Ecological Economics*, *64*(4), 701–715. doi:10.1016/j.ecolecon.2007.08.001

Kotabe, M., Martin, X., & Domoto, H. (2003). Gaining from vertical partnerships: Knowledge transfer, relationship duration and supplier performance improvements in the U.S. and Japanese automotive industries. *Strategic Management Journal*, *24*(4), 293–316. doi:10.1002/smj.297

Kotler, P. (2000). Marketing management: The millennium edition. *Englewood Cliffs, NJ: Prentice Hall. Maxham, JG Jr & Netemeyer, RG (2002) Modeling customer perceptions of complaint handling over time: the effects of perceived justice on satisfaction and intent. Journal of Retailing*, *78*(4), 239–252.

Kotzab, H., Grant, D. B., Teller, C., & Halldorsson, A. (2009). Supply chain management and hypercompetition. *Logistics Research*, *1*(1), 5–13. doi:10.1007/s12159-008-0002-5

Koustelios, A., & Bagiatis, K. (1997). The Employee Satisfaction Inventory (ESI): Development of a Scale to measure satisfaction of Greek employees. *Educational and Psychological Measurement*, *57*(3), 469–476. doi:10.1177/0013164497057003008

Krause, D. R., Vachon, S., & Klassen, R. D. (2009). Special topic forum on sustainable supply chain management: introduction and reflections on the role of purchasing management. *Journal of Supply Chain Management*, *45*(4), 18–25. doi:10.1111/j.1745-493X.2009.03173.x

Kristensen, N., & Westergård-Nielsen, N. (2004). Does Low Job Satisfaction Lead to Job Mobility? *IZA Discussion paper no. 1026*.

Kudla, N. L., & Klaas-Wissing, T. (2012). Sustainability in shipper-logistics service provider relationships: A tentative taxonomy based on agency theory and stimulus-response analysis. *Journal of Purchasing and Supply Management*, *18*(4), 218–231. doi:10.1016/j.pursup.2012.04.001

Kuik, S. S., Nagalingam, S. V., & Amer, Y. (2011). Sustainable supply chain for collaborative manufacturing. *Journal of Manufacturing Technology Management*, *22*(8), 984–1001. doi:10.1108/17410381111177449

Kulkarni, B. (2010). Powered by passion. *Quest. TQMS*, *5*, 41–43.

Kumar, S. (2013).Strategy to implement sustainable public procurement in India. *Encore BCSD India, 7*(1). Retrieved from http://sustainableprocurement.blogspot.in/2013/02/strategy-to-implement-sustainable.html

Kumar, M. S., Udayasuriyan, G., & Vimala, B. (2008). Motivation among the employees of a public sector concern. *Journal of Community Guidance and Research*, *25*(3), 340–351.

Kuosmanen, T., & N. Kuosmanen. (2009). How not to measure Sustainable Value (and how one might). *Ecological Economics*, 69(2), 235-243.

Kyriakidou, O., & Ozbilgin, M. (2004). Individuals, organizations and careers: A relational perspective. *Career Development International*, *9*(1), 7–11. doi:10.1108/13620430410518101

La Londe, B. J., & Masters, J. M. (1994). Emerging logistics strategies: Blueprints for the next century. *International Journal of Physical Distribution & Logistics Management*, *24*(7), 35–47. doi:10.1108/09600039410070975

Labuschagne, C., & Brent, A. C. (2008). An industry perspective of the completeness and relevance of a social assessment framework for project and technology management in the manufacturing sector. *Journal of Cleaner Production*, *16*(3), 253–262. doi:10.1016/j.jclepro.2006.07.028

Labuschagne, C., Brent, A. C., & van Erck, R. P. G. (2005). Assessing the sustainability performances of industries. *Journal of Cleaner Production*, *13*(3), 73–85.

Lachman, R., & Diamant, E. (1987). Withdrawal and restraining factors in teachers' turnover intentions. *Journal of Occupational Behavior*, *8*(3), 219–232. doi:10.1002/job.4030080304

Lafferty, W. M., & Langhelle, O. (Eds.), (1999). Towards sustainable development: on the goals of development-and the conditions of sustainability. Macmillan.

Lagodimos, A. G., Chountalas, P. T., & Chatzi, K. (2007). The state of ISO 14001 certification in Greece. *Journal of Cleaner Production*, *15*(18), 1743–1754. doi:10.1016/j.jclepro.2006.06.015

Lambert, D. M., & Cooper, M. C. (2000). Issues in supply chain management. *Industrial Marketing Management*, *29*(1), 65–83. doi:10.1016/S0019-8501(99)00113-3

Lambert, E. G., Hogan, N. L., & Barton, S. M. (2001). The impact of Job Satisfaction on Turnover Intent: A test of a structural model using a national sample of workers. *The Social Science Journal, Elsevier*, *38*(2), 233–250. doi:10.1016/S0362-3319(01)00110-0

Lamberton, G. (2005, March). Sustainability accounting—a brief history and conceptual framework. *Accounting Forum*, *29*(1), 7–26. doi:10.1016/j.accfor.2004.11.001

Lamming, R., & Hampson, J. (1996). The environment as a supply chain management issue. *British Journal of Management*, *7*(Suppl. 1), S45–S62. doi:10.1111/j.1467-8551.1996.tb00147.x

Lanen, W., & Anderson, S.W. (2002). Economic Transition, Strategy and the Evolution of Management Accounting Practices: The Case of India.

Laosirihongthong, T., & Adebanjo, D., & ChoonTan, K. (2013). Green supply chain management practices and performances. *Industrial Management & Data Systems*, *113*(8), 1088–1109. doi:10.1108/IMDS-04-2013-0164

Laplume, A. O., Sonpar, K., & Litz, R. A. (2008). Stakeholder theory: Reviewing a theory that moves us. *Journal of Management*, *34*(6), 1152–1189. doi:10.1177/0149206308324322

Large, R. O., Kramer, N., & Hartmann, R. K. (2013). Procurement of logistics services and sustainable development in Europe: Fields of activity and empirical results. *Journal of Purchasing and Supply Management*, *19*(3), 122–133. doi:10.1016/j.pursup.2013.05.002

Laroche, M., Bergeron, J., & Barbaro-Forleo, G. (2001). Targeting consumers who are willing to pay more for environmentally friendly products. *Journal of Consumer Marketing*, *18*(6), 503–520. doi:10.1108/EUM0000000006155

Larson, A. (n. d.). Nike: Moving Down the Sustainability Track Through Chemical Substitution and Waste Reduction.

Laufer, W. S. (2003). Social accountability and corporate greenwashing. *Journal of Business Ethics*, *43*(3), 253–261. doi:10.1023/A:1022962719299

Lau, K. H. (2011). Benchmarking green logistics performance with a composite score. *Benchmarking International Journal (Toronto, Ont.)*, *18*(6), 873–896.

Lauri, B., Benson, G., & Cheney, S. (1996). The top ten trends. *Training & Development*, *11*, 28–42.

Lavassani, K. M., & Movahedi, B. (2010). Critical Analysis of the Supply Chain Management Theories: Toward the Stakeholder Theory. *Proceedings of the POMS 21st Annual Conference,* Vancouver.

Lavastre, O., Gunasekaran, A., & Spalanzani, A. (2012). Supply chain risk management in French companies. *Decision Support Systems*, *52*(4), 828–838. doi:10.1016/j.dss.2011.11.017

Law, K. M., & Gunasekaran, A. (2012). Sustainability development in high-tech manufacturing firms in Hong Kong: Motivators and readiness. *International Journal of Production Economics*, *137*(1), 116–125. doi:10.1016/j.ijpe.2012.01.022

Lawler, E. E. (1990). *Compensation management; Strategic planning; Pay-for-knowledge systems*. San Francisco: Jossey-Bass Publishers.

Lee Weisstein, F., Asgari, M., & Siew, S. W. (2014). Price presentation effects on green purchase intentions. *Journal of Product and Brand Management*, *23*(3), 230–239. doi:10.1108/JPBM-06-2013-0324

Lee, H., Duda, S., James, L., Mackwani, Z., Munoz, R., & Volk, D. (2007). Starbucks Corporation: Building a Sustainable Supply Chain.

Lee, C., Leem, C. S., & Hwang, I. (2011). PDM and ERP integration methodology using digital manufacturing to support global manufacturing. *International Journal of Advanced Manufacturing Technology*, *53*(1-4), 399–409. doi:10.1007/s00170-010-2833-x

Lee, L. H. (2010). 'Don't tweak your supply chain - rethink it end to end'. *Harvard Business Review*, *88*, 46–53.

Lee, S. Y. (2008). Drivers for the participation of small and medium-sized suppliers in green supply chain initiatives. *Supply Chain Management: An International Journal*, *13*(3), 185–198. doi:10.1108/13598540810871235

Lee, T., & Mowday, R. (1987). Voluntary leaving an organization: An empirical investigation of Steers and Mowday's model of turnover. *Academy of Management Journal*, *30*(4), 721–743. doi:10.2307/256157

Leidreiter, A. (2014). 100% renewable energy as centerpiece of a climate action plan. *Energy Transition*. Retrieved from http://energytransition.de/2014/10/100-renewable-energy-as-centerpiece-of-a-climate-action-plan/

Leire, C., & Mont, O. (2010). The implementation of socially responsible purchasing. *Corporate Social Responsibility and Environmental Management*, *17*(1), 27–39.

LeMenestrel, M., Le Menestrel, M., & de Bettignies, H.-C. (2002). Processes and Consequences in Business Ethical Dilemmas: The Oil Industry and Climate Change. *Journal of Business Ethics*, *41*(3), 251–266. doi:10.1023/A:1021237629938

Levitt, T. (1960). Marketing myopia. *Harvard Business Review*, *38*(4), 45–56. PMID:15252891

Lewis, R. E., & Heckman, R. J. (2006). Talent management: A critical review. *Human Resource Management Review*, *16*(2), 139–154. doi:10.1016/j.hrmr.2006.03.001

Liang, H., Saraf, N., Hu, Q., & Xue, Y. (2007). Assimilation of enterprise systems: The effect of institutional pressures and the mediating role of top management. *Management Information Systems Quarterly*, 2007, 59–87.

Light, J. N. (2004). Relationships and effects of employee involvement, employee empowerment and employee satisfaction by job-type in a large manufacturing environment [Unpublished PhD. Dissertation].

Lightfoot, H., Baines, T., & Smart, P. (2013). The servitization of manufacturing: A systematic literature review of interdependent trends. *International Journal of Operations & Production Management*, *33*(11/12), 1408–1434. doi:10.1108/IJOPM-07-2010-0196

Linking economy, society and environment. (2008). *OECDInsights*. Retrieved from http://www.oecd.org/insights/sustainabledevelopmentlinkingeconomysocietyenvironment.htm

Lin, L. C., & Li, T. S. (2010). An integrated framework for supply chain performance measurement using six-sigma metrics. *Software Quality Journal*, *18*(3), 387–406. doi:10.1007/s11219-010-9099-2

Linton, J., Klassen, R., & Jayaraman, V. (2007). Sustainable supply chains: An introduction. *Journal of Operations Management*, *25*(6), 1075–1082. doi:10.1016/j.jom.2007.01.012

Lintott, J. (1999). The Environmental Accounting: Useful to Whom and for What? *Ecological Economics*, *16*(3), 179–190. doi:10.1016/0921-8009(95)00097-6

Lintukangas, K. (2011). Supplier relationship management capability in global supply management. *Int. J. of Procurement Management*, *4*(1), 1–19. doi:10.1504/IJPM.2011.037382

Littig, B., & Griessler, E. (2005). Social sustainability: A catchword between political pragmatism and social theory. *International Journal of Sustainable Development*, *8*(1/2), 65–79. doi:10.1504/IJSD.2005.007375

Liu, S., Kasturiratne, D., & Moizer, J. (2012). A hub-and-spoke model for multi-dimensional integration of green marketing and sustainable supply chain management. *Industrial Marketing Management*, *41*(4), 581–588. doi:10.1016/j.indmarman.2012.04.005

Liu, S., Leat, M., & Smith, M. H. (2011). State-of-the-art sustainability analysis methodologies for efficient decision support in green production operations. *International Journal of Sustainable Engineering*, *4*(3), 236–250. doi:10.1080/19397038.2011.574744

Liu, W., & Liang, M. (2008). Multi-objective design optimization of reconfigurable machine tools: A modified fuzzy-Chebyshev programming approach. *International Journal of Production Research*, *46*(6), 1587–1618. doi:10.1080/00207540600943944

Liu, Z., Nouraei, H., Papini, M., & Spelt, J. K. (2014). Abrasive enhanced electrochemical slurry jet micro-machining: Comparative experiments and synergistic effects. *Journal of Materials Processing Technology*, *214*(9), 1886–1894. doi:10.1016/j.jmatprotec.2014.04.003

Lo, C. K. Y., Yeung, A. C. L., & Cheng, T. C. E. (2012). The impact of environmental management systems on financial performance in fashion and textile industries. *International Journal of Production Economics*, *135*(2), 561–567. doi:10.1016/j.ijpe.2011.05.010

Lohmann, L. (2009). Toward a different debate in environmental accounting: The cases of carbon and cost–benefit. *Accounting, Organizations and Society*, *34*(3), 499–534. doi:10.1016/j.aos.2008.03.002

Longo, F., & Mirabelli, G. (2008). An advanced supply chain management tool based on modeling and simulation. *Computers & Industrial Engineering*, *54*(3), 570–588. doi:10.1016/j.cie.2007.09.008

Long-term care facility worker retention. (2004). Ontario Ministry of Health and Long-Term Care.

Loucks, E. S., Martens, M. L., & Cho, C. H. (2010). Engaging Small- and Medium-sized Businesses in Sustainability. *Sustainability Accounting Management and Policy Journal*, *1*(2), 178–200.

Lozano, R. (2008). Envisioning sustainability three-dimensionally. *Journal of Cleaner Production*, *16*(17), 1838–1846. doi:10.1016/j.jclepro.2008.02.008

Lueneburger, C., & Goleman, D. (2010). The change leadership sustainability demands *MIT. Sloan Management Review*, *51*(4), 49–55.

Lu, L., Bock, D., & Joseph, M. (2013). Green marketing: What the Millennials buy. *The Journal of Business Strategy*, *34*(6), 3–10. doi:10.1108/JBS-05-2013-0036

Lum, L., Kervin, J., Clark, K., Reid, F., & Sirola, W. (1998). Explaining nursing turnover intent: Job satisfaction, pay satisfaction or organizational commitment? *Journal of Organizational Behavior*, *19*(3), 305–320. doi:10.1002/(SICI)1099-1379(199805)19:3<305::AID-JOB843>3.0.CO;2-N

Lu, R. X., Lee, P. K., & Cheng, T. C. E. (2012). Socially responsible supplier development: Construct development and measurement validation. *International Journal of Production Economics*, *140*(1), 160–167. doi:10.1016/j.ijpe.2012.01.032

Luthra, S., Garg, D., & Haleem, A. (2014b). Critical success factors of green supply chain management for achieving sustainability in Indian automobile industry. *Production Planning & Control*.

Luthra, S., Garg, D., & Haleem, A. (2014a). Sustainable purchasing in Indian automobile industry: An ISM approach. In *K. Jain, L. Ganapathy, R. Gokhale, & R. Mukundan (Eds.), Manufacturing Excellence: Imperatives for Emerging Economies* (pp. 95–108). Mumbai: Excel India Publishers.

Luthra, S., Kumar, V., Kumar, S., & Haleem, A. (2011). Barriers to implement green supply chain management in automobile industry using interpretive structural modeling technique: An Indian perspective. *Journal of Industrial Engineering and Management*, *4*(2), 231–257. doi:10.3926/jiem.2011.v4n2.p231-257

Lynes, J. K., & Dredge, D. (2006). Going green: Motivations for environmental commitment in the airline industry. A case study of Scandinavian Airlines. *Journal of Sustainable Tourism*, *14*(2), 116–138. doi:10.1080/09669580608669048

Mackau, D. (2003). SME Integrated management system: A proposed experiences model. *The TQM Magazine*, *15*(1), 43–51. doi:10.1108/09544780310454448

Mahler, D. (2007). The sustainable supply chain. *Supply Chain Management Review*, *11*(8), 59–60.

Maignan, I., & Ralston, D. A. (2002). Corporate social responsibility in Europe and the US: Insights from businesses self-presentations. *Journal of International Business Studies*, *33*(3), 497–514. doi:10.1057/palgrave.jibs.8491028

Mainieri, T., Barnett, E. G., Valdero, T. R., Unipan, J. B., & Oskamp, S. (1997). Green buying: The influence of environmental concern on consumer behavior. *The Journal of Social Psychology*, *137*(2), 189–204. doi:10.1080/00224549709595430

Makhijani, N., Rajendran, K., & Creelman, J. (2009). *Best Practices in Aligning People with Strategic Goals*. Azkia Publishers.

Malarvizhi, P., & Yadav, S. (2008, May). Corporate Environmental Reporting On The Internet: An Insight Into Indian Practices. *Proceedings of the11th Annual Convention of the Strategic Management Forum, May* (pp. 8-10).

Malone, D. W. (1975). An introduction to the application of interpretive structural modeling. *Proceedings of the IEEE*, *63*(3), 397–404. doi:10.1109/PROC.1975.9765

Maloni, M. J., & Brown, M. E. (2006). Corporate social responsibility in the supply chain: An application in the food industry. *Journal of Business Ethics*, *68*(1), 35–52. doi:10.1007/s10551-006-9038-0

Maloutas, T. (2003). Promoting social sustainability: The case of Athens. *City*, *7*(2), 167–181. doi:10.1080/1360481032000136732

Mandal, A., & Deshmukh, S. G. (1994). Vendor selection using interpretive structural modeling (ISM). *International Journal of Operations & Production Management*, *14*(6), 52–59. doi:10.1108/01443579410062086

Mani, V., Agrawal, R., Sharma, V. (2015b). Social Sustainability Practices in the Supply Chain of Indian Manufacturing Industries. *International Journal of Automation and Logistics*, 1(2).

Mani, V., Agrawal, R., Sharma, V. (2015c). Social sustainability in the supply chain: Analysis of enablers. *Management Research Review*, 38(9).

Mani, V., Agrawal, R., Sharma, V. (2015d). Supply Chain Social Sustainability: A Comparative Case Analysis in Indian Manufacturing Industries. *Procedia-Social and behavioural science*.

Mani, M., Madan, J., Lee, J. H., Lyons, K. W., & Gupta, S. K. (2014). Sustainability characterization for manufacturing processes. *International Journal of Production Research*, *52*(20), 5895–5912. doi:10.1080/00207543.2014.886788

Mani, M., Madan, J., Lee, J. H., Lyons, K., & Gupta, S. K. (2013). *Review on Sustainability Characterization for Manufacturing Processes*. US Department of Commerce, National Institute of Standards and Technology. doi:10.6028/NIST.IR.7913

Manimaran, G., & Kumar, M. P. (2013). Multi-response Optimization of Grinding AISI 316 Stainless Steel Using Grey Relational Analysis. *Materials and Manufacturing Processes*, *28*(4), 418–423. doi:10.1080/10426914.2012.709347

Mani, V., Agrawal, R., & Sharma, V. (2014a). Supplier selection using social sustainability: AHP based approach in India. *International Strategic Management Review*, *2*(2), 98–112. doi:10.1016/j.ism.2014.10.003

Manning, L., Baines, R., & Chadd, S. (2008). Benchmarking the poultry meat supply chain. *Benchmarking: An International Journal*, *15*(2), 148–165. doi:10.1108/14635770810864866

Mansi, M. (2015). Sustainable procurement disclosure practices in central public sector enterprises: Evidence from India. *Journal of Purchasing and Supply Management*, *21*(2), 125–137. doi:10.1016/j.pursup.2014.12.002

Marcacci, S. (2015). 7 Interesting global renewable energy trends from NREL (Charts Galore!). *Clean Technica*. Retrieved from http://cleantechnica.com/2015/01/26/7-interesting-global-renewable-energy-trends-from-nrel-charts-galore/

Marksberry, P. W., & Jawahir, I. S. (2008). A comprehensive tool-wear/tool-life performance model in the evaluation of NDM (near dry machining) for sustainable manufacturing. *International Journal of Machine Tools & Manufacture*, *48*(7), 878–886. doi:10.1016/j.ijmachtools.2007.11.006

Marquardt, B. (2006). Historia de la sostenibilidad: Un concepto medioambiental en la historia de Europa central (1000-2006). *Historia Critica*, *32*(2), 172–197.

Mathews, M. R. (1997). Twenty-five years of social and environmental accounting research: Is there a silver jubilee to celebrate? *Accounting, Auditing & Accountability Journal*, *10*(4), 481–531. doi:10.1108/EUM0000000004417

Matthews, H. S., Hendrickson, C. T., & Weber, C. L. (2008). The importance of carbon footprint estimation boundaries. *Environmental Science & Technology*, *42*(16), 5839–5842. doi:10.1021/es703112w PMID:18767634

McDonough, W., Braungart, M., Anastas, P. T., & Zimmerman, J. B. (2003). Peer reviewed: Applying the principles of green engineering to cradle-to-cradle design. *Environmental Science & Technology*, *37*(23), 434A–441A. doi:10.1021/es0326322 PMID:14700308

McKenzie, S. (2004). *Social sustainability: towards some definitions. Magill.* Hawke Research Institute, University of South Australia.

McWilliams, A., & Siegel, D. (2001). Corporate social responsibility: A theory of the firm perspective. *Academy of Management Review*, *26*(1), 117–127.

Meadows, D. H. (1972). *The limits to growth: a report of the Club of Rome's project on the predicament of mankind.* New York: Universe Books.

Meadows, M., & Pike, M. (2010). Performance management for social enterprises. *Systemic Practice and Action Research*, *23*(2), 127–141. doi:10.1007/s11213-009-9149-5

Mearian, L. (2015).The new facility is the largest such commercial solar deal in U.S. history. *COMPUTERWORLD*. Retrieved from http://www.computerworld.com/article/2882762/apples-850m-solar-plant-rockets-it-to-first-place-among-us-corporations.html

Meehan, J., & Bryde, D. (2011). Sustainable procurement practice. *Business Strategy and the Environment*, *20*(2), 94–106. doi:10.1002/bse.678

Melnyk, S. A., Sroufe, R. P., & Calantone, R. (2003). Assessing the impact of environmental management systems on corporate and environmental performance. *Journal of Operations Management*, *21*(3), 329–351. doi:10.1016/S0272-6963(02)00109-2

Mentzer, J. T., DeWitt, W., Keebler, J. S., Min, S., Nix, N. W., Smith, C. D., & Zacharia, Z. G. (2001). Defining supply chain management. *Journal of Business logistics, 22*(2), 1-25.

Mentzer, J. T., DeWitt, W., Keebler, J. S., Min, S., Nix, N. W., Smith, C. D., & Zacharia, Z. G.. (2001). Defining supply chain management. *Journal of Business Logistics*, *22*(2), 1–25. doi:10.1002/j.2158-1592.2001.tb00001.x

Mentzer, J. T., & Gundlach, G. (2010). Exploring the relationship between marketing and supply chain management: Introduction to the special issue. *Journal of the Academy of Marketing Science*, *38*(1), 1–4. doi:10.1007/s11747-009-0150-4

Merli, G. (1991). *Co-makership: The new supply strategy for manufacturers*. Cambridge, MA: Productivity Press.

Messmer, M. (2000). Orientations programs can be key to employee retention. Strategic Finance, 81(8), 12-15.

Meybodi, M. Z. (2009). Benchmarking performance measures in traditional and just-in-time companies. *Benchmarking: An International Journal*, *16*(1), 88–102. doi:10.1108/14635770910936531

Michaels, E., Handfield-Jones, H., & Axelrod, B. (2001). *The War for Talent*. Boston: Harvard Business School Press.

Miemczyk, J., Johnsen, T. E., & Macquet, M. (2012). Sustainable purchasing and supply management: A structured literature review of definitions and measures at the dyad, chain and network levels. *Supply Chain Management: An International Journal*, *17*(5), 478–496. doi:10.1108/13598541211258564

Miller, G., Pawloski, J., & Standridge, C. R. (2010). A case study of lean, sustainable manufacturing. *Journal of industrial engineering and management*, 3(1), 11-32.

Milne, M. J., Kearins, K., & Walton, S. (2006). Creating adventures in wonderland: The journey metaphor and environmental sustainability. *Organization*, *13*(6), 801–839. doi:10.1177/1350508406068506

Milne, M. J., & Patten, D. M. (2002). Securing organizational legitimacy: An experimental decision case examining the impact of environmental disclosures. *Accounting, Auditing & Accountability Journal*, *15*(3), 372–405. doi:10.1108/09513570210435889

Min, H., & Galle, W. P. (2001). Green purchasing practices of US firms. *International Journal of Operations & Production Management*, *21*(9), 1222–1238. doi:10.1108/EUM0000000005923

Minimol, M.C., & Makesh, K.G. (2013). Green accounting and reporting practices among indian corporates. *Asia Pacific Journal of Research*, 1(14).

Minitab User Manual Release 14. (2003). State College, PA, USA: MINITAB Inc.

Mirhedayatian, S.M., Azadi, M. & Saen, R.F. (2014). A novel network data envelopment analysis model for evaluating green supply chain management. *International Journal of Production Economics*, 147(1 Part B), 544-554.

Mirhedayatian, S. M., Azadi, M., & Saen, R. F. (2014). A novel network data envelopment analysis model for evaluating green supply chain management. *International Journal of Production Economics*, *147*, 544–554. doi:10.1016/j.ijpe.2013.02.009

Mitlin, D., & Satterthwaite, D. (1996). Chapter One Sustainable Development and Cities. *Sustainability. Environment and Urbanization.*

Mittal, V. K., & Sangwan, K. S. (2014a). Development of a structural model of environmentally conscious manufacturing drivers. *Journal of Manufacturing Technology Management*, *25*(8).

Mittal, V. K., & Sangwan, K. S. (2014b). Development of a model of barriers to environmentally conscious manufacturing implementation. *International Journal of Production Research*, *52*(2), 584–594. doi:10.1080/00207543.2013.838649

Mittal, V. K., & Sangwan, K. S. (2014b). Modeling drivers for successful adoption of environmentally conscious manufacturing. *Journal of Modelling in Management*, *9*(2).

Model of social sustainability, (n. d.). Western Australian Council of Social Services (WACSS). Retrieved from http://www.wacoss.org.au/downloads/socialsustainable.pdf

Moffett, S., Anderson-Gilliespie, K., & McAdam, R. (2008). Benchmarking and performance measurement: A statistical analysis. *Benchmarking: An International Journal*, *15*(4), 368–381. doi:10.1108/14635770810887203

Mohamed, E. K., & Lashine, S. H. (2003). Accounting knowledge and skills and the challenges of a global business environment. *Managerial Finance*, *29*(7), 3–16. doi:10.1108/03074350310768319

Molamohamadi, Z., & Ismail, N. (2013). Developing a New Scheme for Sustainable Manufacturing. *International Journal of Materials. Mechanics and Manufacturing*, *1*(1), 1–5. doi:10.7763/IJMMM.2013.V1.1

Molla, A. (2013). Identifying IT sustainability performance drivers: Instrument development and validation. *Information Systems Frontiers*, *15*(5), 705–723. doi:10.1007/s10796-013-9415-z

Mollenkopf, D., Stolze, H., Tate, W. L., & Ueltschy, M. (2010). Green, lean, and global supply chains. *International Journal of Physical Distribution & Logistics Management*, *40*(1/2), 14–41. doi:10.1108/09600031011018028

Mollenkopf, D., Stolze, H., Tate, W. L., & Ueltschy, M. (2010). Green, lean, and global supply chains. *International Journal of Physical Distribution & Logistics Management*, *40*(1), 14–41.

Möller, A., & Schaltegger, S. (2005). The sustainability balanced scorecard as a framework for eco-efficiency analysis. *Journal of Industrial Ecology*, *9*(4), 73–83. doi:10.1162/108819805775247927

Montalvo, C., & Kemp, R. (2008). Cleaner technology diffusion: Case studies, modeling and policy. *Journal of Cleaner Production*, *16*(1), S1–S6. doi:10.1016/j.jclepro.2007.10.014

Mont, O., & Leire, C. (2009). Socially responsible purchasing in supply chains: Drivers and barriers in Sweden. *Social Responsibility Journal*, *5*(3), 388–407. doi:10.1108/17471110910977302

Moorthy, K., & Yacob, P. (2013). Green accounting: Cost measures. *Open Journal of Accounting*, *2*(01), 4–7. doi:10.4236/ojacct.2013.21002

Morali, O., & Searcy, C. (2013). A review of sustainable supply chain management practices in Canada. *Journal of Business Ethics*, *117*(3), 635–658. doi:10.1007/s10551-012-1539-4

More than two planets needed by 2030. (2012). WWF Living planet report. Retrieved from www.wwf.org.za

Morgan, R. M., & Hunt, S. D. (1994). The commitment-trust theory of relationship marketing. *Journal of Marketing*, *58*(3), 20–38. doi:10.2307/1252308

Moss, P., & Tilly, C. (2001). *Stories Employers Tell. Race, Skill and Hiring in America*. New York: Russell Sage Foundation.

Mudgal, R. K., Shankar, R., Talib, P., & Raj, T. (2010). Modelling the barriers of green supply chain practices: An Indian perspective. *International Journal of Logistics Systems and Management*, *7*(1), 81–107. doi:10.1504/IJLSM.2010.033891

Muduli, K., & Barve, A. (2014). Impact of Soft Factors on GSCM Performance.*Proc. of the Intl. Conf. on Advances in Social Science, Management and Human Behaviour-- SMHB '13*

Muduli, K., Govindan, K., Barve, A., Kannan, D., & Geng, Y. (2013b). Role of behavioral factors in green supply chain management implementation in Indian mining industries. *Resources, Conservation and Recycling*, *76*, 50–60. doi:10.1016/j.resconrec.2013.03.006

Mukherjee-Saha, J. (2011). Business Sustainability amidst global churning paradoxes and dilemmas. *GMJ*, *5*(1/2), 19–24.

Mullerat, R. (2010). *International Corporate Social Responsibility: the role of corporations in the economic order of the 21st century*. Netherlands: Kluwer Law International.

Murphy, J. (2000). Ecological modernisation. *Geoforum*, *31*(1), 1–8. doi:10.1016/S0016-7185(99)00039-1

Murphy, J., & Gouldson, A. (2000). Environmental policy and industrial innovation: Integrating environment and economy through ecological modernization. *Geoforum*, *31*(1), 33–44. doi:10.1016/S0016-7185(99)00042-1

Musa, A., Yusuf, Y., McArdle, L., & Banjoko, G. (2013). Corporate social responsibility in Nigeria' oil and gas industry: The perspectives of the industry. *International Journal of Process Management and Benchmarking*, *3*(2), 101–135. doi:10.1504/IJPMB.2013.057722

Nadler, S. S., & Kros, J. F. (2010). An assessment of supply chain managers' trust in online auctions. *Industrial Management & Data Systems*, *110*(6), 805–822. doi:10.1108/02635571011055063

Nagel, M. H. (2000). Environmental supply-chain management versus green procurement in the scope of a business and leadership perspective. *Proceedings of the 2000 IEEE International Symposium onElectronics and the Environment ISEE '00* (pp. 219-224). IEEE. doi:10.1109/ISEE.2000.857652

Najmi, M., Rigas, J., & Fan, I. S. (2005). A framework to review performance measurement systems. *Business Process Management Journal*, *11*(2), 109–122. doi:10.1108/14637150510591129

Nakano, M. (2010). A conceptual framework for sustainable manufacturing by focusing on risks in supply chains. In Advances in Production Management Systems. New Challenges, New Approaches (pp. 160-167). Springer Berlin Heidelberg. doi:10.1007/978-3-642-16358-6_21

Namakonzi, R., & Inanga, E. L. (2014). Environmental management accounting and environmental management in manufacturing industries in Uganda. *African Journal of Economic and Sustainable Development*, *3*(4), 288–329. doi:10.1504/AJESD.2014.065578

Narasimhan, R., & Carter, J. R. (1998). *Environmental supply chain management*. Center for Advanced Purchasing Studies.

Naveh, E., & Marcus, A. (2005). Achieving competitive advantage through implementing a replicable management standard: Installing and using ISO 9000. *Journal of Operations Management*, *24*(1), 1–26. doi:10.1016/j.jom.2005.01.004

Nayak, K. C., & Tripathy, R. K. (2012). Relevance vector machine based prediction of MRR and SR for electrochemical machining process. *International Journal of Mechanical Engineering and Technology*, *3*(3), 394–403.

Nazli Nik Ahmad, N., & Sulaiman, M. (2004). Environment disclosure in Malaysia annual reports: A legitimacy theory perspective. *International Journal of Commerce and Management*, *14*(1), 44–58. doi:10.1108/10569210480000173

Neely, A., Adams, C., & Kennerley, M. (2002). *The performance prism: The scorecard for measuring and managing stakeholder relationship*. London, UK: Financial Times-Prentice Hall.

Nelson, D., Marsillac, E., & Rao, S. (2012). Antecedents and evolution of the green supply chain. *Journal of Operations and Supply Chain Management*, *1*(1), 29–43.

Netemeyer, R. G., Boles, J. S., McKee, D. O., & McMurrian, R. (1997). An investigation into the antecedents of organizational citizenship behaviors in a personal selling context. *Journal of Marketing*, *61*(3), 85–98. doi:10.2307/1251791

Neto, J., Silva, E., & Silva, M. (2006). Intervening variables in electrochemical machining. *Journal of Materials Processing Technology*, *179*(1-3), 92–96. doi:10.1016/j.jmatprotec.2006.03.105

Nidumolu, R., Prahalad, C. K., & Rangaswami, M. R. (2009). Why sustainability is now the key driver of innovation. *Harvard Business Review*, *87*(9), 56–64.

Nikolopoulou, A., & Ierapetritou, M. G. (2012). Optimal design of sustainable chemical processes and supply chains: A review. *Computers & Chemical Engineering*, *44*, 94–103. doi:10.1016/j.compchemeng.2012.05.006

Noci, G. (1997). Designing 'green' vendor rating systems for the assessment of a supplier's environmental performance. *European Journal of Purchasing & Supply Management*, *3*(2), 103–114. doi:10.1016/S0969-7012(96)00021-4

Noe, R. A. (1999). *Employee training and development* (pp. 212–218). New York: Irwin McGraw-Hill.

O'Connell, L. (2008). Driving change: KPMG, Studies Impact of Climate Change on Irish M & A Activity. *Accountancy Ireland*, *40*(6), 36–38.

Obal, M., & Lancioni, R. A. (2013). Maximizing buyer–supplier relationships in the digital era: Concept and research agenda. *Industrial Marketing Management*, *42*(6), 851–854. doi:10.1016/j.indmarman.2013.06.002

O'Brien, C. (1999). Sustainable production–a new paradigm for a new millennium. *International Journal of Production Economics*, *60*, 1–7. doi:10.1016/S0925-5273(98)00126-1

Olajide, A. (2000). *Getting the best out of the employees in a developing economy. A personal Psychology Guest Lecture Series*. Nigeria: Department of Guidance and Counseling, University of Ibadan.

Oliveira, O. J., & Serra, J. R. (2010). Benefícios e dificuldades da gestão ambiental com base na ISO 14001 em empresas industriais de São Paulo. *Production*, *20*(3), 429–438. doi:10.1590/S0103-65132010005000013

Oliver, N., Delbridge, R., & Lowe, J. (1996). Lean Production Practices: International Comparisons in the Auto Components Industry. *British Journal of Management*, *7*(Suppl. 1), S29–S44. doi:10.1111/j.1467-8551.1996.tb00146.x

Olugu, E. U., Wong, K. Y., & Shaharoun, A. M. (2011). Development of key performance measures for the automobile green supply chain. *Resources, Conservation and Recycling*, *55*(6), 567–579. doi:10.1016/j.resconrec.2010.06.003

Onishi, Y., Kokubu, K., & Nakajima, M. (2009). Implementing material flow cost accounting in a pharmaceutical company. In *Environmental Management Accounting for Cleaner Production* (pp. 395–409). Springer Netherlands.

Ōno, T. (1988). *Toyota production system: beyond large-scale production*. Productivity press.

OReilly, C. A., Chatman, J., & Caldwell, D. F. (1991). People and organizational culture: A profile comparison approach to assessing person organization fit. *Academy of Management Journal*, *34*(3), 487–516. doi:10.2307/256404

Oruezabala, G., & Rico, J.-C. (2012). The impact of sustainable public procurement on supplier management – the case of French public hospitals. *Industrial Marketing Management*, *41*(4), 573–580. doi:10.1016/j.indmarman.2012.04.004

Ostrom, E., Schroeder, L., & Wynne, S. (1993). *Institutional Incentives and Sustainable Development: Infrastructure Policies in Perspective*. Westview Press.

Ouellette, J. (1996). Coagulants and Flocculants Rise. *Chemical Marketing Reporter*, 250(15).

Our Common Future (Brundtland Report). (1987). World Commission on Environment and Development, Oxford University Press, Oxford.

Our common future. (1987). World Commission on Environment and Development. Oxford University Press, Oxford.

Oztemel, E., & Tekez, E. K. (2009). Interactions of agents in performance based supply chain management. *Journal of Intelligent Manufacturing*, *20*(2), 159–167. doi:10.1007/s10845-008-0229-7

Pagell M, Wu Z (2009). Building a more complete theory of sustainable supply chain management using case studies of 10 exemplars. *Journal of supply chain management*, 45(2), 37-56.

Pagell, M., & Wu, Z. (2009). Building a more complete theory of sustainable supply chain management using case studies of 10 exemplars. *Journal of supply chain management, 45*(2), 37-56.

Pagell, M., & Wu, Z. (2009).Building a more complete theory of sustainable supply chain management using case studies of 10 exemplars. *Journal of supply chain management, 45*(2), 37-56.

Pagell, M., & Wu, Z. (2009). Building a more complete theory of sustainable supply chain management using case studies of 10 exemplars. *Journal of Supply Chain Management*, *45*(2), 37–56. doi:10.1111/j.1745-493X.2009.03162.x

Pahuja, S. (2007). Environmental Reporting Verification: A Critical Evaluation of Accountants' Views and Corporate Practices in India. *Social Responsibility Journal*, *3*(2), 22–31. doi:10.1108/17471110710829696

Pahuja, S. (2009). Relationship between environmental disclosures and corporate characteristics: A study of large manufacturing companies in India. *Social Responsibility Journal*, *5*(2), 227–244. doi:10.1108/17471110910964504

Paju, M., Heilala, J., Hentula, M., Heikkila, A., Johansson, B., Leong, S., & Lyons, K. (2010, December). Framework and indicators for a sustainable manufacturing mapping methodology. *Proceedings of the* Simulation Conference (WSC), *2010*(Winter), 3411–3422.

Palisi, F. X. (2012). Sustainable purchasing practices: An investigation into current industry awareness and practice [Master's Thesis College of Technology Paper 65]. Retrieved from http://docs.lib.purdue.edu/techmasters/65

Palmatier, R. W., Dant, R. P., & Grewal, D. (2007). A comparative longitudinal analysis of theoretical perspectives of interorganizational relationship performance. *Journal of Marketing*, *71*(4), 172–194. doi:10.1509/jmkg.71.4.172

Panackal, N., & Singh, A. (2015). Using Interpretive Structural Modeling to Determine the Relation between Youth and Sustainable Rural Development. *IBMRD's Journal of Management & Research*, *4*(1), 58–74. doi:10.17697/ibmrd/2015/v4i1/60352

Panda, M. C., & Yadava, V. (2012). Intelligent Modeling and Multi-objective Optimization of Die Sinking Electrochemical Spark Machining Process. *Materials and Manufacturing Processes*, *27*(1), 10–25. doi:10.1080/10426914.2010.544812

Parikh, K. (2012). Sustainable development and low carbon growth strategy for India. *Energy*, *40*(1), 31–38. doi:10.1016/j.energy.2012.01.013

Parisi, C. (2013). The impact of organisational alignment on the effectiveness of firms' sustainability strategic performance measurement systems: An empirical analysis. *Journal of Management & Governance*, *17*(1), 71–97. doi:10.1007/s10997-012-9219-4

Parker, O., & Wright, L. (2001). Pay and employee commitment: the missing link. In Ivey Business Journal, 65(3), 70-79.

Parkinson, S. (1999). Agile manufacturing. *Work Study*, *48*(4), 134–137. doi:10.1108/00438029910272409

Pasewark, W. R., & Viator, R. E. (2006). Sources of Work- Family Conflict in the Accounting Profession. *Behavioral Research in Accounting*, *18*(1), 147–165. doi:10.2308/bria.2006.18.1.147

Patney, M.B. (2010). *Indian consumers and their mall patronage: Application of cultural-self and the theory of planned behavior to patronage intentions* [Thesis].

Paul, L., & Hiremath, S. S. (2014). Evaluation of process of ECDM using grey relation analysis. *Procedia Materials Science*, *5*, 2273–2282. doi:10.1016/j.mspro.2014.07.446

Paulraj, A. (2011). Understanding the relationships between internal resources and capabilities, sustainable supply management and organizational sustainability. *Journal of Supply Chain Management*, *47*(1), 19–37. doi:10.1111/j.1745-493X.2010.03212.x

Paulraj, A., & de Jong, P. (2011). The effect of ISO 14001 certification announcements on stock performance. *International Journal of Operations & Production Management*, *31*(7), 765–788. doi:10.1108/01443571111144841

Pazirandeh, A., & Jafari, H. (2013). Making sense of green logistics. *International Journal of Productivity and Performance Management*, *62*(8), 889–904. doi:10.1108/IJPPM-03-2013-0059

Pearce, D. W., Markandya, A., & Barbier, E. (1989). *Blueprint for a green economy*. Earthscan.

Pereira, J. V. (2009). The new supply chain's frontier: Information management. *International Journal of Information Management, 29*(5), 372–379. doi:10.1016/j.ijinfomgt.2009.02.001

Perry, M., & Tse Sheng, T. (1999). An overview of trends related to environmental reporting in Singapore. *Environmental Management and Health, 10*(5), 310–320. doi:10.1108/09566169910289667

Pettersen, J. (2009). Defining lean production: Some conceptual and practical issues. *The TQM Journal, 21*(2), 127–142. doi:10.1108/17542730910938137

Pfeffer, J. (1998). Six myths about pay. *Harvard Business Review*, May-June, 109–119. PMID:10179647

Pfeffer, J., & Salancik, G. (1978). *The External Control of Organizations: A Resource Dependence Perspective*. New York, NY: Harper & Row.

Pham, D. T., & Thomas, A. J. (2011). Fit manufacturing: A framework for sustainability. *Journal of Manufacturing Technology Management, 23*(1), 103–123. doi:10.1108/17410381211196311

Pierson, P. (2000). Increasing returns, path dependence, and study of politics. *The American Political Science Review, 94*(2), 251–267. doi:10.2307/2586011

Pinckard, M., & Wendy, V. (2000, October). Environmental Audit–Working within a Restricted Mandate (Draft paper 20).

Pinkse, J., & Kolk, A. (2012). Multinational enterprises and climate change; Exploring institutional failures and embeddedness. *Journal of International Business Studies, 43*(3), 332–341. doi:10.1057/jibs.2011.56

Pittaway, L., Robertson, M., Munir, K., Denyer, D., & Neely, A. (2004). Networking and innovation: A systematic review of the evidence. *International Journal of Management Reviews, 5*(3-4), 137–168. doi:10.1111/j.1460-8545.2004.00101.x

Pittman, K. (1991). Promoting Youth Development: Strengthening the Role of Youth Serving and Community Organizations. *Paper prepared for USDA, Extension Service, National Youth at Risk Initiative Task Force.*

Plambeck, E., Lee, H., & Yatsko, P. (2012). *Improving environmental performance in your Chinese supply chain.* MIT Sloan Management Review.

Poirier, L. (2010). *PWC Survey: Emissions Reductions Top U.S. Business Concern in Climate change Debate. Global Refining & Fuels Reports.* Hart Energy Publishing.

Poist, R. F. (1989). Evolution of conceptual approaches to the design of logistics systems: A sequel. *Transportation Journal, 28*(3), 35–39.

Pojasek, R. (2006). Is your integrated management system really integrated? *Environmental Quality Management, 16*(2), 89–97. doi:10.1002/tqem.20124

Pojasek, R. (2011). ISO 26000 guidance on Social Responsibility. *Environmental Quality Management, 20*(3), 85–93. doi:10.1002/tqem.20292

Polése, M., & Stren, R. (1999). *The social sustainability of cities: diversity and the management of change.* Toronto: University of Toronto Press.

Polonsky, M.J. (1994). An introduction to green marketing. *Electronic Green Journal, 1*(2).

Polonsky, M. J. (1995). A stakeholder theory approach to designing environmental marketing strategy. *Journal of Business and Industrial Marketing, 10*(3), 29–46. doi:10.1108/08858629510096201

Pope, J., Annandale, D., & Morrison-Saunders, A. (2004). Conceptualizing sustainability assessment. *Environmental Impact Assessment Review, 24*(6), 595–616. doi:10.1016/j.eiar.2004.03.001

Porter, M. E. (1985). *Competitive advantage: Creating and sustaining superior performance*. New York, NY: Free Press.

Porter, M. E., & Kramer, M. R. (2011). Creating shared value. *Harvard Business Review*, *89*(1/2), 62–77.

Porter, M. E., & Linde, C. (1995). Green and competitive: Ending the stalemate. *Harvard Business Review*, *73*(5), 120–134.

Post, J. E., Preston, L. E., & Sachs, S. (2002a). Managing the extended enterprise: The new stakeholder view. *California Management Review*, *45*(1), 6–28. doi:10.2307/41166151

Post, J. E., Preston, L. E., & Sachs, S. (2002b). *Redefining the corporation: Stakeholder management and organizational wealth*. Stanford University Press.

Prabhu, V. V., Jeon, H. W., & Taisch, M. (2012, August). Modeling green factory physics—An analytical approach. In *Automation Science and Engineering (CASE), 2012 IEEE International Conference on* (pp. 46-51). IEEE.

Prabhu, S., Uma, M., & Vinayagam, B. K. (2015). Surface roughness prediction using Taguchi-fuzzy logic-neural network analysis for CNT nano fluids based grinding process. *Neural Computation & Application*, *26*(1), 41–55. doi:10.1007/s00521-014-1696-8

Pradhan, B. B., & Bal, R. K. (2002). Corporate Environmental Reporting: Perceptions of Corporate Managers. *Environmental Accounting and Reporting*.

Pramanik, A. K., Shil, N. C., & Das, B. (2007). Environmental accounting and reporting with special reference to India.

Preuss, L. (2007). Buying into our future: Sustainability initiatives in local government procurement. *Business Strategy and the Environment*, *16*(5), 354–365. doi:10.1002/bse.578

Price, J., & Mueller, C. (1986). *Absenteeism and turnover among hospital employees*. Greenwich, CT: JAI Press.

Prince, B. J. (2005). Career-focused employee transfer processes. *Career Development International*, *10*(4), 293–309. doi:10.1108/13620430510609136

Pujari, D. (2006). Eco-innovation and new product development: Understanding the influences on market performance. *Technovation*, *26*(1), 76–85. doi:10.1016/j.technovation.2004.07.006

Pujari, D., Peattie, K., & Wright, G. (2004). Organizational antecedents of environmental responsiveness in industrial new product development. *Industrial Marketing Management*, *33*(5), 381–391. doi:10.1016/j.indmarman.2003.09.001

Pujari, D., Wright, G., & Peattie, K. (2003). Green and competitive: Influences on environmental new product development performance. *Journal of Business Research*, *56*(8), 657–671. doi:10.1016/S0148-2963(01)00310-1

Pumar, E. S. (2005). Social networks and the institutionalization of the idea of sustainable development. *The International Journal of Sociology and Social Policy*, *25*(1/2), 63–86. doi:10.1108/01443330510791298

Putnam, R. (2001). Social capital: Measurement and consequences. *Canadian Journal of Policy Research*, *2*(1), 41–51.

Putti, J. M. (1989). Work Values and Organizational Commitment: A Study in Asian Context. *Human Relations*, *42*(3), 275. doi:10.1177/001872678904200305

Quest (n. d.). The New world order. *Quest*, 3, 35-39.

Rahbar, E., Shwu Shyan, T., & Abdul Wahi, N. (2011). Factors influencing the green purchase behavior of Penang environmental volunteers. *International Business Management*, *5*(1), 38–49. doi:10.3923/ibm.2011.38.49

Ramanathan, K. V. (1976). Toward a Theory of Corporate Social Accounting. *Accounting Review*, *51*(3), 516–528.

Ramani, K., Ramanujan, D., Bernstein, W. Z., Zhao, F., Sutherland, J., Handwerker, C., & Thurston, D. et al. (2010). Integrated sustainable life cycle design: A review. *Journal of Mechanical Design*, *132*(9), 091004. doi:10.1115/1.4002308

Ramlall, S. (2003). Organizational Application Managing Employee Retention as a Strategy for Increasing Organizational Competitiveness. *Applied H.R.M. Research*, *8*(2), 63–72.

Rao, P. (2002). Greening the supply chain: A new initiative in South East Asia. *International Journal of Operations & Production Management*, *22*(6), 632–655. doi:10.1108/01443570210427668

Rao, P. (2004). Greening production: A South-East Asian experience'. *International Journal of Operations & Production Management*, *24*(3), 289–320. doi:10.1108/01443570410519042

Rao, P. H. (2008). *Greening the supply chain: A guide for Asian managers*. SAGE Publications India.

Rao, P. S., Ramji, K., & Satyanarayana, B. (2014). An Experimental Investigation and Multi-Optimization of Wire EDM Parameter for Surface Roughness- Metal Removal Rate and White Layer in Machining of Aluminium Alloy. *Procedia Material Science*, *5*, 2197–2206. doi:10.1016/j.mspro.2014.07.426

Rao, P., & Holt, D. (2005). Do green supply chains lead to competitiveness and economic performance? *International Journal of Operations & Production Management*, *25*(5), 898–916. doi:10.1108/01443570510613956

Rasmussen, J. M. (2007). *Integrated Management Systems* – an analysis of best practice in Danish companies (Thesis). Aalborg University.

Rattray, J., & Jones, M. C. (2007). Essential Elements of Questionnaire Design and Development. *Journal of Clinical Nursing*, *16*(2), 234–243. doi:10.1111/j.1365-2702.2006.01573.x PMID:17239058

Raynolds, L. T., Murray, D., & Heller, A. (2007). Regulating sustainability in the coffee sector: A comparative analysis of third-party environmental and social certification initiatives. *Agriculture and Human Values*, *24*(2), 147–163. doi:10.1007/s10460-006-9047-8

Redclift, M. (2005). Sustainable development (1987–2005): An oxymoron comes of age. *Sustainable Development*, *13*(4), 212–227. doi:10.1002/sd.281

Reefke, H., & Trocchi, M. (2013). Balanced scorecard for sustainable supply chains: design and development guidelines. *International Journal of Productivity and Performance Management, 62*(8), 805-826.

Reefke, H., & Trocchi, M. (2013). Balanced scorecard for sustainable supply chains: Design and development guidelines. *International Journal of Productivity and Performance Management*, *62*(8), 805–826. doi:10.1108/IJPPM-02-2013-0029

Reinhardt, F. L. (1999). Bringing the environment down to earth. *Harvard Business Review*, *77*(4), 149–157. PMID:10539206

Report, M. (2003). Mercer study raises red flags for employer pay and benefits plans (findings of the 2002 people at work survey). In Human resource department management report (pp. 8-15).

Reuter, C., Goebel, P., & Foerstl, K. (2012). The impact of shareholder orientation on sustainability and cost prevalence in supplier selection decisions. *Journal of Purchasing and Supply Management*, *18*(4), 270–281. doi:10.1016/j.pursup.2012.06.004

Ricart, J. E., Rodríguez, M. Á., & Sánchez, P. (2005). Sustainability in the boardroom: An empirical examination of Dow Jones Sustainability World Index leaders. *Corporate Governance*, *5*(3), 24–41. doi:10.1108/14720700510604670

Richardson, J. (2014). 100% Renewable Electricity Will Be Achieved In German State Soon. *Clean Technica*. Retrieved from http://cleantechnica.com/2014/06/23/100-renewable-electricity-will-achieved-german-state-soon/

Rimanoczy, I., & Pearson, T. (2010). Role of HR in the new world of sustainability. *Industrial and Commercial Training, 42*(1), 11–17. doi:10.1108/00197851011013661

Ritzman, L. P., Krejewski, L. J., & Klassen, R. D. (2004). *Foundations of operations management*. Upper Saddle River, NJ: Prentice-Hall.

Robertson, A., & Abbey, G. (2003). *Managing Talented People*. Harlow: Pearson Education Limited.

Roberts, S. (2003). Supply chain specific? Understanding the patchy success of ethical sourcing initiatives. *Journal of Business Ethics, 44*(2-3), 159–170. doi:10.1023/A:1023395631811

Robinson, J. (2004). Squaring the circle? Some thoughts on the idea of sustainable development. *Ecological Economics, 48*(4), 369–384. doi:10.1016/j.ecolecon.2003.10.017

Robinson, J., Bradley, M., Busby, P., Connor, D., Murray, A., Sampson, B., & Soper, W. (2006). Climate Change and Sustainable Development: Realizing the opportunity. *Ambio, 35*(1), 2–7. doi:10.1579/0044-7447-35.1.2 PMID:16615692

Robinson, S. (2000). Key survival issues: Practical steps toward corporate environmental sustainability. *Corporate Environmental Strategy, 7*(1), 92–105. doi:10.1016/S1066-7938(00)80118-9

Rodrigue, J. P., Slack, B., & Comtois, C. (2001). Green logistics (the paradoxes of). The Handbook of Logistics and Supply-Chain Management, 2.

Rof, L. M., & Danuletiu, D. C. (2010). Environmental Accounting and Corporate Social Responsibility. *Annals of Faculty of Economics, 1*(2), 885–891.

Rogers, E. M. (2003). *Diffusion of Innovations* (5th ed.). New York, NY: Free Press.

Romar, E. J. (2009). Snapshots of the Future: Darfur, Katrina, and Maple Sugar (Climate Change, the Less Well –off and Business Ethics). *Journal of Business Ethics, 85*(Suppl. 1), 121–132. doi:10.1007/s10551-008-9932-8

Romilly, P. (2007). Business and climate change risk: A regional time series analysis. *Journal of International Business Studies, 38*(3), 474–480. doi:10.1057/palgrave.jibs.8400266

Rondinelli, D. A., & Berry, M. A. (2000). Environmental citizenship in multinational corporations: Social responsibility and sustainable development. *European Management Journal, 18*(1), 70–84. doi:10.1016/S0263-2373(99)00070-5

Roos, D., Womack, J. P., & Jones, D. (1990). *The Machine That Changed the World: The Story of Lean Production*. Harper Perennial, New York: Rawson.

Roth, A. V., Giffi, C. A., & Seal, G. M. (1992). *Operating strategies for the 1990s: elements comprising world-class manufacturing. In Manufacturing Strategy. Process and Content* (pp. 133–165). London: Chapman & Hall.

Rothenberg, S. (2007). Environmental managers as institutional entrepreneurs: The influence of institutional and technical pressures on waste management. *Journal of Business Research, 60*(7), 749–757. doi:10.1016/j.jbusres.2007.02.017

Rowley, T. (1997). Moving beyond dyadic ties: A network theory of stakeholder influences. *Academy of Management Review, 22*(4), 887–910.

Roznowski, M., & Hulin, C. (1992). The scientific merit of valid measures of general constructs with special reference to job satisfaction and job withdrawal. In C. J. Cranny, P. C. Smith, & E. F. Stone (Eds.), *Job satisfaction: How people feel about their jobs and how it affects their performance*. New York: Lexington Books.

Rugman, A. M., Li, J., & Oh, C. H. (2009). Are supply chains global or regional? *International Marketing Review, 26*(4-5), 384–395.

Rusinko, C. A. (2007). Green Manufacturing: An evaluation of environmentally sustainable manufacturing practices and their impact on competitive outcomes. *IEEE Transactions on Engineering Management, 54*(3), 445–454. doi:10.1109/TEM.2007.900806

Saberi, S., Nookabadi, A. S., & Hejazi, S. R. (2012). Applying agent-based system and negotiation mechanism in improvement of inventory management and customer order fulfilment in multi echelon supply chain. *Arabian Journal for Science and Engineering, 37*(3), 851–861. doi:10.1007/s13369-012-0197-2

Sachs, I. (1999). Social sustainability and whole development: exploring the dimensions of sustainable development. *Sustainability and the social sciences: a cross-disciplinary approach to integrating environmental considerations into theoretical reorientation* (pp. 25-36).

Sahay, A. (2004). Environmental reporting by Indian corporations. *Corporate Social Responsibility and Environmental Management, 11*(1), 12–22. doi:10.1002/csr.51

Sahu, A. K., Datta, S., & Mahapatra, S. S. (2013). Green supply chain performance benchmarking using integrated IVFN-TOPSIS methodology. *International Journal of Process Management and Benchmarking, 3*(4), 511–551. doi:10.1504/IJPMB.2013.058272

Salomone, R. (2008). Integrated management systems: Experiences in Italian organizations. *Journal of Cleaner Production, 16*(16), 1786–1806. doi:10.1016/j.jclepro.2007.12.003

Sambasivan, M., & Fei, N. Y. (2008). Evaluation of critical success factors of implementation of ISO 14001 using analytic hierarchy process (AHP): A case study from Malaysia. *Journal of Cleaner Production, 16*(13), 1424–1433. doi:10.1016/j.jclepro.2007.08.003

SAMDA report. (n. d.). Retrieved from www.samda.co.za

Sánchez, A. M., & Pérez, M. P. (2001). Lean indicators and manufacturing strategies. *International Journal of Operations & Production Management, 21*(11), 1433–1452. doi:10.1108/01443570110407436

Sanchez, X., A., Lehtiranta, L., D. Hampson, K., & Kenley, R. (. (2014). Evaluation framework for green procurement in road construction. *Smart and Sustainable Built Environment, 3*(2), 153–169. doi:10.1108/SASBE-05-2013-0028

Sanders, N. R., & Wagner, S. M. (2011). Multidisciplinary and multimethod research for addressing contemporary supply chain challenges. *Journal of Business Logistics, 32*(4), 317-323.

Sarkis, J. (2003). A strategic framework for green supply chain management. *Journal of Cleaner Production, 11*(4), 397–409. doi:10.1016/S0959-6526(02)00062-8

Sarkis, J. (2006). *Greening the supply chain.* Berlin, Germany: Springer-Verlag. doi:10.1007/1-84628-299-3

Sarkis, J. (2012). A boundaries and flows perspective of green supply chain management. *Supply Chain Management: An International Journal, 17*(2), 202–216. doi:10.1108/13598541211212924

Sarkis, J., Gonzalez-Torre, P., & Adenso-Diaz, B. (2010). Stakeholder pressure and the adoption of environmental practices: The mediating effect of training. *Journal of Operations Management, 28*(2), 163–176. doi:10.1016/j.jom.2009.10.001

Sarkis, J., Zhu, Q., & Lai, K.-H. (2011). An organizational theoretic review of green supply chain management literature. *International Journal of Production Economics, 130*(1), 1–15. doi:10.1016/j.ijpe.2010.11.010

Satolo, E. G. (2012). *Measurement model for assessment of organizational sustainability* [Thesis in Portuguese]. Santa Bárbara d'Oeste: Unimep.

Satolo, E.G., Calarge, F.A. & Miguel, P.A.C. (2013). Experience with an integrated management system in a sugar and ethanol manufacturing unit: possibilities and limitations. *Management of Environmental Quality: an international journal*, 24(6), 710-725.

Schaltegger, S., Bennett, M., & Burritt, R. (Eds.). (2006). *Sustainability accounting and reporting* (Vol. 21). Springer. doi:10.1007/978-1-4020-4974-3

Schiemann, W. A. (2009). *Reinventing Talent Management*. Hoboken, NJ: John Wiley & Sons.

Schlossberg, M., & Zimmerman, A. (2003). Developing state wide indices of environmental, economic, and social sustainability: a look at Oregon and the Oregon Benchmarks. *Local Environment*, *86*, 641–660.

Schneider, F., Kall-is, G., & Martinez-Alier, J. (2010). Crisis or opportunity? Economic degrowth for social equity and ecological sustainability. Introduction to this special issue. *Journal of Cleaner Production*, *18*(6), 511–518. doi:10.1016/j.jclepro.2010.01.014

Schneider, L., & Wallenburg, C. M. (2012). Implementing sustainable sourcing – does purchasing need to change? *Journal of Purchasing and Supply Management*, *18*(4), 243–257. doi:10.1016/j.pursup.2012.03.002

Schonberger, R. J. (1986). *World Class Manufacturing: The Lessons of Simplicity Applied*. New York: Free Press.

Schone, R., & Molka, D. (2014). Integrating performance analysis and energy efficiency optimizations in a unified environment. *Computer Science - Research for Development*, *29*(3-4), 231–239. doi:10.1007/s00450-013-0243-7

Schrettle, S., Hinz, A., Scherrer-Rathje, M., & Friedli, T. (2014). Turning sustainability into action: Explaining firms' sustainability efforts and their impact on firm performance. *International Journal of Production Economics*, *147*, 73–84. doi:10.1016/j.ijpe.2013.02.030

Schroeder, R. G., & Flynn, B. B. (Eds.). (2002). *High performance manufacturing: Global perspectives*. NY: Wiley.

Schuster, R., Kirchiner, V., Allongue, P., & Etrl, G. (2000). Electrochemical micro machining. *Inder Science*, *289*(5476), 98–101. doi:10.1126/science.289.5476.98 PMID:10884233

Schwarz, J., Beloff, B., & Beaver, E. (2002). Use sustainability metrics to guide decision-making. *Chemical Engineering Progress*, *98*(7), 58–63.

Scott, J. (1990). *A matter of Record – Documentary Sources in Social Research*. Cambridge: Polity.

Searcy, C., Karapetrovic, S., & McCartney, D. (2008). Application of a systems approach to sustainable development performance measurement. *International Journal of Productivity and Performance Management*, *57*(2), 182–197. doi:10.1108/17410400810847429

Seliger, G. (2007). *Sustainability in manufacturing: recovery of resources in product and material cycles*. Springer. doi:10.1007/978-3-540-49871-1

Seliger, G., Kim, H. J., Kernbaum, S., & Zettl, M. (2008). Approaches to sustainable manufacturing. *International Journal of Sustainable Manufacturing*, *1*(1), 58–77. doi:10.1504/IJSM.2008.019227

Sen, A. (2013). The ends and means of sustainability. *Journal of Human Development and Capabilities*, *14*(1), 6–20. doi:10.1080/19452829.2012.747492

Sen, S., & Bhattacharya, C. B. (2001). Does doing good always lead to doing better? Consumer reactions to corporate social responsibility. *JMR, Journal of Marketing Research*, *38*(2), 225–243. doi:10.1509/jmkr.38.2.225.18838

Senthil Kumar, C., Ganesan, G., & Karthikeyan, R. (2012). Optimization of ECM Process Parameters using NSGA-II. *Journal of Minerals and Materials Characterization and Engg, 11*(10), 931–937. doi:10.4236/jmmce.2012.1110091

Seppanen, R., Blomqvist, K., & Sundqvist, S. (2007). Measuring inter-organizational trust: A critical review of the empirical research in 1990–2003. *Industrial Marketing Management, 36*(2), 249–265. doi:10.1016/j.indmarman.2005.09.003

Seuring, S. (2013). A review of modeling approaches for sustainable supply chain management. *Decision Support Systems, 54*(4), 1513–1520. doi:10.1016/j.dss.2012.05.053

Seuring, S., & Gold, S. (2012). Conducting content-analysis based literature reviews in supply chain management. *Supply Chain Management: An International Journal, 17*(5), 544–555. doi:10.1108/13598541211258609

Seuring, S., & Müller, M. (2008). Core issues in sustainable supply chain management–a Delphi study. *Business Strategy and the Environment, 17*(8), 455–466. doi:10.1002/bse.607

Seuring, S., & Müller, M. (2008). From a literature review to a conceptual framework for sustainable supply chain management. *Journal of Cleaner Production, 16*(15), 1699–1710. doi:10.1016/j.jclepro.2008.04.020

Shah, J. (2009). *Supply chain management: Text and Cases*. Pearson Education India.

Shang, G. Q., & Sun, C. H. (2008). Application of BP neural Network for Predicting Anode Accuracy in ECM. *Proceedings of theInternational Symposium on Information Science and Engineering* (pp. 20-25). doi:10.1109/ISISE.2008.55

Shank, J. K., & Govindarajan, V. (1993). *Strategic cost management*. New York, NY: Free Press.

Sharma, M. K., Bhagwat, R., & Dangayach, G. S. (2005). Practice of performance measurement: Experience from Indian SMEs. *International Journal of Globalisation and Small Business, 1*(2), 183–213. doi:10.1504/IJGSB.2005.008014

Sharma, M., & Kodali, R. (2008). TQM implementation elements for manufacturing excellence. *The TQM Journal, 20*(6), 599–621. doi:10.1108/17542730810909365

Sharma, S., & Jain, V. K. (2014). Drivers Affecting the Green Supply Chain Adaptation: A Review. *The IUP Journal of Operations Management, 9*(1), 54–63.

Sharma, S., & Ruud, A. (2003). On the path to sustainability: Integrating social dimensions into the research and practice of environmental management. *Business Strategy and the Environment, 12*(4), 205–214. doi:10.1002/bse.366

Sharma, S., & Vredenburg, H. (1998). Proactive corporate environmental strategy and the development of competitively valuable organizational capabilities. *Strategic Management Journal, 19*(8), 729–753. doi:10.1002/(SICI)1097-0266(199808)19:8<729::AID-SMJ967>3.0.CO;2-4

Shepherd, C., & Gunter, H. (2006). Measuring supply chain performance: Current research and future directions. *International Journal of Productivity and Performance Management, 55*(3-4), 242–258. doi:10.1108/17410400610653219

Sheth, J. N. (2008). *Chindia rising: How India and China will benefit your business*. New Delhi: Tata McGraw Hill.

Sheu, J. B., & Chen, Y. J. (2012). Impact of government financial intervention on competition among green supply chains. *International Journal of Production Economics, 138*(1), 201–213. doi:10.1016/j.ijpe.2012.03.024

Shi, V.G., Koh, S.C.L., Baldwin, J. & Cucchiella, F. (2012). *Natural resource based green supply chain management*, 12(1), 54-67.

Shields, M. A., & Wheatley Price, S. W. (2002). Racial Harassment, Job Satisfaction and Intentions to Quit: Evidence from the British Nursing Profession. *Economica, 69*(274), 295–326. doi:10.1111/1468-0335.00284

Shi, X., & Liao, Z. (2013). The mediating effects of interfirm business process integration and joint teamwork on firm performance in supply chains. *Asia Pacific Journal of Management*, *30*(4), 1243–1264. doi:10.1007/s10490-012-9308-6

Shoushtari, K. D. (2013). Redesigning a large supply chain management system to reduce the government administration: A socio-functional systems approach. *Systemic Practice and Action Research*, *26*(2), 195–216. doi:10.1007/s11213-012-9244-x

Shrimp sold at Walmart Costco tied to slave labour. (2014, June 11). CBS News report. Retrieved from http://www.cbsnews.com/news/shrimpsold-at-walmart-costco-tied-to-slave-labor/

Shrivastava, P. (1995). Environmental technologies and competitive advantage. *Strategic Management Journal*, *16*(Suppl. 1), 183–200. doi:10.1002/smj.4250160923

Shukla, A., & Vyas, N. (2000). Environmental Accounting & Reporting in India. *Bharat Petroleum Company Limited & Oil & Natural Gas Company Limited.*

Siebertz, L., Van Bebber, D., & Hochkirchen, T. (2010). *Design of Experiments (DoE).* London: Springer Heidelberg Dordrecht.

Siena, O. (2008). Método para avaliar desenvolvimento sustentável: Técnicas para escolha e ponderação de aspectos e dimensões. *Production*, *18*(2), 359–374. doi:10.1590/S0103-65132008000200012

Sikdar, S. K., & Howell, S. G. (1998). On developing cleaner organic unit processes. *Journal of Cleaner Production*, *6*(3-4), 253–259. doi:10.1016/S0959-6526(98)00026-2

Silbert, L. T. (2005). The effect of Tangible Rewards on Perceived Organizational Support. Management Sciences. Retrieved from http://etd.uwaterloo.ca/etd/lsilbert2005.pdf?origin=publication_detail

Silva, A. K. M., & McGeough, J. (1998). Process monitoring of electrochemical micromachining. *Journal of Materials Processing Technology*, *76*(1-3), 165–169. doi:10.1016/S0924-0136(97)00334-8

Simon, A., Karapetrovic, S., & Casadesus, M. (2012). Evolution of Integrated Management Systems in Spanish firms. *Journal of Cleaner Production*, *23*(1), 8–19. doi:10.1016/j.jclepro.2011.10.025

Simpson, D. F., & Power, D. J. (2005). Use the supply relationship to develop lean and green suppliers. *Supply chain management:International Journal (Toronto, Ont.)*, *10*(1), 60–68.

Simpson, D., Power, D., & Samson, D. (2007). Greening the automotive supply chain: A relationship perspective. *International Journal of Operations & Production Management*, *27*(1), 28–48. doi:10.1108/01443570710714529

Simpson, J. A. (2006). *The measurement and recognition of soft skills-Developing a Common Standard?* Brussels: European Union.

Singh, V.K. (2009). PDCA Cycle: a quality approach. *Utthan – The Journal of Management Sciences*, 1(1), 89-96.

Singh, A., Narain, R., & Yadav, R. C. (2006). Benchmarking and performance measurement of supply chain management practices: A survey of Indian organizations. *International Journal of Services and Operations Management*, *2*(4), 313–334. doi:10.1504/IJSOM.2006.010250

Singh, G., & Joshi, M. (2009). Environment management and disclosure practices of Indian companies. *International Journal of Business Research*, *9*(2), 116–128.

Singh, J., & Gill, S. S. (2009). Fuzzy Modelling and Simulation of Ultrasonic Drilling of Porcelain Ceramic with Hollow Stainless Steel Tools. *Materials and Manufacturing Processes*, *24*(4), 468–475. doi:10.1080/10426910802714407

Singh, R. K., Murty, H. R., Gupta, S. K., & Dikshit, A. K. (2009). An overview of sustainability assessment methodologies. *Ecological Indicators*, *15*(1), 189–212. doi:10.1016/j.ecolind.2008.05.011

Sinha, D., & Sinha, S. (2008). Talent Retention in Academics: A Case Study on Management.

Sisaye, S. (2011). Ecological systems approaches to sustainability and organizational development emerging trends in environmental and social accounting reporting systems. *Leadership and Organization Development Journal*, *32*(4), 379–398. doi:10.1108/01437731111134652

Sjoberg, A., & Sverke, M. (2000). The interactive effect of job involvement and organizational commitment on job turnover revisited: A note on the mediating role of turnover intention. *Scandinavian Journal of Psychology*, *41*(3), 247–252. doi:10.1111/1467-9450.00194 PMID:11041307

Slack, N. (2005). The flexibility of manufacturing systems. *International Journal of Operations & Production Management*, *25*(12), 1190–1200. doi:10.1108/01443570510633594

Smeraldi, R. (2009). *The new manual of sustainable business (in Portuguese)*. São Paulo: Publifolha.

Smilansky, J. (2006). *Developing Executive Talent*. Chichester: Jossey-Bass.

Smith, M. K. (2001) Young people, informal education and association *Informal Education*. Retrieved from www.infed.org/youthwork/ypandassoc.htm

Smith, L., & Ball, P. (2012). Steps towards sustainable manufacturing through modelling material, energy and waste flows. *International Journal of Production Economics*, *140*(1), 227–238. doi:10.1016/j.ijpe.2012.01.036

Smyth, E. (2011), LED vs CFL — Which Light Bulb is More Efficient? *Clean Technica*. Retrieved from http://cleantechnica.com/2011/09/01/led-vs-cfl-which-light-bulb-is-more-efficient/

Sobel, M. E. (1982). Asymptotic confidence intervals for indirect effects in structural equation models. *Sociological Methodology*, *13*, 290–312. doi:10.2307/270723

Sodano, V., & Hingley, N. (2013). The food system, climate change and CSR from business to government case. *British Food Journal*, *115*(1), 75–91. doi:10.1108/00070701311289885

Solar Means Business 2015: Top U.S. Corporate Solar Users. (2015). *Solar Energy Industries Association*. Retrieved from http://www.seia.org/research-resources/solar-means-business-2015-top-us-corporate-solar-users

Soni, G., & Kodali, R. (2010). Internal benchmarking for assessment of supply chain performance. *Benchmarking: An International Journal*, *17*(1), 44–76. doi:10.1108/14635771011022316

Souza, G. C. (2013). Closed-loop supply chains: A critical review, and future research. *Decision Sciences*, *44*(1), 7–38. doi:10.1111/j.1540-5915.2012.00394.x

Spangenberg, J. H., Omann, I., & Hinterberger, F. (2002). Sustainable growth criteria: Minimum benchmarks and scenarios for employment and the environment. *Ecological Economics*, *42*(3), 429–443. doi:10.1016/S0921-8009(02)00125-8

Spekman, R. E., Kamauff, J. W. Jr, & Myhr, N. (1998). An empirical investigation into supply chain management: A perspective on partnerships. *Supply Chain Management: An International Journal*, *3*(2), 53–67. doi:10.1108/13598549810215379

Spence, L., & Bourlakis, M. (2009). The evolution from corporate social responsibility to supply chain responsibility: The case of Waitrose. *Supply Chain Management: An International Journal*, *14*(4), 291–302. doi:10.1108/13598540910970126

Srai, J. S., & Gregory, M. (2008). A supply network configuration perspective on international supply chain development. *International Journal of Operations & Production Management*, *28*(5), 386–411. doi:10.1108/01443570810867178

Srinivas, N., & Deb, K. (1995). Multi-objective optimization using non-dominated sorting in genetic algorithms. *Evolutionary Computation*, *3*, 221–248.

Srivastava, S. K. (2007). Green supply chain management: A state-of-the-art- literature review. *International Journal of Management Reviews*, *9*(1), 53–80. doi:10.1111/j.1468-2370.2007.00202.x

Stavins, R. N., Wagner, A. F., & Wagner, G. (2003). Interpreting sustainability in economic terms: Dynamic efficiency plus intergenerational equity. *Economics Letters*, *79*(3), 339–343. doi:10.1016/S0165-1765(03)00036-3

Steel, R. P., Griffeth, R. W., & Hom, P. W. (2002). Practical retention policy for the practical manager. *The Academy of Management Executive*, *18*(2), 149–169. doi:10.5465/AME.2002.7173588

Steel, R., & Ovalle, N. (1984). A Review and meta-analysis of research on the relationship between behavioral intentions and employee turnover. *The Journal of Applied Psychology*, *69*(4), 673–686. doi:10.1037/0021-9010.69.4.673

Steers, R. M. (1997). Antecedents and outcomes of organizational commitment. *American Scientist*, *53*, 36–39.

Stefan, S., & Roger, B. (2014). Measuring and managing sustainability performance of supply chains. *Supply Chain Management International Journal (Toronto, Ont.)*, *19*(3).

Steudel, H. J., & Desruelle, P. (1992). *Manufacturing in the nineties, how to become a mean, lean, world-class competitor*. New York: Von Nostrand Reinhold.

Stevenson, W. (2011). *Operations management*. McGraw-Hill/Irwin.

Stieb, J. A. (2009). Assessing Freeman's stakeholder theory. *Journal of Business Ethics*, *87*(3), 401–414. doi:10.1007/s10551-008-9928-4

Stock, J. R., Boyer, S. L., & Harmon, T. (2010). Research opportunities in supply chain management. *Journal of the Academy of Marketing Science*, *38*(1), 32–41. doi:10.1007/s11747-009-0136-2

Stock, J. R., & Lambert, D. M. (2001). *Strategic logistics management* (Vol. 4). Boston, MA: McGraw-Hill/Irwin.

Stonebraker, P. W., & Afifi, R. (2004). Toward a contingency theory of supply chains. *Management Decision*, *42*(9), 1131–1144. doi:10.1108/00251740410565163

Storey, J., & Sisson, K. (1993). *Managing Human Resources and industrial Relations* (pp. 22–31). Buckingham: Open University Press.

Stoughton, A. M., & Ludema, J. (2012). The driving forces of sustainability. *Journal of Organizational Change Management*, *25*(4), 501–517. doi:10.1108/09534811211239191

Strecker, S., Frank, U., Heise, D., & Kattenstroth, H. (2012). MetricM: A modeling method in support of the reflective design and use of performance measurement systems. *Information Systems and e-Business Management, 10*(2), 241-276.

Strong, C. (1997). The problems of translating fair trade principles into consumer purchase behaviour. *Marketing Intelligence & Planning*, *15*(1), 32–37. doi:10.1108/02634509710155642

Sundarakani, B., De Souza, R., Goh, M., Wagner, S. M., & Manikandan, S. (2010). Modeling carbon footprints across the supply chain. *International Journal of Production Economics*, *128*(1), 43–50. doi:10.1016/j.ijpe.2010.01.018

Sun, S., Hsu, M., & Hwang, W.-J. (2009). The impact of alignment between supply chain strategy and environmental uncertainty on SCM performance. *Supply Chain Management: An International Journal*, *14*(3), 201–244. doi:10.1108/13598540910954548

Sushil. (2001). Framework. *Global Journal of Flexible Systems Management.*

Sushil. (2005a). Interpretive matrix: A tool to aid interpretation of management and social research. *Global Journal of Flexible Systems Management*, 6(2), 27-30

Sushil. (2005b). A flexible strategy framework for managing community and change. International Journal of Global Business and Competitiveness, 1(1), 22-32

Sushil. (2009). Interpretive ranking process. *Global Journal of Flexible Systems Management*, 10(4), pp.1-10

Sushil. (2012). Interpreting the Interpretive Structural Model. *Global Journal of Flexible Systems Management*, 13(2), 87-106

Sushil, A. (2005a). Interpretive matrix: A tool to aid interpretation of management and social research. *Global Journal of Flexible Systems Management*, *6*(2), 27–30.

Sushil, A. (2005b). A flexible strategy framework for managing continuity and change. *International Journal of Global Business and Competitiveness*, *1*(1), 22–32.

Sustainability report. (2009). IISD. Retrieved from http://www.iisd.org/about/sdreporting/2002/eco_dim.asp

Sustainability reporting guidelines. (2000Global Reporting Initiative. Amsterdam: Global Reporting Initiative.

Sustainable Public Procurement: Towards a low carbon economy (2007GL01). (2007). TERI, New Delhi.

Sustaining Employment Growth: The Role of Manufacturing and Structural Change (Industrial Development Report). (2013). United Nation Industrial Development Organization (UNIDO).

Szekely, F., & Knirsch, M. (2005). Responsible leadership and corporate social responsibility: Metrics for sustainable performance. *European Management Journal*, *23*(6), 628–647. doi:10.1016/j.emj.2005.10.009

Tachizawa, E. M., & Wong, C. Y. (2014). Towards a theory of multi-tier sustainable supply chains: A systematic literature review. *Supply Chain Management: An International Journal, 19*(5/6).

Tagesson, T., Blank, V., Broberg, P., & Collin, S. O. (2009). What explains the extent and content of social and environmental disclosures on corporate websites: A study of social and environmental reporting in Swedish listed corporations. *Corporate Social Responsibility and Environmental Management*, *16*(6), 352–364. doi:10.1002/csr.194

Taguchi, G. (1990). *Introduction to Quality Engineering*. Tokyo: Asian Productivity Organization.

Talluri, S., & Sarkis, J. (2002). A model for performance monitoring of suppliers. *International Journal of Production Research*, *40*(16), 4257–4269. doi:10.1080/00207540210152894

Tan, K. C., Kannan, V. R., Handfield, R. B., & Ghosh, S. (1999). Supply chain management: An empirical study of its impact on performance. *International Journal of Operations & Production Management*, *19*(10), 1034–1052. doi:10.1108/01443579910287064

Tapiero, C. S., & Kogan, K. (2008). Sustainable infrastructure investment with labor-only production. *International Journal of Production Economics*, *113*(2), 876–886. doi:10.1016/j.ijpe.2007.12.004

Taplin, J.R.D., Bent, D., & Aeron-Thomas, D. (2006). Developing a Sustainability Accounting Framework to Inform Strategic Business Decisions: a Case Study from the Chemicals.

Tassabehji, R., & Moorhouse, A. (2008). The changing role of procurement: Developing professional effectiveness. *Journal of Purchasing and Supply Management*, *14*(1), 55–68. doi:10.1016/j.pursup.2008.01.005

Tata. (2015). Leadership with trust. Retrieved from www.tarta.co.in/aboutus/sub-index/Leadership–with–trust

Tate, W. L., Ellram, L. M., & Dooley, K. J. (2012). Environmental purchasing and supplier management: Theory and practice. *Journal of Purchasing and Supply Management*, *18*(3), 173–188. doi:10.1016/j.pursup.2012.07.001

Taticchi, P., Balachandran, K., & Tonelli, F. (2012). Performance measurement and management systems: State of the art, guidelines for design and challenges. *Measuring Business Excellence*, *16*(2), 41–54. doi:10.1108/13683041211230311

Taticchi, P., Tonelli, F., & Cagnazzo, L. (2009). A decomposition and hierarchical approach for business performance measurement and management. *Measuring Business Excellence*, *13*(4), 47–57. doi:10.1108/13683040911006783

Taticchi, P., Tonelli, F., & Pasqualino, R. (2013). Performance measurement of sustainable supply chains: A literature review and a research agenda. *International Journal of Productivity and Performance Management*, *62*(8), 782–804. doi:10.1108/IJPPM-03-2013-0037

Tecnologia, R. (2007). Especificação de requisitos comuns de sistemas de gestão com estrutura para a integração. São Paulo: Risk Tecnologia Editora.*Sistemas Integrados de Gestão PAS, 99.*

Terrin Report. (2012). Retrieved from http://www.teriin.org/policybrief/docs/spp2013.pdf

Terziovski, M., & Power, D. (2007). Increasing ISO 9000 certification benefits: A continuous improvement approach. *International Journal of Quality & Reliability Management*, *24*(2), 141–163. doi:10.1108/02656710710722266

Testa, F., & Iraldo, F. (2010). Shadows and lights of GSCM (Green Supply Chain Management): Determinants and effects of these practices based on a multi-national study. *Journal of Cleaner Production*, *18*(10), 953–962. doi:10.1016/j.jclepro.2010.03.005

The emerging Indian economy. (2013). *Center for Strategic & International Studies.* Retrieved from http://csis.org/files/publication/130206_Khambatta _IndiaEmergEcon_Web.pdf

The Handbook of National Accounting, Studies in Methods Series F, No. 78 Integrated Environmental and Economic Accounting: An Operational Manual. (2000). United Nations, New York.

The high performance work force: separating the digital economy's winners from losers. (2001). *Accenture: In the battle for retention accentures study.*

The Institute for Sustainability. (2011). Retrieved from http://theinstituteforsustainability.com/

The more things change: The World Bank, TATA and enduring abuses on India's tea plantations. (2014). *World Bank report.* Retrieved from https://web.law.columbia.edu/sites/default/files/microsites/human-rights-institute/files/tea_report_final_draft-smallpdf.pdf

Thomas, A., Francis, M., John, E., & Davies, A. (2012). Identifying the characteristics for achieving sustainable manufacturing companies. *Journal of Manufacturing Technology Management*, *23*(4), 426–440. doi:10.1108/17410381211230376

Thompson, C. A., & Prottas, D. J. (2005). Relationships among Organizational Family Support, Job Autonomy, Perceived Control, and Employee well-being. *Journal of Occupational Health Psychology*, *11*(1), 100–118. doi:10.1037/1076-8998.10.4.100 PMID:16551178

Thompson, P., & Zakaria, Z. (2004). Corporate social responsibility reporting in Malaysia. *Journal of Corporate Citizenship*, *2004*(13), 125–136. doi:10.9774/GLEAF.4700.2004.sp.000014

Toffel, M.W. (2014, March). Implementing Environmentally Sustainable Operations. *Harvard Business School Module Note 613-090.*

Topping, N. (2015).Transparency + a price on carbon can radically accelerate decarbonization. *The World Bank*. Retrieved From http://www.acs.org/content/acs/en/greenchemistry/what-is-green-chemistry/principles/sandestin-declaration.html

Tough choices facing the South African mining industry. (n. d.). *Deloitte Blog*. Retrieved http://deloitteblog.co.za/wp-content/uploads/2013/08

Tower, P. (2003). Rewards: The not-so-secret ingredient for managing talent Retention. *HRFocus*, *80*(1), 3–10.

Tracey, S., & Anne, B. (2008). *OECD Insights Sustainable Development Linking Economy, Society, Environment: Linking Economy, Society, Environment*. OECD Publishing.

Tranfield, D., Denyer, D., & Smart, P. (2003). Towards a methodology for developing evidence-informed management knowledge by means of systematic review. *British Journal of Management*, *14*(3), 207–222. doi:10.1111/1467-8551.00375

Tsahalis, J., Moussas, V. C., & Tsahalis, H. T. (2012, July 4-7). Towards a holistic optimization of production sustainability: a multi-level approach. *Proceedings of the5th International Conference from Scientific Computing to Computational Engineering (5th IC-SCCE)*, Athens.

Tsai, K. M., & Wang, P. J. (2001). Predictions on Surface Finish in Electrical Discharge Machining based upon Neural Network Models. *International Journal of Machine Tools & Manufacture*, *41*(10), 1385–1403. doi:10.1016/S0890-6955(01)00028-1

Tsai, W.-H., & Chou, W.-C. (2009). Selecting management systems for sustainable development in SMEs: A novel hybrid model based on DEMATEL, ANP, and ZOGP. *Expert Systems with Applications*, *36*(2), 1444–1458. doi:10.1016/j.eswa.2007.11.058

Tybout, A. M., & Roehm, M. (2009). Let the response fit the scandal. *Harvard Business Review*, *87*(12), 82–88.

Ugrasen, G., Ravindra, H. V., Naveen Prakash, G. V., & Keshavamurthy, R. (2014). Estimation of machining performances using MRA, GMDH and Artificial Neural Network in Wire EDM of EN-31. *Procedia. Materials Science*, *6*, 1788–1797.

UNE 66177. *Management systems: guide for the integration of management systems.* (2005). AENOR, Madrid.

UNEP Report. (2011). Retrieved from http://www.unep.org/greeneconomy/greeneconomyreport/tabid/29846/default.aspx

United Nations Commission on Sustainable Development. Measuring Changes in Consumption and Production Patterns. (1998UNCSD, Division for Sustainable Development, Department of Economic and Social Affairs, New York.

United Nations Information Centre (UNIC). (n. d.). New Delhi. Retrieved from http://www.unic.org.in/items/India_and_the_MDGs_small_web.pdf

United States Environmental Protection Agency. (2015). Retrieved from http://www3.epa.gov/

Unruh, G. (2010). *Earth, Inc.: Using Nature's Rules to Build Sustainable Profits*. Boston, Massachusetts: Harvard Business Press.

Unruh, G., & Ettenson, R. (2011). Growing green: Three smart paths to developing sustainable products. *Harvard Business Review*, *87*(9), 23–39.

Upadhyay, D. K. ((2009). Promoting Sustainable Development from the Grassroots Level: Indian Perspectives. *Madhya Pradesh Journal of Social Sciences*.

Vachon, S. (2007). Green supply chain practices and the selection of environmental technologies. *International Journal of Production Research*, *45*(18-19), 4357–4379. doi:10.1080/00207540701440303

Vachon, S., Halley, A., & Beaulieu, M. (2009). Aligning competitive priorities in the supply chain: The role of interactions with suppliers. *International Journal of Operations & Production Management*, *29*(4), 322–340. doi:10.1108/01443570910945800

Vachon, S., & Klassen, R. D. (2006a). Extending green practices across the supply chain: The impact of upstream and downstream integration. *International Journal of Operations & Production Management, 26*(7), 795–821. doi:10.1108/01443570610672248

Vachon, S., & Klassen, R. D. (2006b). Green project partnership in the supply chain: The case of the package printing industry. *Journal of Cleaner Production, 14*(6-7), 661–671. doi:10.1016/j.jclepro.2005.07.014

Vachon, S., & Klassen, R. D. (2007). Supply chain management and environmental technologies: The role of integration. *International Journal of Production Research, 45*(2), 401–423. doi:10.1080/00207540600597781

Vachon, S., & Klassen, R. D. (2008). Environmental management and manufacturing performance: The role of collaboration in the supply chain. *International Journal of Production Economics, 111*(2), 299–315. doi:10.1016/j.ijpe.2006.11.030

Vachon, S., & Mao, Z. (2008). Linking supply chain strength to sustainable development: A country-level analysis. *Journal of Cleaner Production, 16*(15), 1552–1560. doi:10.1016/j.jclepro.2008.04.012

Vallance, S., Perkins, H. C., & Dixon, J. E. (2011). What is social sustainability? A clarification of concepts. *Geoforum, 42*(3), 342–348. doi:10.1016/j.geoforum.2011.01.002

Van Aken, J. E. (2005). Management research as a design science: Articulating the research products of mode 2 knowledge production in management. *British Journal of Management, 16*(1), 19–36. doi:10.1111/j.1467-8551.2005.00437.x

Van de Ven, B., & Jeurissen, R. (2005). Competing responsibly. *Business Ethics Quarterly, 15*(2), 299–317. doi:10.5840/beq200515216

Van Hoek, R. I. (1999). From reversed logistics to green supply chains. *Supply Chain Management: An International Journal, 4*(3), 129–135. doi:10.1108/13598549910279576

van Hoof, B., & Lyon, T. P. (2013). Cleaner production in small firms taking part in Mexico's Sustainable Supplier Program. *Journal of Cleaner Production, 41*, 270–282. doi:10.1016/j.jclepro.2012.09.023

Van Weele, A. J. (2005). *Purchasing & supply chain management: analysis, strategy, planning and practice*. London: Thomson learning.

Varsei, M., Soosay, C. A., Fahimnia, B., & Sarkis, J. (2014). Framing sustainability performance of supply chains with multidimensional indicators. *Supply Chain Management: An International Journal, 19*(3), 3–3.

Vasconcelos, D. S. C. (2011). Integrated Management System and the PAS 99.

Vasileiou, K., & Morris, J. (2006). The sustainability of the supply chain for fresh potatoes in Britain. *Supply Chain Management: An International Journal, 11*(4), 317–327. doi:10.1108/13598540610671761

Vasques, e. (2007). How thing in sustainability (in portuguese). Revista B2B.

Veleva, V., & Ellenbecker, M. (2001). Indicators of sustainable production: Framework and methodology. *Journal of Cleaner Production, 9*(6), 519–549. doi:10.1016/S0959-6526(01)00010-5

Veleva, V., Hart, M., Greiner, T., & Crumbley, C. (2001). Indicators of sustainable production. *Journal of Cleaner Production, 9*(5), 447–452. doi:10.1016/S0959-6526(01)00004-X PMID:17208696

Verma, M. & Roy. (1988). Work Motivation and Job Satisfaction as Determinants of Organizational Commitment. *Indian Journal of Applied Psychology, 25*(2), 21.

Verma, O. P. (1986). Organizational Commitment as a Function of Managerial Respect. Asian. *Journal of Psychology and Education, 17*(1), 1–6.

Vikas, S., Shashikant, , Roy, A. K., & Kumar, K. (2014). Effect and Optimization of Machine Process Parameters on MRR for EN19 & EN41 materials using Taguchi. *Procedia Technology*, *14*, 204–210. doi:10.1016/j.protcy.2014.08.027

Vimal, K. E. K., & Vinodh, S. (2013). Development of checklist for evaluating sustainability characteristics of manufacturing processes. *International Journal of Process Management and Benchmarking*, *3*(2), 213–232. doi:10.1504/IJPMB.2013.057726

Vinodh, S., & Rajanayagam, D. (2010). CAD and DFM: Enablers of sustainable product design. *International Journal of Sustainable Engineering*, *3*(4), 292–298. doi:10.1080/19397038.2010.501870

Vinodkumar, M. N., & Bhasi, M. (2011). A study on the impact of management system certification on safety management. *Safety Science*, *49*(3), 498–507. doi:10.1016/j.ssci.2010.11.009

Viscusi, W. (1986). Kip (1979).The Impact of Occupational Safety and Health Regulation. *The Bell Journal of Economics*, *10*(1), 117–140. doi:10.2307/3003322

Vonderembse, M. A., Uppal, M., Huang, S. H., & Dismukes, J. P. (2006). Designing supply chains: Towards theory development. *International Journal of Production Economics*, *100*(2), 223–238. doi:10.1016/j.ijpe.2004.11.014

Wackernagel, M. (2001). Shortcomings of the Environmental Sustainability Index. Retrieved from http://www.anti-Iomborg.com/ESI%20critique.rtf

Waddock, S., & Bodwell, C. (2004). Managing responsibility: What can be learned from the quality movement? *California Management Review*, *471*(1), 25–37. doi:10.2307/41166285

Walker, H., Mayo, J., Brammer, S., Touboulic, A., & Lynch, J. (2012). Sustainable procurement: An international policy analysis of 30 OECD countries. *Proceedings of the5th International Public Procurement Conference,* August, Seattle, USA (p. 19).

Walker, J.W. (2001). Perspectives. *Human resource planning*, 24(1), 6-10.

Walker, H., DiSisto, L., & McBain, D. (2008). Drivers and barriers to environmental supply chain management practices: Lessons from the public and private sectors. *Journal of Purchasing and Supply Management*, *14*(1), 69–85. doi:10.1016/j.pursup.2008.01.007

Walker, H., & Brammer, S. (2009). Sustainable procurement in the United Kingdom public sector. *Supply Chain Management: An International Journal*, *14*(2), 128–137. doi:10.1108/13598540910941993

Walker, H., & Brammer, S. (2012). The relationship between sustainable procurement and e-procurement in the public sector. *International Journal of Production Economics*, *140*(1), 256–268. doi:10.1016/j.ijpe.2012.01.008

Walker, H., Miemczyk, J., Johnsen, T., & Spencer, R. (2012). Sustainable procurement: Past, present and future. *Journal of Purchasing and Supply Management*, *18*(4), 201–206. doi:10.1016/j.pursup.2012.11.003

Walker, H., & Preuss, L. (2008). Fostering sustainability through sourcing from small businesses: Public sector perspectives. *Journal of Cleaner Production*, *16*(15), 1600–1609. doi:10.1016/j.jclepro.2008.04.014

Walmart Expands Online Sustainable Shopping. (2015). *Supplychain247*. Retrieved from http://www.supplychain247.com/article/walmart_expands_online_sustainable_shopping

Walton, S. V., Handfield, R. B., & Melnyk, S. A. (1998). The green supply chain: Integrating suppliers into environmental management processes. *Journal of Supply Chain Management*, *34*(2), 2.

Walton, S., & Huey, J. (1993). *Sam Walton, Made in America: My Story*. Bantam.

Wang, C.-H., & Tsai, D.-R. (2009). Integrated installing ISO 9000 and ISO 27000 management systems on an organization. *Proceedings of International Carnahan Conference On Security Technology*, (265–267). doi:10.1109/CCST.2009.5335527

Wang, F., Lai, X., & Shi, N. (2011). A multi-objective optimization for green supply chain network design. *Decision Support Systems*, *51*(2), 262–269. doi:10.1016/j.dss.2010.11.020

Warfield, J. N. (1974). Structuring complex systems. Columbus, O.H: Battele Memorial Ins. 4.

Warfield, J. N. (1974). Toward interpretation of complex structural models. *IEEE Transactions on Systems, Man and Cybernetics*, 5, 405-417.

Warfield, J. N. (1994). *A science of generic design: Managing complexity through systems design*. Iowa: Iowa State University press.

Warfield, J. N. (1994). *Science of Generic Design: Managing Complexity through Systems Design*. Iowa State Press.

Warfield, J. N. (1999). Twenty laws of complexity: Science applicability in organizations. *Systems Research and Behavioral Science*, *16*(1), 3–40. doi:10.1002/(SICI)1099-1743(199901/02)16:1<3::AID-SRES241>3.0.CO;2-F

Watkinson, J. (2009). WCED (1987)–Copenhagen (2009) Will We Ever Take the Environment Seriously. Retrieve from myliberaldemocratpoliticalramblings.wordpress.com/2009/08/27lwced-1987-copenhagen-2009-wilI-we-ever-take-theenvironment-seriously

Watson, R. H. (1978). Interpretive structural modeling—A useful tool for technology assessment? *Technological Forecasting and Social Change*, *11*(2), 165–185. doi:10.1016/0040-1625(78)90028-8

Watson, W. (1999). *Work USA 2000: Employee commitment and the bottom line*. Bethesda, MD: Watson Wyatt.

Webster, J., & Watson, R. T. (2002). Analyzing the past to prepare for the future: Writing a literature review. *Management Information Systems Quarterly*, *26*(2), 3.

Westkamper, E. (2008). Manufuture and sustainable manufacturing. In *Manufacturing Systems and Technologies for the New Frontier* (pp. 11–14). Springer London. doi:10.1007/978-1-84800-267-8_3

Westkamper, E., Alting, L., & Arndt, G. (2001). Life cycle management and assessment: Approaches and visions towards sustainable manufacturing. *Proceedings of the Institution of Mechanical Engineers. Part B, Journal of Engineering Manufacture*, *215*(5), 599–626. doi:10.1243/0954405011518557

Westphal, I., Thoben, K. D., & Seifert, M. (2010). Managing collaboration performance to govern virtual organizations. *Journal of Intelligent Manufacturing*, *21*(3), 311–320. doi:10.1007/s10845-008-0182-5

Wetland, D. (2003). The strategic training of employee's model: balancing organizational constraints and training content. S.A.M. Advanced Management Journal, winter, 103-107.

Whitelegg, J. (1993). *Transport for a sustainable future: the case for Europe*. Belhaven Pr.

Whooley, N. (2004). Social Responsibility in Europe. *PWC.com*. Retrieved from www.pwc.com/extweb/newcolth.nsf/0/503508DDA107A61885256F35005C1E35S

Wiener, Y. (1988). Forms of value system: A focus on organizational effectiveness and cultural change and maintenance. *Academy of Management Review*, *13*(4), 534–545.

Wiengarten, F., Pagell, M., & Fynes, B. (2012). Supply chain environmental investments in dynamic industries: Comparing investment and performance differences with static industries. *International Journal of Production Economics*, *135*(2), 541–551. doi:10.1016/j.ijpe.2011.03.011

Wildman, T. M., & Niles, J. A. (1987). Essentials of professional growth. *Educational Leadership*, *44*(5), 4–10.

Wilkes, R. S. (2008). Low linolenic soybeans and beyond. *Lipid Technology*, *20*(12), 277–279. doi:10.1002/lite.200800072

Wilkinson, G., & Dale, B. G. (1999). Integrated management systems: An examination of the concept and theory. *The TQM Magazine*, *11*(2), 95–104. doi:10.1108/09544789910257280

Williams, L., & Hazer, J. (1986). Antecedents and consequences of satisfaction and commitment in turnover models: A re-analysis using latent variable structural equation models. *The Journal of Applied Psychology*, *71*(2), 219–231. doi:10.1037/0021-9010.71.2.219

Williams, M., & Dreher, G. (1992). Compensation systems attributes and applicant pool characteristics. *Academy of Management Journal*, *35*(3), 571–595. doi:10.2307/256487

Williamson, O. E. (1981). The economics of organization: The transaction cost approach. *American Journal of Sociology*, *87*(3), 548–577. doi:10.1086/227496

Williams, P. (2010). SA 8000 - The first decade: Implementation, influence and impact. *Development in Practice*, *20*(2), 298–300. doi:10.1080/09614520903564280

Williams, Z., Lueg, J. E., Taylor, R. D., & Cook, R. L. (2009). Why all the changes?: An institutional theory approach to exploring the drivers of supply chain security (SCS). *International Journal of Physical Distribution & Logistics Management*, *39*(7), 595–618. doi:10.1108/09600030910996279

Willis, C. (2000). Go for your goals. *Working Woman*, 2000, 6–7.

Winston, A. (2014). Resilience in a hotter world. *Harvard Business Review*, *92*(4), 56–64. PMID:24830282

Wolfe, A. (2002). Questionnaire design. In R. Birn (Ed.), *The international hand book of marketing research techniques* (pp. 87–102). London: Market Research Society.

Wolf, J. (2014). The relationship between sustainable supply chain management, stakeholder pressure and corporate sustainability performance. *Journal of Business Ethics*, *119*(3), 317–328. doi:10.1007/s10551-012-1603-0

Womack, J., & Jones, D. T. (1996). *Lean thinking: Banish waste and create wealth in your corporation*. London, UK: Simon and Schuster.

Womack, J., & Jones, D. T. (2005). *Lean solutions: How companies and customers can create value and wealth together*. London, UK: Simon and Schuster.

Womack, J., Jones, D. T., & Roos, D. (1990). *The machine that changed the world*. New York, NY: Rawson Associates.

Womack, J., Jones, D., & Roos, D. (1990). *The Machine that Changed the World*. New York, NY: Rawson Associates.

Wong, S., Liu, J., & Culpan, T. (2010, June 3). Why Apple and others are nervous about Foxconn. *Bloomberg Business Week*.

Wood, D. J. (1991). Corporate social performance revisited. *Academy of Management Review*, *16*(4), 691–718.

Woodruffe, C. (1999). Winning the talent war: A strategic approach to attracting, developing and retaining the best people. Chichester, UK: John Wiley & Sons. Retrieved from www.cpscccsp.ca/PDFS/CPSC%20Final%20Report%20June28%20%207%20case%20studis2%20oct%207%2004.pdf

World conservation strategy: living resource conservation for sustainable development. (1980). IUCN/UNEP/WWF, Gland, Switzerland.

Wu, G. C. (2013). The influence of green supply chain integration and environmental uncertainty on green innovation in Taiwan's IT industry. *Supply Chain Management: An International Journal, 18*(5), 539–552. doi:10.1108/SCM-06-2012-0201

Wu, H. J., & Dunn, S. C. (1995). Environmentally responsible logistics systems. *International Journal of Physical Distribution & Logistics Management, 25*(2), 20–38. doi:10.1108/09600039510083925

Wu, Z., & Pagell, M. (2011). Balancing priorities: Decision-making in sustainable supply chain management. *Journal of Operations Management, 29*(6), 577–590. doi:10.1016/j.jom.2010.10.001

Xu, F. L., Zhao, S. S., Dawson, R. W., Hao, J. Y., Zhang, Y., & Tao, S. (2006). A triangle model for evaluating the sustainability status and trends of economic development. *Ecological Modelling, 195*(3-4), 327–337. doi:10.1016/j.ecolmodel.2005.11.023

Xu, H. B., Zhang, Y., Li, Z. H., Zheng, S. L., Wang, Z. K., Qi, T., & Li, H. Q. (2006). Development of a new cleaner production process for producing chromic oxide from chromite ore. *Journal of Cleaner Production, 14*(2), 211–219. doi:10.1016/j.jclepro.2004.09.001

Yakhou, M., & Dorweiler, V. P. (2004). Environmental accounting: An essential component of business strategy. *Business Strategy and the Environment, 13*(2), 65–77. doi:10.1002/bse.395

Yakovleva, N., Sarkis, J., & Sloan, T. (2012). Sustainable benchmarking of supply chains: The case of the food industry. *International Journal of Production Research, 50*(5), 1297–1317. doi:10.1080/00207543.2011.571926

Yanadori, Y., & Kato, T. (2010). Work and Family Practices in Japanese Firms: Their Scope, Nature and Impact on Employee Turnover. *International Journal of Human Resource Management, 20*(2), 439–456. doi:10.1080/09585190802673494

Yee, L. M., Devi, S., & Khin, E. W. S. (2011). Kyoto Protocol and Social Accounting Implication on Global-Warming in Malaysia. *African Journal of Agricultural Research, 6*(6), 1489–1499.

Yen, Y. X., & Yen, S. Y. (2012). Top-management's role in adopting green purchasing standards in high-tech industrial firms. *Journal of Business Research, 65*(7), 951–959. doi:10.1016/j.jbusres.2011.05.002

York, J., & Larson, A. (2006). *Nike: Moving down the sustainability track through chemical substitution and waste reduction. UVA-ENT-0098.* Charlottesville, VA: Darden Business Publishing.

Yu, W., Chavez, R., Feng, M., & Wiengarten, F. (2014). Integrated green supply chain management and operational performance. *Supply Chain Management: An International Journal, 19*(5/6).

Yuan, C., Zhai, Q., & Dornfeld, D. (2012). A three dimensional system approach for environmentally sustainable manufacturing. *CIRP Annals-Manufacturing Technology, 61*(1), 39–42. doi:10.1016/j.cirp.2012.03.105

Zadeh, F. O., & Eskandari, A. (2012). Firm Size as Company's Characteristic and Level of Risk Disclosure: Review on Theories and Literatures. *International Journal of Business and Social Science, 3*(17), 9–17.

Zeldin, S., Camino, L., & Calvert, M. (2007). Toward an understanding of youth in community governance: Policy priorities and research directions. *Análise Psicológica*, 2007, 77–95.

Zeng, S. X., Shi, J. J., & Lou, G. X. (2007). A synergetic model for implementing an integrated management system: An empirical study in China. *Journal of Cleaner Production, 15*(18), 1760–1767. doi:10.1016/j.jclepro.2006.03.007

Zeng, S. X., Tam, V. W. Y., & Tam, C. M. (2008). Towards occupational health and safety systems in the construction industry of China. *Safety Science, 46*(8), 1155–1168. doi:10.1016/j.ssci.2007.08.005

Zeppel, H., & Beaumont, N. (2012). Climate change and tourism futures: Responses by Australian tourism agencies. *Tourism and Hospitality Research*, *12*(2), 73–88. doi:10.1177/1467358412444807

Zeytinoglu, I. U. & Denton, M. (2005). Satisfied workers, Retained Workers: Effects of work and work environment on Homecare worker's Job satisfaction, Stress, Physical Health, and Retention. *Canadian Health Services Research Foundation, Foundation canadienne de la recherché sur les services de sante.*

Zhang, C., & Dhaliwal, J. (2009). An investigation of resource-based and institutional theoretic factors in technology adoption for operations and supply chain management. *International Journal of Production Economics*, *120*(1), 252–269. doi:10.1016/j.ijpe.2008.07.023

Zhang, H. C., Kuo, T. C., Lu, H., & Huang, S. H. (1997). Environmentally conscious design and manufacturing: A state-of-the-art survey. *Journal of Manufacturing Systems*, *16*(5), 352–371. doi:10.1016/S0278-6125(97)88465-8

Zheng, J., Knight, L., James, K., Harland, C., & Humby, S. (2007). An analysis of research into the future of purchasing and supply management. *Journal of Purchasing and Supply Management*, *13*(1), 69–83. doi:10.1016/j.pursup.2007.03.004

Zhiyong, L., & Hua, J. (2009). Machining accuracy prediction of Aero-engine Blade in electrochemical Machining based on BP Neural network. *Proceedings of theInternational Workshop on Information Security and Application (IWISA)* (Vols. 21-22, pp. 244-247).

Zhu, Q., & Cote, R. P. (2004). Integrating green supply chain management into an embryonic eco-industrial development: A case study of the Guitang Group. *Journal of Cleaner Production*, *12*(8-10), 1025–1035. doi:10.1016/j.jclepro.2004.02.030

Zhu, Q., Dou, Y., & Sarkis, J. (2010). A portfolio-based analysis for green supplier management using the analytical network process. *Supply Chain Management: An International Journal*, *15*(4), 306–319. doi:10.1108/13598541080001414

Zhu, Q., Geng, Y., Fujita, T., & Hashimoto, S. (2010). Green supply chain management in leading manufacturers: Case studies in Japanese large companies. *Management Research Review*, *33*(4), 380–392. doi:10.1108/01409171011030471

Zhu, Q., & Sarkis, J. (2004). Relationships between operational practices and performance among early adopters of green supply chain management practices in Chinese manufacturing enterprises. *Journal of Operations Management*, *22*(3), 265–289. doi:10.1016/j.jom.2004.01.005

Zhu, Q., & Sarkis, J. (2006). An inter-sectoral comparison of green supply chain management in China: Drivers and practices. *Journal of Cleaner Production*, *14*(5), 472–486. doi:10.1016/j.jclepro.2005.01.003

Zhu, Q., Sarkis, J., Cordeiro, J. J., & Lai, K. H. (2008). Firm-level correlates of emergent green supply chain management practices in the Chinese context. *Omega*, *36*(4), 577–591. doi:10.1016/j.omega.2006.11.009

Zhu, Q., Sarkis, J., & Geng, Y. (2005). Green supply chain management in China: Pressures, practices, and performance. *International Journal of Operations & Production Management*, *25*(5), 449–468. doi:10.1108/01443570510593148

Zhu, Q., Sarkis, J., & Lai, K. H. (2007). Green supply chain management: Pressures, practices and performance within the Chinese automobile industry. *Journal of Cleaner Production*, *15*(11), 1041–1052. doi:10.1016/j.jclepro.2006.05.021

Zhu, Q., Sarkis, J., & Lai, K. H. (2012). Examining the effects of green supply chain management practices and their mediations on performance improvements. *International Journal of Production Research*, *50*(5), 1377–1394. doi:10.1080/00207543.2011.571937

Zhu, Q., Sarkis, J., Lai, K. H., & Geng, Y. (2008). The role of organizational size in the adoption of green supply chain management practices in China. *Corporate Social Responsibility and Environmental Management*, *15*(6), 322–337. doi:10.1002/csr.173

Zhu, Q., Sarkis, J., & Lai, K.-H. (2008a). Confirmation of a measurement model for green supply chain management practices implementation. *International Journal of Production Economics*, *111*(2), 261–273. doi:10.1016/j.ijpe.2006.11.029

Zhu, Q., Sarkis, J., & Lai, K.-H. (2008b). Green supply chain management implications for closing the loop. *Transportation Research Part E, Logistics and Transportation Review*, *44*(1), 1–18. doi:10.1016/j.tre.2006.06.003

Zhu, Q., Sarkis, J., & Lai, K.-H. (2013). Institutional-based antecedents and performance outcomes of internal and external green supply chain management practices. *Journal of Purchasing and Supply Management*, *19*(2), 106–117. doi:10.1016/j.pursup.2012.12.001

Zokaei, K., & Hines, P. (2007). Achieving consumer focus in supply chains. *International Journal of Physical Distribution & Logistics Management*, *37*(3), 223–247. doi:10.1108/09600030710742434

Zokaei, K., & Simons, D. W. (2006). Value chain analysis in improvement of customer focus: A case study of UK red meat industry. *International Journal of Logistics Management*, *17*(2), 141–162. doi:10.1108/09574090610689934

Zsidisin, G. A. (2003). Managerial perceptions of supply risk. *Journal of supply chain management, 39*(4), 14-26.

Zsidisin, G. A., & Ellram, L. (1999). Supply risk assessment analysis. *PRACTIX. Best Practices in Purchasing & Supply Chain Management*, *2*(4), 9–12.

Zsidisin, G. A., & Siferd, S. P. (2001). Environmental purchasing: A framework for theory development. *European Journal of Purchasing and Supply Management*, *7*(1), 61–73. doi:10.1016/S0969-7012(00)00007-1

Zubir, A. F. M., & Habidin, N. F. (2012). The development of sustainable manufacturing practices and sustainable performance in Malaysian automotive industry. *Journal of Economics and Sustainable Development*, *3*(7), 130–138.

Zucker, L. G. (1987). Institutional theories of organization. *Annual Review of Sociology*, *13*(1), 443–464. doi:10.1146/annurev.so.13.080187.002303

Zuckerman, A. (2000). Quality Assurance through ISO 9000. *School Administrator*, *57*(6), 12–16.

Zutshi, A., & Sohal, A. S. (2004). Adoption and maintenance of environmental management systems: Critical success factors. *Management of Environmental Quality: An International Journal*, *15*(4), 399–419. doi:10.1108/14777830410540144

About the Contributors

Rameshwar Dubey is currently working as an Associate Professor with Symbiosis International University and Visiting Professor & Research Fellow with South University of Science and Technology of China. He is also associated with various institutions of repute as a visiting scholar which includes institutes of national importance and some global universities like DePaul University, University of Massachusetts Dartmouth and Anhui School of Economics and Finance, China and South University of Science and Technology, China. He is also serving as Associate Editor of Global Journal of Flexible Systems Management (Springer) and International Journal of Innovation Science (Emerald). He has over 66 papers indexed in Scopus and SCI/SSCI and over 210 citations.

Angappa Gunasekaran is a Professor and Dean at the Charlton College of Business, University of Massachusetts, Dartmouth. Dr. Gunasekaran has held academic positions in UK, Australia, Finland, India and Canada. He has over 250 articles published in peer-reviewed journals. He has presented about 50 papers and published 50 articles in conferences and given a number of invited talks in many countries. He is on the Editorial Board of several journals. He has organized several international workshops and conferences in the emerging areas of operations management and information systems. He is currently interested in researching logistics and supply chain management. Dr. Gunasekaran has been the founding Director of Business Innovation Research Center (BIRC) since 2006.

***

Rajat Agrawal is currently an Assistant Professor at the Department of Management Studies IIT-Roorkee. His research areas include general management and supply chain management. He has contributed many research papers, book chapters—in top international publications—and has presented his research works at several national and international conferences. He teaches courses on supply chain management, operations management and quantitative techniques to students of the Masters in Business Administration program at IIT-Roorkee.

Padmanabha Aital is a Professor of Information Technology and Operations Management at the Symbiosis Institute of Operations Management, Nashik, India. He holds a PhD from IIT Bombay, and has over 18 years of teaching experience. He has published over 20 research papers in refereed international journals and conference proceedings.

Surajit Bag holds a PhD degree in Supply Chain Management from the University of Petroleum & Energy Studies and is actively involved in research. Dr. Bag has trained in Case Study Teaching from the Indian Institute of Management, Calcutta. He has also trained in Total Quality Management from the Indian Institute of Technology, Kanpur. His research interests are in Green Supply Chain, Green Manufacturing, Environmental Management and Total Quality Management. He has more than six years of industry experience with exposure in the areas of international procurement and logistics management. He is a member of several professional bodies such as the Indian Rubber Institute, the Society of Operations Management, the Operational Research Society of India, and AIMS International.

Sonali Bhattacharya is an associate professor at SCMHRD, a constituent of the Symbiosis International University. She has a PhD in statistics from the Delhi University and also has a Master's degree in insurance. Her research interests include theoretical and applied statistics, social media analytics, and social science and management. She has 40 published research papers to her credit.

Sergio Silva Braga Jr. is a doctor in Business Administration from Universidade Nove July (UNINOVE) has a Master's in Organizational Management from the University of São Paulo (USP). He is a graduate in Business Administration from Integrated Faculties of Toledo (UniToledo). Currently an Assistant Professor at UNESP, Tupã Campus, in Business Administration course, lecturing the disciplines of Materials Management.

Susheela Girisaballa is a Research Scholar at the Symbiosis International University. She holds a Bachelor's Degree in Agriculture and a Master's Degree in Management. Susheela has been working as a faculty member for the past 8 years in various management schools in India. Her research interests include sustainable supply chains, operations management, and statistics and operations research.

Shirish Jeble has a MBA in Operations Management and B.E. (Mechanical) from Mumbai University, currently pursuing his PHD from Symbiosis International University, Pune, India. He has over 19 years' industry experience and over 3 years of experience in academics. His industry experience includes imports, materials management, Global ERP implementation projects, program management and information technology projects. He has taught industry-oriented subjects such as supply chain management, operations management, enterprise resource planning, project management and industrial management in a reputed business school in Pune. His research interests are in sustainable supply chain management, Big Data and Lean Production Systems.

Shibin K. T. has a M Tech in Industrial Engineering from the National Institute of Technology Calicut. He has 3 years of experience in the automotive industry. He is currently working with Tata Technology Limited, Pune, as a Simulation Engineer for 2 years and 6 months.

Kijpokin Kasemsap received his BEng degree in Mechanical Engineering from King Mongkut's University of Technology Thonburi, his MBA degree from Ramkhamhaeng University, and his DBA degree in Human Resource Management from Suan Sunandha Rajabhat University. He is a Special Lecturer for the Faculty of Management Sciences, Suan Sunandha Rajabhat University based in Bangkok, Thailand. He is a Member of the International Association of Engineers (IAENG), the International

Association of Engineers and Scientists (IAEST), the International Economics Development and Research Center (IEDRC), the International Association of Computer Science and Information Technology (IACSIT), the International Foundation for Research and Development (IFRD), and the International Innovative Scientific and Research Organization (IISRO). He also serves on the International Advisory Committee (IAC) for the International Association of Academicians and Researchers (INAAR). He has published numerous original research articles in top international journals, conference proceedings, and book chapters on business management, human resource management, and knowledge management published internationally.

Kaushik Kumar received his B.Tech. in Mechanical Engineering from REC, Warrangal, India, MBA from IGNOU and a PhD from Jadavpur University, Kolkata. He has over 25 years of teaching, research and industrial experience. Presently, he is an Associate Professor in the Department of Mechanical Engineering, Birla Institute of Technology, Mesra, Ranchi, India. His areas of interests are optimization, non-conventional machining, CAD/CAM, rapid prototyping and composites. He has 3 Patents, 71 international journals, 18 internationals and 8 national conference publications to his credit. He is on the editorial board and review panel of 7 international and 1 national journals of repute. He is a Mentor of the Organising body of the Institute for Research and Development India (IRD India) and has been felicitated with many awards and honours.

Arunavo Mukherjee is currently Vice President of Advisory Services, Tata Cleantech Capital Ltd, a joint venture of Tata Capital and International Finance Corporation (IFC). He has extensive experience in energy industry-spanning coal, natural gas and other renewable sources. He has been with the Tata group since 1991, and has worked in India, USA, Europe and Africa. In July 2008, Mr. Mukerjee was selected to lead the Tata group effort to combat climate change across its 100 companies. In early 2013, he was transferred to Tata Cleantech Capital Ltd to establish the advisory business. His academic credentials include a BS in Engineering from the Indian Institute of Technology and a MBA from Carlson School of Management, University of Minnesota, USA. Mr. Mukerjee is a member of the IEEE, USA, and the Society of Petroleum Engineers (SPE), USA. He is a Steering Committee member of the UN's Caring for Climate Initiative, and has been a speaker for events organized by the IEEE (USA), the Confederation of Indian Industries (CII) and the UN Global Compact, across the globe

Sujata Mukherjee has a PH.D in Social Sciences from the Tata Institute of Social Sciences, Mumbai. She is a full-time faculty at SVKM's NMIMS, School of Business, in the Area of Business Strategy and Environment. She teaches subjects like corporate social responsibility and corporate governance, business research methodology, livelihood management and has done research in the area of entrepreneurship management, and strategic reviews of NGOs. She has experience in working with national and international NGOs in India. Dr. Mukherjee is a member of the CII-Indian Women Network and the Society for Human Resource Management.

Nehajoan Panackal has completed her Bachelor's in mechanical engineering and is a post-graduate in business administration-human resource management. Her research areas include sustainability, human resource management and business administration.

Supriyo Roy, M. Sc (Statistics), M. Tech (OR&IBM), Ph. D (Optimization Tech.) has been working as Associate Professor, BIT-Mesra, Ranchi since 1st July, 2012. He has completed his Post-Doctoral Fellowship from EADS-SMI Center for Supply Chain, IIM - Bangalore. He has an experience of 4 years in industry and 14 years of post-graduate teaching. His teaching interests comprise the interdisciplinary areas of management science, production and operations management, decision science, and soft-computing for manufacturing. His publications have appeared in leading journals in INFOR-Canada, AMSE-France, BJO&PM - Brazil, IGI-Global, etc. His research interests cover development of optimization models relating to management science problems via structural equation modeling.

Anil Sankhi is a Manager – System and Audits with Constantia Flexibles, Ahmadabad, Gujarat, India and has a rich 7 years of experience in manufacturing organization. He also has expertise in the financial health analysis for maintaining quality standards and audits for organizations. He has successfully implemented various quality modules in organizations including the AX module. He has a zest to improve the processes within the organization and thus, also facilitates need-based training for the internal departments of the organization.

Vinay Sharma is currently an Associate Professor at the Department of Management Studies IIT-Roorkee. His research areas include marketing management, rural marketing, advertising and business strategy. He has contributed many research papers, book chapters in top international publications and has presented his research works at several national and international conferences. He teaches courses in marketing management, advertising and business strategy to students of the Masters in Business Administration program at IIT-Roorkee.

Eduardo Guilherme Satolo is a doctor in Industrial Engineering from the Universidade Metodista de Piracicaba (UNIMEP). Has a master's in Production Engineering from UNIMEP is a graduate in Production Engineering from UNIMEP is an Assistant Professor at UNESP, Tupã Campus, in Business Administration course, lecturing on the disciplines of Production Management.

Alexandre T Simon is a doctor in Industrial Engineering from UNIMEP, has a Master's in Mechanical Engineering from the Universidade Estadual de Campinas (UNICAMP) is a Graduate in Mechanical Engineering from UNESP is a Professor at UNIMEP, in the Industrial Engineering Post Graduation Program.

Archana Singh has a double post-graduate in business administration and cost, and management accounting. She has 9 years' teaching experience and is currently pursuing her Ph.D. Her areas of research include sustainability, accounts, finance and business administration.

Rupali Singh is a sponsored part-time research scholar with the Motilal Nehru National Institute of Technology Allahabad, Uttar Pradesh, India.

Kavitha. T. N. is a Senior Lecturer in the Department of Automobile Engineering, Government Polytechnic, Tumkur. She has contributed many research articles to reputed, peer-reviewed journals. Her research interests include: designing, CAD and CAM, robotics, and sustainable manufacturing.

Abhishek Tiwari received his BTech in Mechanical Engineering from AIMT, Greater Noida. He is currently pursuing a Master's in Engineering with a specialization in Design from the Department of Mechanical Engineering, Birla Institute of Technology, Mesra, Ranchi, India. His areas of interests are optimization and non-conventional machining. He has 6 international Journals and 1 International Conference publication to his credit.

Mani V. is a full time Research Scholar in the Department of Management Studies, Indian Institute of Technology, Roorkee, and is pursuing research in the field of supply chain management and operations. He has contributed many research papers in refereed journals by leading publishers that include Elsevier, Emerald, Inderscience, and Indian Management (AIMA). Prior to joining his PhD, he served for over a decade in several Fortune 500 companies in various capacities. He has also presented research articles in International conferences conducted by the Indian Institute of Management, Bangalore (IIMB), the Indian Institute of Technology, Roorkee (IITR) in association with Society of Operations Management (SOM), and the Glogift society in association with Curtin University (Glowgift 2014-Singapore). His research is titled 'It is time to change the system' published by the Indian Management –Journal of AIMA, which won the best article of the year in 2014.

Bhama Venkataramani has more than 25 years of teaching experience, teaching the subjects of Financial Accounting, Management Accounting and Financial Management. She has to her credit, a monograph on Factoring service published by Think line publishers, Nashik. She has co-authored three textbooks in Accountancy for the syllabus prescribed by University of Pune. She is currently supervising four PhD students registered under Symbiosis International University. Her research interests are in areas of financial inclusion and inclusive education and she has published papers in topics relating to these areas. Dr. Bhama Venkataramani is the Dean of Academics and Administration at Symbiosis International University.

V.G. Venkatesh is working as a faculty member at the Symbiosis Institute of Business Management, in the area of supply chain and logistics management and has also qualified as a certified supply chain professional from APICS (CSCP). He has international apparel industry experience from Honduras, Hong Kong, Bangladesh, and Sri Lanka. He has presented papers in overseas conferences in the USA, Taiwan and Hong Kong and published as a contributing author in journals including: the International Journal of Procurement Management, the International Journal of Operational Research, and Benchmarking –an International Journal. He is a regular contributor for industry magazines related to supply chain education and business areas. He has also published a book on merchandising concepts. His research interests are in reverse logistics systems and international business in the Indian apparel domain.

Prince Vijai is a Research Scholar in the Department of Operations and Systems at the ICFAI Business School of IFHE University, Hyderabad, India. He teaches under-graduate and graduate level courses related to Operations and Supply Chain Management. His research focuses on manufacturing, service, and supply chain operations and studies both strategic questions and tactical execution. He has published research articles in refereed international journals.

Index

I

J

L

M

N

O

P

R

S

T

U

V

www.ingramcontent.com/pod-product-compliance
Lightning Source LLC
LaVergne TN
LVHW060400300926
849162LV00006B/1316